BARRON'S

E-Z

ORGANIC CHEMISTRY

Bruce A. Hathaway, Ph.D.
Professor of Chemistry
LeTourneau University
(Formerly at Southeast Missouri State University)

BARRON'S

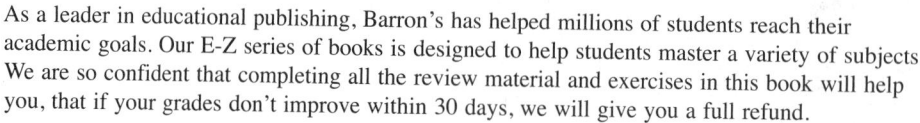

Better Grades or Your Money Back!

As a leader in educational publishing, Barron's has helped millions of students reach their academic goals. Our E-Z series of books is designed to help students master a variety of subjects. We are so confident that completing all the review material and exercises in this book will help you, that if your grades don't improve within 30 days, we will give you a full refund.

To qualify for a refund, simply return the book within 90 days of purchase and include your store receipt. Refunds will not include sales tax or postage. Offer available only to U.S. residents. Void where prohibited. Send books to **Barron's Educational Series, Inc., Attn: Customer Service** at the address on this page.

All inquiries should be addressed to:
Barron's Educational Series, Inc.
250 Wireless Boulevard
Hauppauge, New York 11788
www.barronseduc.com

ISBN-13: 978-0-7641-4467-7

Library of Congress Catalog Card No.: 2011000113

Library of Congress Cataloging-in-Publication Data
Hathaway, Bruce A.
   Barron's E-Z organic chemistry / Bruce A. Hathaway.
      p. cm.
   Rev. ed. of: Organic chemistry the easy way. c2006.
   Includes index.
   ISBN 978-0-7641-4467-7
 1. Chemistry, Organic—Problems, exercises, etc.
2. Chemistry, Organic—Outlines, syllabi, etc. I. Hathaway,
Bruce A. Organic chemistry the easy way. II. Title. III. Title:
E-Z organic chemistry.
   QD257.H38 2011
   547—dc22   2011000113

Printed in the United States of America
9 8 7 6 5 4 3 2

CONTENTS

INTRODUCTION

I have found organic chemistry to be a fascinating subject for over thirty years. Part of this fascination is due to some of the professors I had in undergraduate and graduate school. My first undergraduate organic chemistry professor, Philip L. Stotter, had a genuine infectious enthusiasm for the subject. He encouraged discussion and forced us students to integrate information from the text, lecture, and lab to do well on his exams. I had a chance to do a year of undergraduate research under his direction, and even though it was ultimately unsuccessful, I learned a lot about perseverance and creativity in problem solving. My major professor in graduate school, David E. Nichols, had a creative approach to teaching and mentoring and allowed me considerable freedom in the pursuit of my graduate research. Initially, I thought my project would require six months to a year to finish. Three years later, I had learned that even though a reaction is known and published in the chemical literature, it doesn't mean that it will work when I do it, and it doesn't mean that it will work in the molecules to which I was trying to apply it.

Over the past twenty-nine years, I have had the opportunity to teach and conduct research in organic chemistry at Southeast Missouri State University in Cape Girardeau, Missouri, and now at LeTourneau University. I have tried to impart the same enthusiasm for the subject that I experienced to my students. I have enjoyed involving undergraduates in a variety of research projects. Both the students and I have learned a lot about organic reactions in research and have had our share of successes and failures.

As you approach the study of organic chemistry, you will have to study differently for this course than you have for many others. There is so much to the subject that you must work on it daily to master it. Take advantage of whatever resources you have available to you to help you learn: your professor, teaching assistants, lab assistants, other students, your textbook, credible on-line resources, and whatever else you can find. There is no substitute for spending time with your text and working problems. If you can work the problems in your text (without looking at the answers), then you should do well on your exams. I have provided some problems with this text. Your text will have other problems. Work as many as you can. Work problems in groups if possible and contribute to the solutions and the discussion of the solutions. Explaining why you think a solution is right is one of the best ways to cement what you know into your memory.

I'd like to make a few comments about the content of this book. I have tried to summarize a huge amount of material in a book one-third the size of a typical organic chemistry book. I have tried to be complete, but I have omitted some things due to space constraints, such as the Hoffman Elimination and Nucleophilic Aromatic Substitution, and my perception of what is important in the typical organic chemistry course. I have not included any specific biochemistry chapters in this book because I could not do it justice, and many organic courses do not cover this material anyway. I have included some biochemical examples where relevant. In the discussion of spectroscopy, I have omitted ultraviolet and visible (UV-Vis) spectroscopies, because they are not of much use in solving the structures of organic molecules. In my twenty-nine years of teaching, I have run only two UV-Vis spectra. Nuclear magnetic resonance (NMR) spectroscopy has become the most powerful technique, and I have focused most of my efforts on NMR, along with infrared spectroscopy and mass spectrometry.

In the discussion of organic reactions, I have tried to include the mechanisms to most of the reactions. To really learn the reactions, you need to know how they work. Practice writing the mechanisms because this will aid your understanding immensely. As you plan a synthesis, you also need to know the mechanisms of the reactions, so you can tell which of the functional groups in the molecule will survive the reaction conditions.

I have chosen to write water as HOH in many places in the book. I do this to emphasize that the hydrogens are bonded to the oxygen, and to show that water is similar to alcohols in structure. It is perhaps a little subtle, but I hope it helps in your understanding as you see it written a little differently than you usually do.

A large number of individuals deserve to be acknowledged for their contributions and encouragement in the writing of this book. In addition to Phil Stotter and David Nichols, several others contributed to my development as a chemist and teacher and have significantly encouraged me as I have been writing this book. I sincerely appreciate my former colleagues in the Chemistry Department at Southeast Missouri State University for their willingness to share their expertise and help me explain better what I wanted to say. I would particularly like to acknowledge Sharon Coleman, for her encouragement to write this book, and Bjorn Olesen, for his constant encouragement, help, and support. I would also like to acknowledge the late Ronald Popham, who was my department chairman and friend for twenty years. He encouraged me to be the best teacher and researcher I could be and was very creative in scheduling my classes and allocating departmental resources to help me achieve those goals. My former chairman, Philip Crawford, has continued in this tradition, and I am very grateful for his support in this project. The editorial team at Barron's Educational Publishing has been very helpful in all aspects of this project, and I am grateful to them for this opportunity. My mother, Marian Hathaway, and my brother, Wayne Hathaway, have been encouraging during this project. My wife, Ruth, has had to put up with my strange schedule as I have worked on this project, and I appreciate all of her support through the past year and a half. Finally, I would like to acknowledge the support of God, to whom all the glory is due.

Important Concepts That Underlie Most of Organic Chemistry

WHAT YOU WILL LEARN

In this chapter, you will learn:

- how do draw correct chemical structures of organic molecules;
- what the difference is between isomers and resonance structures, and how to draw them;
- how to use electronegativity to predict which molecules will be more water-soluble and have a higher boiling point;
- how to recognize common functional groups.

SECTIONS IN THIS CHAPTER

- Electronegativity
- Ionic and Covalent Bonding
- Writing Chemical Structures
- Isomerism
- Resonance
- Hydrogen-Bonding and Boiling Points
- Water-Solubility
- Acid-Base Concepts
- Functional Groups

A s you embark on your study of organic chemistry, several concepts from general chemistry will form the groundwork for much of what you will learn. I hope that this chapter will refresh your memory of these topics and allow you to apply them as you study the reactivity and properties of organic compounds. My students would tell you that I almost beat these concepts to death when I teach them, with the goal that they will become ingrained in my students' minds.

A. Electronegativity

Electronegativity is defined as the tendency of an element to draw electrons in a covalent bond toward itself. As you look at the periodic table, there are two general trends in electronegativity of elements. Ignoring the noble gases in the rightmost column, electronegativity generally increases going from left to right. Second, electronegativity increases going from bottom to top. Some electronegativity values are shown in the following table.

GROUP NUMBER: PREFERRED U.S. (CURRENT IUPAC)							
Period	IA (1)	IIA (2)	IIIA (13)	IVA (14)	VA (15)	VIA (16)	VIIA (17)
1	H 2.1						
2	Li 1.0	Be 1.5	B 2.0	C 2.5	N 3.0	O 3.5	F 4.0
3	Na 0.9	Mg 1.2	Al 1.5	Si 1.8	P 2.1	S 2.5	Cl 3.0
4	K 0.8	Ca 1.0			As 2.0	Se 2.4	Br 2.8
5	Rb 0.8						I 2.5
6	Cs 0.7						

The most electronegative element is fluorine, and the least electronegative element is cesium. Hydrogen is a little anomalous with its value of 2.1. This reflects hydrogen's ability to form compounds with virtually every element in the periodic table, other than the noble gases.

B. Ionic and Covalent Bonding

There are two general types of bonds between pairs of atoms. One type is an **ionic bond**, which is defined as the attraction between oppositely charged ions. Ionic bonds are most commonly formed between atoms that have a large difference in electronegativity. NaCl, sodium chloride, or common table salt, is a large crystal lattice of alternating Na^+ and Cl^- ions. Most ionic bonds are formed between a metal cation and a nonmetal anion.

The other type of bond is a **covalent bond**. In a covalent bond, electrons are shared between a pair of atoms. In general, a pair of atoms can share one or more pairs of electrons if their electronegativities are not too different. In general, covalent bonds are formed between two nonmetals.

C. Writing Chemical Structures

Ionic bonds are usually not formally shown. For example, if you were asked to write a representation of sodium chloride, you would just write a sodium cation and a chloride anion. This is shown here.

$$\overset{\oplus}{Na} \quad \overset{\ominus}{\underset{\cdot\cdot}{\overset{\cdot\cdot}{:Cl:}}}$$

The dots represent unshared valence electrons. A chlorine atom has seven unshared valence electrons, while a chloride ion has eight unshared valence electrons. The extra electron gives the chloride ion the negative charge. Similarly, a sodium atom has one valence electron, while the sodium ion has none, so it has a positive charge. It is important to include the charges when writing structures of compounds. The charges are a key factor in recognizing an ionic compound.

In contrast to ionic bonds, covalent bonds are explicitly shown. There are two commonly used ways to do this. One is to use a pair of dots for each pair of electrons shared by two atoms. The other is to use a dash to indicate a shared pair of electrons. These are illustrated here.

Molecule	Structure Using Dots	Structure Using Dashes
H_2	H : H	H——H
CO_2	$\overset{\cdot\cdot}{O} :: C :: \overset{\cdot\cdot}{O}$	$\overset{\cdot\cdot}{\underset{\cdot\cdot}{O}} = C = \overset{\cdot\cdot}{\underset{\cdot\cdot}{O}}$
N_2	: N ::: N :	:N≡N:

Most organic chemistry structures are shown using dashes for shared pairs of electrons because they are easier to write. There is some jargon you should know about types of covalent bonds. If two atoms are sharing one pair of electrons, this is called a **single bond**. If two atoms are sharing two pairs of electrons, this is called a **double bond**. If two atoms are sharing three pairs of electrons, this is called a **triple bond**.

In most organic compounds, none of the atoms have a formal charge; that is, the atoms are neutral. One of the most important concepts in organic chemistry is writing structures so that each atom is making the correct number of bonds. If you don't learn this early in the game, you will miss points all through the year on homework and exams. One way to learn this is to go back to the periodic table, and use it to help you.

Group Number: Preferred U.S.	IVA	VA	VIA	VIIA
Period 2 Atom	C	N	O	F

The preferred U.S. group number tells you the number of valence electrons an atom has. For a stable electron configuration, a period 2 atom needs to own or share eight electrons (the "octet rule"). Because carbon has four valence electrons, and it needs eight, it must have access to four more electrons. If carbon shared four pairs of electrons with other atoms, it would fulfill the octet rule. Hence, a neutral carbon atom shares four pairs of electrons. A neutral nitrogen shares three pairs of electrons; an oxygen, two pairs; and a fluorine, one pair. Since hydrogen is in period 1, a full outer shell is two valence electrons, so hydrogen only shares one pair of electrons.

Another important concept in writing structures of organic compounds is using the proper numbers of electrons. You can use only the number of valence electrons that the atoms possess in your structure, no more and no less! The two most common errors students make involve these two concepts.

Let's do an example. Ethanol (ethyl alcohol, the "alcohol" of alcoholic beverages) has the molecular formula C_2H_6O. It is known that the two carbons are covalently bonded together. How do we write a structure for ethanol?

First, count up the total valence electrons you have. Each hydrogen has one, each carbon has four, and the oxygen has six. This adds up to $(6 \times 1) + (2 \times 4) + 6 = 20$ valence electrons. Twenty electrons implies some combination of ten shared or unshared pairs of electrons (occasionally we will run into structures with unpaired electrons, but we will ignore this possibility for the moment). I'll walk us through the process next.

We start off with what we know: The two carbons are covalently bonded, which means that they are sharing at least one pair of electrons. We can have them share more if we need to later. We have used only two of the twenty electrons so far.	C—C
Carbon needs to share four pairs of electrons, so let's just add three single bonds to each atom. This now gives us a total of fourteen electrons. We still need six more electrons, as well as the six hydrogens and the oxygen.	—C—C—
Oxygen shares two pairs of electrons, so we can bond it to one of the carbons, and then add another single bond to the oxygen, so it will make two bonds.	—C—C—O—
Now let's add the six hydrogens to the ends of the six single bonds. At this point, we have used all of our atoms, but have only used sixteen electrons with the eight single bonds. We still have four electrons to use. Where do they go?	H—C—C—O—H (with H's on carbons)

The oxygen is not fulfilling the octet rule until it has eight electrons around it, so we need to add two unshared pairs of electrons to the oxygen. Now, each atom has a full outer shell of electrons, and is making the correct number of bonds for a neutral atom. We have also used all of the atoms, so we are done!	*(structure: ethanol, H—C—C—O—H with H atoms and two lone pairs on oxygen)*

Let's do another example. Propene (or propylene), a raw material used in production of plastics, has the formula C_3H_6. Therefore, we have eighteen valence electrons to work with, six from the hydrogens, and four each from the three carbons ($6 + 4 + 4 + 4 = 18$).

I usually start off by bonding the carbons together with single bonds. If this doesn't work, I can change it later.	C—C—C
Let's add enough single bonds to each carbon so that each carbon is making four bonds. If we count electrons, we have used twenty! We can't do this because we have only eighteen electrons. Also, we only have six hydrogens, so we don't have enough atoms for the carbons to make eight more single bonds.	*(structure: three carbons each with four single bonds)*
If we can't use all single bonds, we can try a double bond between two of the carbons, and then add enough single bonds so the carbons are making four bonds. This gives us eighteen electrons (four for the double bond, and fourteen for the seven single bonds). This also gives us six places for hydrogens to be attached.	*(structure: —C—C═C— with single bonds)*
Adding the six hydrogens gives us a structure where each atom has a full outer shell of electrons and is making the correct number of bonds for a neutral atom. We have also used all of the atoms, so we are done! The bottom structure more closely reflects what the structure actually looks like.	*(structure: propene, H—C—C═C—H with H atoms; and an alternate angled drawing below)*

There are other ways to represent organic compounds as well. This can be illustrated with ethanol and propene.

Structure using dashes and unshared electron pairs	H—C—C—O—H (ethanol)	H—C—C=C—H (propene)
Condensed structure not showing bonds to hydrogens or unshared electron pairs	CH_3—CH_2—OH	CH_3—CH=CH_2
Condensed structure not showing any single bonds or unshared electron pairs	CH_3CH_2OH	CH_3CH=CH_2
Line-angle structure	OH	

The first type of condensed structure assumes you know that carbon makes four bonds, and hydrogen makes only one. It also assumes you know that oxygen has unshared pairs of electrons. The second type of condensed structure assumes you will recognize when atoms are connected by single bonds. The line-angle formula represents structures somewhat differently. Carbons are at the ends of line segments, if there is not another atom shown. Carbons are also at the corner of an angle. A line angle formula assumes that each carbon is bonded to enough hydrogens to make four bonds, unless that carbon is charged or has unshared electrons shown. See the drawing below.

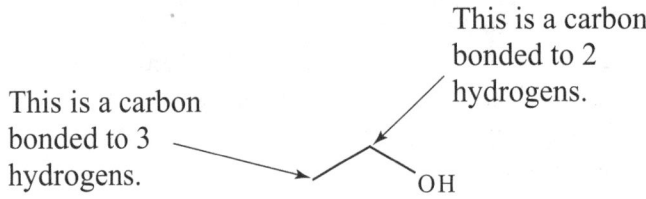

This is a carbon bonded to 3 hydrogens.

This is a carbon bonded to 2 hydrogens.

OH

There is also another shorthand way to represent some molecules. This can be illustrated with heptane, as shown here. The fourth type assumes you know that the five CH_2 groups are bonded in row. It saves space, especially when you are typing.

Type of Structural Formula	Structure
Structure using dashes	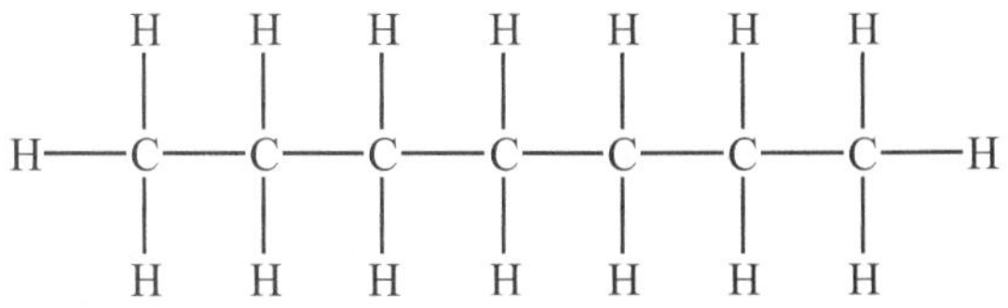
Condensed structure not showing bonds to hydrogens	$CH_3—CH_2-CH_2-CH_2-CH_2-CH_2-CH_3$
Condensed structure not showing any single bonds or unshared electron pairs	$CH_3CH_2CH_2CH_2CH_2CH_2CH_3$
Condensed structure grouping the five CH_2's together	$CH_3(CH_2)_5CH_3$
Line-angle structure	

Why do chemists use such a variety of formulas? Most of the time, we use different types of formulas to emphasize different aspects of the molecule. Other times, it is to save time and space when drawing a large molecule. For example, the following drawing of cholesterol (molecular formula $C_{27}H_{46}O$), shows every atom, bond, and unshared pair of electrons.

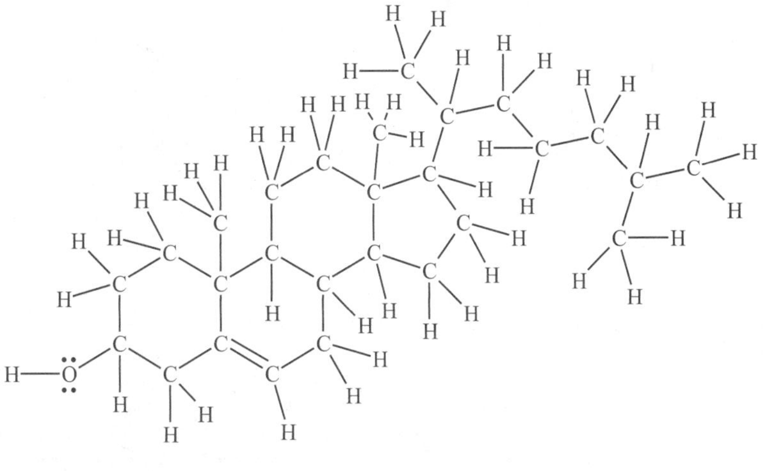

As you can imagine, it takes a long time to draw the structure of such a large molecule, showing every atom and bond. The shortest time I can do it in class is about two minutes. In contrast, the line-angle drawing of cholesterol is shown below. I can draw it in about 20 seconds. It assumes you know each angle and end of a line segment is a carbon bonded to enough hydrogens so that carbon is making four bonds. It is much easier to draw, especially if you can draw hexagons and pentagons quickly. Remember, line segments always end in carbons.

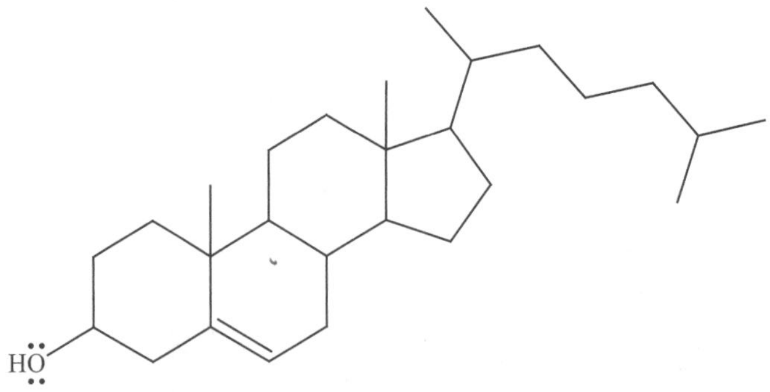

Some organic molecules have atoms that are formally charged. One example is acetylcholine, a molecule that is involved in transmitting electrical impulses between certain nerve cells. Here is the structure of acetylcholine.

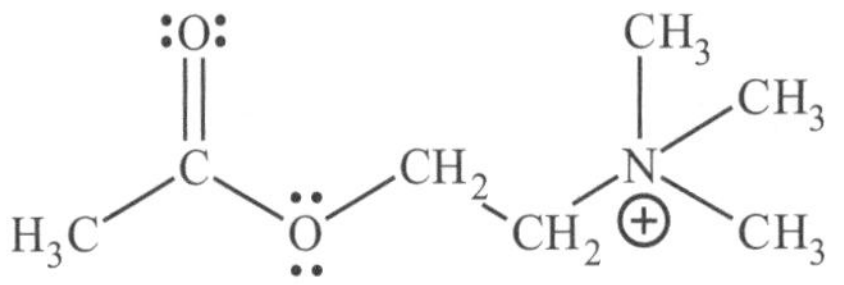

Acetylcholine

Why is the nitrogen positively charged? The nitrogen in acetylcholine is sharing four pairs of electrons, whereas a neutral nitrogen has an unshared pair of electrons and is sharing three pairs of electrons. By sharing an electron pair that the nitrogen would otherwise have to itself, the nitrogen is, in effect, giving away an electron to another atom. Because electrons are negative, giving away an electron makes the nitrogen positive.

There is a simple formula you can use to calculate the charge on an atom. The formula is

$$\text{Charge} = \text{No. of valence electrons} - \text{No. of unshared electrons} - \text{No. of shared pairs of electrons}$$

Let's look at the nitrogen of acetylcholine. It has five valence electrons (from the periodic table). It has no unshared electrons (no dots around it). It is sharing four pairs of electrons (each dash is a pair of electrons). So, to calculate the charge,

$$\text{Charge} = 5 - 0 - 4 = +1$$

Let's do another example. Consider the following structure.

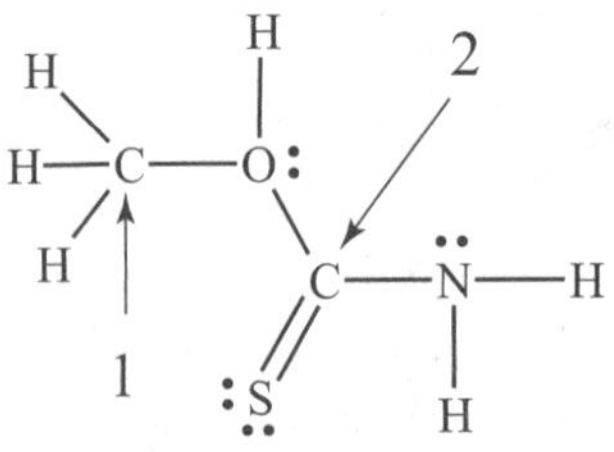

We can calculate the charges on all of the atoms, using the preceding charge formula. I won't do the hydrogens because they are all making one bond.

Atom	No. of Valence Electrons	No. of Unshared Electrons	No. of Shared Pairs of Electrons	Charge
C_1	4	0	4	0
C_2	4	0	4	0
O	6	2	3	+1
S	6	4	2	0
N	5	2	3	0

So if you can read the periodic table, and count dashes and dots, you can calculate the charge on an atom.

D. Isomerism

In Section 1.C, we drew structures for ethanol and propene. However, it is possible to draw other reasonable structures with those same atoms. For example, in addition to the structure of ethanol, another correct structure can be drawn with the formula C_2H_6O. This molecule is called methoxymethane (or commonly, dimethyl ether).

Ethanol	Methoxymethane
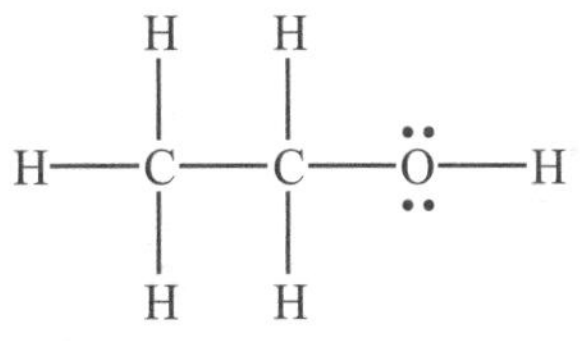	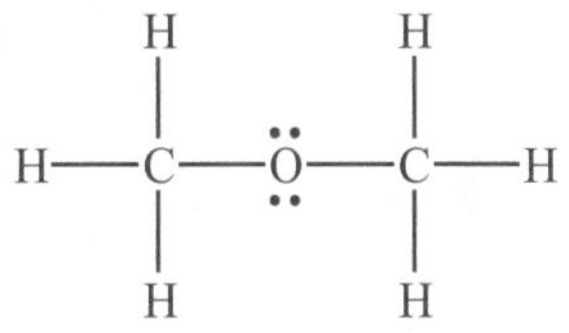

Both structures follow the octet rule for all atoms and use the correct number of electrons. They differ in the order or sequence in which the atoms are bonded together. Structures that differ in this way are called **structural isomers** or **constitutional isomers**.

There is also a structural isomer that can be drawn for propene, C_3H_6. This molecule, cyclo-propane, is shown next.

Propene	Cyclopropane
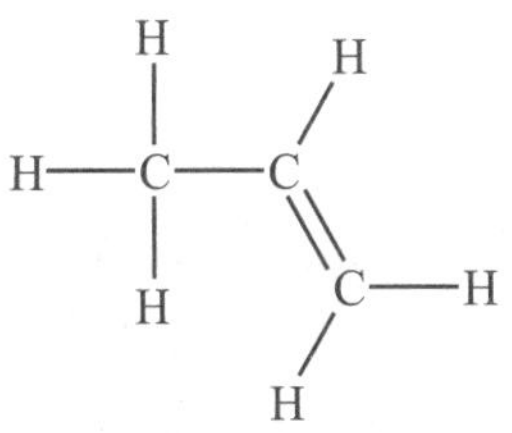	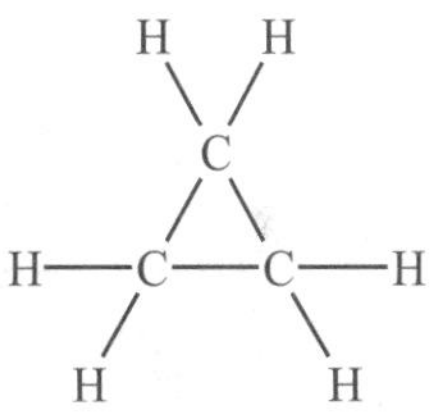

As you can see, the carbons in cyclopropane are arranged like a triangle, with two hydrogens bonded to each carbon. When carbons are arranged in polygons like triangles, squares, and hexagons, the carbons are described as being in a **ring**. Rings are quite common in organic molecules, such as the four rings in cholesterol in Section 1.C.

E. Resonance

Some molecules cannot be adequately represented by just one structure. I'm not talking about molecular formulas, like C_2H_6O, but actual molecules. One of these is nitromethane (CH_3NO_2), which is sometimes used as a fuel for race cars. Nitromethane can be drawn as shown here at the left.

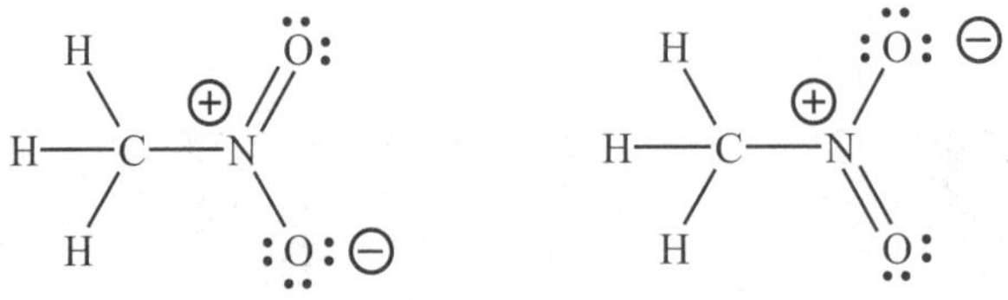

After looking at the structure, you may be thinking, "Can't you write it a different way, such as the structure on the right?" Yes, you can! The left structure has the N=O double bond to the top oxygen, and the right structure has the N=O double bond to the bottom oxygen. So, which one is correct? The answer is, neither one! The two nitrogen–oxygen bonds in nitromethane are the same length and are intermediate in length between a single and a double bond. The "real" structure is something in between these two; it looks something like this:

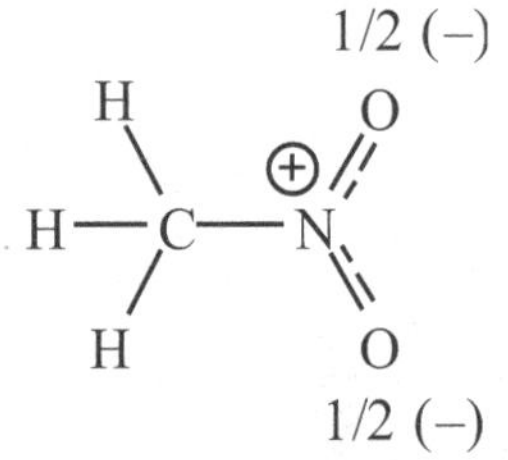

The two nitrogen–oxygen bonds are intermediate between a single and a double bond (hence, the dotted second line), and the oxygens are sharing equally the negative charge (which is why I gave each oxygen one-half of a negative charge). Structures like this are sometimes difficult to understand and can be hard to draw (notice I didn't put in any unshared electrons). The two structures I drew previously tell us something about the real structure (each has a negative oxygen, and each shows an N=O), but individually, they don't give us the whole story. Structures like these, which aren't real, and only differ in the locations of electrons, are called **resonance structures**. Resonance structures are drawn with a double-headed arrow between them, as shown here.

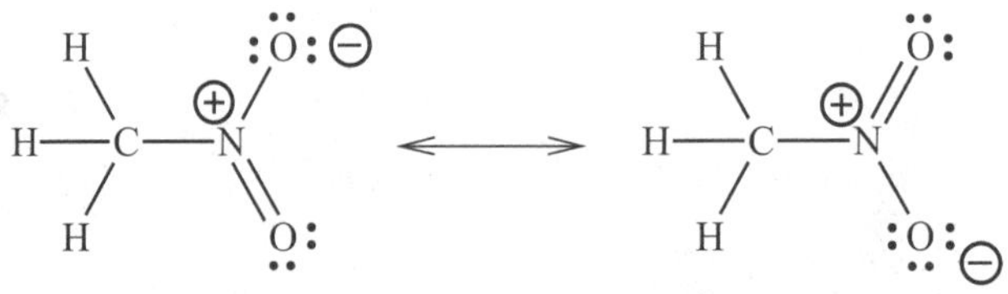

The double-headed arrow signifies that the structures represent the extremes, with reality somewhere in the middle. It does not mean equilibrium! Nitromethane is not oscillating between the left structure and the right structure, nor is it sometimes one structure or the other.

Another common example of resonance structures is the case of benzene, C_6H_6. The six carbons of benzene are in the shape of a hexagon, each carbon–carbon bond is the same length, and each carbon is bonded to one hydrogen only. The two resonance forms of benzene are shown next, along with the "real" structure.

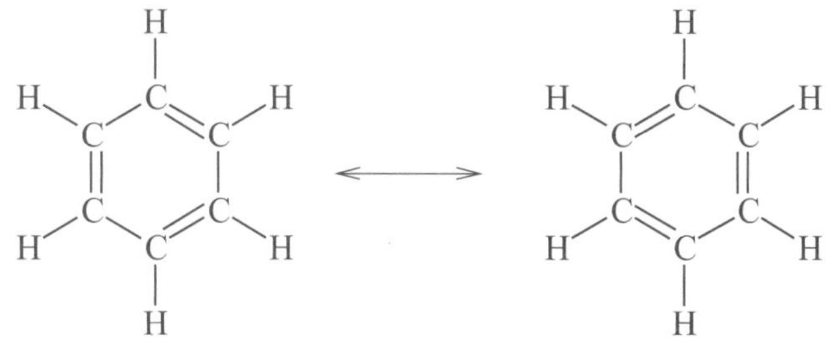

| **Resonance Structures of Benzene** | **Real Structure** |

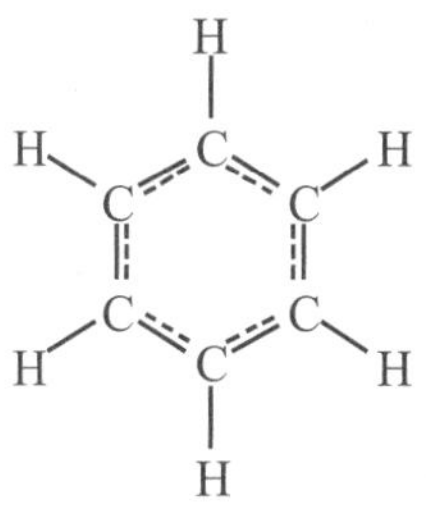

The resonance structures of benzene show alternating single and double bonds between pairs of carbon atoms, while the real structure shows "1½" bonds between each pair of carbon atoms.

One thing professors like to have students do is to draw resonance forms of structures. For example, you might be given the following resonance form of allyl cation, and be asked to draw another one. What do you do?

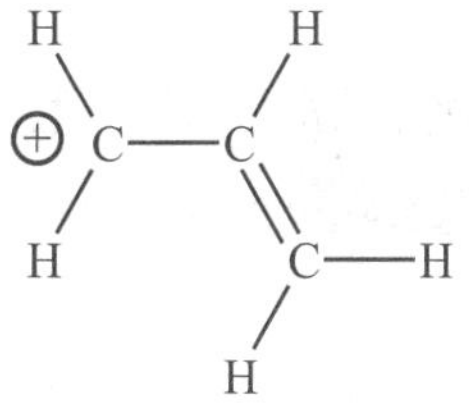

Allyl cation

First, we need to know what the properties of a "good" or stable resonance structure are. I've summarized them here.

1. No atoms can be moved, only electrons. Usually, only double bonds, triple bonds, and unshared pairs of electrons are involved.
2. Try to keep the same number of bonds. The more bonds, the more stable, so if you draw a structure with fewer bonds, it is not likely to be as good as what you started with.
3. In general, don't increase the number of charges. This is really the same as rule 2, just from another point of view.
4. Remember how many bonds a neutral atom makes! A neutral carbon makes four bonds. Nothing else will drive your organic chemistry professor more crazy than drawing carbons with five or six bonds, hydrogens with two bonds, and oxygens with four or six bonds (I've seen all of these and more)!

To keep track of where to put electrons, chemists use "curved arrows" to show where electrons are going, in a formal sense, when we draw a resonance structure. We are not implying that one resonance structure is being converted into another: We are just trying to keep track of electrons. A curved arrow shows where an electron pair is coming from, and where it ends up. For example, with allyl cation, we can "move" a pair of electrons from the double bond to the bond between the middle carbon and the positively charged carbon, as shown here.

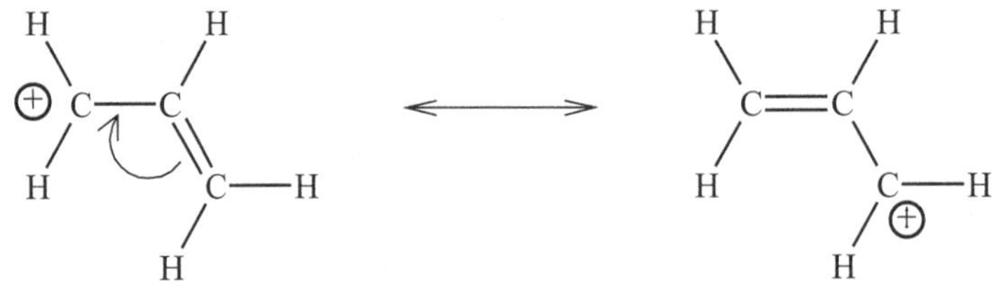

The structure on the right has the same number of bonds and charges as the one on the left, so it is equivalent to it in stability.

To convert from one of the nitromethane structures into the other, we need to move two pairs of electrons simultaneously, as shown here.

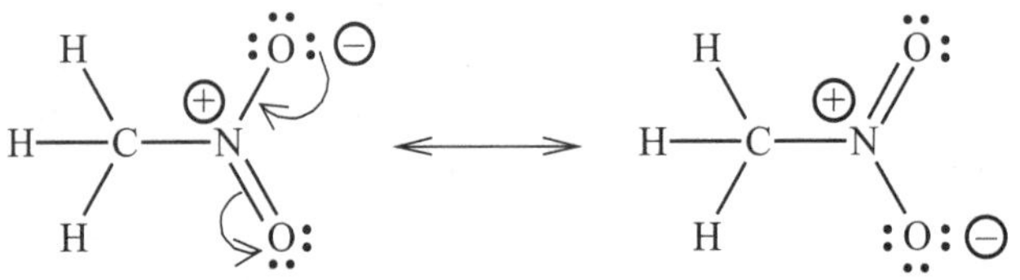

We will encounter resonance quite often in organic chemistry, so get used to it now, and you won't be surprised by it later.

F. Hydrogen-Bonding and Boiling Points

Water is an amazing substance. All of life on earth depends upon it. Chemically, it has some unusual properties. One of these is its unusually high boiling point for a molecule of such small molecular weight, as shown in the following table.

Molecule	Molecular Weight	Boiling Point (°C)
CH_4 (methane)	16	−161.4
NH_3 (ammonia)	17	−33.4
H_2O (water)	18	100
$CH_3CH_2CH_2OH$ (propan-1-ol)	60	97.2
$CH_3CH_2CH_2CH_2CH_2CH_2CH_3$ (heptane)	100	98.4

Water has a much higher boiling point than both methane and ammonia, even though it is very similar in molecular weight to them. It has essentially the same boiling point as propan-1-ol and hexane, even though these molecules have much larger molecular weights. Why? Consider the structure of water shown next.

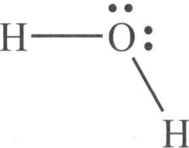

Water is a bent molecule. The hydrogens and the oxygen are covalently bonded. However, oxygen is more electronegative than hydrogen, so it has a greater attraction for the electrons in the covalent bonds than the hydrogens do. As a result, the electrons in those bonds are closer to the oxygen than they are to the hydrogens. This causes the oxygen to have a slight negative charge, and the hydrogens to have a slight positive charge. In organic chemistry, a slightly positive atom is shown by writing "δ^+" (the small Greek letter delta with a plus sign) near the atom. Similarly, a slightly negative atom is designated "δ^-". Therefore, we can draw water with the partial charges as shown here.

Now, if there are many water molecules together, the slightly positive H's of some molecules will be attracted to the partially negative O's of other molecules. These attractions are usually indicated by a dotted line between a δ^+ H and a δ^- O, as shown here.

This is really a three-dimensional network of water molecules attracted to each other. The outer water molecules are attracted to other water molecules as well.

An attraction between a partially positive atom and a partially negative atom is called a **dipole-–dipole attraction**. If the partially positive atom is a hydrogen, then it is also called a **hydrogen bond**. These attractive forces are much weaker than covalent bonds. An O–H covalent bond in water has a bond dissociation energy of 119 kcal/mol. A hydrogen bond between a partially positive H to the partially negative O has a bond dissociation energy of about 7 kcal/mol, so it is a much weaker attractive force than a covalent bond.

How does this relate to boiling point? In general, molecules in the liquid state are closer together that molecules in the vapor state, and attractive forces between molecules in the liquid must be overcome to separate those molecules. How are these attractive forces overcome? One way is by heating them. Groups of molecules with stronger attractive forces between them require more heat to overcome these attractive forces, so they have higher boiling points.

Why does water have a higher boiling point than ammonia, and why does ammonia have a higher boiling point than methane? Going back to the periodic table with electronegativity values on it, we can see what the electronegativity differences are between hydrogen and carbon, nitrogen, and oxygen are. See below.

Atom Pair	Electronegativity Difference
C–H	0.4
N–H	0.9
O–H	1.4

With an electronegativity difference of 1.4, the O–H bond in water is highly polar, with a significant partial negative charge on oxygen and partial positive charge on hydrogen. An N–H bond is not as polar because the electronegativity difference is smaller. C–H bonds are not very polar at all, and there is little partial charge character on carbon or hydrogen. Therefore, ammonia molecules have some dipole–dipole attraction for each other, but not as much as water, so less heat is required to separate ammonia molecules than water molecules. Because C–H bonds have little polar character, even less heat is required to separate methane molecules from each other. This simplified analysis of polarity works well when comparing molecules of similar shape and molecular weight.

There is one other type of attractive force that all organic molecules have for one another, and that is called the **induced dipole-induced dipole** interaction, sometimes referred to in different texts as London forces, dispersion forces, or van der Waals forces. What this refers to is the attraction of nuclei in one molecule for electrons in another molecule, due to distortions in the electron clouds around molecules. London forces are very short range and weak attractive forces, compared to dipole–dipole attractions. All molecules have these attractive forces for other molecules. That's why I didn't mention them when I discussed methane, ammonia, and water. However, London forces are important when comparing molecules with different molecular weights, such as methane and heptane. Because heptane has more atoms than methane, it has more nuclei to be attracted to more electrons than in methane, so there are more London forces holding heptane molecules together.

The amount of London forces is also dependent upon the shape of a molecule. The more surface area that can interact with another molecule, the more London forces there are, and the higher the boiling point. The more linear a molecule is, the higher the boiling point. I will discuss this more in Chapter 2.

Heptane and propan-1-ol have similar boiling points, due to differing amounts of London and dipole–dipole attractions. Heptane is bigger than propan-1-ol, so it has more London forces. Since propan-1-ol has an O–H group, it has hydrogen-bonding attractive forces that heptane doesn't have (see the following drawing). So the hydrogen-bonding attractive force makes up for a lot of London forces, because hydrogen-bonding is a much stronger attractive force.

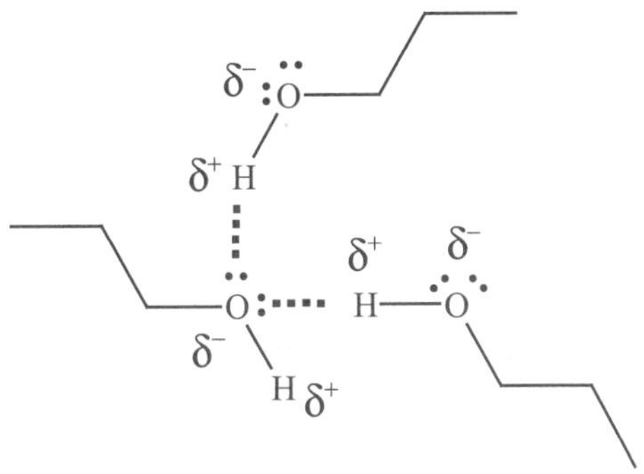

Normally, if your instructor asks you to tell which compound has a higher boiling point, he/she won't ask you to compare molecules like heptane and propan-1-ol. Instead, the molecules will differ in polarity, size, or shape, but not in more than one of these at a time.

This is admittedly a very simplified view of how attractive forces relate to boiling points, but it does work well for many cases.

G. Water-Solubility

You have probably known that sodium chloride is soluble in water since childhood. Why is it soluble? Because sodium chloride is a salt, it is composed of sodium and chloride ions. Because water is polar, the partially positive hydrogens can be attracted to the chloride anions, and the partially negative oxygens are attracted to the sodium cations. Eventually, the crystal lattice of the sodium chloride is broken down, and the sodium and chloride ions are surrounded by water molecules. This is shown next.

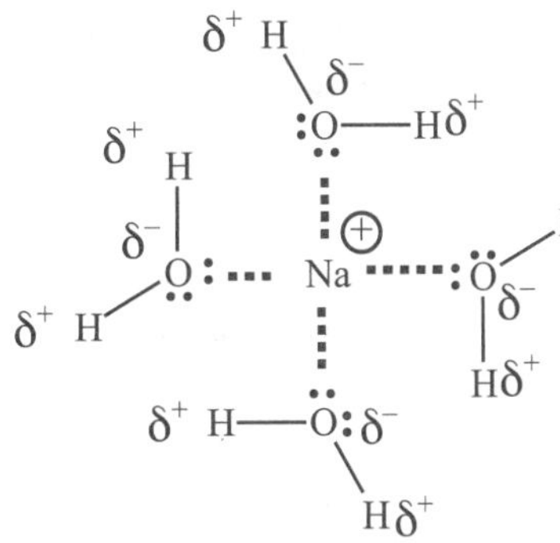

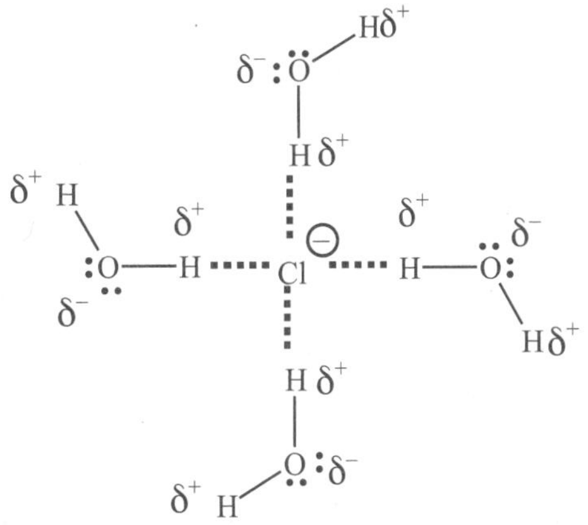

Organic molecules with partial positive and negative charges can also be attracted to water. Many smaller organic compounds with oxygens or nitrogens are significantly water-soluble. The attractions of propan-1-ol with water are shown here.

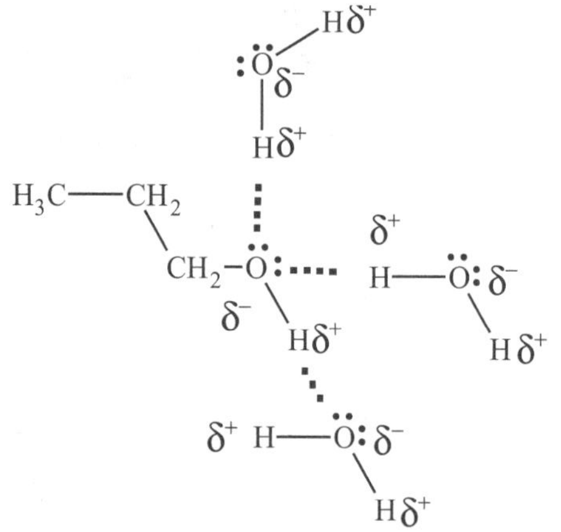

Notice that the water molecules are only attracted to the O–H group of propan-1-ol. The CH_2's and the CH_3 are not significantly polar because carbon and hydrogen are similar in electronegativity, so the water is not attracted to that part of propan-1-ol.

If you look up propan-1-ol in the *Merck Index* (13th edition), it is listed as being miscible in water. My old 51st edition of the *CRC Handbook of Chemistry and Physics* lists its solubility as "∞." Both of these imply that propan-1-ol and water form a homogenous solution no matter what proportions we mix of them. The newer editions of the *CRC Handbook of Chemistry and Physics* just list water as a solvent that propan-1-ol is soluble in, which is not very helpful.

Heptane is listed as being insoluble in water in both handbooks. This makes sense because heptane has only carbon and hydrogen, so there are no polar bonds for water to be attracted to. It is actually soluble to some small extent, due to London forces. One material safety data sheet for heptane lists the solubility as 0.5 g/liter of water. This is pretty close to insoluble.

H. Acid–Base Concepts

You may have seen or made a demonstration of a volcano. You pour some baking soda into the cone of the volcano and then pour in some vinegar; very shortly, you have a bunch of fizzy foam. If you were really creative, you added red food coloring to make the foam look more like lava. But why does it fizz? You have just done a reaction of an acid with a base, which ultimately produced carbon dioxide bubbles as the fizzy by–product.

Let's review some definitions. The Brønsted–Lowry acid is something that can donate H^+ ions. The Brønsted–Lowry base accepts H^+ from something. For most acids and bases, these definitions work pretty well. In the preceding reaction, acetic acid from the vinegar reacted with sodium bicarbonate. The acetic acid donated an H^+ ion (which is also often called a proton), and bicarbonate accepted the proton. When the acetic acid gives up a proton, it becomes an acetate ion. The sodium ion followed along to balance the charge. The bicarbonate ion accepted a proton to become carbonic acid. Carbonic acid is in equilibrium with CO_2 and H_2O. The equation is shown here.

$$CH_3-\overset{\overset{\displaystyle :O:}{\|}}{C}-\overset{..}{\underset{..}{O}}-H \quad + \quad \overset{Na \oplus}{\underset{..}{:\underset{..}{O}}}-\overset{\overset{\displaystyle :O:}{\|}}{\underset{\ominus}{C}}-\overset{..}{\underset{..}{O}}H \quad \rightleftharpoons \quad CH_3-\overset{\overset{\displaystyle :O:}{\|}}{C}-\overset{Na \oplus}{\underset{..}{\underset{\ominus}{O}:}} \quad + \quad H-\overset{..}{\underset{..}{O}}-\overset{\overset{\displaystyle :O:}{\|}}{C}-\overset{..}{\underset{..}{O}}H$$

| Acetic acid | Sodium bicarbonate | Sodium acetate | Carbonic acid |

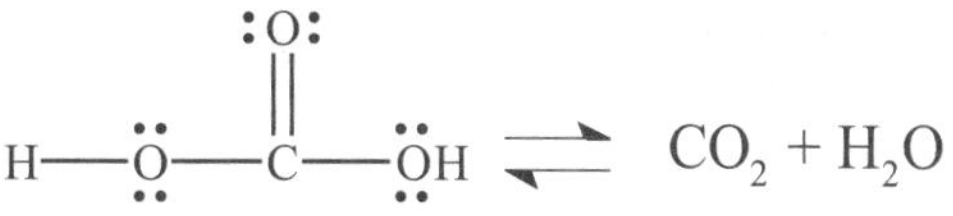

$$H-\overset{..}{\underset{..}{O}}-\overset{\overset{\displaystyle :O:}{\|}}{C}-\overset{..}{\underset{..}{O}}H \quad \rightleftharpoons \quad CO_2 + H_2O$$

There is another way of defining acids and bases that complements the Brønsted–Lowry definition. A Lewis acid accepts an electron pair, and a Lewis base donates an electron pair. The bicarbonate ion donates an electron pair to the hydrogen of acetic acid and forms a new bond. Acetic acid accepted the electron pair, which resulted in breaking its O–H bond. I have redrawn the equation here, using curved arrows to show the formal electron movements.

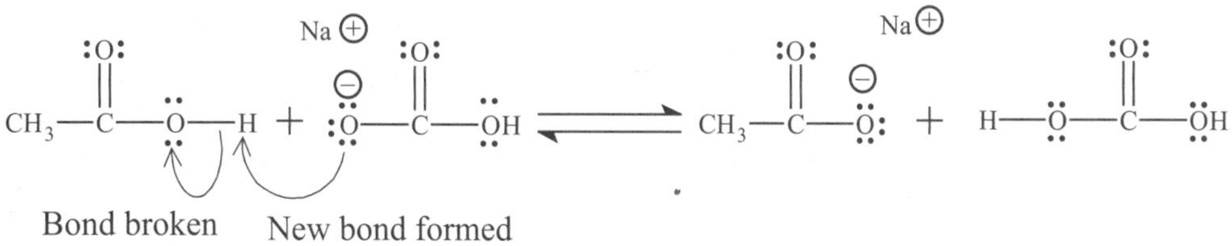

$$CH_3-\overset{\overset{\displaystyle :O:}{\|}}{C}-\overset{..}{\underset{..}{O}}-H \quad + \quad \overset{Na \oplus}{\underset{..}{:\underset{..}{O}}}-\overset{\overset{\displaystyle :O:}{\|}}{\underset{\ominus}{C}}-\overset{..}{\underset{..}{O}}H \quad \rightleftharpoons \quad CH_3-\overset{\overset{\displaystyle :O:}{\|}}{C}-\overset{Na \oplus}{\underset{..}{\underset{\ominus}{O}:}} \quad + \quad H-\overset{..}{\underset{..}{O}}-\overset{\overset{\displaystyle :O:}{\|}}{C}-\overset{..}{\underset{..}{O}}H$$

Bond broken   New bond formed

We will be using the Lewis concepts of acids and bases often in organic chemistry. One further example is in the following reaction of BH_3 (borane) with ammonia.

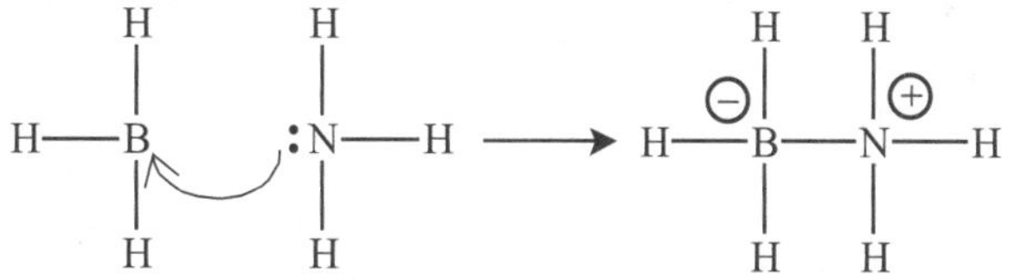

$$H-\overset{\overset{\displaystyle H}{|}}{\underset{\underset{\displaystyle H}{|}}{B}} \quad :\overset{\overset{\displaystyle H}{|}}{\underset{\underset{\displaystyle H}{|}}{N}}-H \quad \longrightarrow \quad H-\overset{\overset{\displaystyle H}{|}}{\underset{\underset{\displaystyle H}{|}}{\underset{\ominus}{B}}}-\overset{\overset{\displaystyle H}{|}}{\underset{\underset{\displaystyle H}{|}}{\underset{\oplus}{N}}}-H$$

In this case, the boron accepts the electron pair from the nitrogen. Therefore, borane is the Lewis acid, and ammonia is the Lewis base. Borane is not a Brønsted–Lowry acid because it is not donating H^+.

I. Functional Groups

You may have picked up on the idea that there are many different organic compounds because there can be many possible structures for the same molecular formula. For example, there are 366,319 possible different structures with the formula $C_{20}H_{42}$! At present, there are over 6 million known organic compounds, and more are being made every day. Now, one possible way to teach organic chemistry would be to discuss each organic compound separately. Assuming I only spent 1 page on each compound, this book would have over 6 million pages, and would be a little heavy for you to carry around!

There is a better way to teach you about organic compounds. This is to teach you about the properties and reactivity of different classes of compounds. These classes are referred to as functional groups. A **functional group** may be defined as an atom of group of atoms that undergoes certain reactions under prescribed sets of conditions. The following table contains a list of common functional groups, some simple examples, and the IUPAC endings we will use as we learn how to name organic compounds.

Name	Functional Group Structure*	Simple Example	IUPAC Name Ending
Alkane		H_3C——CH_3 Ethane	*-ane*
Alkene		H_2C==CH_2 Ethene (Ethylene)	*-ene*
Alkyne		H——C≡C——H Ethyne (Acetylene)	*-yne*
Aromatic Ring		 Benzene	Varies
Alkyl halide		H_3C——Cl Chloromethane (Methyl chloride)	None
Alcohol		H_3C——OH Methanol (Methyl alcohol)	*-ol*

Name	Functional Group Structure*	Simple Example	IUPAC Name Ending
Ether		H_3C—O—CH_3 Methoxymethane (Dimethyl ether)	None
Amine		H_3C—NH_2 Methanamine (Methylamine)	*-amine*
Aldehyde		Ethanal (Acetaldehyde)	*-al*
Ketone		2-Propanone (Acetone)	*-one*
Carboxylic Acid		Ethanoic acid (Acetic acid)	*-oic acid*
Acid Chloride		Ethanoyl chloride (Acetyl chloride)	*-oyl chloride*

Name	Functional Group Structure*	Simple Example	IUPAC Name Ending
Anhydride		 Ethanoic anhydride (Acetic anhydride)	*-oic anhydride*
Ester		 Ethyl ethanoate (Ethyl acetate)	*-oate*
Amide		 Ethanamide (Acetamide)	*-amide*
Nitrile	—C≡N	H_3C—C≡N Ethanenitrile (Acetontrile)	*-nitrile*

*The bonds whose connections aren't specified are assumed to be attached to carbon or hydrogen atoms in the rest of the molecule.

How does using functional groups help us learn organic chemistry? Remember the 366,319 possible isomers of $C_{20}H_{42}$? All of those compounds are alkanes. So instead of spending my time teaching you the same types of reactivity over and over again for 366,319 pages, I can spend a much smaller number of pages (see Chapter 2) teaching you the characteristic reactivity and properties of a whole class of molecules.

REVIEW EXERCISES FOR CHAPTER 1

1. Draw *two* different structures that correspond to each of the following molecular formulas. Each atom must be neutral and must fulfill the octet rule. Show all bonds, all atoms, and all unshared pairs of electrons. One structure for each formula must contain a ring.
   a. $C_4H_6O_2$ b. $C_3H_3F_3N_2$

2. In the following bonds, label one atom as δ^+ and one atom as δ^-.
   a. C–Br   b. Cl–P   c. H–N   d. S–O

3. Draw curved arrows to show how the structure on the left is converted into the structure on the right.

   a.

4. Draw a line formula for the expanded structure in a, and expanded formulas for the line structures in b and c.

a.   b. 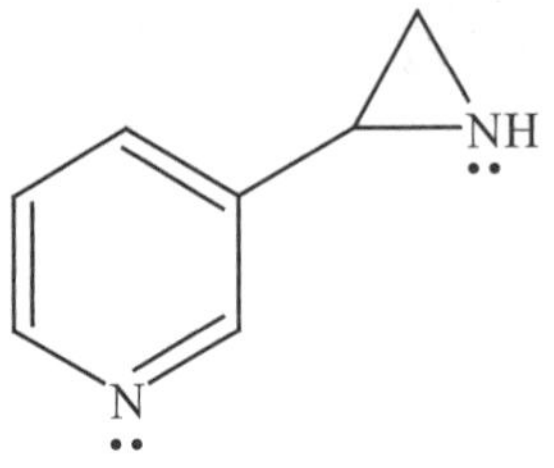   c.

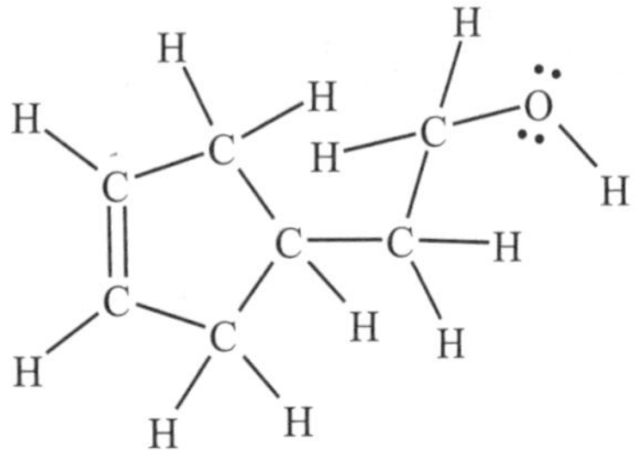

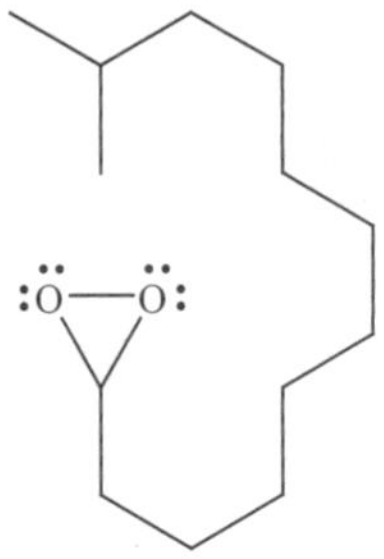

5. Explain why 1-pentanol has a higher boiling point and is more water-soluble than hexane. Describe the types of intermolecular forces involved in each case.

1-Pentanol 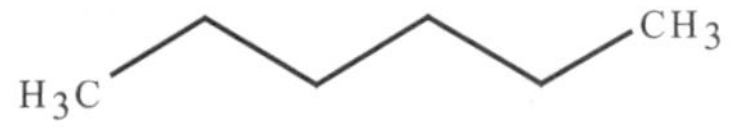   Hexane

6. Circle and label neatly the functional groups listed in Chapter 1 in the following structures. Do **not** include alkanes.

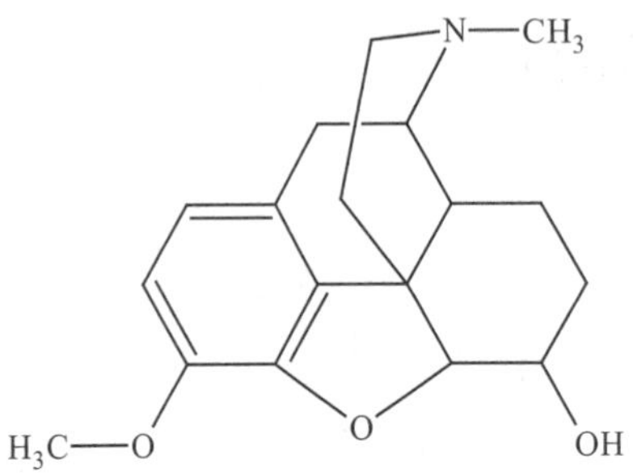

Codeine	Aspirin

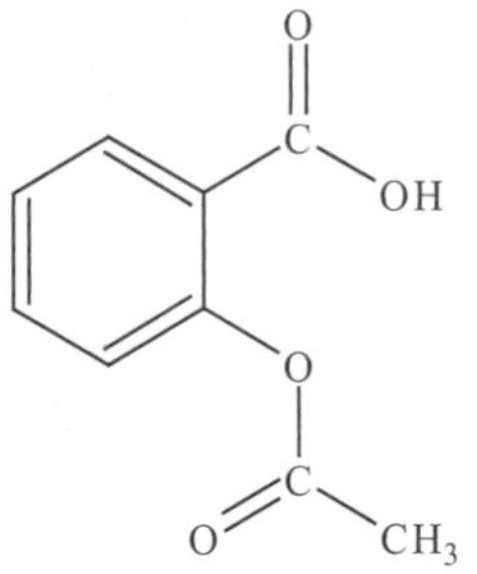

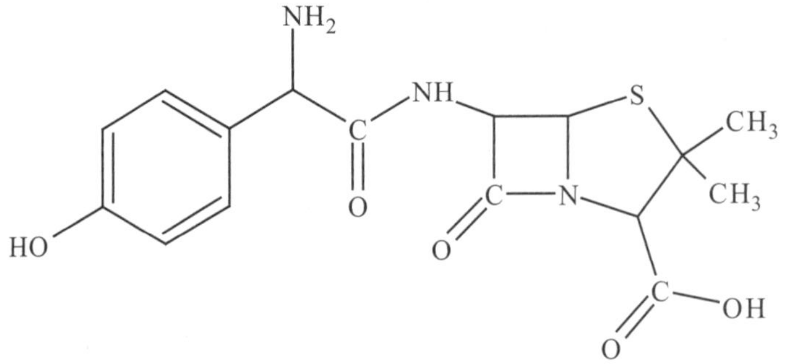

Amoxicillin

7. Consider the following acid–base reactions. Label the compound on the left side of the equation that is the acid, and the one that is the base. Draw arrows to show how bonds were broken and formed to produce the products of the reaction.

a.

$$CH_3—C\equiv\overset{K\oplus}{\underset{\ominus}{C}}\colon\ \ +\ \ H—\ddot{\underset{\cdot\cdot}{C}}l\colon\ \ \rightleftharpoons\ \ CH_3—C\equiv C—H\ \ +\ \ \overset{\oplus}{K}\ \ \overset{\ominus}{\colon}\ddot{\underset{\cdot\cdot}{C}}l\colon$$

b.

$$H_3C—\ddot{\underset{\cdot\cdot}{S}}—H\ \ +\ \ \colon\!\overset{H}{\underset{H}{N}}\!—CH_3\ \ \rightleftharpoons\ \ H_3C—\overset{\ominus}{\ddot{\underset{\cdot\cdot}{S}}}\colon\ +\ \ H—\overset{H}{\underset{H}{\overset{\oplus}{N}}}\!—CH_3$$

Alkanes

WHAT YOU WILL LEARN

In this chapter, you will learn:

- how to use IUPAC nomenclature to name alkanes;
- how to figure out the most stable shapes of simple non-cyclic and cyclic alkanes;
- how the structure of alkanes influences their boiling points and water-solubilities.

SECTIONS IN THIS CHAPTER

- Structure and Bonding

- IUPAC Nomenclature

- Shapes of Simple Alkanes

- Shapes of Cyclic Alkanes

- Physical Properties of Alkanes

The simplest class of organic compounds is the alkanes. Alkanes contain only carbon and hydrogen, and all the bonds between atoms are single bonds. Alkanes are commonly found in petroleum (gasoline, kerosene, and diesel fuel) and natural gas. Methane is also a product of anaerobic respiration of some prokaryotes.

A. Structure and Bonding

The simplest alkane is methane, CH_4. The hydrogens are arranged around the central carbon atom so that each hydrogen atom is as far apart from the other hydrogens as possible. This arrangement is shown here, where the lighter ball in the middle is a carbon, and the four darker balls are hydrogens.

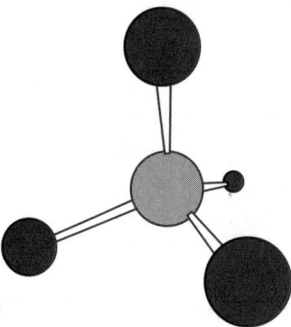

The H–C–H bond angles are 109.5°, and all the C–H bonds are the same length. How do we account for this geometry? To explain this, we need to go back to the general chemistry and look at atomic orbitals.

Hydrogen only has one electron, and it is in a $1s$ orbital. The $1s$ orbital is spherical in shape. Carbon's valence electrons are in $2s$ and $2p$ orbitals. The $2s$ orbital is spherical in shape, and the $2p$ orbital is sort of dumbbell-shaped.

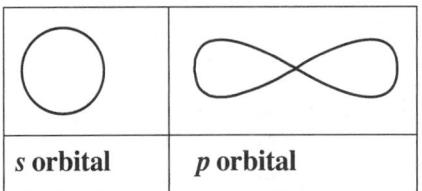

s orbital	*p* orbital

If you try to draw an orbital picture of how the valence orbitals are arranged around the carbon nucleus, you end up with a picture the looks sort of like this.

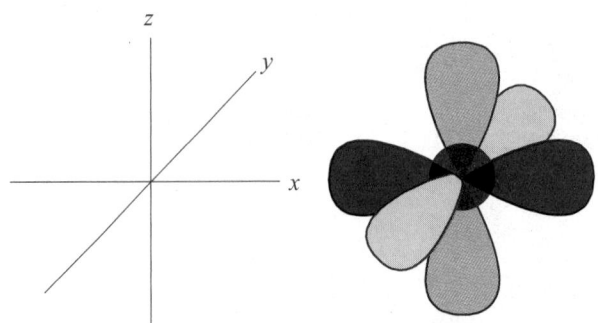

The *s* orbital is the dark sphere, and the *p* orbitals are all perpendicular to each other, on the *x*-, *y*-, and *z*-axes. Because the angles between the *p* orbitals are 90°, it is difficult to see how we can use these individual orbitals for bonding, so that we could obtain 109.5° bond angles.

Remember that orbitals are really mathematic functions that describe where an electron would probably be. We could imagine mathematically combining the wave functions for the 2*s* and the three 2*p* orbitals and producing four equivalent orbitals from them. Because these new orbitals were derived from one *s* and three *p* orbitals, they would have one-fourth *s* character and three-fourths *p* character. Such orbitals are called **sp³ hybrid orbitals**. These orbitals are arranged so that they are as far apart form each other as possible, that is, they have 109.5° bond angles. Since the carbon in methane uses *sp³* hybrid orbitals to describe its bonding, the carbon is referred to as *sp³* hybridized.

What do *sp³* hybrid orbitals look like? They generally look like this.

It looks kind of like a *p* orbital, with one very small lobe, and one larger lobe that is a little fatter than the lobe of a *p* orbital. The larger lobe is the one used for bonding. So a hybrid orbital picture of methane might look something like the following, with the carbon *sp³* orbitals (distorted ovals) overlapping with H *s* orbitals (spheres), as shown in the following drawing on the left. The smaller lobes of the *sp³* orbitals have been omitted for clarity. The *sp³* orbital on the lower left is larger, to try to show that it is coming toward you. The smaller *sp³* orbital is pointing away from you.

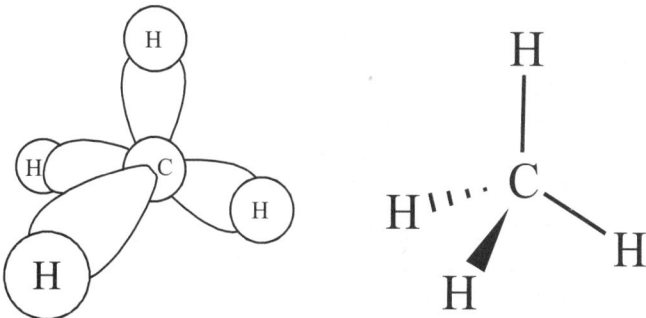

Rather than drawing orbitals, we usually draw bonds as lines. To try to show a three-dimensional molecule in a two-dimensional medium, there is a standard convention used. If a bond is shown as a solid wedge, it is coming at you out of the paper. If it is a dashed line, it is going away from you. So, we write the three-dimensional structure of methane as shown in the preceding drawing on the right.

B. IUPAC Nomenclature

The International Union of Pure and Applied Chemistry (IUPAC, for short) has developed a set of rules for giving a unique name for each organic compound. The complete rules for IUPAC naming, with examples, can be found on the following web site: *http://www.acdlabs.com/iupac/ nomenclature/*

You will not have to learn to name every possible organic compound, but by the end of the course, you should be able to name many thousands. This will allow you to read chemical articles, and even product labels, and have some idea what the structures of the molecules are that are discussed.

B.1. NAMING CHAINS

Alkanes come in all shapes and sizes. The simplest alkanes have all of the carbons bonded together in a row (which is sometimes referred to as a chain of carbons). Consider the following structure, whose IUPAC name is heptane:

$$CH_3–CH_2–CH_2–CH_2–CH_2–CH_2–CH_3$$

There are two parts to the name. *Hept-* means that there are seven carbons bonded together in a continuous chain. I call this the **base part** of the name, and *-ane* means the compound is an alkane. This is the **functional group suffix**. Put them together and you have the name. To name other simple alkanes, you need to know the appropriate base part for other numbers of carbons in a continuous chain. See the following table.

No. of Carbons	Base	No. of Carbons	Base
1	*Meth*	11	*Undec*
2	*Eth*	12	*Dodec*
3	*Prop*	13	*Tridec*
4	*But*	14	*Tetradec*
5	*Pent*	15	*Pentadec*
6	*Hex*	16	*Hexadec*
7	*Hept*	17	*Heptadec*
8	*Oct*	18	*Octadec*
9	*Non*	19	*Nonadec*
10	*Dec*	20	*Icos*

The base names for between one and four carbons come from old names that people have used for a long time, and so they were sort of "grandfathered in." The names for five carbons and up make more sense. If you know these twenty base names, you can name twenty alkanes. See the following table.

Structure	Name
 H_3C [chain structure] CH_3	Nonane: nine carbons in a row.

H_3C—CH_2–CH_2–CH_2
              |
         H_2C——CH_3
	Hexane: there are six carbons in a row, even if I didn't draw them on the same horizontal line.

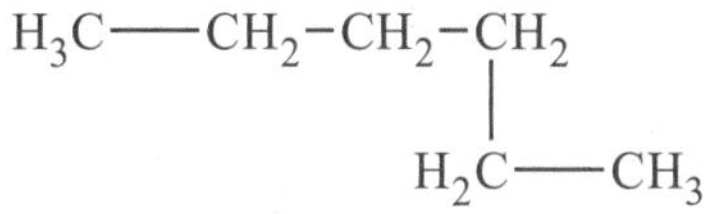

B.2. NAMING RINGS

Organic compounds can also exist as rings. If a compound is a ring, the prefix *cyclo-* is placed in front of the base part of the name. Rings can be written in differently. See the following table.

Structure	Name
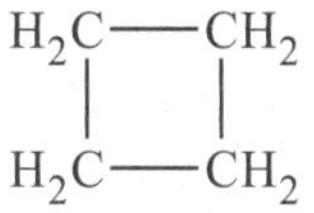 H_2C——CH_2 | | H_2C——CH_2	Cyclobutane: four carbons in the ring
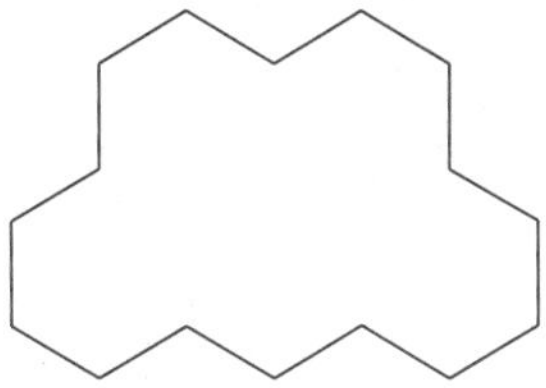 [sixteen-membered ring structure]	Cyclohexadecane: sixteen carbons in the ring

B.3. ONE GROUP ON A CHAIN

Unfortunately for organic students, alkanes can exist in more forms than simple chains or rings. See the following structure. It has seven carbons, but they are not arranged in a chain. Six carbons are in a chain, with a CH_3 attached.

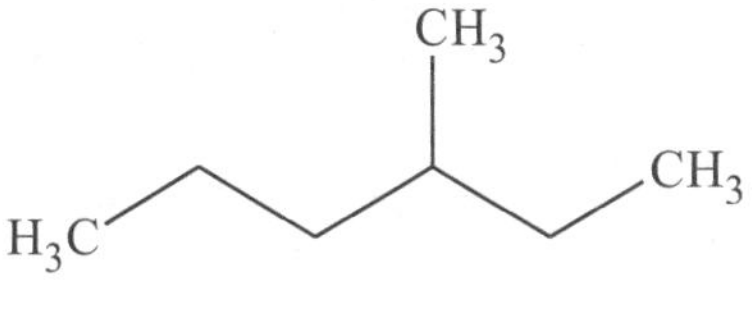

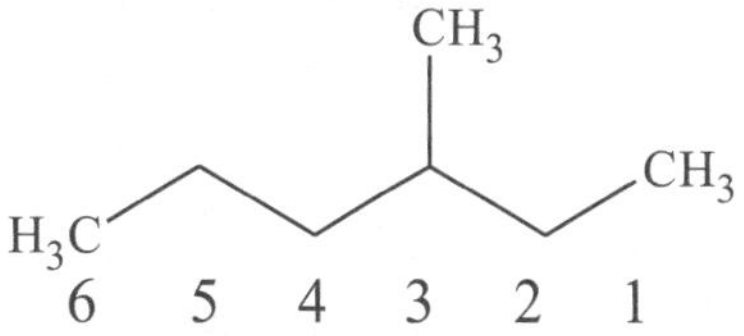

First we name that longest continuous chain. An alkane with six carbons in a row is hexane. The base for one carbon is *meth*. IUPAC defines the suffix *-yl* to indicate a carbon group attached to the longest chain, so the CH_3 group is called a methyl group. Methyl goes in front of hexane in the name, so we have a methylhexane. We are not done yet. We have to indicate to which carbon of the hexane chain the methyl group is attached. To do this, we number the carbons in the chain from the end of the chain closest to the methyl group, as shown above right. This would place the methyl group on the third carbon, so a number is placed before the methyl to indicate where on the longest chain the methyl is attached. So the complete name for this structure is **3-methyl-hexane**. Numbers are separated from words with dashes.

The longest chain is not always written straight across, as in the next structure.

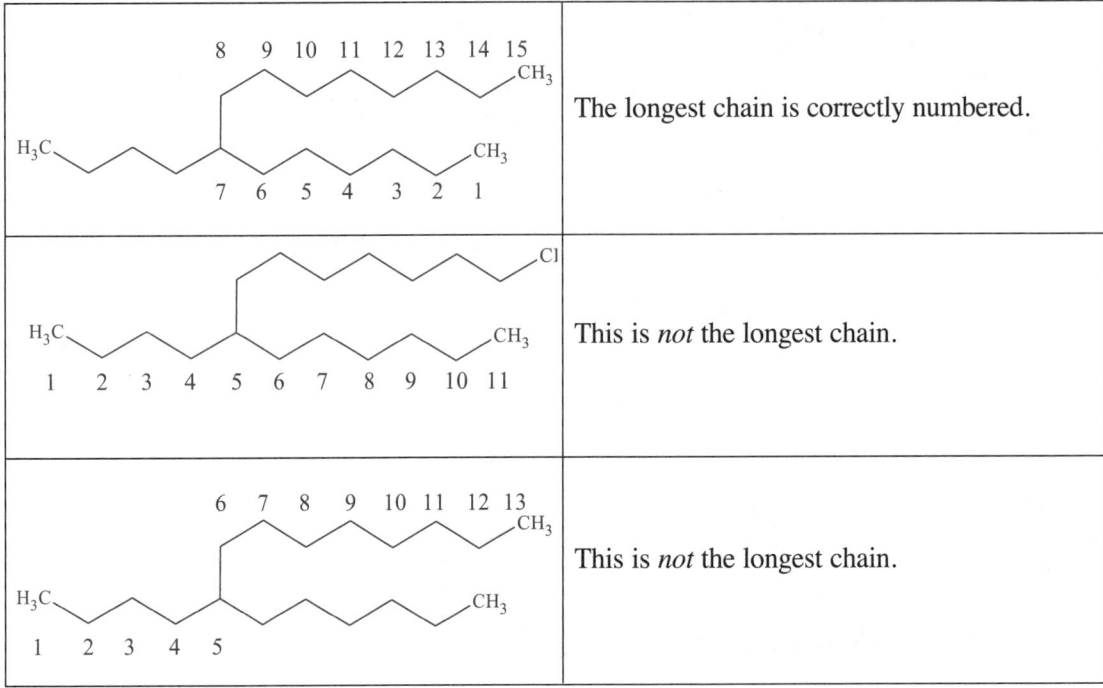

Remember, the first rule of naming is to find the longest continuous chain of carbons. This is fifteen, not eleven or thirteen.

The longest chain is fifteen, so it is a pentadecane. There is a four-carbon group attached to the pentadecane, so it is a butylpentadecane. Numbering the longest chain from the end closest to the butyl group gives the number 7, so the correct name is **7-butylpentadecane**.

B.4. ONE GROUP ON A RING, ONE RING ON A CHAIN, AND TWO RINGS

Groups can also be attached to rings. If the chain contains fewer carbons than the ring, or the same number of carbons, the compound is named as an **alkylcycloalkane**. If there is only one group attached to a ring, there is no need to use a number to locate the group on the ring. If the chain contains more carbons than the ring, the compound is named as a **cycloalkylalkane**. A number is used to tell where the cycloalkyl group is. If there are two rings, the smaller ring is named as a substituent on the larger ring. See the following examples.

	Propylcyclopentane
	Pentylcyclodecane
	1-Cyclobutylheptane
	Cyclopropylcycloheptane

B.5. COMPLEX GROUPS

Consider the following two structures. What names would you give them?

According to the rules we covered previously, you might name them both propylcyclopentane because they both have rings of five carbons, with a three-carbon group attached. However, the way the three-carbon group has been attached to the ring differs. In the molecule on the left, the three-carbon chain is attached at one of the end carbons of the three-carbon chain. In the molecule on the right, the three-carbon chain is attached at the middle carbon of the three-carbon chain. In both molecules, the group attached has three carbons and seven hydrogens, but the manner in which the group is attached differs.

IUPAC allows two different ways to overcome our difficulty in naming these groups. The first is to use what might be called common names for the two groups. A $CH_2CH_2CH_3$ group is named propyl. The name for the group on the right compound is called isopropyl (*iso* reflecting the word "isomer," which we defined as having a different structure or arrangement of atoms). So the compound on the left is propylcyclopentane, and the compound on the right is isopropylcyclopentane.

The other way to name it is more systematic. You start numbering at the carbon of the group attached to the cyclopentane group. If this carbon is number 1, then the next carbon is carbon 2, and we've come to the end of the chain, so it is an ethyl group. But the ethyl group has a methyl group attached to carbon one of the ethyl group, so this is a 1-methylethyl group, as shown here.

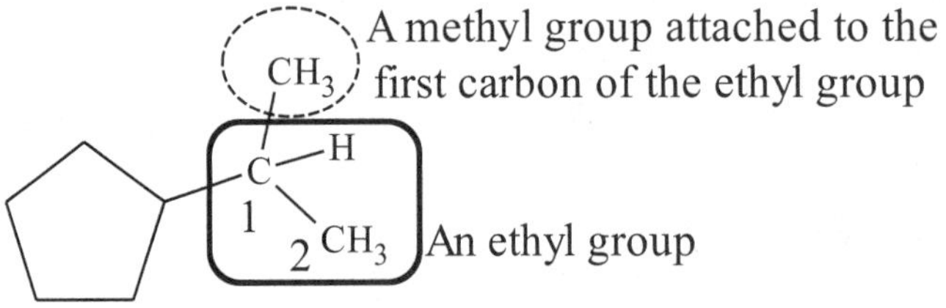

A methyl group attached to the first carbon of the ethyl group

An ethyl group

Therefore, the other IUPAC allowed name for this compound is (1-methylethyl)cyclopentane; "1-methylethyl" is in parentheses to keep us from thinking there is a methyl and an ethyl group in the structure.

There are four types of four-carbon groups. These are illustrated and named in the following table.

Stucture	Common Name	Systematic Name
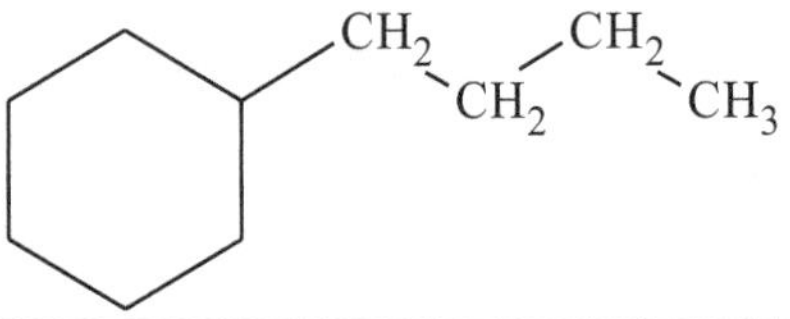	Butylcyclohexane (sometimes called *n*-butyl, where *n* stands for normal)	Butylcyclohexane
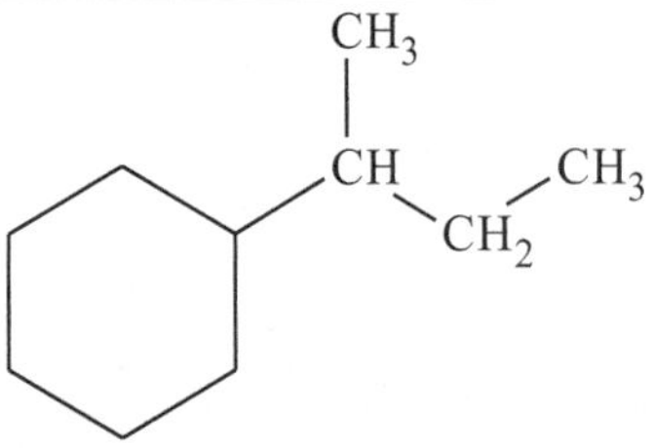	*sec*-Butylcyclohexane (*sec* is short for secondary, and is sometimes abbreviated as *s*)	(1-Methylpropyl)cyclohexane
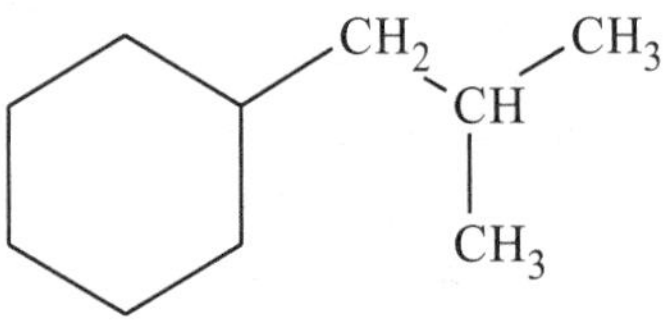	Isobutylcyclohexane	(2-Methylpropyl)cyclohexane
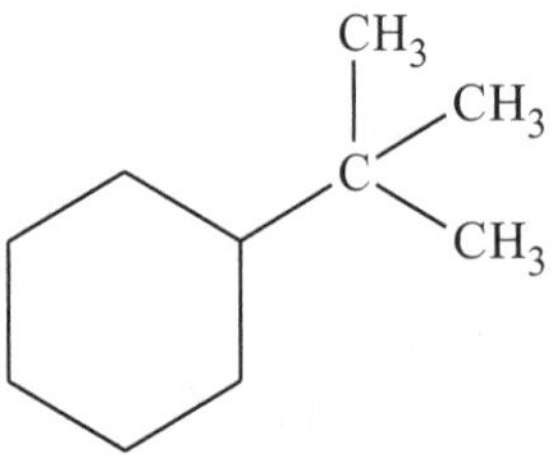	*tert*-Butylcyclohexane (*tert* is short for tertiary, and is sometimes abbreviated as *t*)	(1,1-Dimethylethyl)cyclohexane

"Secondary" implies that there are *two* carbons in the group that are bonded to the carbon attached to the cyclohexane. "Tertiary" implies that there are *three* carbons in the group that are bonded to the carbon attached to the cyclohexane. Both "sec" and "tert" are italicized, and separated from "butyl" by a hyphen. "Iso," on the other hand, is not italicized nor separated by a hyphen. I don't know why.

If a group has more than four carbons, it is generally named systematically. See the following examples.

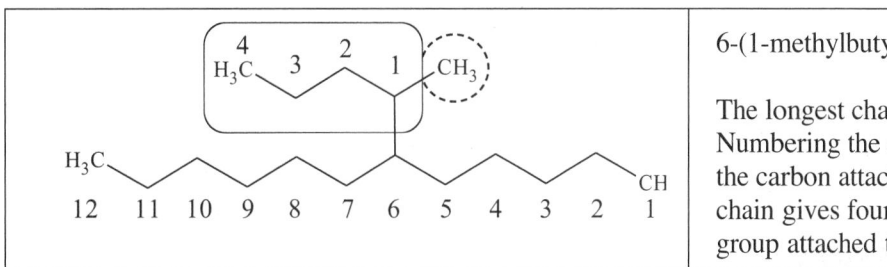

6-(1-methylbutyl)dodecane	

The longest chain is twelve. Numbering the side chain from the carbon attached to the main chain gives four, with a methyl group attached to the first carbon.

B.6. TWO OR MORE OF THE SAME GROUP ON A CHAIN

Organic compounds can have more than one of a group attached to the longest chain. If they do, we apply the rules for naming as follows:

1. Find the longest continuous chain of carbons.
2. Add a prefix before the group name to tell how many of the group there are. There are different prefixes for simple groups (such as methyl, butyl, or isopropyl), and for the complex groups, such as (1-methylethyl) or (2-ethylhexyl).

Number of Groups	Simple Group Prefix	Complex Group Prefix
2	*di*	*bis*
3	*tri*	*tris*
4	*tetra*	*tetrakis*
5	*penta*	*pentakis*
6	*hexa*	*hexakis*

3. Number the chain from the end with a group closest to it.
4. Assign a number to each group, so we know where each group is.

See the following examples.

Numbered incorrectly	1. The longest chain is seven, so it is a heptane. 2. There are two methyl groups, so it is a dimethylheptane. 3. We number from the right end, since we run into a methyl group attached to carbon 2. If we numbered from the left end, we wouldn't hit a methyl group until carbon 3. 4. Numbering from the right, the methyl groups are on carbons 2 and 5 of the heptane, so the name is **2,5-dimethylheptane**.
	1. The longest chain is ten, so it is a decane. 2. There are two isopropyl or (1-methylethyl) groups. 3. Numbering from the left gives us the lowest number for the first group. 4. Either **4,5-diisopropyldecane**, or **4,5-bis(1-methylethyl)decane** are acceptable.

B.7. MORE THAN ONE OF THE SAME GROUP ON A RING

This variation is similar to those in the preceding examples, except now we have to use numbers to tell where the groups are attached. Because a ring doesn't have ends, we start numbering at the carbon that one of the groups is attached to. Then we number in the direction of the closest group. See the following examples.

	This is a trimethylcyclopentane. If we start numbering at the carbon with the methyl attached on the bottom right, and go counterclockwise, we get the numbers 1, 2, and 4. If we start at the carbon with the methyl attached on the top right, and go clockwise, we also get 1, 2, and 4, as shown in italics. So this is **1,2,4-trimethylcyclopentane**.

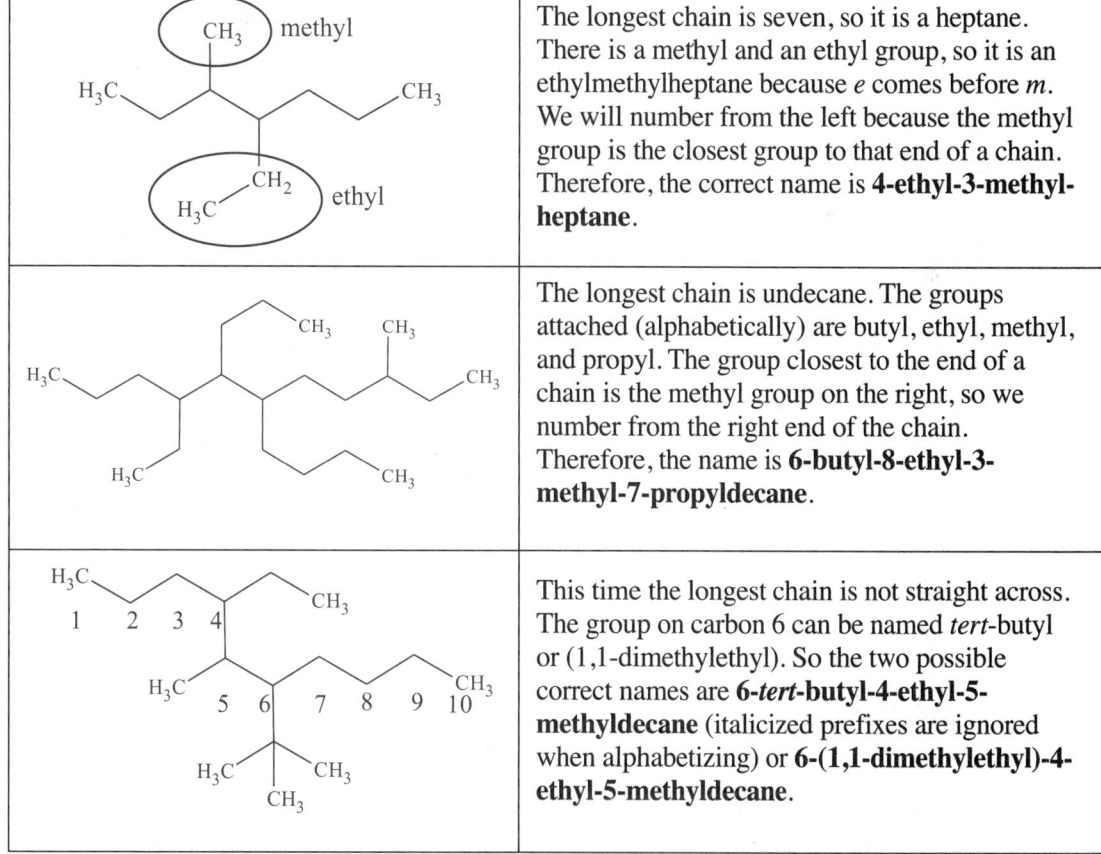	This is a six-membered ring with two ethyl groups attached, so it is a diethylcyclohexane. You start numbering at one of the carbons that has an ethyl group attached, and number toward the other one, so it is 1,2-diethylcyclohexane. Notice that the two ethyl groups are attached to the cyclohexane with wedged bonds. This means that the two ethyl groups are on the same side of the ring. Two groups on the same side of a ring are referred to as "cis." So the correct name is *cis*-**1,2-diethylcyclohexane**. If the two groups were on opposite sides of the ring, it would be referred to as "trans."

"Cis" and "trans" are used when there are only two groups on a ring, and if the bonds are indicated with dashes and/or wedges. If the bonds are just normal lines, then you can't tell if it is cis or trans.

B.8. TWO OR MORE DIFFERENT GROUPS ON A CHAIN

If there are two or more different groups on a chain, the groups are arranged alphabetically in the name. You still number from the end of the chain closest to the first group. See the following examples.

	The longest chain is seven, so it is a heptane. There is a methyl and an ethyl group, so it is an ethylmethylheptane because *e* comes before *m*. We will number from the left because the methyl group is the closest group to that end of a chain. Therefore, the correct name is **4-ethyl-3-methyl-heptane**.
	The longest chain is undecane. The groups attached (alphabetically) are butyl, ethyl, methyl, and propyl. The group closest to the end of a chain is the methyl group on the right, so we number from the right end of the chain. Therefore, the name is **6-butyl-8-ethyl-3-methyl-7-propyldecane**.
	This time the longest chain is not straight across. The group on carbon 6 can be named *tert*-butyl or (1,1-dimethylethyl). So the two possible correct names are **6-*tert*-butyl-4-ethyl-5-methyldecane** (italicized prefixes are ignored when alphabetizing) or **6-(1,1-dimethylethyl)-4-ethyl-5-methyldecane**.

B.9. TWO OR MORE DIFFERENT GROUPS ON A RING

In general, this is the same as with any ring. You start numbering at a carbon on the ring where a group is attached, and go in the direction that gets you to another group first. If there are only two groups on the ring, then the one that comes first alphabetically will be numbered as 1. See the following examples.

	This is a cyclopentane, with an ethyl and a (2-methylpropyl) or an isobutyl group attached. Ethyl is first alphabetically in either case, so the possible names are **1-ethyl-3-isobutylcyclopentane** or **1-ethyl-3-(2-methylpropyl) cyclopentane.**
	1-isopropyl-1-methylcyclohexadecane or **1-methyl-1-(1-methylethyl)cyclohexadecane**. Both groups are on the same carbon, so we start there. We have to use numbers because the groups could be on different carbons.

B.10. ANY NUMBER OF AS MANY DIFFERENT GROUPS ON A RING OR A CHAIN AS YOU WANT

When you have more than one of a kind of group, you use the appropriate prefix to tell how many of the groups you have. However, IUPAC ignores the numerical prefix when alphabetizing. So, for example, "dimethyl" is alphabetized as "methyl," and "tris(2-ethylpentyl)" is alphabetized as "(ethylpentyl)." Also, "methyl" comes before "(2-methylpropyl)" when alphabetizing, just like "Smith" comes before "Smithson" in the phone book. See the following example.

	This is an octane with three ethyl and two methyl groups attached. Because *e* comes before *m*, it is a triethyldimethyloctane. We start numbering from the left, so the name is **3,4,5-triethyl-3,6-dimethyloctane**.

B.11. OTHER ASPECTS OF ALKANE NAMING

In the previous set of examples, the longest continuous chain of carbons should have been clear. Sometimes, there is more than one way to pick a longest chain of the same length, which could potentially give you more than one possible set of substituent groups. See the following example.

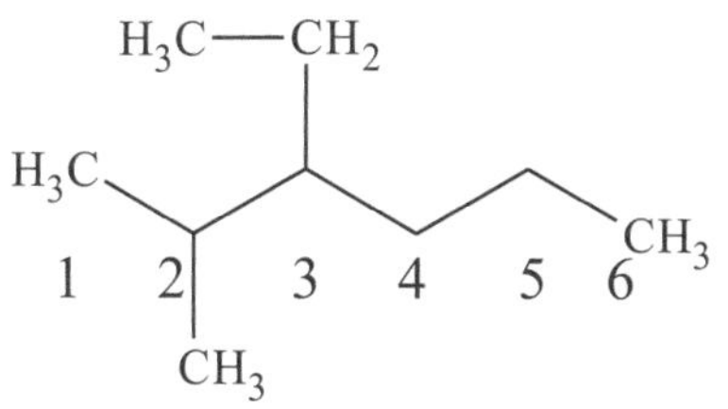 H₃C—CH₂ ... (structure)	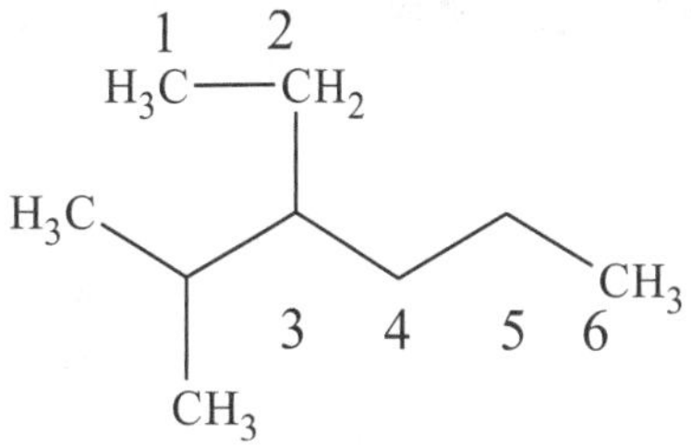 1 2 H₃C—CH₂ ... (structure)
Choosing this as the longest chain gives us **3-ethyl-2-methylhexane** as the name.	Choosing this as the longest chain gives us **3-isopropylhexane**.

Both of these names seem reasonable, but only one is correct. IUPAC says that, in this case, we choose the longest chain that has the most groups attached. So the name on the left is the correct one. If there are choices between longest chains with the same number of groups, we pick the longest chain with the most of the smallest groups. See the following examples.

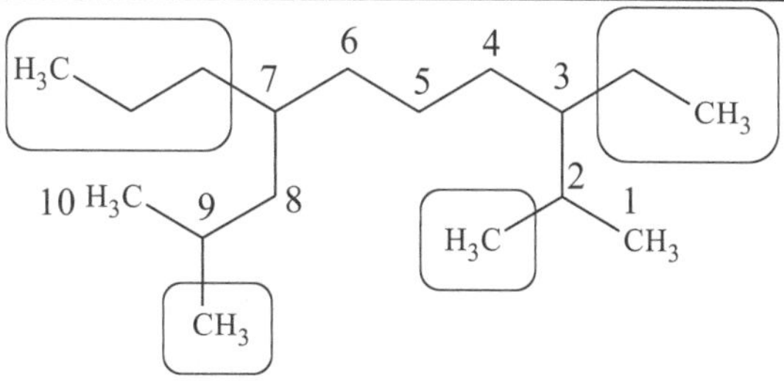 (structure)	Correct. The longest chain has four groups attached. The name is **3-ethyl-2,9-dimethyl-7-propyldecane**.
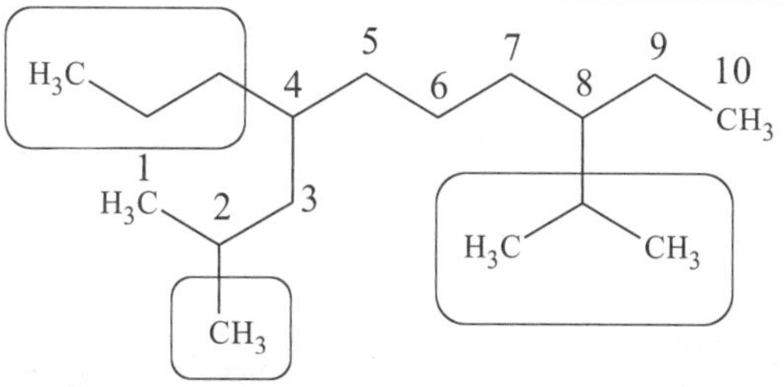 (structure)	Incorrect. The longest chain has only three groups attached.
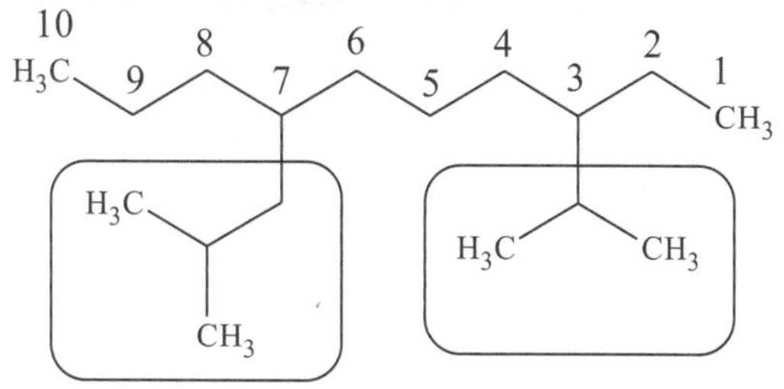 (structure)	Incorrect. The longest chain has only two groups attached.

As you can imagine, there are many possible structures that could have really complex names. If you can handle these examples, and the ones in the review problems, you should do pretty well. Again, consult your instructor on his or her preference in naming the three- and four-carbon groups.

B.12. NAMING ALKYL HALIDES

Alkyl halides are alkanes that have one or more halogens (F, Cl, Br, or I) replacing of one or more of the hydrogens. Alkyl halides are convenient starting materials to prepare lots of other compounds, so we should know how to name them. The halogens are always named as substituents. To make a halogen name into a substituent name, you drop the -ine from the halogen name, and add an o. So fluorine becomes fluoro, chlorine becomes chloro, and so on. They are alphabetized within the name of the compound. See the following examples.

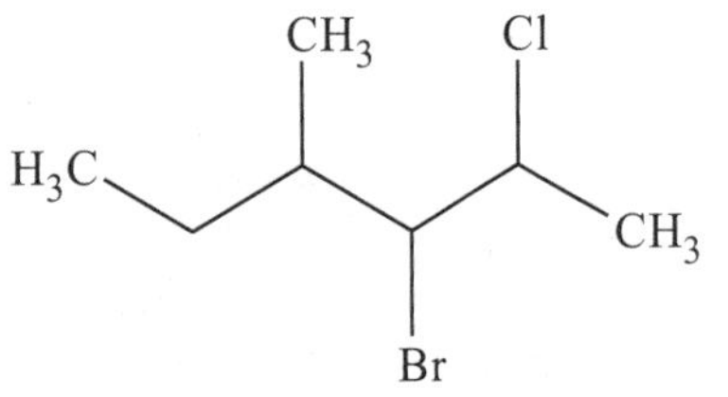	The longest chain is hexane, with bromo, chloro, and methyl attached, so the name is **3-bromo-2-chloro-4-methylhexane**.
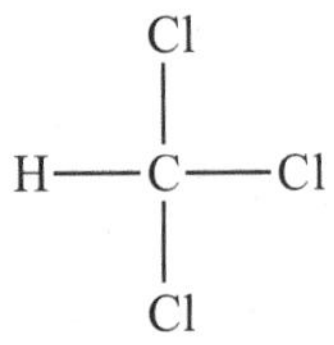	There are three chlorines on a one-carbon alkane, so this is **trichloromethane** (no numbers are needed because there is only one carbon). The common name for this is chloroform. It was one of the first general anesthetics used in surgery, but it is not used now because it is toxic.
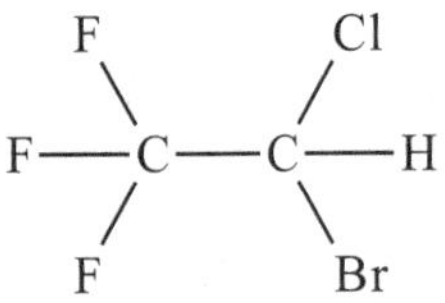	This is **2-bromo-2-chloro-1,1,1-trifluoroethane**. This is a general anesthetic with the generic name of halothane. Halothane is much safer than chloroform.

C. Shapes of Simple Alkanes

We want to consider two aspects of alkane shapes. One is the shape of the chain of carbons. The other is the positions of the hydrogens on a carbon relative to the hydrogens on the adjacent carbons.

Normally when we draw alkane structures, we draw the carbons in a horizontal line, as shown here for propane.

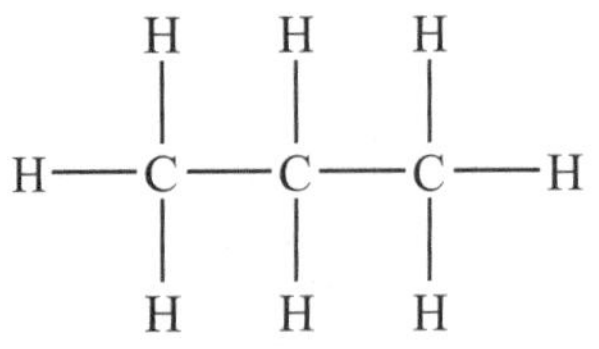

Remember, however, that the carbons are sp^3-hydridized, so they have bond angles of about 109.5°. This means that carbon chain of propane is bent, and "really" looks more like these two drawings.

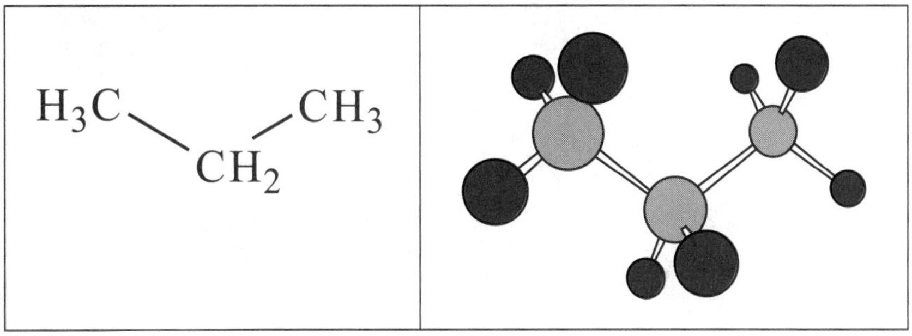

The second aspect of the shape of an alkane is a little more subtle. Let's look at two structures of ethane shown here. How do they differ in shape?

Projection	Ethane #1	Ethane #2
Ball-and-stick model	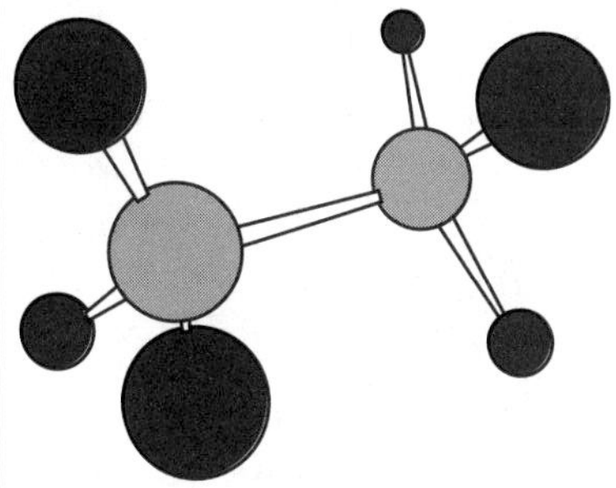	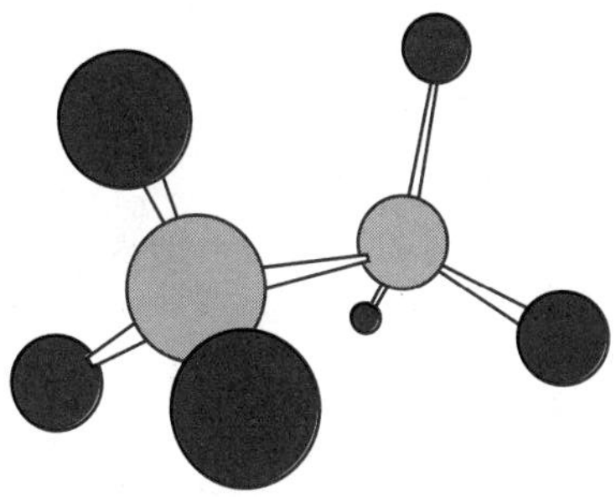
Saw-horse projection	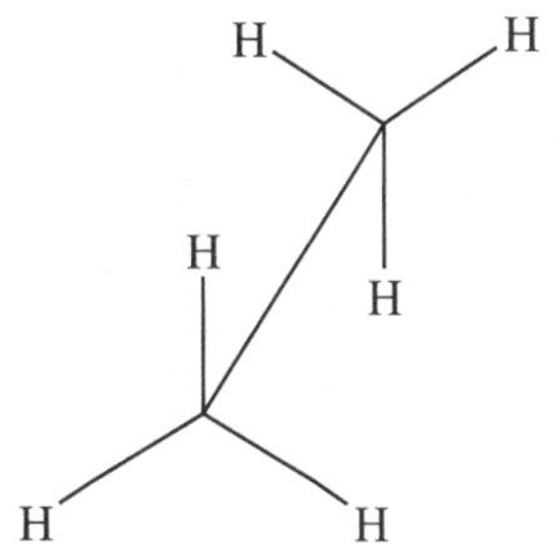	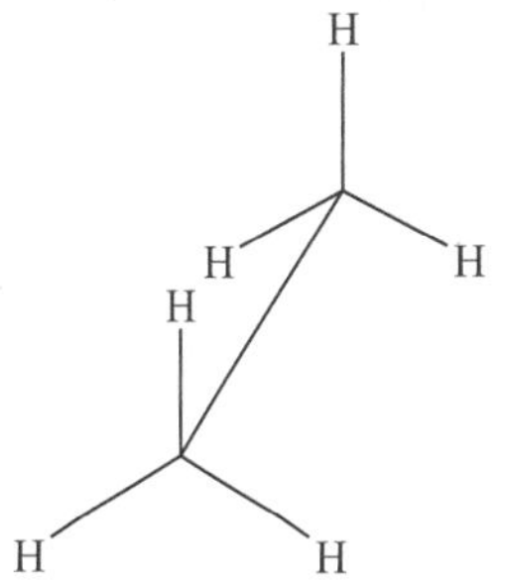

This is easier to see if you have a set of molecular models to play with. If you don't, you can improvise with toothpicks and gumdrops or colored marshmallows. If you look down the carbon–carbon bond, here is what you see.

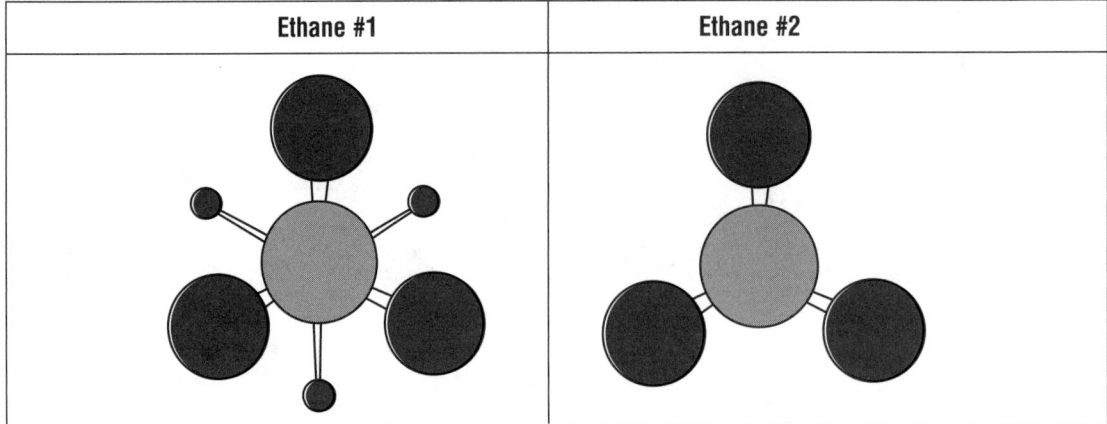

Ethane #1	Ethane #2

In ethane #1, you see the carbon in the front (the gray ball), attached to its three hydrogens (the big red balls). You can't see the carbon in the back; it is covered up by the front carbon. You can see the hydrogens on the back carbon (the small red balls). In ethane #2, the carbon and the hydrogens on the front carbon are covering up the back carbon and hydrogens. In organic chemistry jargon, hydrogens on the front carbon are "eclipsing" the hydrogens on the back carbon, so ethane #2 is in the **eclipsed conformation**. Ethane #1 is described as being in the **staggered conformation** because the hydrogens on the front and back carbons are as far as possible from each other as they can be.

Instead of drawing balls and sticks, there are other ways to represent shapes of alkanes. One of these is the Newman projection, first proposed by Melvin Newman (for more information about him see *http://www.nap.edu/html/bio73h/newman.html*). In a Newman projection, the carbon in the front is at the intersection of the bonds to the hydrogens, and the carbon in the back is represented by a circle. Newman projections of staggered and eclipsed ethane are shown next. Normally, the hydrogens in the eclipsed form are slightly offset so that you can see the hydrogens in the back more easily.

Ethane #1 Staggered	Ethane #2 Eclipsed	Ethane #2 Eclipsed, with H's slightly offset

There is one more jargon term you need to know: **dihedral angle** (DA, sometimes called torsion angle). This is the apparent angle between a hydrogen on the front carbon, and a hydrogen on the back carbon. For example, in the staggered form of ethane, the dihedral angle between a hydrogen at the 12 o'clock position on the front carbon, and the hydrogen at the 2 o'clock position on the back carbon is 60°. This is shown next. Hydrogens that are eclipsed have a dihedral angle of 0°.

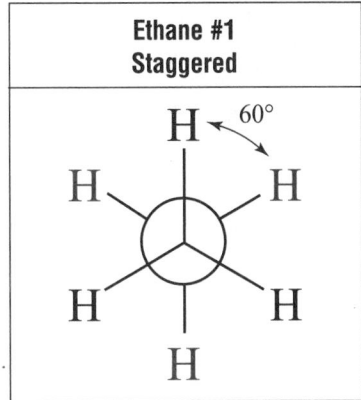

Why is the dihedral angle important? It allows us to describe the relative position of two groups without drawing the structure. By why is the shape of ethane important? Because the potential energy of ethane is dependent upon its shape. If the hydrogens are eclipsed, the electron clouds around the hydrogens are as close together as possible. Because electrons are negative, the electron clouds repel each other, and want to be as far apart as possible. The way they can be as far apart as possible is for the hydrogens to be staggered. Therefore, the staggered form of ethane is somewhat more stable (lower in potential energy) than the eclipsed form of ethane. The energy difference is not that great (2.9 kcal/mol), but it is measurable.

Actually, there is a whole range of dihedral angles by which two hydrogens can be separated, anywhere between 0 and 360°. The energies of these shapes are illustrated in the following graph. The dihedral angle is the angle between the two circled hydrogens. You can envision this as rotating the back carbon to the right in 60° increments.

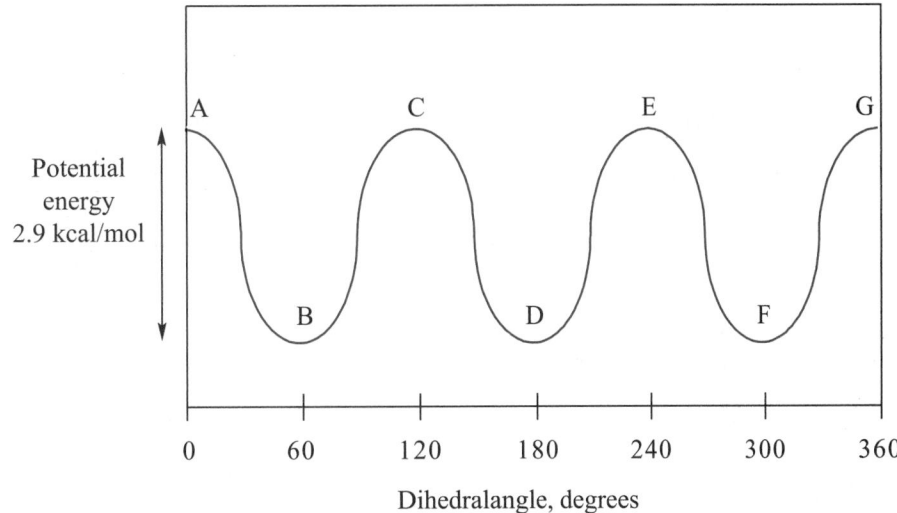

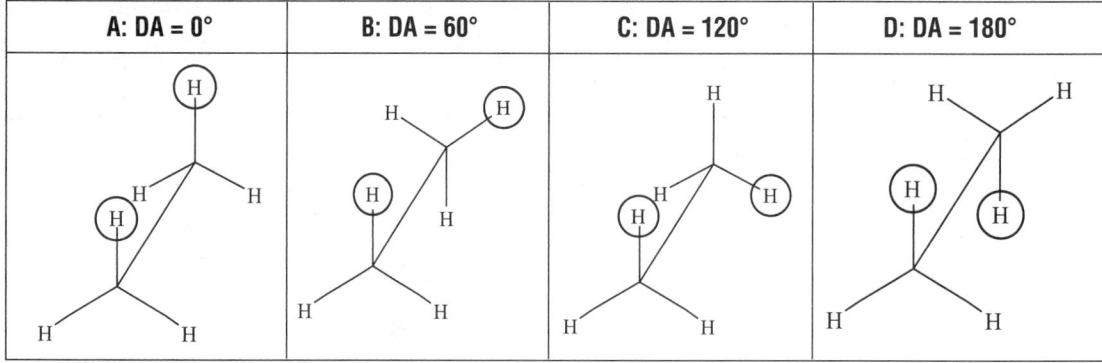

A standard problem in organic chemistry texts relates to the conformations of butane. Look down the bond between carbons 2 and 3, and rotate the back carbon by 60°, as we did with ethane. DA measures the angle between the two methyl groups.

A: DA = 0°	B: DA = 60°	C: DA = 120°	D: DA = 180°
CH_3 CH_3 H H H H	CH_3 H CH_3 H H H	CH_3 H H H CH_3	CH_3 H H H H CH_3
Methyls eclipsed with each other	Methyls gauche	Methyls eclipsed with H's	Methyls anti

E: DA = 240°	F: DA = 300°	G: DA = 360° = 0°	
CH_3 H H_3C H H H	CH_3 H_3C H H H H	CH_3 CH_3 H H H H	
Methyls eclipsed with H's	Methyls gauche	Methyls eclipsed with each other	

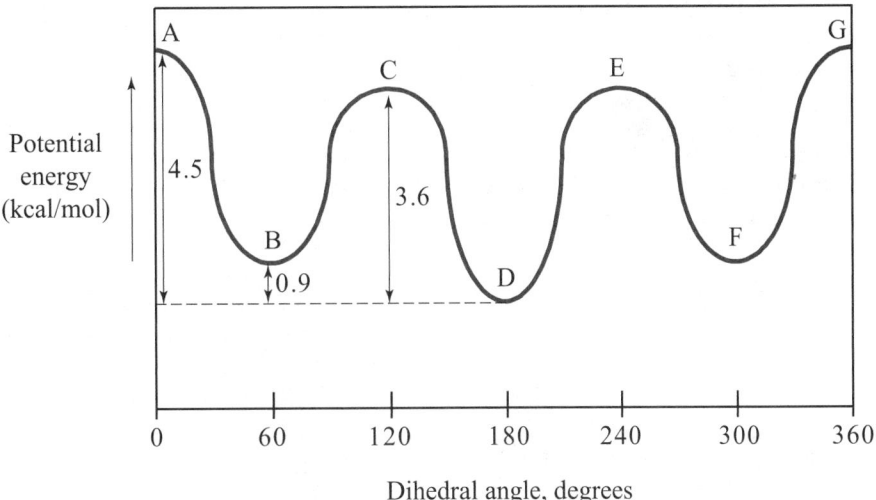

The most stable form has the groups staggered and the methyls as far from each other as possible (DA = 180°). This form is referred to as the **anti configuration**. The next most stable form has the groups staggered, but the methyls are only 60° apart, the so-called **gauche form**. It is slightly higher in energy because the methyls are closer to each other, and the electron clouds repel each other. The eclipsed forms are even higher in energy. The one with the two methyls eclipsed is the highest in energy, since the methyls are as close to each other as they can be, so repulsive forces are maximized.

D. Shapes of Cyclic Alkanes

D.1. INTRODUCTION

We have been drawing the cycloalkanes as the simple geometric figures, but this is not necessarily how they look. The angles are the internal angles of the polygons, assuming that they are flat. Because alkanes use sp^3 hybrid orbitals to form bonds, we would expect the bond angles to be close to 109.5°, if at all possible. The conclusion we can draw from this is that many of these molecules are not flat. Except for cyclopropane, where the three carbons have to be in the same plane, all of the cycloalkanes are not planar.

Many organic compounds contain cyclohexane rings. Therefore, we will consider the shape of cyclohexane and substituted cyclohexanes in more detail.

D.2. CYCLOHEXANE AND SUBSTITUTED CYCLOHEXANE SHAPES

Cyclohexane is not a planar hexagon. The shape shown in the first figure below is called a **chair** and is the most stable shape. If you have a good imagination, you can envision the four carbons in the middle as the seat, the carbon in the upper left as the backrest, and the carbon on the lower right as the footrest. It can be drawn as a "boat," as shown below, as well.

Chem3D Structure of a Chair	Drawing Showing the H's	Standard Line Drawing
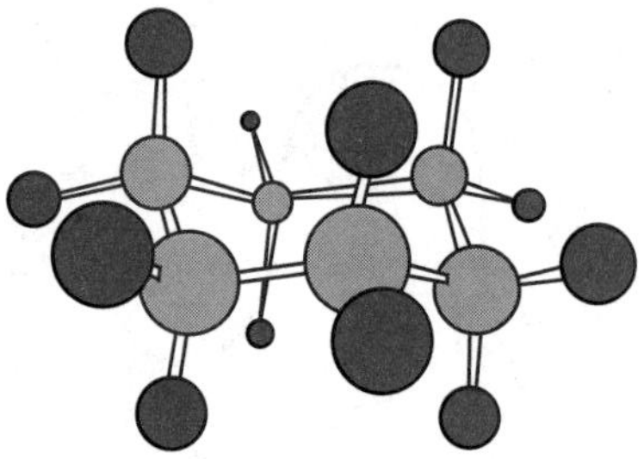	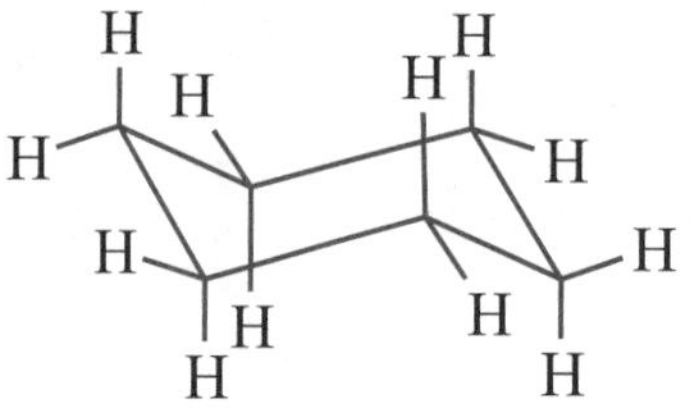	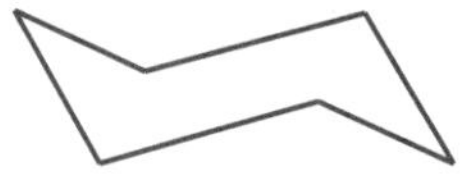

There are other possible shapes for cyclohexane as well. Some are shown below.

A Boat Showing the H's	Boat as a Line Drawing	The Twist Boat
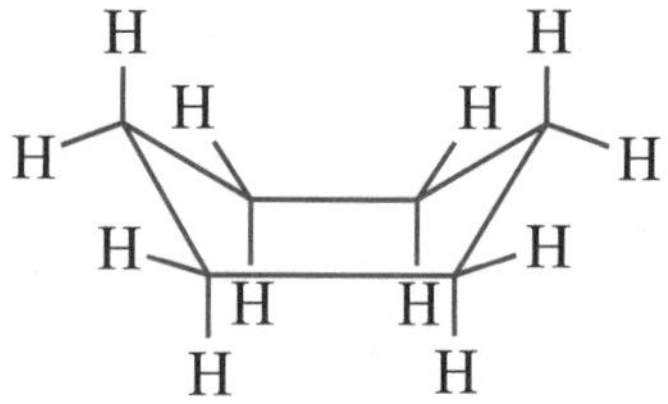	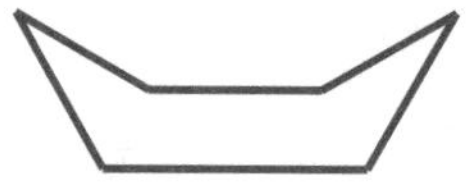	

The boat is much less stable than the chair, by about 6.5 kcal/mol. In the boat shape, there are eclipsing interactions, as well as interactions between hydrogens across the ring from one another. A twist boat is about 5.5 kcal/mol less stable than the chair. By twisting it a little, some of the eclipsing interactions are reduced. Since the chair is the most stable form, we will deal with it now.

D.2.a. DRAWING CHAIRS

You will have to learn how to draw chairs: It is one of the rites of passage in organic chemistry, and you will have to do it on exams and homework. There are lots of ways to do it; here is one that works for me.

Step 1. Draw a line that is not quite horizontal. Draw a second line parallel to the first, but below it and slightly to the left.	
Step 2. Connect the right ends of the lines with a V.	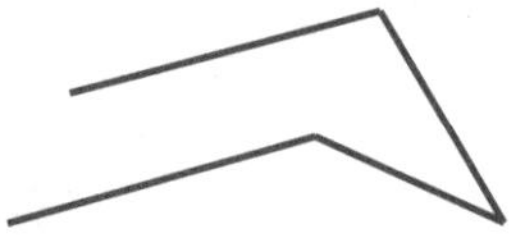
Step 3. Connect the left ends of the lines with an inverted V	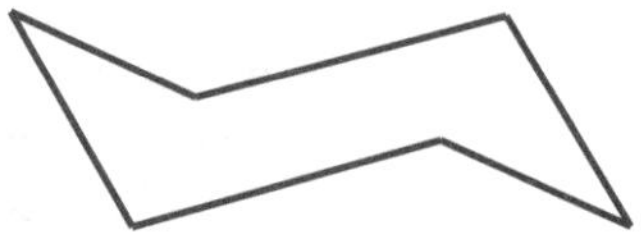

Practice drawing chairs so you can do it quickly. It will help you on exams. You also need to know where the hydrogens are on each carbon. This is shown below.

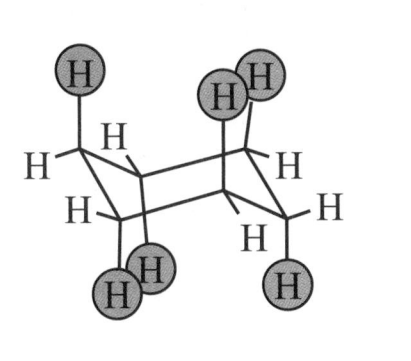

There are two different sets of hydrogens. The circled hydrogens are referred to as **axial**. They are either pointed straight up or straight down. Notice that as you go around the ring, the directions of the axial hydrogens alternate. If the axial hydrogen is pointed up on one carbon, it is pointed down on the next carbon, then up on the following carbon. The hydrogens that are not circled are called **equatorial** hydrogens (from "equator"), since they are around the middle of the molecule.

There is a simple way to tell where the axial hydrogens are. If the carbon in the chair is pointed up, the axial hydrogen is up. If the carbon is pointed down, the axial hydrogen is down. This is shown below.

This carbon is pointed up . . .

so the axial H is pointed up.

This carbon is pointed down . . .

so the axial H is pointed down.

Equatorial hydrogens are generally pointed out to the side. Technically, the C–H bond of an equatorial hydrogen is parallel to the C–C bond in the ring two bonds away. This is illustrated below.

This equatorial C–H is parallel to this C–C bond

This equatorial C–H is parallel to this C–C bond.

D.2.b. INTERCONVERTING CHAIR FORMS

At normal temperatures, the C–C bonds of cyclohexane are twisting and bending. Cyclohexane primarily exists as two chair forms that are in equilibrium with each other. If we could label the axial hydrogens in one cyclohexane molecule and then let it equilibrate, we would discover that half of the time, all of the labeled hydrogens are in axial positions, and the other half of the time, they are all in equatorial positions. This is shown below.

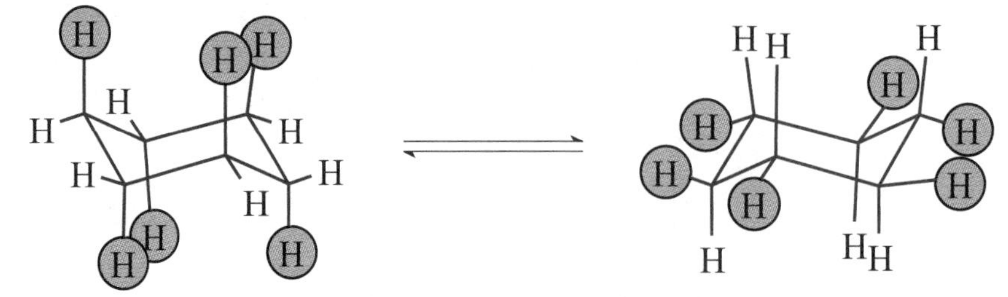

All circled hydrogens are axial.	All circled hydrogens are equatorial.

Let's look at a substituted cyclohexane, such as methylcyclohexane. Draw the two chair forms. This is pretty easy; just substitute a CH_3 group for a hydrogen. Remember, when you draw the other chair form, whatever was axial in one form is equatorial in the other. See below.

Chair with the Methyl Axial 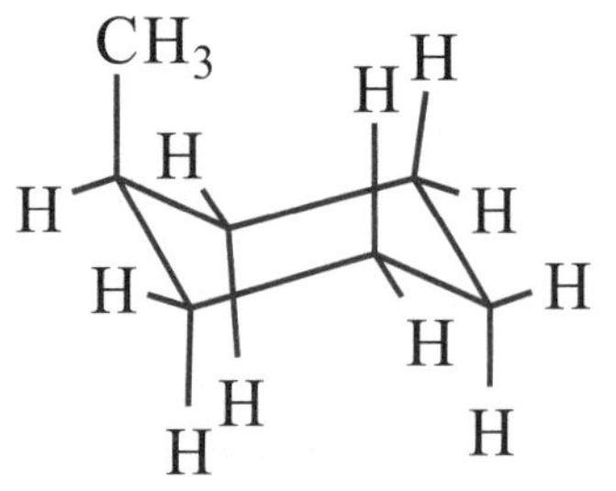	Chair Flipped so the Methyl Is Equatorial 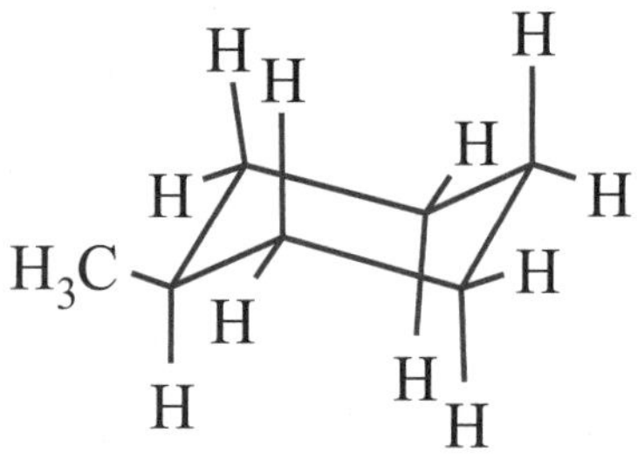

Which one is more stable? Almost any group is more stable when it is in the equatorial position. When the methyl is axial, its hydrogens can be close to the other axial hydrogens on the same side of the ring. The electron clouds of the methyl hydrogens repel those of the other axial hydrogens. When the methyl is equatorial, there are no similar repulsive interactions. We can look at cyclohexanes in Newman projections as well, as shown below. I have numbered the carbons in the ring in both projections. In the Newman projections, we are looking down the bond from carbon 1 to carbon 2, and from carbon 5 to carbon 4. Carbons 1 and 5 are at the intersections of the lines, and carbons 2 and 4 are the circles.

Chair with the Methyl Axial	Chair Flipped so the Methyl Is Equatorial

From the Newman projection on the left, we can see that when the methyl is axial on carbon 1, it is gauche to carbon 3. From the Newman projection on the right, where the methyl is equatorial on carbon 1, it is anti to carbon 3. Therefore, we can use the same argument for rationalizing the more stable form of methylcyclohexane that we used to form butane conformations.

As I mentioned earlier, almost any group is more stable in the equatorial position than in the axial position. Some representative groups, the values of $\Delta G°$ (in kcal/mol) by which each group is more stable in the equatorial position, and the corresponding % equatorial are shown in the following table.

Group	$-\Delta G°$ (kcal/mol): How Much of Each Group Is More Stable in the Equatorial Position	Percentage of the Group That Is in the Equatorial Position
CH_3	1.8	95
CH_2CH_3	1.9	96
$CH(CH_3)_2$	2.1	97
$C(CH_3)_3$	>4.5	>99.9
C_6H_5 (phenyl)	2.9	99.3
$CH=CH_2$	1.7	94
$C\equiv CH$	0.5	70
F	0.24	60
Cl	0.53	71
Br	0.48	69
I	0.47	69

D.2.c. CHAIR FORMS OF CYCLOHEXANES WITH MORE THAN TWO GROUPS

Now let's look at a cyclohexane with two groups attached, such as *cis*-1,2-diethylcyclohexane. We want to draw the two chair forms and figure out which one is the more stable. The first part of this problem often gives students the most trouble. Here is how I approach it.

Step 1. First, draw the molecule normally, using dashes and wedges as needed. It really doesn't matter where you put the groups, as long as they are "1,2," and whether the groups are both wedged or dashed. When you are looking at this structure, you are looking at the molecule from over the ring. Arbitrarily label the atoms with the ethyls 1 and 2.	
Step 2. Now draw a chair and arbitrarily label two of the atoms 1 and 2. Draw the positions for the two groups that will be attached to those two ring carbons. When you are looking at a chair, you are really looking at the ring from the side, not from the top.	
Step 3. Look at carbon 1 in both structures. The ethyl is coming at you in the top structure, or coming up toward you. The "up" position in the chair is the position that is closest to the top of the box, or the axial position of carbon 1. So put the ethyl on the axial position, and the H on the equatorial position of carbon 1.	
Step 4. Look at carbon 2 in both structures. The ethyl is coming at you in the top structure, or coming up toward you. The "up" position in the chair is the position that is closest to the top of the box, or the equatorial position of carbon 2. So put the ethyl on the equatorial position, and the H on the axial position of carbon 2.	
Step 5. Congratulations! You have drawn one correct chair form of *cis*-1,2-diethylcyclohexane. Now you need to draw the other one. Start with the initial chair you drew above. Pull carbon 1 down, and pull carbon 4 up, to generate the other chair. Draw the positions for the two groups that will be attached to carbons 1 and 2.	
Step 6. Now repeat steps 3 and 4 to position the two ethyls on the correct positions. In this case, the ethyl on carbon 1 is equatorial, and the ethyl on carbon 2 is axial. Or you could have just remembered that the groups would be in the opposite positions as in the first chair because when you flip a chair, whatever is equatorial becomes axial, and vice versa.	

E. Physical Properties of Alkanes

Two important physical properties of organic compounds are water-solubility and boiling point. We will discuss these for each functional group. In general, alkanes have the lowest water-solubilities and boiling points of any functional groups we will discuss. The table below gives some examples.

Compound	Water-solubility (g/100 mL of water)	Boiling point (°C)
Methane	0.0025	–161
Ethane	0.0064	–88
Propane	0.013	–42
Butane	0.039	–0.5
Pentane	0.036	36
Hexane	0.0095	69
Heptane	0.0025	98.4

In general, alkanes are not very soluble in water. Because carbon and hydrogen are similar in electronegativity, alkanes have no significant partially positive or negative atoms to be attracted to the partially positive hydrogens and partially negative oxygens of water molecules. Therefore, there is little attraction of alkanes for water molecules. The increase in solubility going from methane to butane is attributed to small hydrocarbon molecules being able to fit into spaces between water molecules.

The boiling points of alkanes increase with increasing chain length. This is due to increasing van der Waals forces between molecules. Another factor that can influence boiling points of alkanes is the shape. Consider the table on the next page.

Compound	3-D Structure	Boiling point (°C)
H_3C CH_2 CH_2 CH_2 CH_3 Pentane		36
CH_3 CH CH_3 H_3C CH_2 2-Methylbutane		28
CH_3 H_3C—C—CH_3 CH_3 2,2-Dimethylpropane		10

The more linear pentane molecule has the highest boiling point, whereas the more spherical 2,2-dimethylpropane has the lowest boiling point, and 2-methylbutane is in between. This is because the more linear molecules have more surface area that can be involved in van der Waals attractive forces between molecules than the more branched alkanes have.

REVIEW EXERCISES FOR CHAPTER 2

1. Give IUPAC names for the following structures.

   a.

   b.

   c.

   d.

2. Draw acceptable structures that correspond to the following names.
   a. 5-*sec*-butyltridecane
   b. 1-*t*-butyl-5-heptyl-2-propylcyclononane
   c. 2,7-dicyclobutyl-1-cyclopropyl-4,6-bis(2-methylpropyl)cyclooctane
   d. 2,3,3,4,4-pentamethylhexane
   e. 2,2,4-trifluoro-3-methylheptane
   f. Draw both chair forms of *trans*-1-methyl-3-propylcyclohexane. Indicate which one is the more stable, and why.

3. Draw Newman projections for all the staggered forms of the following molecule. Put the carbon labeled 1 in the front and the carbon labeled 2 in the back. Which one is less stable than the other two? Explain.

Stereochemistry

WHAT YOU WILL LEARN

In this chapter, you will learn:

- what stereogenic atoms are, why they are important, and how to name them;
- how enantiomers and diastereomers differ from structural isomers and from each other;
- how the properties of stereoisomers are potentially different.

SECTIONS IN THIS CHAPTER

- Stereogenic Atoms

- Naming Molecules with Stereogenic Atoms:
  The *R*,*S* (Cahn–Ingold–Prelog) System

- Possible Relationships Between Two Structures
  with the Same Molecular Formula

- Optical Rotation

- Implications of Stereoisomerism

Stereochemistry is a fancy word for looking at structures in three dimensions. If you have access to a set of molecular models, it will make this topic much easier to understand. Again, toothpicks and gumdrops or marshmallows will work, but they get sticky quickly.

A. Stereogenic Atoms

An atom at which the interchange of two groups gives a stereoisomer is called a **stereogenic atom**. Often, but not always, a stereogenic atom is bonded to four different types of groups of atoms. Several other terms are used to describe this: stereocenter, stereogenic center, center of chirality, or chiral atom. Find out what your instructor prefers. Most people use the term "chiral" (from the Greek word for hand) to describe a molecule, rather than an atom. A chiral molecule does not have a plane of symmetry and is not superimposible on its mirror image.

Let's look at some examples of molecules, and see if they contain stereogenic atoms.

$H-\overset{\overset{\displaystyle H}{\mid}}{\underset{\underset{\displaystyle H}{\mid}}{C}}-\overset{\overset{\displaystyle H}{\mid}}{\underset{\underset{\displaystyle H}{\mid}}{C}}-H$	There are no stereogenic atoms in this molecule. A hydrogen cannot be a stereogenic atom because it is bonded to only one group. Neither carbon can be a stereogenic atom because each carbon is bonded to three hydrogens. Therefore, no atoms are bonded to four different types of groups.
$Cl-\overset{\overset{\displaystyle H}{\mid}}{\underset{\underset{\displaystyle H}{\mid}}{C}}-\overset{\overset{\displaystyle H}{\mid}}{\underset{\underset{\displaystyle Br}{\mid}}{C}}-Cl$	The right carbon is a stereogenic atom because it is bonded to four different types of groups: a hydrogen, a chlorine, a bromine, and a CH_2Cl group.
$H_3C-\overset{\overset{\displaystyle OH}{\mid}}{\underset{\underset{\displaystyle H}{\mid}}{C}}-CH_2-CH_2-CH_3$	The second carbon from the left is a stereogenic atom because it is bonded to a methyl group, a hydrogen, an OH group, and a propyl group. Remember, my definition said different types of *groups*, not different types of *atoms*. The other carbons are not stereogenic atoms because they are each bonded to two or three hydrogens.
$\begin{array}{c} CH_2 \\ H_2C \qquad CH-Br \\ CH_2-CH_2 \end{array}$	There are no stereogenic atoms in this molecule. This is sometimes difficult to see in a ring. The CH_2's can't be stereogenic because they are bonded to two hydrogens. The only possible stereogenic atom could be the CH bonded to the bromine. To see why it is not, let's envision doing the following.

Let's suppose the molecule was really big, and you and a friend decided to walk around it, starting from carbon 1. You go clockwise, and your friend goes counterclockwise. When you get to carbon 5, you call out to your friend, "I found a CH_2." Your friend gets to carbon 2, and says, "I found a CH_2, too." You both keep going, and when you get to carbon 4, you call out, "I found a CH_2." Your friend gets to carbon 3, and says, "I found a CH_2, too." You both keep going, and meet each other. Since you both found the same groups of atoms going around the ring, it means carbon one is attached to two of the same group, so it is not a stereogenic atom.

Another way that you could have seen that carbon 1 is not a stereogenic atom is to notice the plane of symmetry through the molecule, as shown by the line at the left. The line goes right through carbon 1. Let's look at another example.

The only possible stereogenic atoms are carbons 1 and 3. If you start at carbon 1, and go around the ring each way like we did before, you discover CH_2's at carbons 2 and 5, so you keep going. When you get to carbons 3 and 4, there is now a difference: Carbon 3 is a CHBr, while carbon 4 is a CH_2. Therefore, it is now different going around the ring each way, so carbon 1 is a stereogenic atom. If you do a similar thing starting at carbon 3, you discover it is also a stereogenic atom.

You may have noticed that there is a plane of symmetry through this molecule as well. However, it does not go through either carbon 1 or carbon 3, so it doesn't affect their being stereogenic atoms.

The only stereogenic atom is the starred carbon. The carbons making double bonds or triple bonds in this molecule cannot be stereogenic atoms because they are not bonded to four different groups.

If you want to give yourself a challenge, find all of the stereogenic atoms in cholesterol. There are eight!

Atoms other than carbon can be stereogenic atoms if they are bonded to four different groups. You will run across relatively few examples of these in your organic chemistry course. Some of these are shown below.

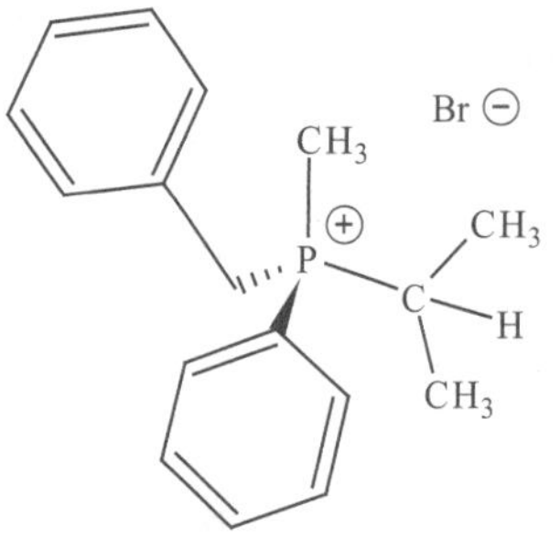	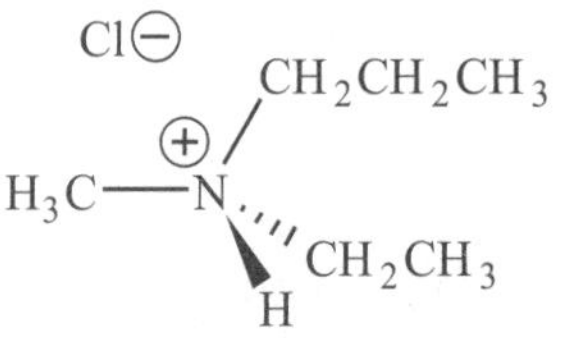
The phosphorus is a stereogenic atom.	The nitrogen is a stereogenic atom.

B. Naming Molecules with Stereogenic Atoms: The *R,S* (Cahn–Ingold–Prelog) System

Let's consider a simple molecule with one stereogenic atom, 1-bromo-1-chloroethane. If we draw the molecule in a three-dimensional representation, using dashes and wedges, we could come up with two different structures, as shown below.

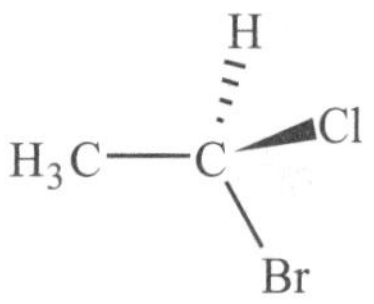

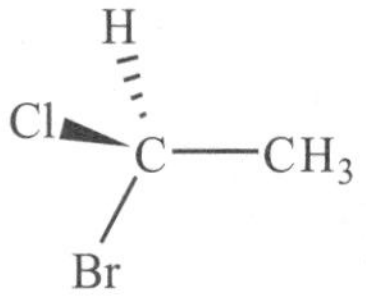

These two structures are not identical because they cannot be superimposed onto each other. In fact, these two structures are mirror images of each other. However, according to the naming rules we know so far, we would give them the same name. We can't have two different molecules with the same name, and IUPAC provides a way to distinguish them. This involves using the rules developed by three chemists, Cahn, Ingold, and Prelog. These rules are as follows:

1. Find the stereogenic atom. It is starred.	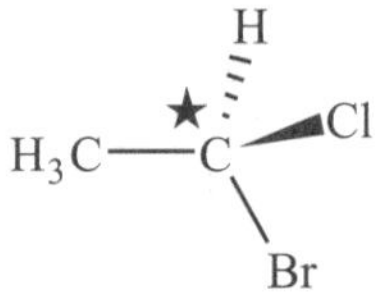
2. Locate the atoms bonded to the stereogenic atom.	C, H, Br, Cl
3. Rank the atoms from highest to lowest in atomic number.	Br > Cl > C > H

4. Label the atoms to indicate their priority in atomic number. Many texts use numbers to indicate priority (#1 being the highest, to #4 being the lowest). I prefer to use the letters A, B, C, and D because we all know A is the highest grade in school.	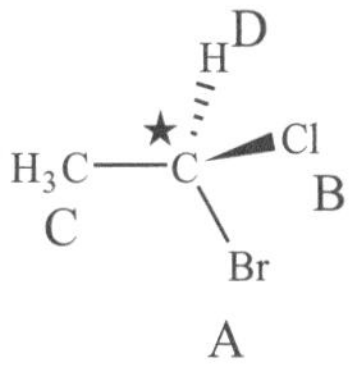
5. This step is critical. Orient the molecule so that the lowest priority group is pointing away from you. In this case, I drew the molecule so that the hydrogen was pointing away from you.	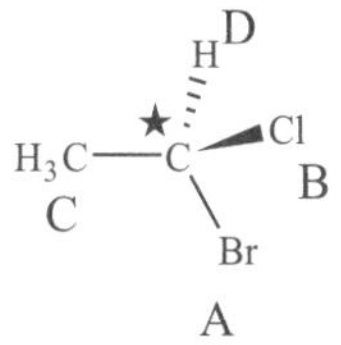
6. Starting from the A group, draw a curved arrow going from the A group to the B group to the C group.	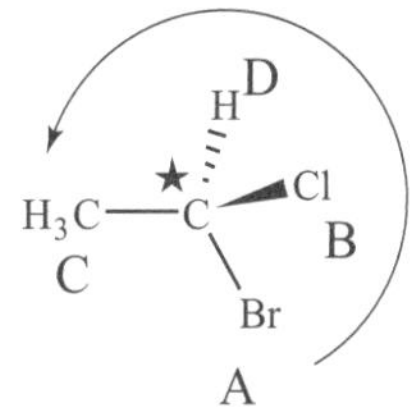
7. In this case, the arrow is in the counterclockwise direction. In the Cahn–Ingold–Prelog system, the stereogenic atom is designated as *S*, which comes from the Latin word, *sinister*, which means left. Had it been clockwise, it would have been designated *R*, from the Latin, *rectus*, for right. The total name for this molecule is (*S*)-1-bromo-1-chloroethane. *R* and *S* are written as italics.	
The mirror-image molecule is shown at the right. In this case, the arrow is in a clockwise direction, so the stereogenic atom is "*R*," and the name of the molecule is (*R*)-1-bromo-1-chloroethane.	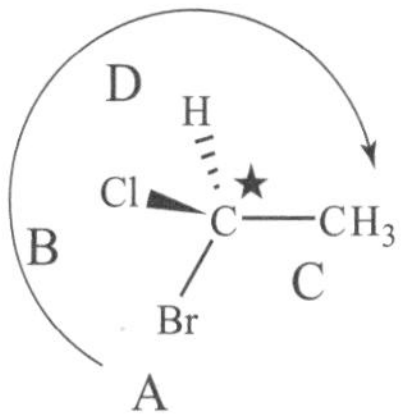

Use models when you practice this. That way, you can orient the molecule however you need to to get the lowest priority group to go away from you.

What do you do if the lowest priority group on the stereogenic atom is not pointed away from you? One way is to rotate a bond, so the lowest priority group goes away from you. See the example on the next page.

1. Find the stereogenic atom. It is starred.	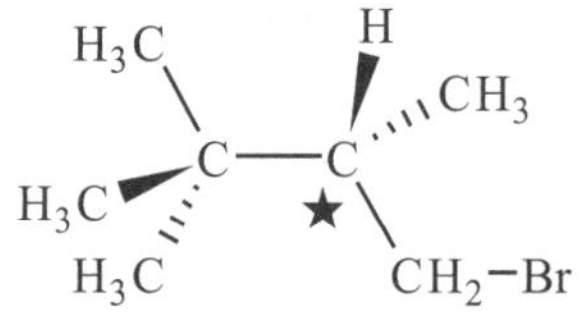 H_3C H_3C C C H CH_3 H_3C ★ CH_2-Br
2 & 3. Locate the atoms bonded to the stereogenic atom, and rank the atoms from highest to lowest in atomic number. The three carbons win out over the H.	$C = C = C > H$
3a & 4. We look at the atoms attached to the three carbons bonded to the stereogenic atom. A CH_3 is always lower in priority to any carbon group because it is bonded to three atoms with atomic number one. The CH_2-Br is higher in priority than the $(CH_3)_3C$ group because Br has a higher atomic number than C. So the priorities are as shown.	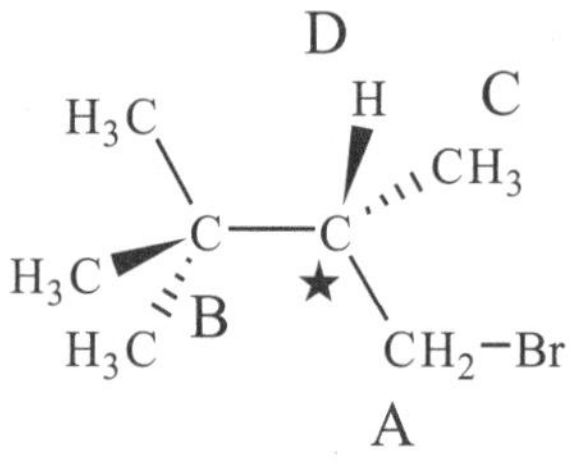 D H_3C H C H_3C C C CH_3 H_3C B ★ CH_2-Br A
5. Orient the molecule so the lowest priority group is pointing away from you. Because the hydrogen is pointing at you, we need to rotate the bond between the starred carbon and the carbon labeled B to get the hydrogen going away from you. You rotate a bond by interchanging any *three* groups on an atom, as shown to the right.	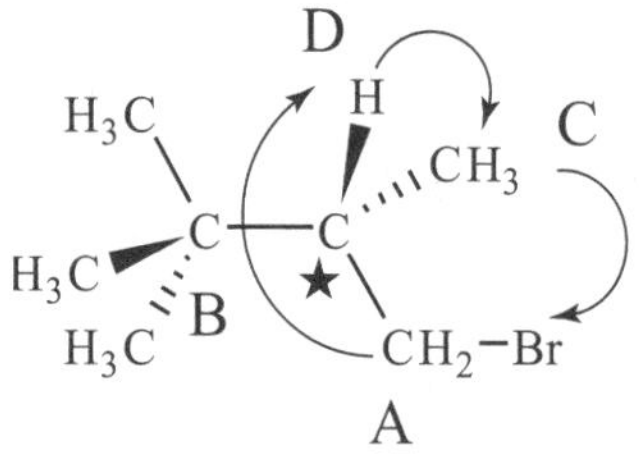 D H_3C H C H_3C C C CH_3 H_3C B ★ CH_2-Br A
5a. Here's how the molecule looks after the bond rotation. Now the hydrogen is going away from you.	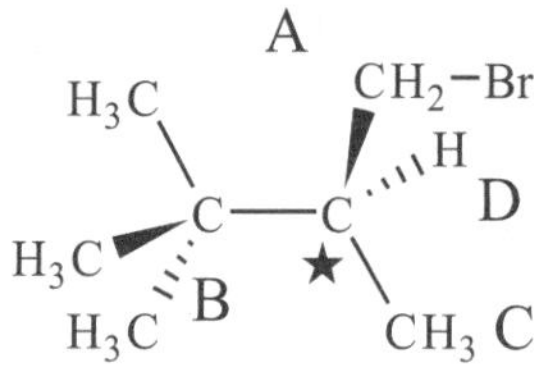 A H_3C CH_2-Br H_3C C C H D H_3C B ★ CH_3 C
6 & 7. Starting from the A group, draw a curved arrow going from the A group to the B group to the C group. In this case, the arrow is in the counterclockwise direction. Therefore, the stereogenic atom is designated as *S*.	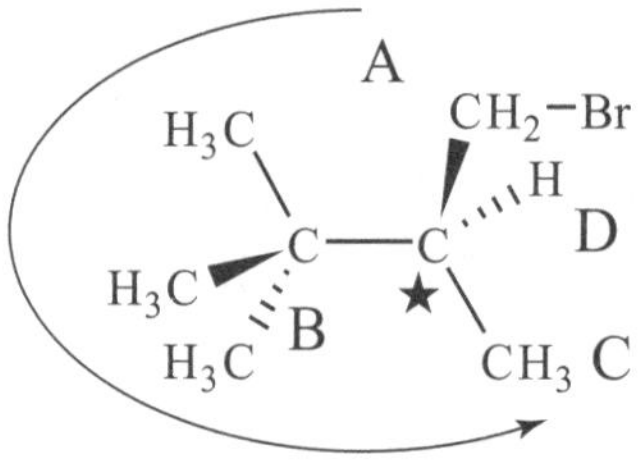 A H_3C CH_2-Br H_3C C C H D H_3C B ★ CH_3 C

Interchanging three groups is the same as a bond rotation. You can use Newman projections to see this.

Three-Dimensional Drawing	Newman Projection	
— You		To view this as a Newman projection, you need to look down the indicated carbon–carbon bond. That makes the stereogenic atom the front carbon of the Newman projection, and the carbon with the three methyl groups attached to the back carbon.
		Interchanging three groups is like using the front carbon of the Newman projection as a handle, and turning the handle counterclockwise 120°. You are just rotating the carbon–carbon bond, and going from one staggered conformation to another.
		Here we have our final conformations.

Another way to deal with a molecule that has the lowest priority group coming at you is to recognize that the configuration is the opposite of what it looks like. See the example below.

Here is the original structure with the assigned priorities. If we draw the arrow going from A to B to C, it is going clockwise, so it looks like it should be _R_. However, because the lowest priority group is coming at you, we have to reverse it to _S_.	
Here is the structure "flipped over," so the hydrogen is going away from you. Now you can see that the stereocenter is _S_ because A to B to C is counterclockwise.	

If your instructor lets you use models on exams, then you can turn the molecule so that the lowest priority group is going away from you. Otherwise, practice doing the rotations on paper.

The priority rules as described above cover most of the cases you will probably encounter. Here are a few more that cover other situations.

1. When comparing only hydrocarbon groups, the carbon group with the most carbons directly attached to the carbon attached to the stereogenic atom has the higher priority. This is illustrated below.

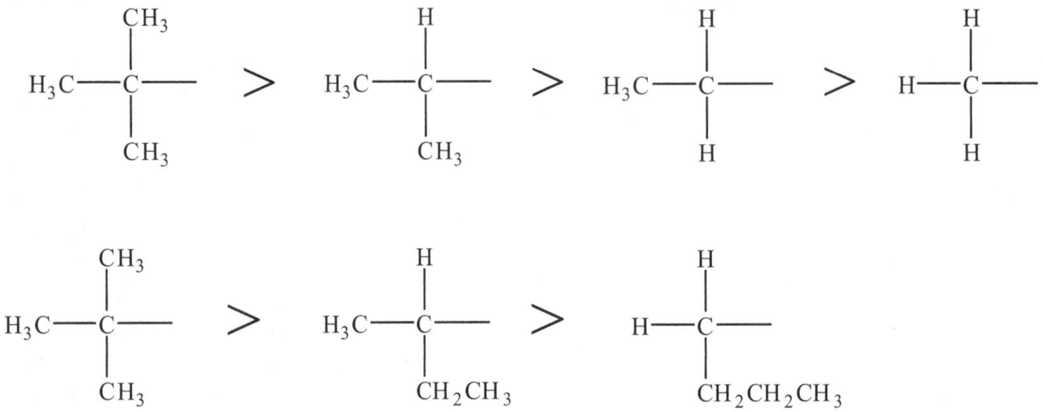

2. Double and triple bonds are treated as multiple single bonds, but the atoms at each end of the bond are duplicated for the purpose of determining priority. However, the duplicated atoms only make one bond. For example,

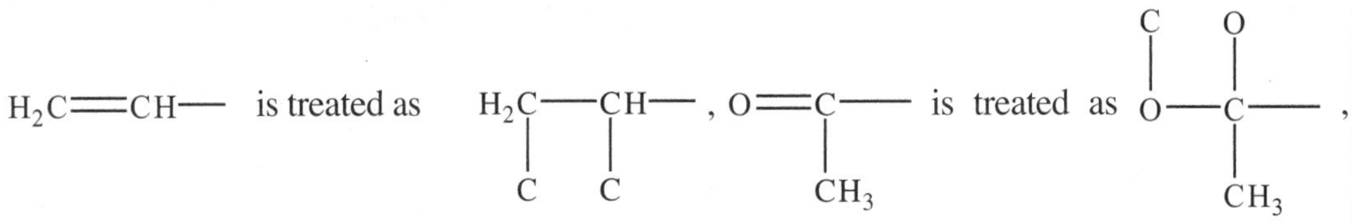

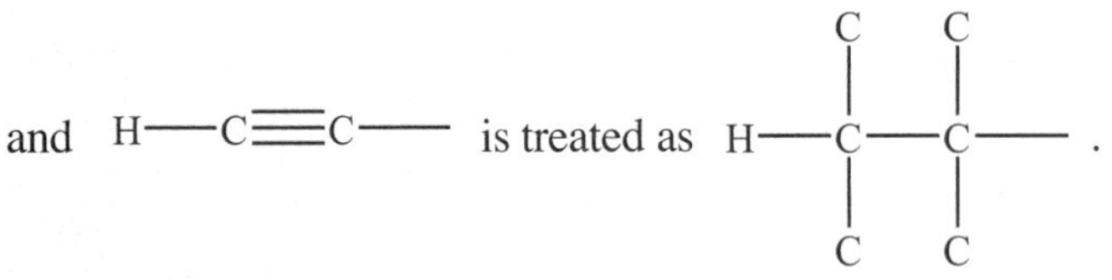

The net result of this is that $H_2C=CH$ is higher in priority than $(CH_3)_2CH$, and $HC\equiv C$ is higher in priority than $(CH_3)_3C$.

3. If you are comparing two isotopes of the same element, the one with the higher atomic weight has the higher priority. Hence, ^3H (tritium, T) > ^2H (deuterium, D) > ^1H.

C. Possible Relationships Between Two Structures with the Same Molecular Formula

If two structures have the same molecular formula, there are only four possible ways that they can be related: They can be identical, constitutional isomers, enantiomers, or diastereomers. We will take these in order.

C.1. IDENTICAL

If two structures are identical, it means that every atom of one structure can be superimposed on every atom of the other structure. Another way to tell if two structures are identical is to name both of them (correctly): If you come up with the same name, the structures are identical.

   Professors and textbook authors like to modify structures to try to make them look different. Here are some common examples.

Original Structure	Modified Structure	What Was Done
H_3C-C with H, F, Cl	H_3C-C with F, Cl, H	The C–C single bond was rotated.
H_3C-C with H, F, Cl	$C-CH_3$ with H, F, Cl	The molecule was flipped over. Notice that these are not mirror images; the F is coming toward you in the left structure, but the hydrogen is coming toward you in the right structure.
H_3C chain with CH_2, CH_2, CH_2, CH_3, H, CH_3	bent chain with H, CH_3, H_3C, CH_2, CH_2, H_3C, CH_2, CH_2	The ever-popular "bend the carbon chain back on itself" approach was used.
$H_3C-C-CH_3$ with H, OH	$H_3C-C-CH_3$ with H, OH	A molecule without a stereogenic atom is flipped over to make it look different.

C.2. CONSTITUTIONAL ISOMERS

Constitutional isomers are structures that have a different bonding sequence of atoms. This kind of isomer is also referred to as a structural isomer and is probably what you think of when you hear the word isomer. The following structures are all constitutional isomers with the formula C_4H_8O. There are more isomers than are shown here.

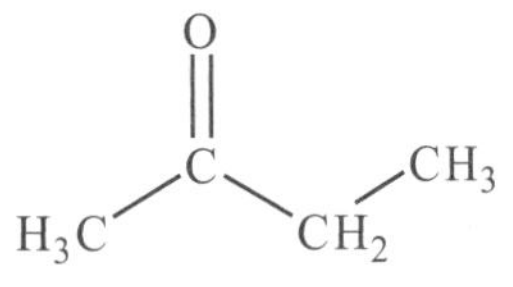

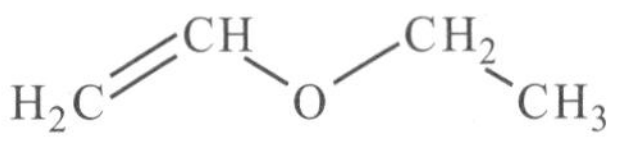

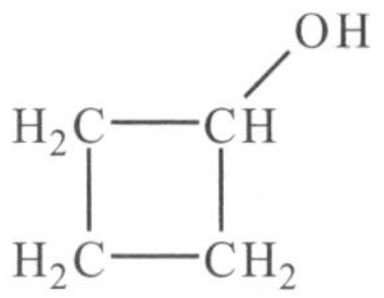

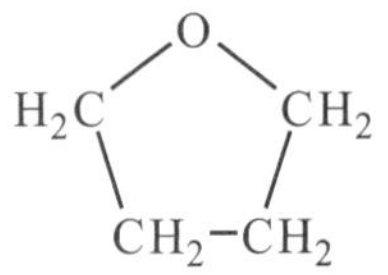

C.3. ENANTIOMERS

Enantiomers are one of two types of molecules called **stereoisomers**, the other being diastereomers, which are coming up later. In general, **stereoisomers** have the same bonding arrangement of atoms, but differ in the three-dimensional arrangement of atoms in space. Most stereoisomers contain one or more stereogenic atoms. An **enantiomer** is a structure that is not superimposible on its mirror image. An example of a pair of enantiomers is shown below.

| 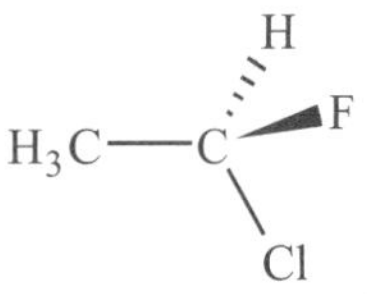 (S)-1-chloro-1-fluoroethane | 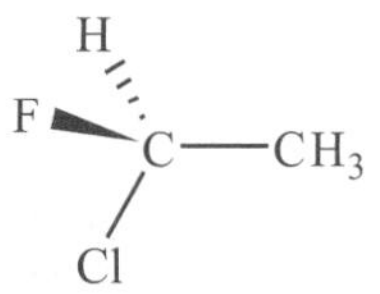 (R)-1-chloro-1-fluoroethane |

The two structures are mirror images. The line between them is like a mirror, and one structure is the reflection of the other. They cannot be superimposed, that is, every atom on one structure cannot overlay each atom on the other structure: This is best shown with models. If they did superimpose, they would be identical. The two structures are not identical because we name them differently: one is S, and the other is R. This is an easy way to tell if you have enantiomers: The names are the same, except that the configuration around each stereogenic atom is the opposite.

Molecules with more than one stereogenic atom can also be enantiomers, as the examples below indicate.

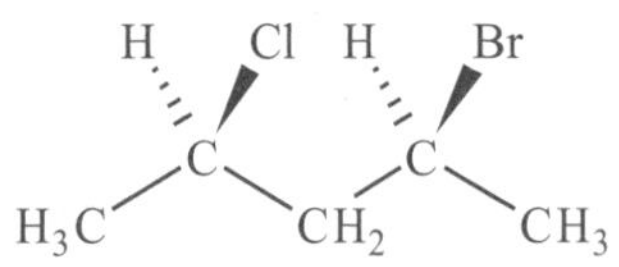

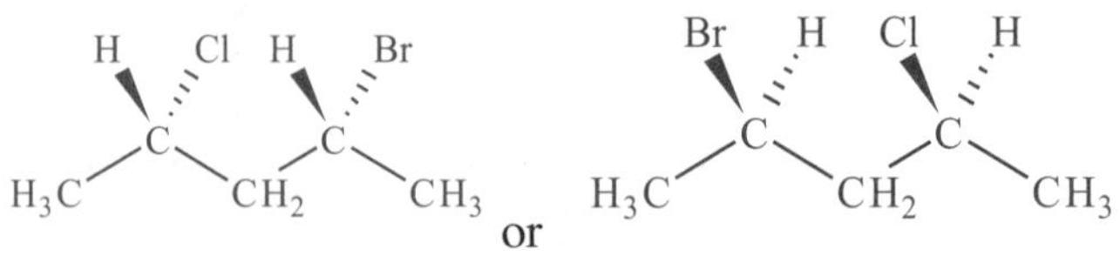

| (2S,4R)-2-Bromo-4-chloropentane | (2R,4S)-2-Bromo-4-chloropentane |

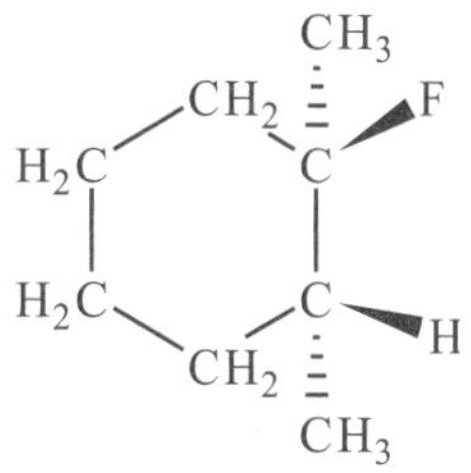

(1R,2R)-1-Fluoro-1,2-dimethylcyclohexane

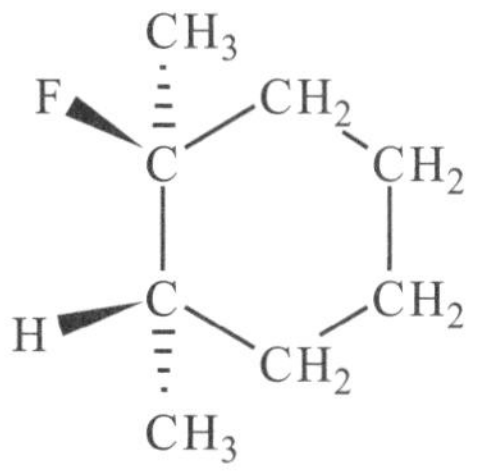

or

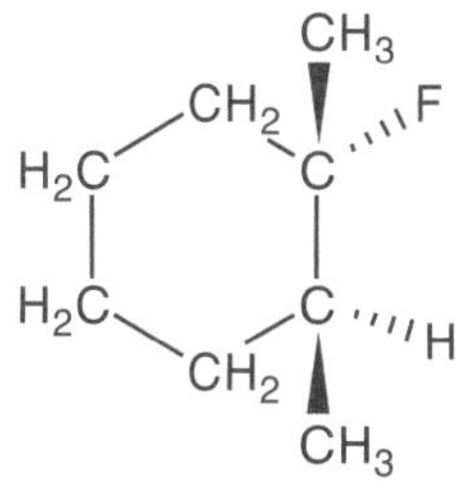

(1S,2S)-1-Fluoro-1,2-dimethylcyclohexane

However, just because a structure has stereogenic atoms doesn't mean it has an enantiomer. If a structure has a plane of symmetry through it, it will be superimposible on its mirror image, as the examples below show.

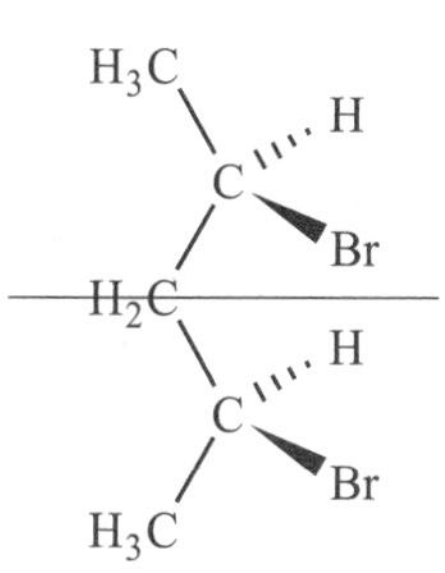

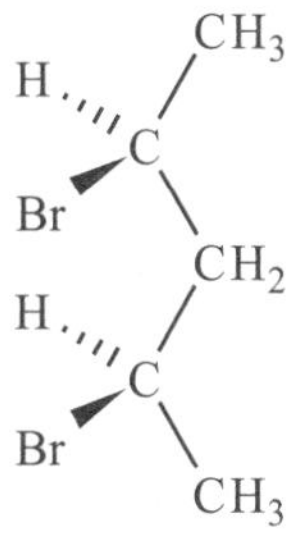

which, if you rotate 180°, is

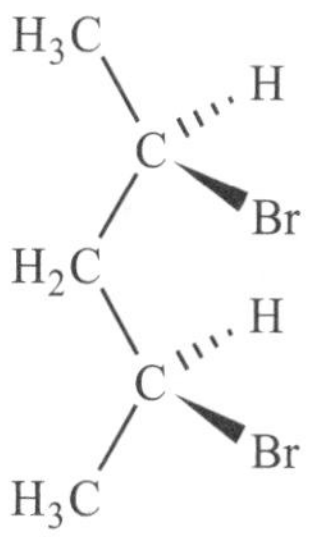

There is a plane of symmetry through the middle of the structure. Rotating the structure 180° gives you the other structure. They are both (2R,4S)-2,4-dibromopentane.

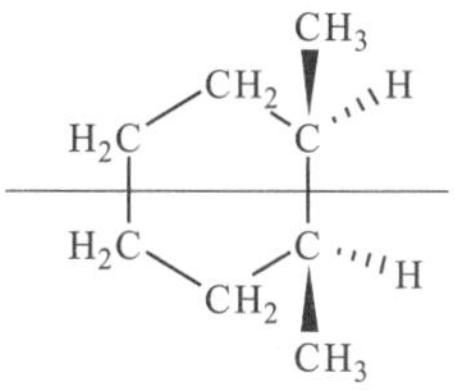

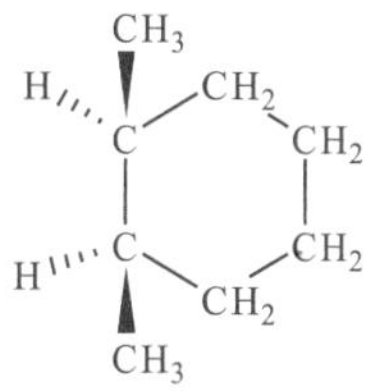

Again, there is a plane of symmetry through the middle of each of these structures. They are both (1R,2S)-1,2-dimethylcyclohexane. Compounds such as the examples above are called **meso compounds**.

A few structures do not have stereogenic atoms, but still have an enantiomer. A couple of examples are shown below.

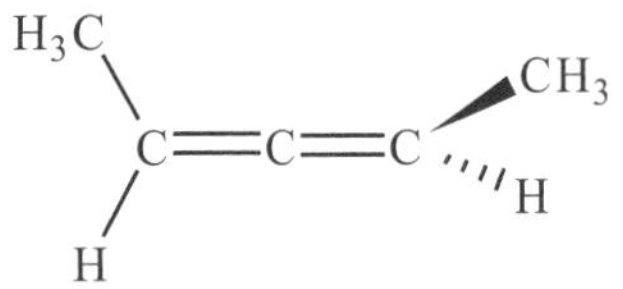

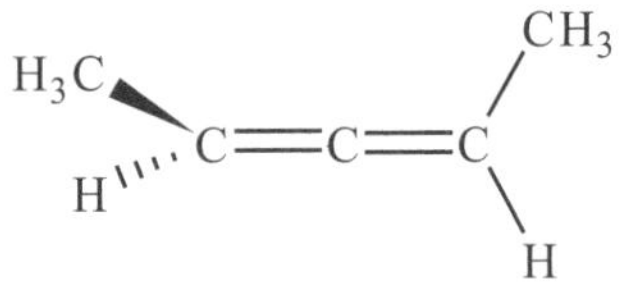

These are enantiomers because the overlapping *p* orbitals of the two C=C's are perpendicular, forcing the groups on the ends to also be perpendicular.

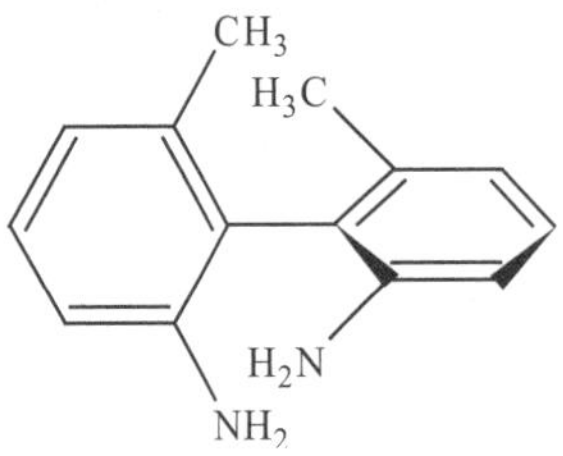

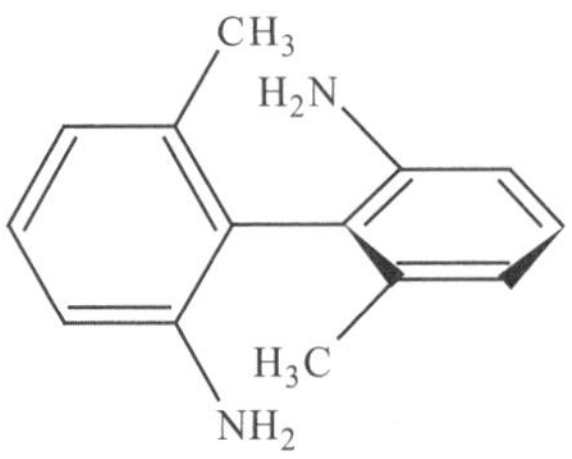

These two structures are enantiomers because the single bond between the two benzene rings is not free to rotate. The methyl groups and the amino groups cannot get past each other.

C.4. DIASTEREOMERS

A few pages ago, I made the statement, "If two structures have the same molecular formula, there are only four possible ways that they can be related: They can be identical, constitutional isomers, enantiomers, or diastereomers." So if you are examining a pair of structures, and they are not identical, constitutional isomers, or enantiomers, they must be diastereomers! I know this sounds like defining something by what it is not. However, the technical definition of diastereomers is "stereoisomers that are not enantiomers." More specifically, diastereomers are stereoisomers that are not superimposible and are not mirror images. Some examples of diastereomers are shown below.

Original Structure	**Diastereomer**
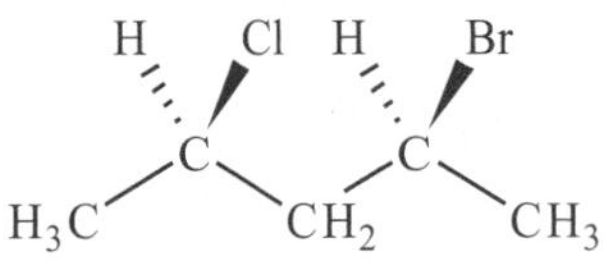	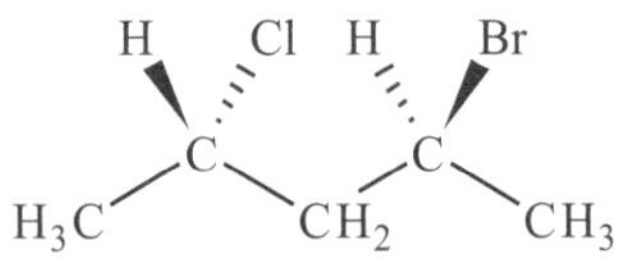
(2*S*,4*R*)-2-Bromo-4-chloropentane	(2*S*,4*S*)-2-Bromo-4-chloropentane

They are not identical because they do not have the same configurations around both stereogenic atoms. They are not constitutional isomers because the same atoms are bonded together in the same order. They are not enantiomers because they do not have opposite configurations around both stereogenic atoms. Therefore, they must be diastereomers.

Original Structure	Diastereomer
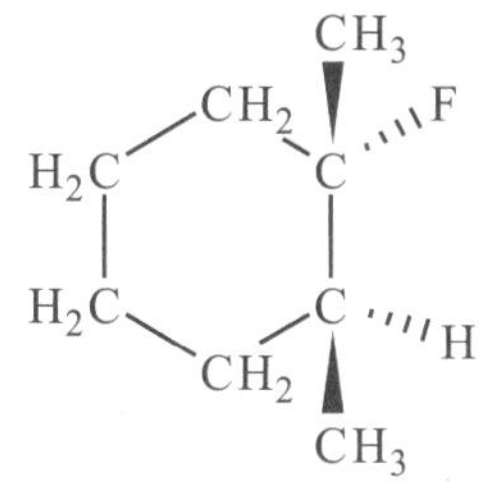	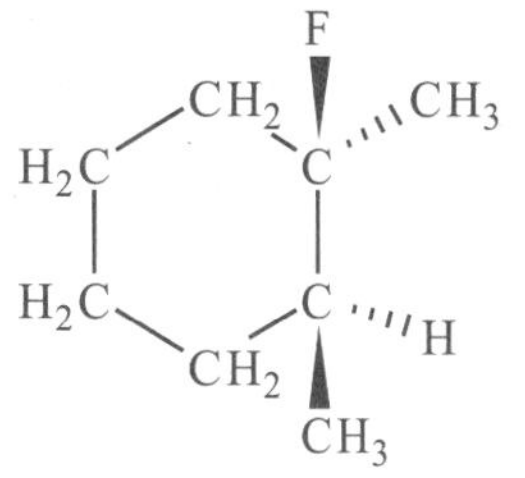
(1*S*,2*S*)-1-Fluoro-1,2-dimethylcyclohexane	(1*R*,2*S*)-1-Fluoro-1,2-dimethylcyclohexane

They are not identical because they do not have the same configurations around both stereogenic atoms. They are not constitutional isomers because the same atoms are bonded together in the same order. They are not enantiomers because they do not have opposite configurations around both stereogenic atoms. Therefore, they must be diastereomers.

One point needs to be made about the relationship between the number of stereocenters and the number of possible stereoisomers. If *n* is the number of stereocenters, then 2^n is the maximum number of stereoisomers. So with two stereocenters, the maximum number of stereoisomers is $2^2 = 4$. With five stereocenters, you could have $2^5 = 32$ stereoisomers. Most of the time, you will have the maximum number. The only time you don't is when there is a plane of symmetry through a structure, and its mirror image ends up being identical. This is illustrated below. The molecule has two stereocenters, but there are only three different stereoisomers.

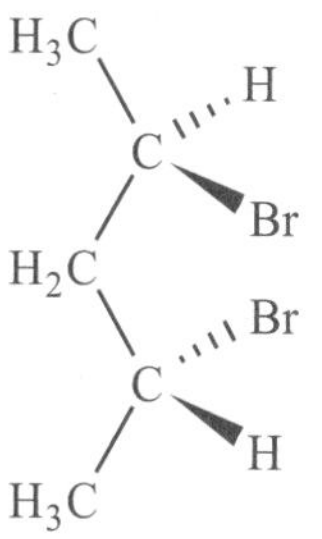	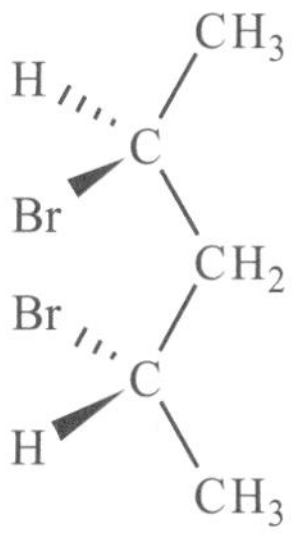	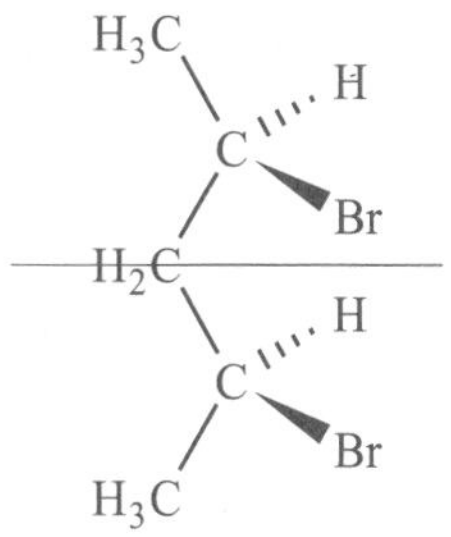	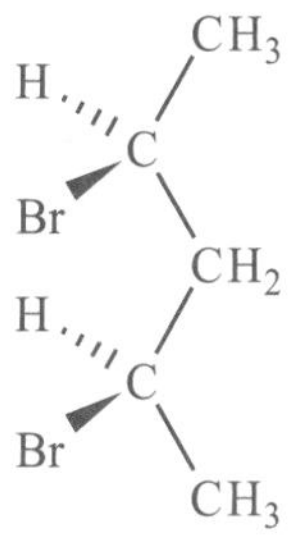
(2*R*,4*R*)	(2*S*,4*S*)	(2*R*,4*S*)	(2*R*,4*S*)
These are enantiomers.		These are identical because there is a mirror plane through the middle of the structure. This is a meso compound.	

D. Optical Rotation

Let's consider the enantiomers of 2-butanol, shown below. How do they differ in common physical properties? According the table below, they don't differ in all common physical properties, except in something called optical rotation.

Name	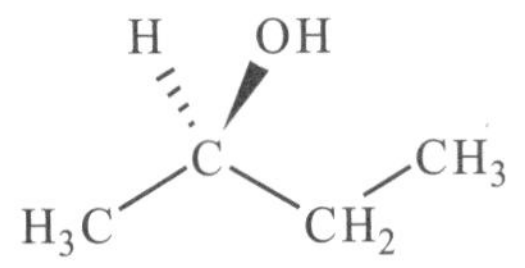 (R)-2-Butanol	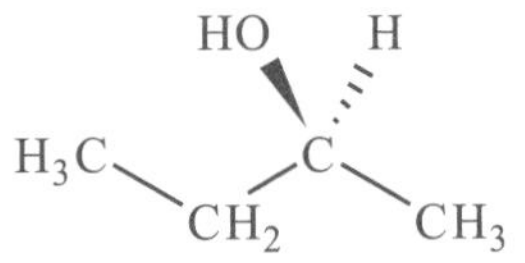 (S)-2-Butanol
Boiling point (°C)	98	98
Density (g/mL)	0.804	0.804
Refractive index	1.395	1.395
Water-solubility (g/100 mL)	8	8
Optical rotation, $[\alpha]_D$	−13.5	+13.5

Enantiomers have equal but opposite optical rotations. **Optical rotation** is the ability of a compound to rotate a plane of polarized light. The term, $[\alpha]_D$, refers to the specific rotation of a given material. Specific rotation is calculated using the following formula:

$$[\alpha]_D = \alpha/(C \cdot L)$$ where α = the observed rotation of the plane of polarized light in degrees, C = the concentration of the solution in grams per milliliter, and L = the length of the sample tube in decimeters.

When polarized light travels through a solution of a chiral material, the plane of polarization gets rotated (if the plane of polarization is rotated in a clockwise direction, it is called positive). Because enantiomers are mirror images, they rotate polarized light in opposite directions, but in equal amounts. If you rearrange the preceding equation to the following form:

$$\alpha = [\alpha]_D \cdot C \cdot L$$

you can see that the observed rotation, α, is proportional to the concentration of the solution and the length of the sample tube.

Optical rotation sounds a little weird, but it is the *only* simple physical property in which a pair of enantiomers differs. On the other hand, diastereomers differ in all of their physical properties because they are not mirror images.

One other point should be made. There is no correlation between (R) and (S) configurations with (+) and (−) optical rotations. Some (R) molecules rotate polarized light clockwise (+), and some rotate polarized light counterclockwise (−). If you can come up with a way to predict the direction of rotation of a structure, you can be famous!

E. Implications of Stereoisomerism

If you buy generic ibuprofen at the store for relief of headache pain, you are buying the following material. It has one stereocenter, so it has two stereoisomers, which are enantiomers.

| (S)-(+)-Ibuprofen | (R)-(−)-Ibuprofen |

The pain-relieving and anti-inflammatory effects of ibuprofen are solely due to the *S*-enantiomer: the *R*-enantiomer has no effect. However, the *R*-enantiomer is slowly metabolized to the *S*-enantiomer in the body, so it is not a total waste.

Why is the *R*-enantiomer inactive, while the *S*-enantiomer is active? It has to do with how well the two enantiomers bind to the receptor. A crude example of this is trying to put your hands in a baseball glove. If you throw with your right hand, you put the glove on your left hand. Your left hand fits very nicely into the glove, and you can manipulate the glove to catch balls with it, hopefully. If you try to use the glove on your right hand, your right hand doesn't fit into it very well, and you probably can't catch balls that way. In this example, the *S*-ibuprofen is your left hand, which fits into the receptor (the glove) in such a way that it produces the desired effect (the ability to catch balls). The *R*-ibuprofen (your right hand) just doesn't work.

REVIEW EXERCISES FOR CHAPTER 3

1. Star (*) all of the stereogenic atoms in the following structures.

2. Classify the stereogenic atoms in the structures below as either *R* or *S*.

a.

b.

c.

d.

3. Classify the following pairs of structures as identical, structural isomers, enantiomers, or diastereomers.

Alkenes

WHAT YOU WILL LEARN

In this chapter, you will learn:

- what alkenes look like and how to name them;
- how alkenes react with a variety of reagents;
- how to draw the arrow-pushing mechanism of many alkene reactions;
- how to synthesize alkenes from alkyl halides and alcohols.

SECTIONS IN THIS CHAPTER

- Structure and Bonding

- IUPAC Nomenclature

- Physical Properties

- The General Reaction of Alkenes: Addition to the C = C

- Introduction to Reaction Mechanisms

- Addition Reactions of Alkenes, Classified by Mechanistic Similarities

- Diels–Alder Reaction

- Preparations of Alkenes

Alkenes are organic compounds that contain one or more C=C's. Many naturally occurring compounds contain C=C's, including those shown below.

beta-Carotene is found in carrots and tomatoes and was the precursor to vitamin A.

(R)-(+)-Limonene is found in oranges and gives fruit the "citrus" odor.	alpha-Pinene beta-Pinene Constituents of turpentine that are found in evergreen trees and give the trees their characteristic "pine" odor.

A. Structure and Bonding

The C=C double bond is shorter (1.34 Å) than a C–C single bond (1.54 Å). This is partly due to the shorter sp^2 orbitals that the alkene carbons use for the sigma bond, but also to the overlap of the p orbitals.

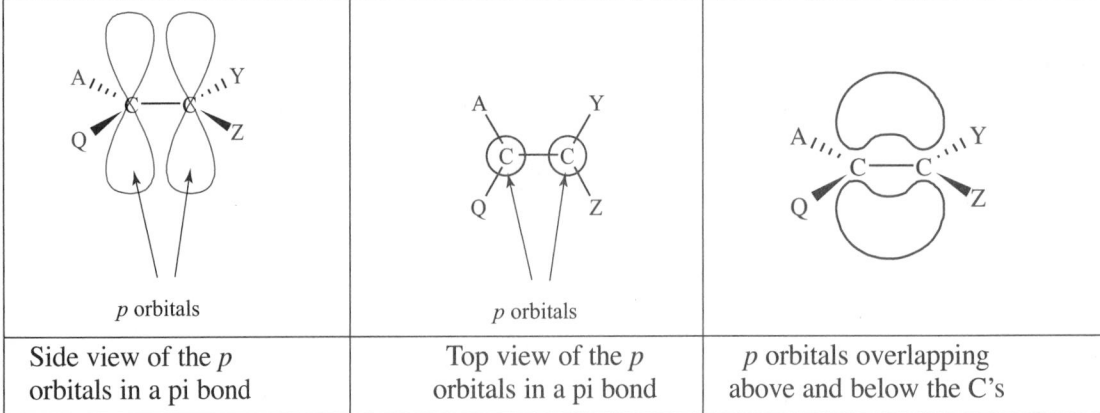

Side view of the p orbitals in a pi bond	Top view of the p orbitals in a pi bond	p orbitals overlapping above and below the C's

The other bond of the C=C is a pi bond, formed by the sideways overlap of the p orbitals on each carbon. For the pi bond to form, the p orbitals must be parallel to each other. As a result, the two carbons, and the four atoms attached to those two carbons (A, Q, Y, and Z), *must all lie in the same plane*. Also, there is no rotation around a C=C. To rotate around this bond would require you to break the overlap of the *p* orbitals, thus breaking the bond: This is not a good thing to do!

Ethene
(Ethylene)

According to VSEPR theory, the bond angles around an atom bonded to three things should be approximately 120°. As the example shows, the actual angles are pretty close to 120°.

For a nifty web site that will calculate bond lengths and bond angles of simple molecules, as well as lots of other information, visit *http://www.colby.edu/chemistry/webmo/mointro.html*.

B. IUPAC Nomenclature

The complete rules for IUPAC naming, with examples, can be found on *http://www.acdlabs.com/iupac/nomenclature/*.

The functional group ending for an alkene is *-ene*. The longest carbon chain is numbered from the end that gives the C=C the lowest number. See the following examples.

1   2   3   4 numbering from the left

H_2C=CH—CH_2—CH_3 But-1-ene

4   3   2   1 numbering from the right

The longest carbon chain contains four carbons, making the base part of the name *but*. Numbering from the left has the C=C start at carbon 1. The *1* is placed before the *ene*, and separated from the letters by dashes. *Note*: Some textbooks put the *1* before the base part of the name *but*, making the name 1-butene. Current IUPAC rules place the number for the functional group as close to the functional group as possible.

H_3C——CH_2—CH_2-CH=C——CH_3 2-Methylhex-2-ene
6   5   4   3   2   1

with CH_3 branch on carbon 2

The longest carbon chain is 6, which is hex. Because the molecule contains a C=C, it is a hexene. Numbering from the right gives the C=C the lower number, 2, so it is a hex-2-ene. There is a one-carbon group attached to the longest chain, so it is a methylhex-2-ene. The methyl group is attached to carbon 2, so the complete name is **2-methylhex-2-ene**.

Cyclopentene

When there is only one C=C in a ring, no location for the C=C is specified: It is defined to be at number 1. Remember the *cyclo-* prefix to indicate that there is a ring.

1,4,6-Trimethylcyclohexene

The trimethylcyclohexene is the easy part. Numbering it looks tricky. Remember that IUPAC says if we can have a substituent at the lowest numbered carbon, that is what we want. So if there is one substituent group on a C=C in a ring, that group must be at carbon 1. Carbon 2 must be the other carbon of the C=C. So the numbering is as shown.

H_3C CH_2-CH_3 C=C $H_3C-CH_2-CH_2$ H	(*E*)-4-Methylhept-3-ene
H_3C CH_2-CH_3 C≡C $H_3C-CH_2-CH_2$ H	Numbering from the right gives 4-methylhept-3-ene. The groups attached to each carbon of the C=C are prioritized using the Cahn–Ingold–Prelog rules (see Chapter 3). Propyl is higher in priority than methyl, and ethyl is higher in priority than H. Because the two higher priority groups are on opposite sides of the C=C, the compound is in the *E* configuration. *E* stands for *entgegen*, which in German means apart.
$H_3C-CH_2-CH_2$ CH_2-CH_3 C≡C H_3C H	(*Z*)-4-Methylhept-3-ene. *Z* stands for *zusammen*, which in German means together.

C. Physical Properties

Because alkenes contain only carbon and hydrogen, they are relatively nonpolar. As a result, their physical properties, such as boiling points and water-solubilities, are similar to those of alkanes. The boiling points of pentane and the isomeric pentenes are shown below.

Compound	Boiling Point (°C)
Pentane	36
Pent-1-ene	30
(*E*)-Pent-2-ene	36
(*Z*)-Pent-2-ene	37
2-Methylbut-2-ene	38

C.1. STABILITY OF ALKENES

The position of the C=C in an alkene can affect the relative stability of the alkene. When we talk about stability here, we are talking about thermodynamic stability, that is, how much energy can be derived from a compound. For example, pent-1-ene, (*E*)-pent-2-ene, and (*Z*)-pent-2-ene will each react with H_2 in the presence of a palladium catalyst to form pentane. These reactions are shown below. The amount of heat given off by each reaction is different. We can use the amount of heat given off to establish an order of stability for the alkenes. See the table below.

Reaction	ΔH (kcal/mol)
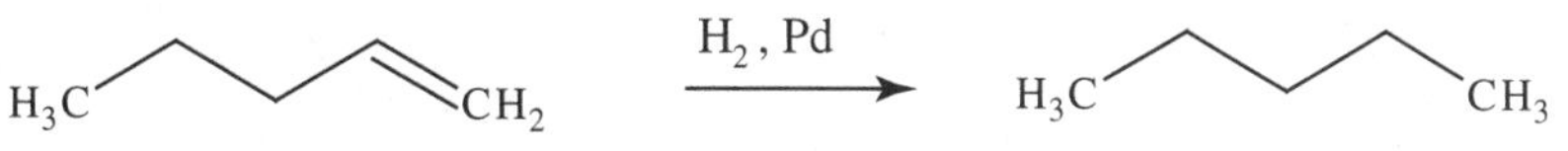 Pent-1-ene	−30.1
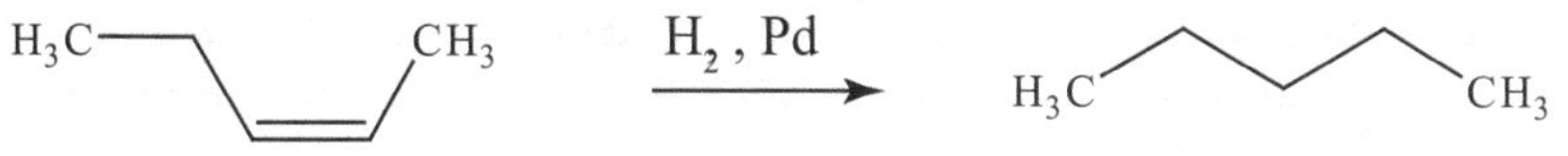 (*Z*)-Pent-2-ene	−28.6
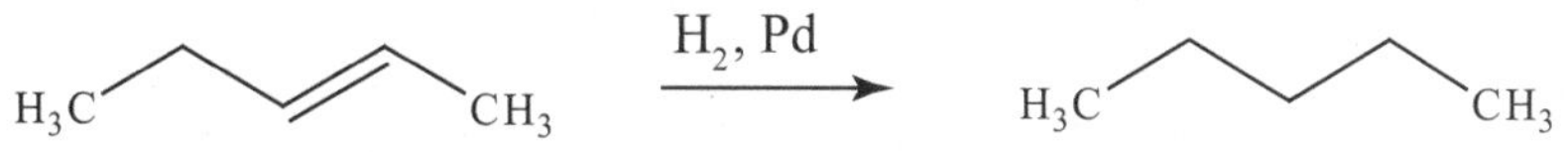 (*E*)-Pent-2-ene	−27.6

The more negative ΔH is, the more exothermic the reaction is, or the greater the difference in energy the alkene starting material is from pentane. We can use these ΔH values to establish an order of stabilities for these alkenes. See Figure 1.

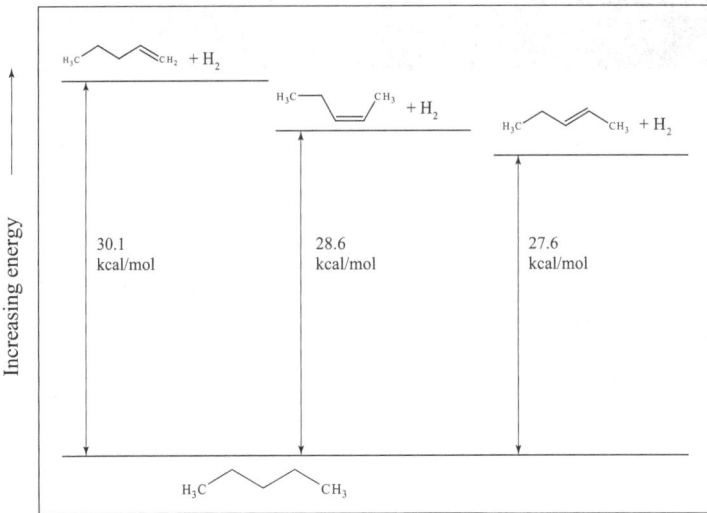

Figure 1. Energy changes in hydrogenation of pentene isomers.

As Figure 1 shows, pent-1-ene is higher in energy than (*Z*)-pent-2-ene, which is higher in energy than (*E*)-pent-2-ene. Higher in energy means that it is less stable thermodynamically. So, the order of stability (most stable to least stable) is (*E*)-pent-2-ene > (*Z*)-pent-2-ene > pent-1-ene.

In general, the trend of stabilities of alkenes is the following: the more C's that are directly bonded to the C=C, the more stable the alkene is. Also, (*E*)-alkenes are more stable than (*Z*)-alkenes.

D. The General Reaction of Alkenes: Addition to the C=C

The general reaction of an alkene is addition to the pi bond of the C=C. This is illustrated below.

$$\text{C}=\text{C} \;+\; \text{Y}-\text{Z} \;\longrightarrow\; \text{C}-\text{C} \begin{smallmatrix}\text{Y}\;\;\text{Z}\end{smallmatrix}$$

I have shown a generic alkene reacting with a generic reagent, Y–Z. One bond of the C=C is broken, and the Y part of the generic reagent becomes bonded to one carbon of the double bond, and the Z part ends up bonded to the other carbon. A specific example, with an actual alkene and a real reagent, is shown below.

5%   95%

Notice that there are two possible ways to add H–Br to this alkene, and that the products are formed in different amounts. We will look at why the two products are formed in different amounts and how to predict the major product of a reaction in later sections.

E. Introduction to Reaction Mechanisms

When you look at the preceding reaction, you might ask, "How are the products formed? When are bonds broken and formed?" Answering these questions has been the goal of chemists over the last 150 years. There is much experimental evidence to indicate how these reactions take place. Even though much of the evidence will not be presented in this text, be assured there is good evidence.

In any reaction, covalent bonds are formed and broken. Sometimes atoms lose electrons, and others gain them. Sometimes an atom loses electrons initially and then regains them later in the process. We need to keep track of the electrons as bonds are broken and formed. Chemists use curved arrows to show the "movement" of electrons in a reaction. This sequence of electron movements that shows how bonds are broken and formed is called a **reaction mechanism**.

In many reactions, polarity of the bonds in the molecules plays an important part. If you can apply the concept of electronegativity to a bond and determine which atom is partially positive and which is partially negative, this will help you to move electrons in ways that make chemical sense. For example, when considering the previous reaction of an alkene with H–Br, we know that Br is more electronegative than H. So, in H–Br, the Br is partially negative and the H is partially positive. The H will want to attract electrons to itself. One place it can get some electrons is from the pi bond of the C=C. We show this attraction of electrons to the H from the C=C, and the resulting cleavage of the H–Br bond using curved arrows. See below.

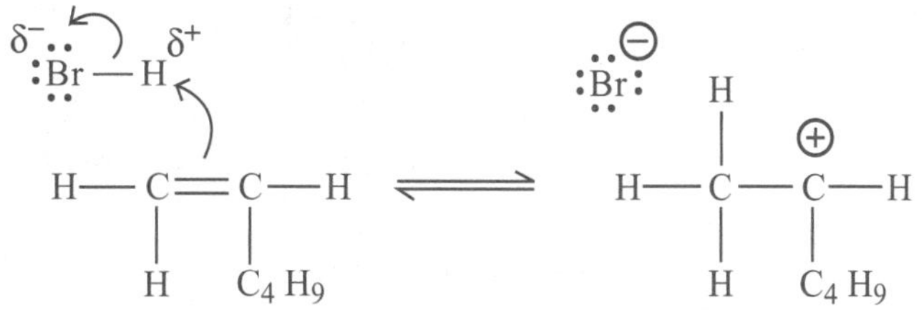

One arrow shows a pair of electrons being attracted to the partially positive H and forming a new bond between the C and the H. Simultaneously, the bond between the H and the Br is broken, and the pair of electrons from that bond ends up on the Br. This makes sense because Br is more electronegative than H. Because the Br gained electrons, it is now negative. Likewise, because electrons were removed from the right C of the C=C, that C is now positive. The positive carbon is called a **carbocation**. The carbocation is bonded to two other carbons and is referred to as a **secondary carbocation**.

To complete the reaction, we need to form a bond between the negative Br and the positive C. Because the Br has the electrons we need to form the bond, we draw an arrow from the Br to the C, as shown below.

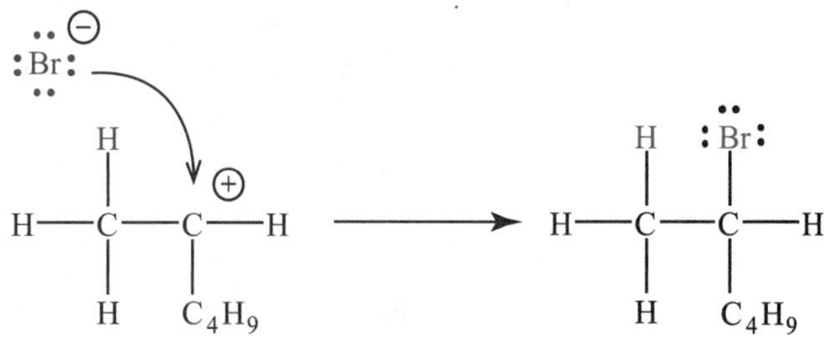

A similar series of arrows can be drawn to show how the minor product is formed.

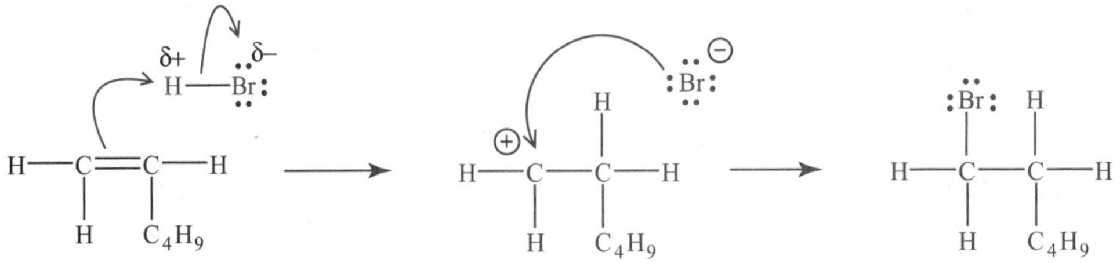

In this mechanism, the positively charged carbon in the second structure is bonded to only one other carbon. This carbon is referred to as a **primary carbocation**.

F. Addition Reactions of Alkenes, Classified by Mechanistic Similarities

Sometimes students feel overwhelmed by the mechanisms of reactions. Often this is because they don't see the similarities between the mechanisms. I have grouped the reactions below by similarities in the types of mechanisms, to try and overcome this.

F.1. ADDITIONS OF H–X AND H₂O (ACID-CATALYZED)

All of these reagents are similar, in that they have a partially positive H bonded to an electronegative atom. As a result, they react in similar fashions with an alkene.

F.1.a. ADDITION OF H–CL, H–BR, AND H–I
General Reaction:

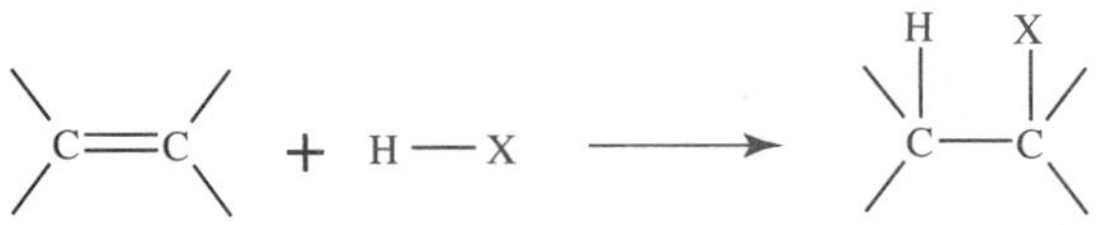

Specific Examples:

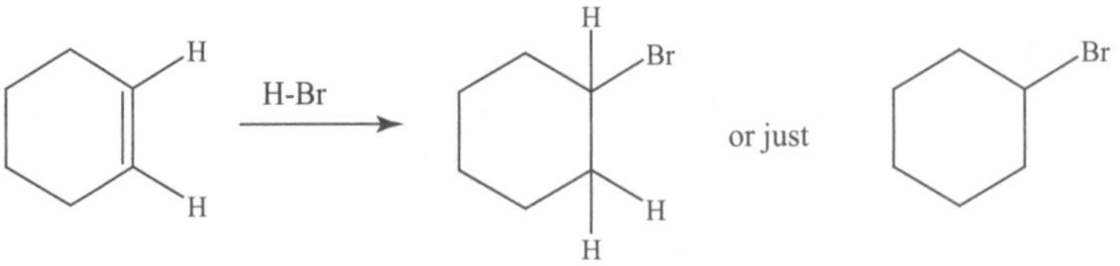

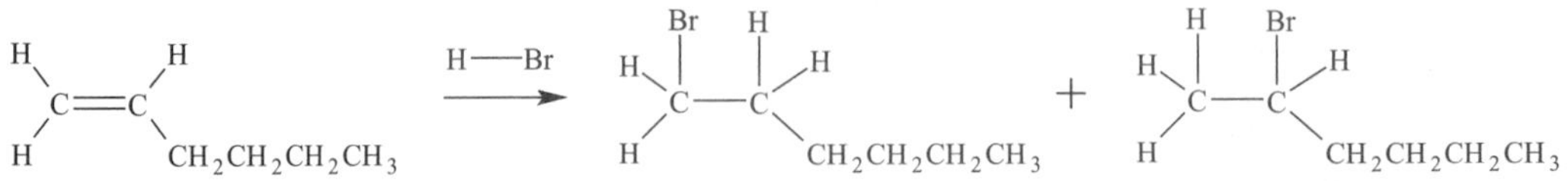

In 1869, Vladimir Vasilevich Markovnikov formulated a rule to rationalize how hydrogen halides would add to unsymmetrical alkenes. Markovnikov's rule states that the major product of the reaction has the H of the H–X bonded to the carbon of the C=C that is bonded directly to the most H's, and the X ends up bonded to the other carbon. He did not know why the reaction happened this way; he simply observed the general trend.

The reason for Markovnikov's rule has to do with the stability of the possible carbocations formed. In general, the more carbons attached to the carbocation, the more stable it is, so the order of stability of carbocations is tertiary > secondary > primary. When we say "more stable," we mean lower in energy. Figure 2 illustrates what happens when H⁺ adds to the C=C of 2-methylpropene.

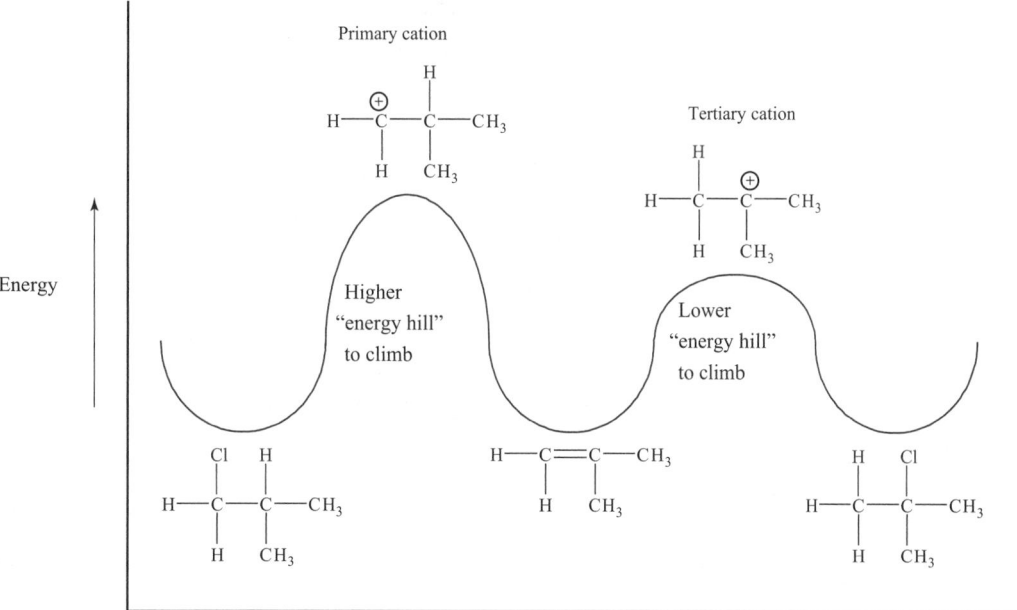

Figure 2. Reaction coordinate for addition of H–Cl to 2-methylpropene.

If the H adds to carbon 2, then carbon 1 becomes a primary carbocation. If the H adds to carbon 1, then carbon 2 becomes a tertiary carbocation. Because the tertiary carbocation is more stable, then less energy is needed to get to it from the alkene starting material (there is a lower energy hill to climb, so to speak). If it is easier to get to the tertiary carbocation intermediate, then chloride is more likely to react with the tertiary carbocation, to form 2-chloro-2-methylpropane.

F.1.b. ADDITION OF WATER (ACID-CATALYZED)

General Reaction:

Specific Examples:

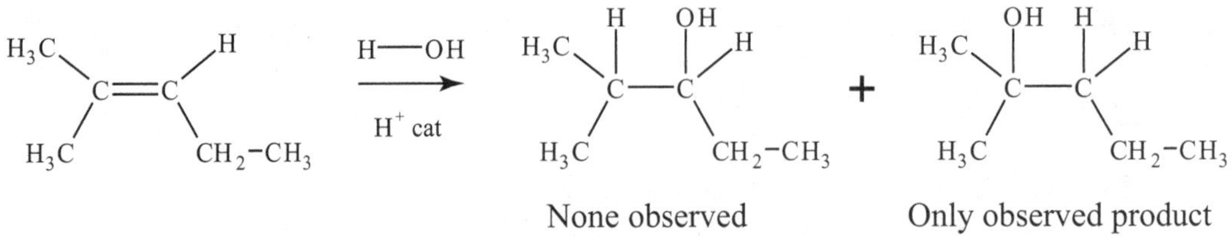

None observed   Only observed product

As with the addition of H–X, the addition of water also follows Markovnikov's Rule. Because water is not as strong of an acid as H–X, an acid catalyst (usually sulfuric acid) is used to attract the electron pair from the C=C. The oxygen in water then attacks the carbocation. H^+ is lost to regenerate the H^+ catalyst. The mechanism is shown below.

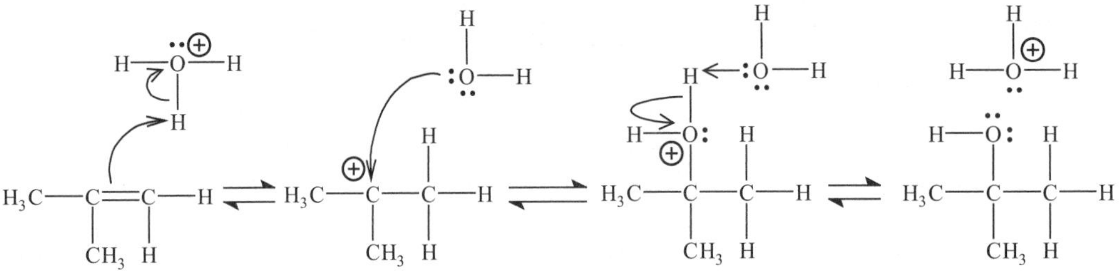

F.1.c. POLYMERIZATION OF ALKENES WITH ACID

Suppose you have 1 million molecules of methylpropene (commonly called isobutylene), and you add one H^+X^-. What will happen? Initially, the H^+ will attract an electron pair from the C=C and form a carbocation, as we did above. Now the carbocation is looking for electrons. The most likely source of electrons is the C=C of another molecule of isobutylene, so it attacks the carbocation. This is shown below.

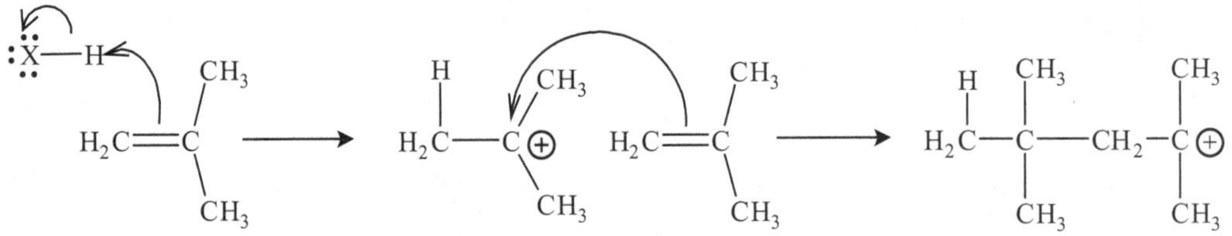

This could repeat over and over again until all of the isobutylene is used up.

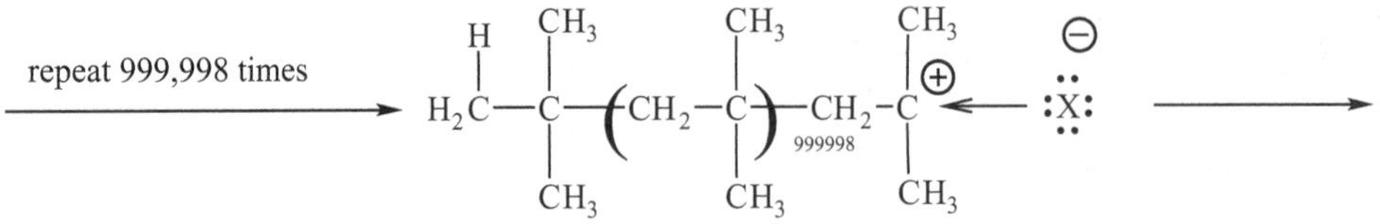

Finally, the anion (X⁻) could attack the cation, and we would have the following.

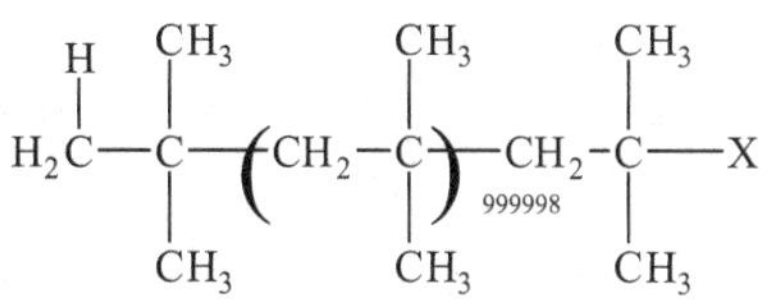

or just

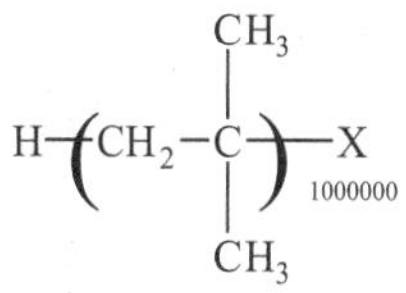

We now have a very large molecule! Very large molecules formed from smaller repeating units are called **polymers**. We polymerized isobutylene above, and the polymer we made is called poly-isobutylene. A more general way to write the structure of the polymer is shown below: n represents a very large number. Some other monomers and polymers are shown in the table below.

Monomer	Polymer	Uses
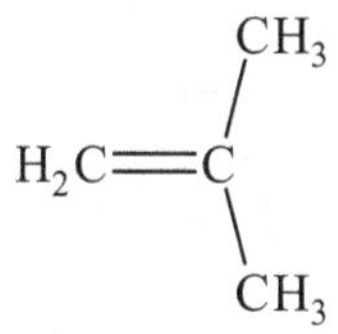 "Isobutylene" or methylpropene	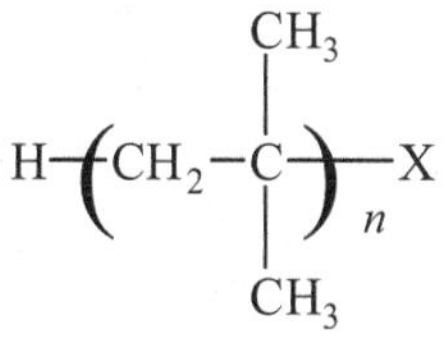 Polyisobutylene	Synthetic rubber
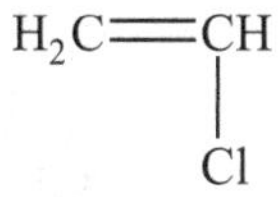 Chloroethene or "vinyl chloride"	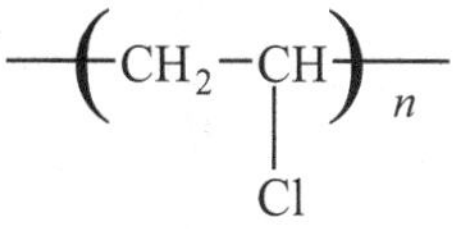 Polyvinyl chloride (PVC)	Rubber substitutes, electrical wire and cable coverings, pliable thin sheeting, raincoats, tubing, gaskets, shoe soles.
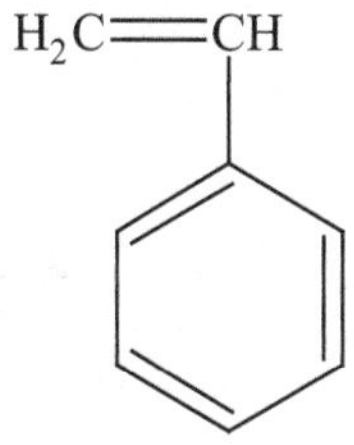 "Styrene"	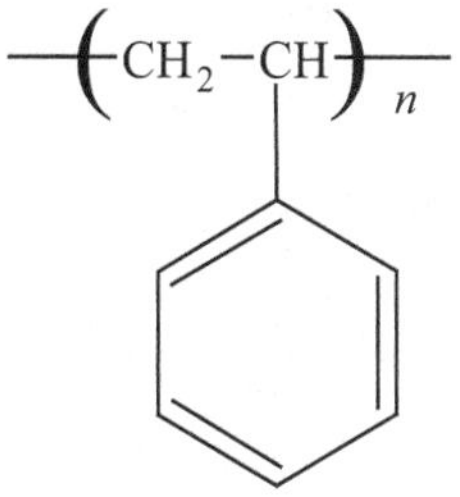 Polystyrene	Plastics, Styrofoam containers

F.2. HYDROBORATION/OXIDATION

General Reaction:

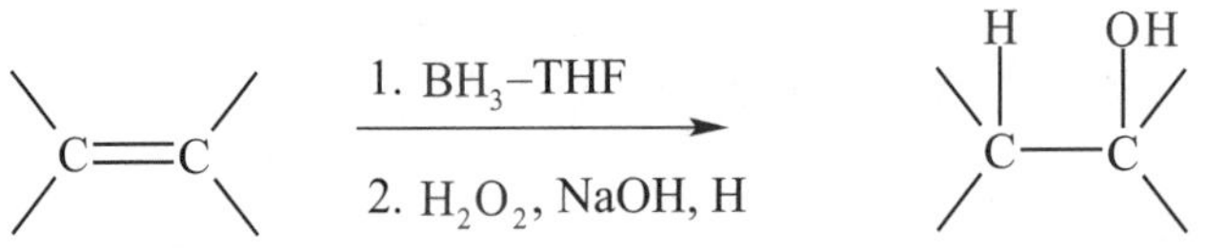

Specific Examples:

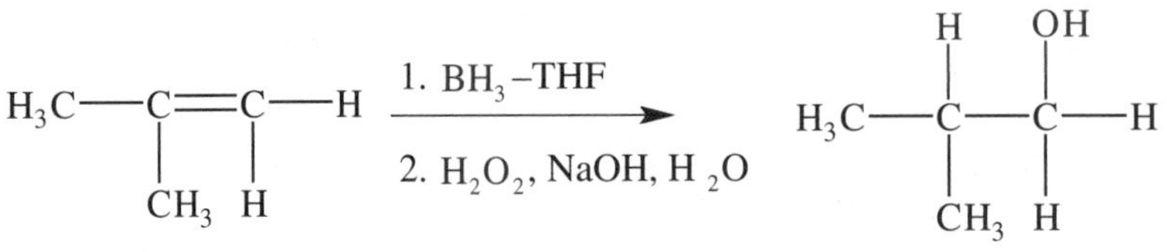

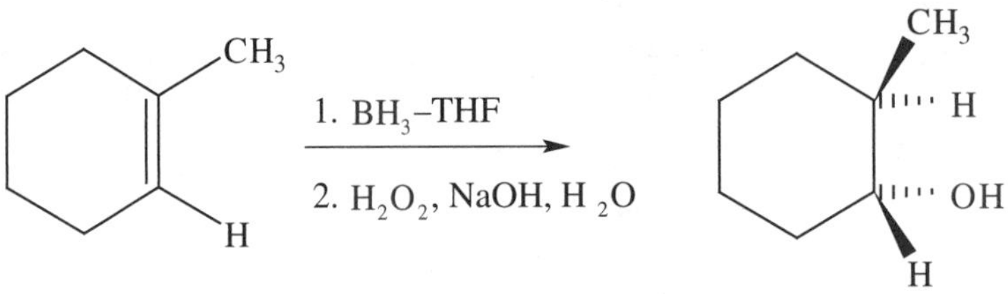

In 1959, Herbert C. Brown published the first of many papers dealing with reductions using borane (BH_3) and related compounds. This two-step procedure results in a net "anti-Markovnikov" addition of water to a C=C: The H ends up bonded to the more substituted C of the C=C, and the OH ends up bonded to the least substituted carbon. In addition, the H and the OH are added to the same side of the C=C, which is referred to as **syn addition**, as illustrated in the reaction of 1-methylcyclohexene.

The addition of BH_3 is called hydroboration. The mechanism of the hydroboration reaction is shown below. The boron is bonded to three H's and has no unshared pair of electrons, so it is electron deficient. Also, boron is less electronegative than hydrogen. The combination of these two effects results in the boron attracting the electron pair from the C=C toward itself. Simultaneously, an H with a pair of electrons (a hydride) is transferred to the more substituted carbon. The result of this is the organoborane intermediate, with a carbon–boron single bond.

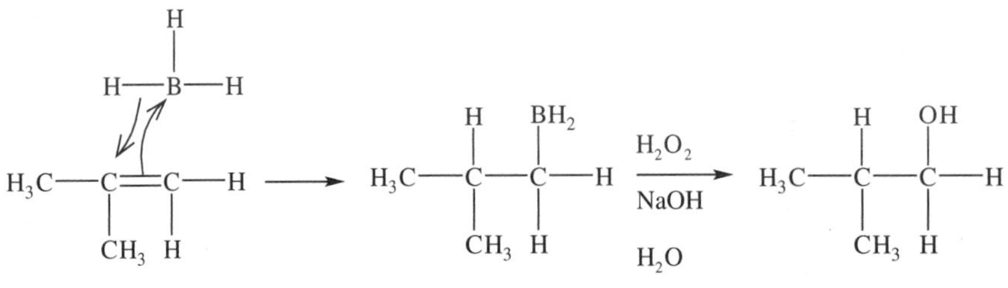

Organoborane

The second step is the oxidation. In a series of steps, hydrogen peroxide adds to the boron, and a rearrangement occurs to make a C–OH. The boron ends up as Na_3BO_3, sodium borate.

F.3. ADDITIONS OF HALOGENS (X₂), CARBENES, EPOXIDATION, AND OXYMERCURATION

F.3.a. ADDITION OF HALOGENS

General Reaction:

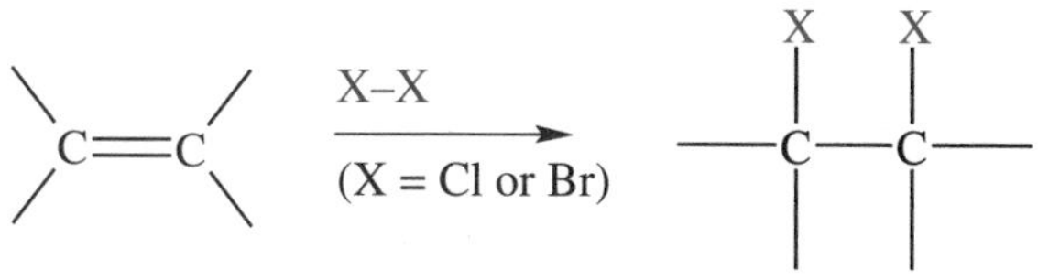

Specific Example:

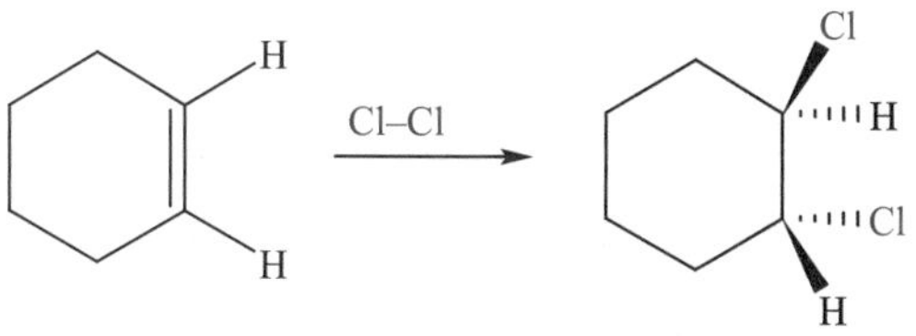

Chlorines are trans

Halogens, most commonly chlorine or bromine, add to C=C to form 1,2-dihalo addition compounds. As shown with the cyclohexene above, the halogens add to opposite sides of the C=C, which is referred to as **anti addition**.

The mechanism is shown below. When a molecular halogen interacts with a C=C, the halogen become polarized, so one atom reacts as if it were partially positive, and the other being partially negative. The electron pair from the C=C attacks the partially positive Br and pushes the electron pair from the Br–Br bond onto the other Br. However, because the partially positive bromine also has unshared pairs of electrons, the electrons are attracted to the developing carbocation. This forms a cyclic bromonium ion shown below. The positive charge is shared by the bromine and the more stable carbocation. The bromide ion then attacks the carbocation from the opposite side from the bromine in the bromonium ion, to form the dibromo product.

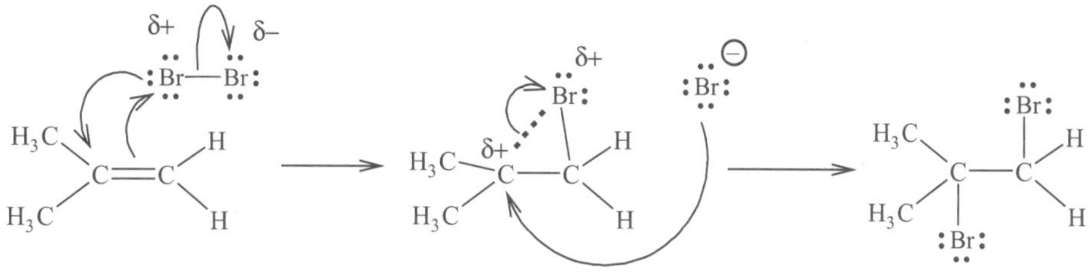

Bromonium ion

Treatment of an alkene or an alkyne with a dilute solution of bromine is a simple qualitative test for a C=C or C≡C, because the red–orange color of the bromine solution disappears after each drop of the bromine solution reacts.

F.3.b. OXYMERCURATION

General Reaction:

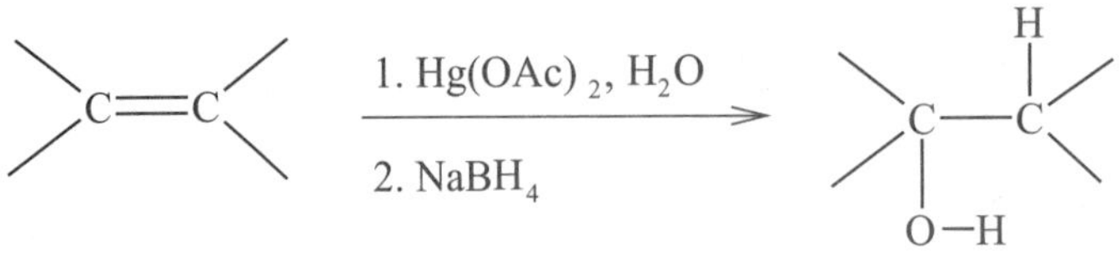

Specific Examples:

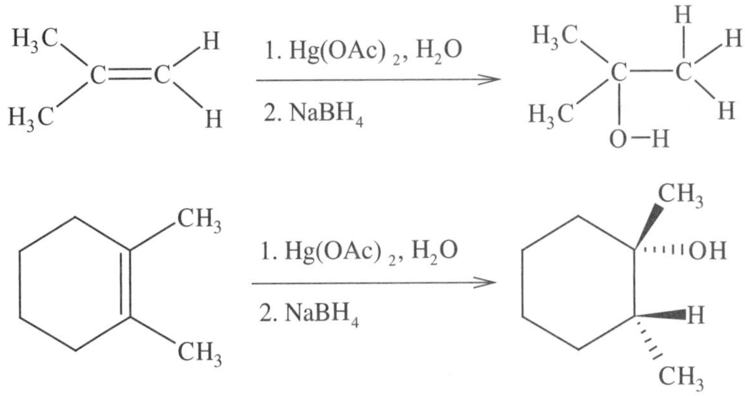

The oxymercuration/demercuration procedure is a two-step sequence that adds water to a C=C in a Markovnikov fashion. The yields are often higher than direct acid-catalyzed addition of water. Also, this procedure is less prone to side reactions common in acid-catalyzed addition of water. These side reactions will be discussed in Chapter 6.

The mechanism of the reaction is shown below. It is very similar to the halogenation mechanism. Mercuric acetate, $Hg(OAc)_2$, is a commonly used catalyst because it has some solubility in water and polar organic solvents. Mercuric trifluoroacetate, $Hg(O_2CCF_3)_2$, is also used. The mercury atom is partially positive, so it attracts electrons from the C=C. An electron pair is pushed onto one of the acetates. Simultaneously, an electron pair on the mercury is attracted to the developing carbocation to form a cyclic mercurium ion. Water attacks the more positive carbon and displaces the mercury. Loss of H^+ forms the intermediate organomercury-alcohol.

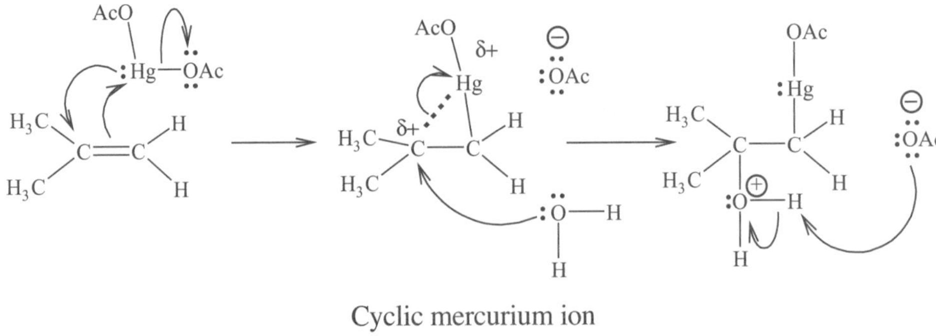

Cyclic mercurium ion

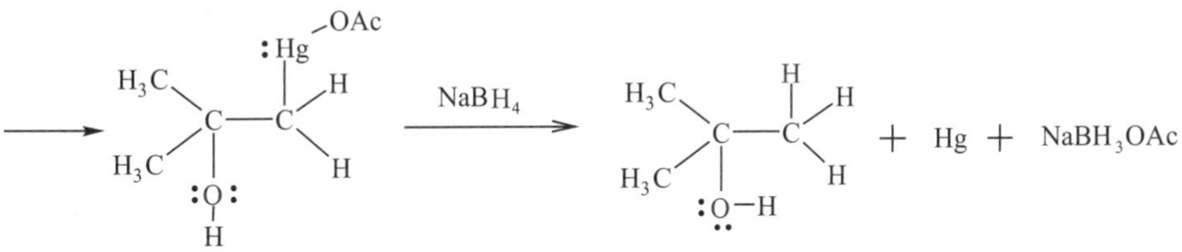

Organomercury-alcohol

In the second step, sodium borohydride ($NaBH_4$) reduces the organomercury-alcohol to produce the alcohol product. The H replaces the Hg with retention of configuration, probably through a mechanism involving radicals.

A variation on this reaction uses an alcohol instead of water. This variation results in the addition of the alcohol to the C=C to form an ether as a final product. An example is shown below.

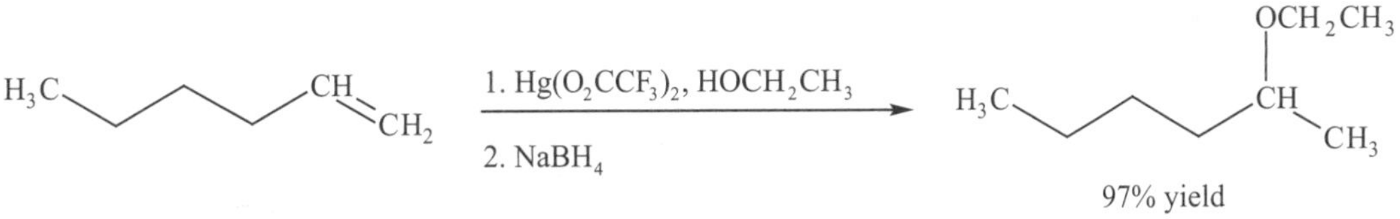

97% yield

F.3.c. EPOXIDATION

General Reaction:

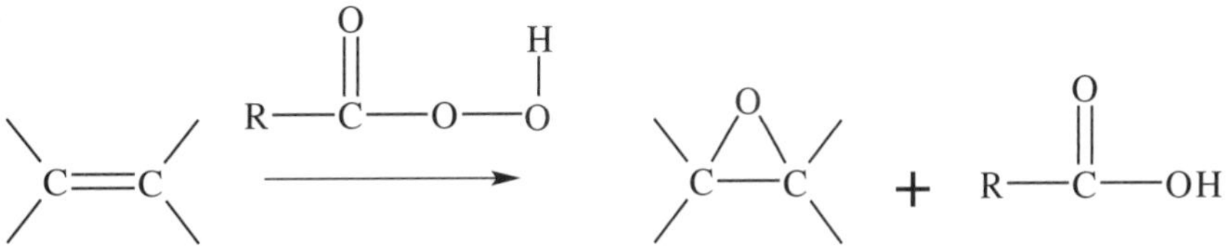

Specific Example:

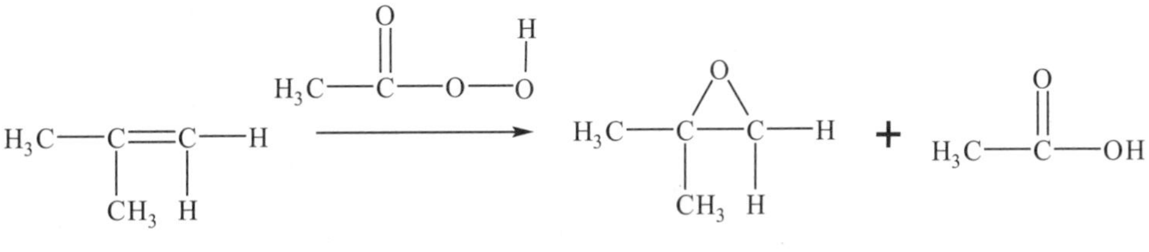

Epoxides are ethers in which the oxygen is part of a three-membered ring. Epoxides are conveniently prepared by reacting alkenes with peracids, whose general formula is RCO_3H. Some peracids are commercially available. Others can be prepared by reacting a carboxylic acid with hydrogen peroxide.

The mechanism is similar to the two preceding ones. The electron pair from the C=C attacks the O, and the weak O–O bond is broken. The indicated oxygen is attacked because the relatively weakly basic acetate ion is formed as a by–product. If the other oxygen were attacked, then hydroxide, a strong base, would be formed. Simultaneously, an electron pair on the oxygen is attracted to the developing carbocation to form a cyclic oxonium ion. Also simultaneously, acetate then pulls the H off the oxonium ion to give the epoxide.

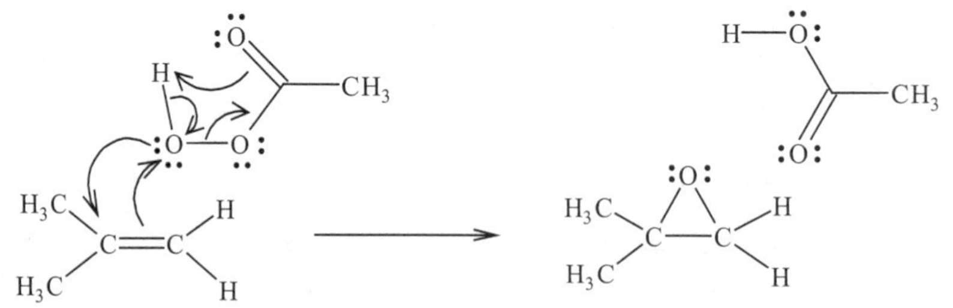

F.3.d. Addition of Carbenes

General Reaction:

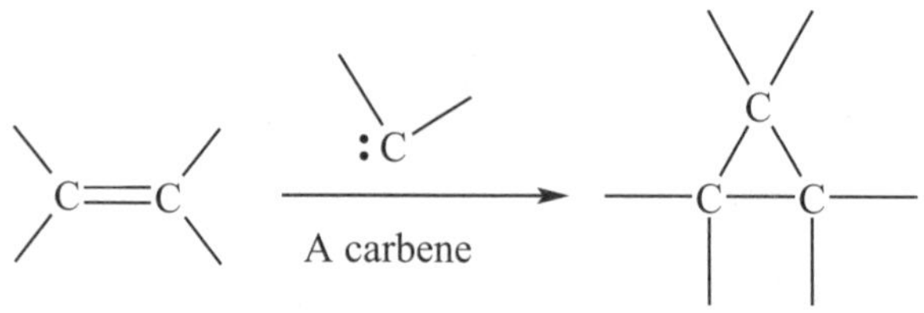

Specific Examples:

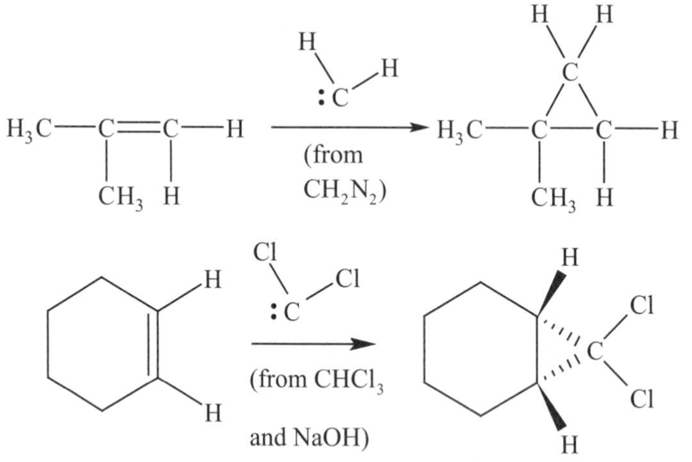

Carbenes are neutral, highly reactive species. They are highly reactive because the carbon is electron deficient: It does not have an octet of outer shell electrons. They exist only for short periods of time and are generated and used immediately. The reaction of carbenes with alkenes is a convenient way to form cyclopropane rings.

Carbenes can be generated in a variety of ways. One way is treatment of trihalomethanes with hydroxide to produce a dihalocarbene.

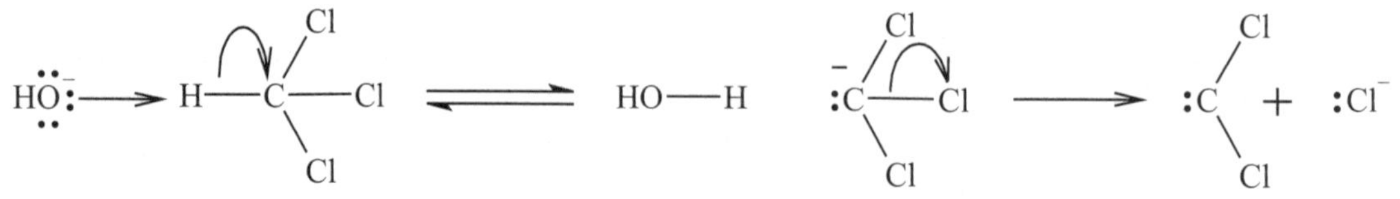

Another is treatment of an *N*-methyl-*N*-nitroso compound with potassium hydroxide to form diazomethane, CH_2N_2. Treatment of diazomethane with ultraviolet light produces a carbene.

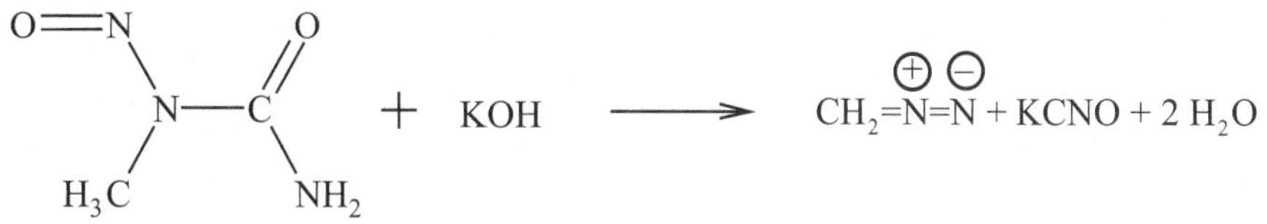

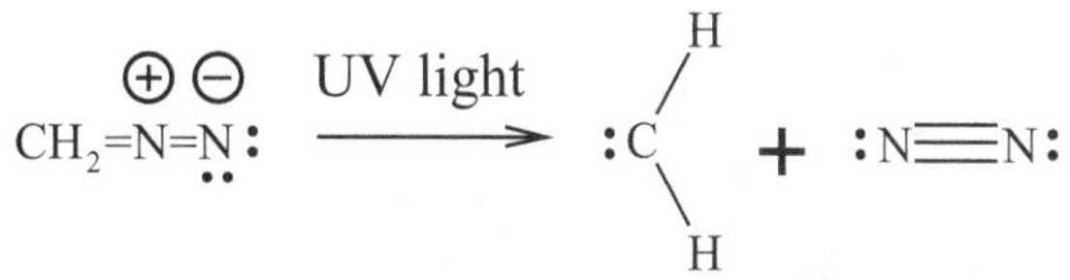

The mechanism should now be familiar. An electron pair from the C=C attacks the electron deficient C of the carbene. The unshared pair of electrons on the carbene C attacks the other carbon of the C=C to form the cyclopropane ring.

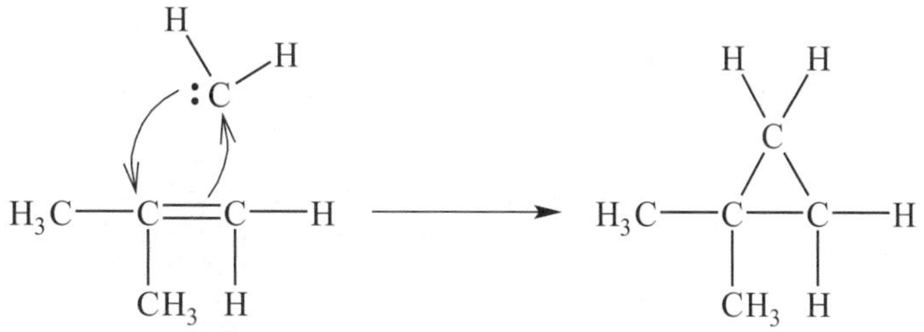

F.4. OXIDATIONS OF ALKENES WITH OZONE AND KMnO$_4$

The C=C of an alkene can be oxidized by a number of reagents. Two of the more common ones are ozone (O_3) and aqueous potassium permanganate ($KMnO_4$). Although the products produced by each reagent are different, the initial addition of the reagents follows a similar mechanism.

F.4.a. OXIDATION WITH DILUTE AQUEOUS POTASSIUM PERMANGANATE (KMnO$_4$)

General Reaction:

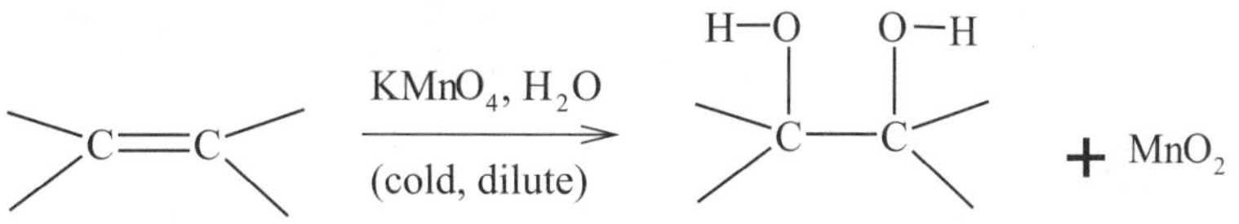

Specific Examples:

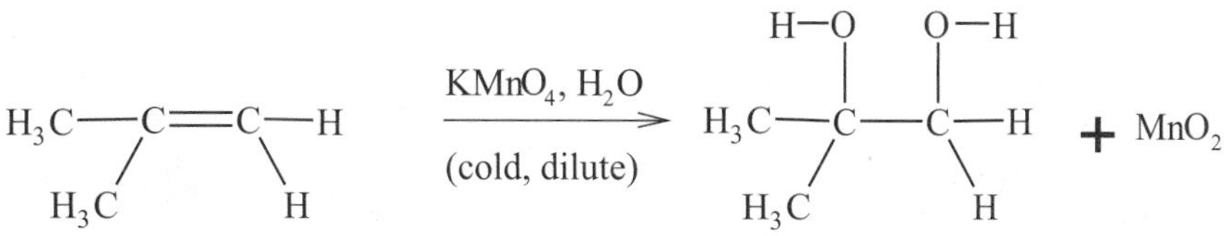

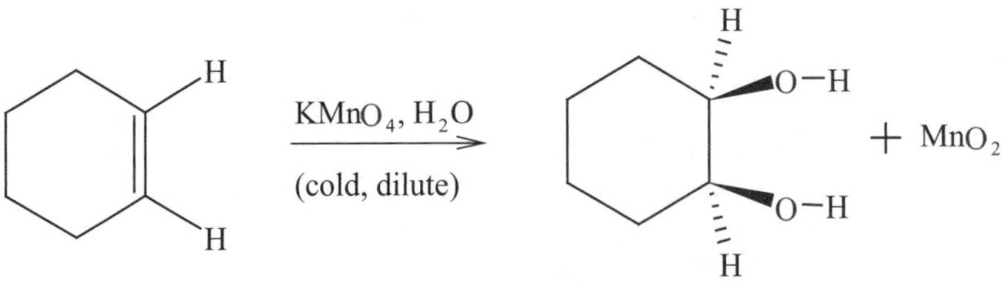

A dilute solution of potassium permanganate in water oxidizes a C=C to a diol. The net reaction breaks the pi bond of the C=C and adds an OH to each carbon. The first step of the reaction mechanism has permanganate ion adding to the C=C, to form a five-membered ring intermediate. Water then hydrolyzes the Mn–O bonds to form the diol. This addition is shown below. Because permanganate must add from one side of the C=C, the two OH groups end up on the same side of the molecule, or a **syn addition**.

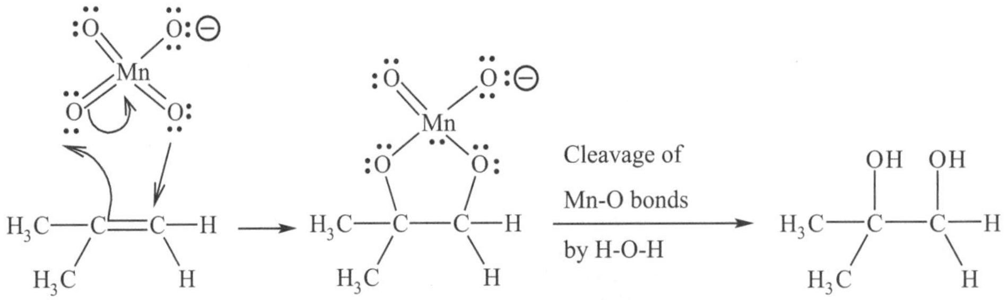

As an aside, you probably noticed that I didn't show the manganese-containing by–product of the above reaction. Although MnO_2 is the ultimate manganese-containing by–product, it is not formed directly from this reaction. After reviewing a number of inorganic chemistry texts, it appears that an Mn^{5+} oxide is formed initially. This disproportionates to MnO_2 and MnO_4^{2-} (manganate). MnO_4^{2-} reacts with water to form MnO_2 and MnO_4^{-} (permanganate). This is all very complex, which is why most organic chemistry texts tend to ignore inorganic by–products of many reactions, and especially those of oxidation reactions!

This reaction is a simple qualitative test for the presence of a C=C or a C≡C, because the purple color of the permanganate ion disappears and the black–brown precipitate of MnO_2 is formed.

Another reagent will do this same oxidation is osmium tetroxide, OsO_4. Osmium tetroxide is more toxic and expensive than $KMnO_4$, but it can be used in catalytic amounts in the presence of other oxidants, such as hydrogen peroxide. The osmium tetroxide oxidizes the alkene, then the hydrogen peroxide reoxidizes the reduced osmium species back to OsO_4.

F.4.b. OZONOLYSIS

General Reaction:

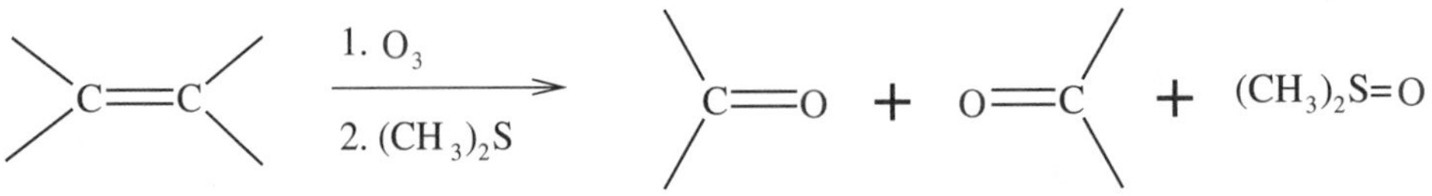

Specific Examples:

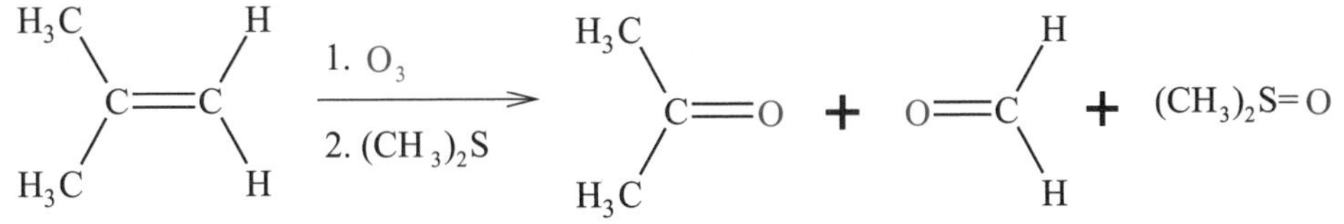

Ozone, O_3, has the following structural formula, which is represented by a pair of resonance structures.

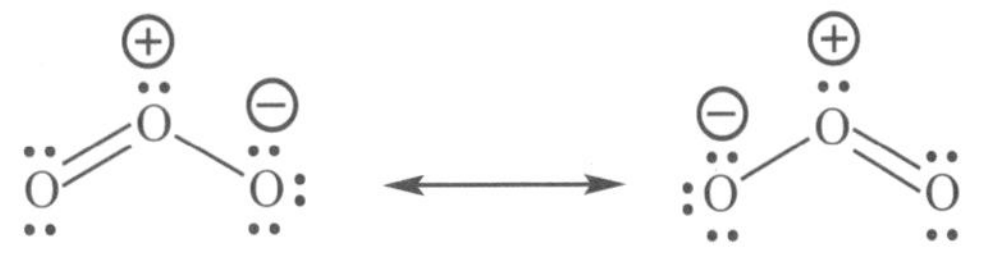

Ozone initially adds to a C=C to produce a cyclic intermediate, referred to as a primary ozonide, or molozonide. The molozonide rearranges to the ozonide. This is shown below.

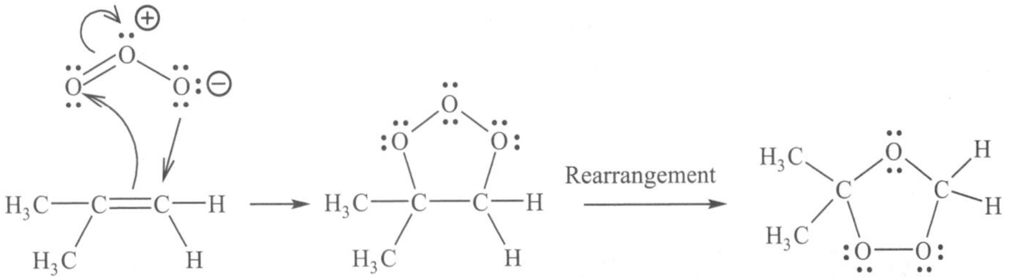

Reduction of the ozonide by dimethylsulfide, $(CH_3)_2S$, produces the C=O compounds, with dimethylsulfoxide, $(CH_3)_2S=O$, as a by–product. The net reaction is cleavage of the alkene C=C and formation of aldehydes and/or ketones.

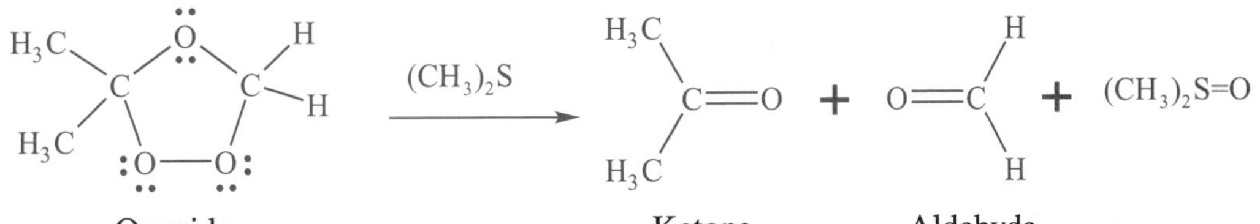

| Ozonide | Ketone | Aldehyde |

F.4.c. OXIDATIVE CLEAVAGE OF ALKENES BY TRANSITION METAL OXIDANTS AND RELATED COMPOUNDS

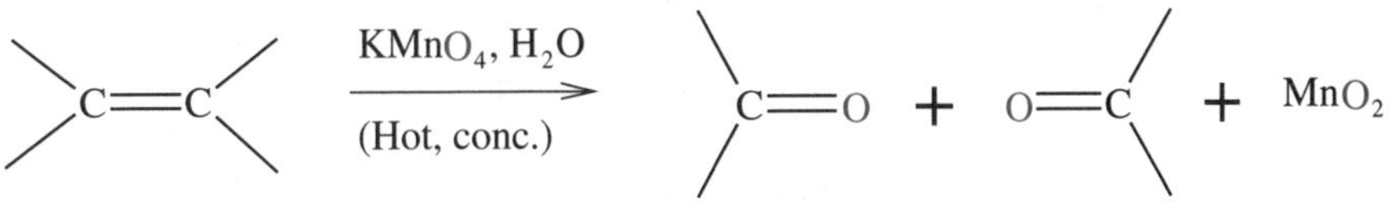

$$H_3C \quad \backslash \quad / \quad H$$
$$C = C \xrightarrow[\text{(Hot, conc.)}]{KMnO_4, H_2O} \quad H_3C \backslash C = O \quad + \quad O = C = O \quad + \quad MnO_2$$
$$H_3C \quad / \quad \backslash \quad H \qquad\qquad\qquad H_3C /$$

(cyclohexene) $\xrightarrow[\text{(Hot, conc.)}]{KMnO_4, H_2O}$ (diacid with O—H, C=O, C=O, O—H groups) $+ \; MnO_2$

A variety of transition metal oxides and similar reagents can cleave C=C's to give C=O containing products. Generally, ketones and/or carboxylic acids are formed, depending on the starting material. One-carbon pieces are oxidized to CO_2. Most commonly, hot concentrated solutions of potassium permanganate or hot, concentrated chromic acid solutions have been used, but $KMnO_4$ with a phase transfer catalyst, catalytic $KMnO_4$ with HIO_4, $RuO_4/NaIO_4$, and catalytic OsO_4 in $CrO_3/H_2SO_4/H_2O$ also do this reaction.

The mechanism of the reaction probably involves formation of the diol, as in Section F.4.a, followed by C–C bond cleavage and further oxidation.

F.5. CATALYTIC HYDROGENATION

General Reaction:

$$\backslash \quad / \qquad\qquad\qquad H \quad H$$
$$C = C \xrightarrow[\text{catalyst}]{H - H} \quad -C - C-$$
$$/ \quad \backslash \qquad\qquad\qquad | \quad |$$

Specific Examples:

(1,2-dimethylcyclohexene) $\xrightarrow[\text{catalyst}]{H - H}$ (cis-1,2-dimethylcyclohexane)

This is one of the simplest addition reactions: A hydrogen adds to each carbon of the C=C. The traditional catalysts are palladium or platinum metal, which are supported in a solid material, such as finely powdered carbon. A number of other metals such as rhodium, ruthenium, and specially prepared nickel-containing catalysts are also used.

Generally, addition of the two hydrogens occurs from the same side of the C=C. Hydrogen appears to absorb onto the metal surface, which breaks the H–H single bond. The alkene complexes to the metal as well, and the pi bond of the C=C is broken. Hydrogens are then transferred to the alkene carbons. No arrow-pushing mechanism will be given here.

G. Diels–Alder Reaction

General Reaction:

Specific Examples:

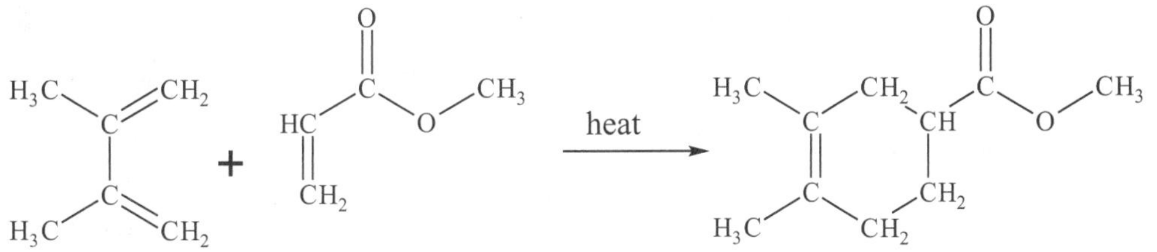

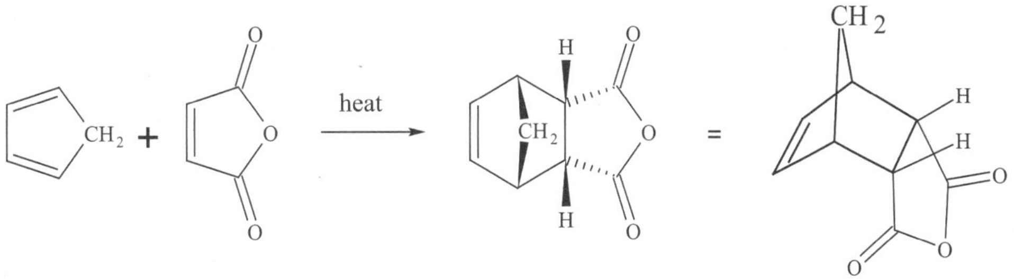

In 1928, Otto Diels and Kurt Alder reported the first examples of the reaction that came to bear their names. In 1950, they received the Nobel Prize in chemistry for their years of work.

As seen in the preceding examples, the Diels–Alder reaction forms a six-membered ring from two starting materials: a conjugated diene and an alkene, which is referred to as the **dienophile** because it reacts with the diene. By conjugated, we mean that the C=C's in the diene are separated by exactly one C–C single bond. The diene and/or the dienophile can be parts of other rings, so relatively complex structures with several rings can be formed in one step.

In general, the C=C of the dienophile is attached to other electron-withdrawing functional groups, such as nitriles, nitro groups, esters, or other C=O containing groups. These groups increase the reactivity of the dienophile toward most dienes. The dienes often have electron-donating groups attached to them to increase their reactivity.

The mechanism of the reaction involves a concerted movement of three pairs of electrons, similar to that of the initial addition of ozone to a C=C that we talked about previously. This is shown below. One of my colleagues refers to this as a "whirl of electrons."

There are no intermediates in this mechanism. The "movement" of electrons is similar to that which we did when we talked about resonance of benzene in Chapter 1. See below.

Notice that the arrows we drew here to show how we formally convert one resonance structure of benzene into the other are exactly the same as the arrows we drew earlier for the mechanism of the Diels–Alder reaction!

There is usually a certain stereochemistry associated with these reactions. If there are two groups on the dienophile, and they are trans, then they will end up trans in the product. If the two groups are cis, they will be cis in the product. See the examples below.

Ester groups are trans   Ester groups are trans

Ester groups are cis   Ester groups are cis

H. Preparations of Alkenes

H.1. DEHYDRATION OF ALCOHOLS

General Reaction:

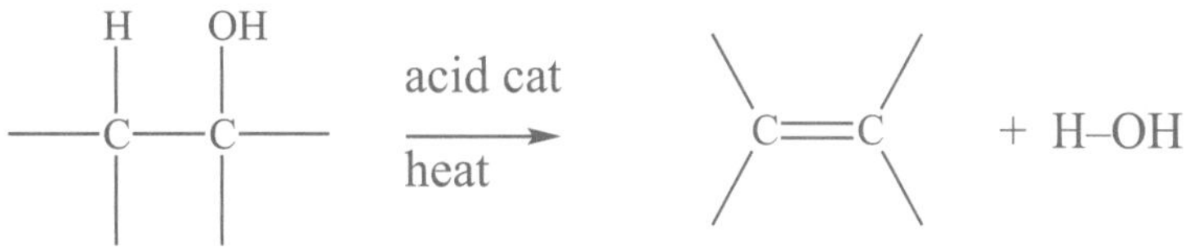

Specific Examples:

Dehydration of cyclohexanol

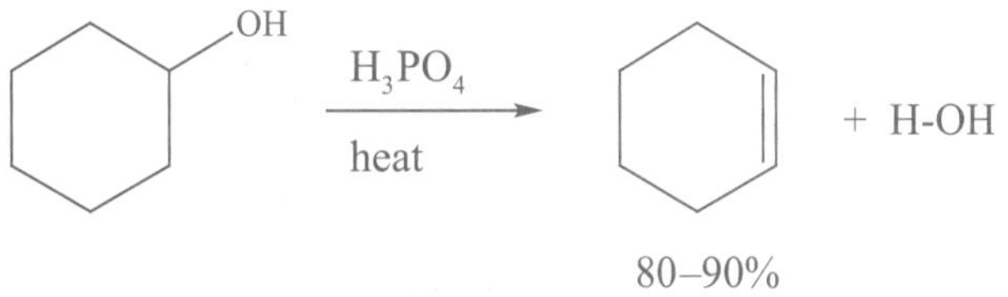

80–90%

Dehydration of butan-2-ol

4%   55%   41%

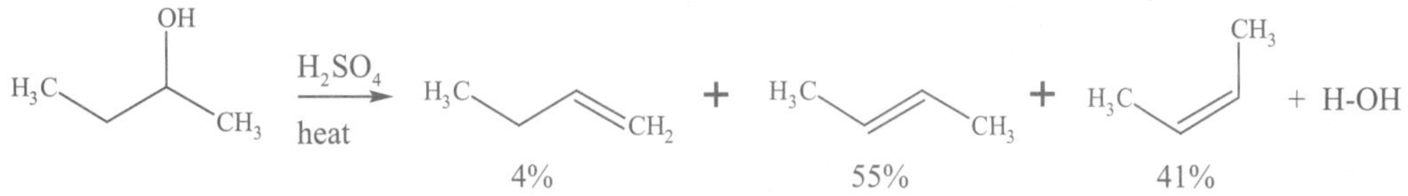

Heating an alcohol with a catalytic amount of a strong acid (usually sulfuric or phosphoric acid) results in the formation of an alkene, with water as a by–product. Because water is formed, the reaction is called a **dehydration reaction**. This is formally the reverse reaction of addition of water to an alkene (see Section F.1.b). The mechanism of the reaction is also the reverse of addition of water. See below.

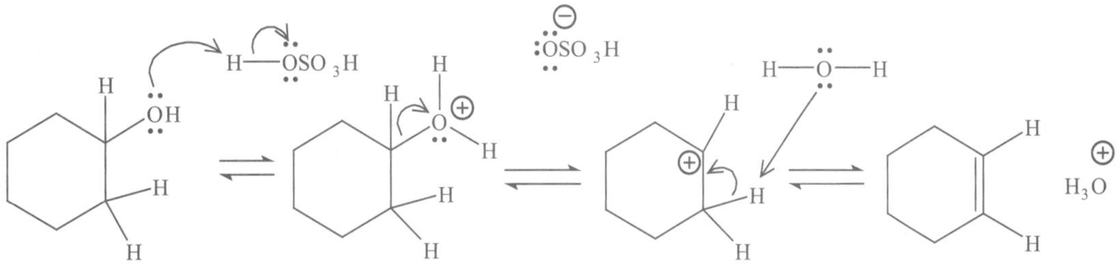

Protonation of the alcohol forms an oxonium ion. Loss of water produces a carbocation. Loss of H$^+$ from an adjacent carbon forms the C=C. In this case, the carbocation intermediate formed is symmetrical, so only one alkene product is formed.

If the carbocation formed is not symmetrical, then more than one alkene product can form. See the mechanism for the dehydration of butan-2-ol on the next page.

Just like the first mechanism, the alcohol is protonated, and water is lost, to form a cation. This cation can lose H⁺ in three different ways to form the different alkene products. See below.

Why is (*E*)-but-2-ene the major product? In Section C, you read that (*E*)-disubstituted alkenes are more stable than (*Z*)-disubstituted alkenes, which are more stable than monosubstituted alkenes. In general, the more stable alkene is the major product from a dehydration reaction.

Because this is a carbocation mechanism, rearrangement of the carbocation is possible, if a more stable carbocation can form. This is demonstrated by the dehydration of butan-1-ol.

The only reasonable way to form the two but-2-ene products is if a secondary cation is produced as an intermediate. This could happen as follows.

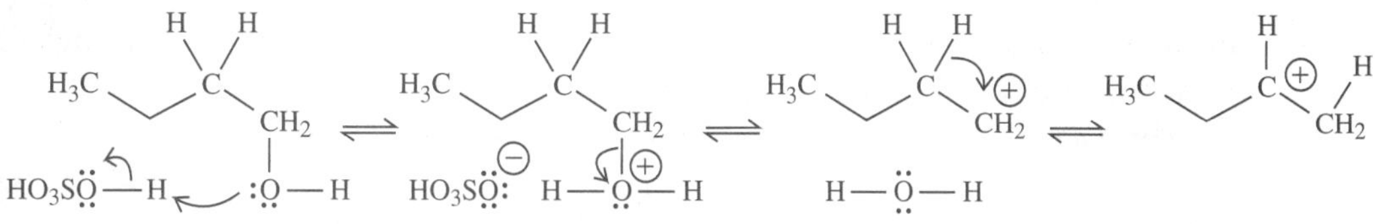

Protonation of the alcohol, followed by loss of water, would give a primary cation. Primary cations are very unstable, so a 1,2-hydride shift gives the more stable secondary cation.

Most people believe that primary cations are too unstable to be formed under these reaction conditions. Instead of forming a primary cation, the 1,2-hydride shift occurs simultaneously with the loss of water, as shown below.

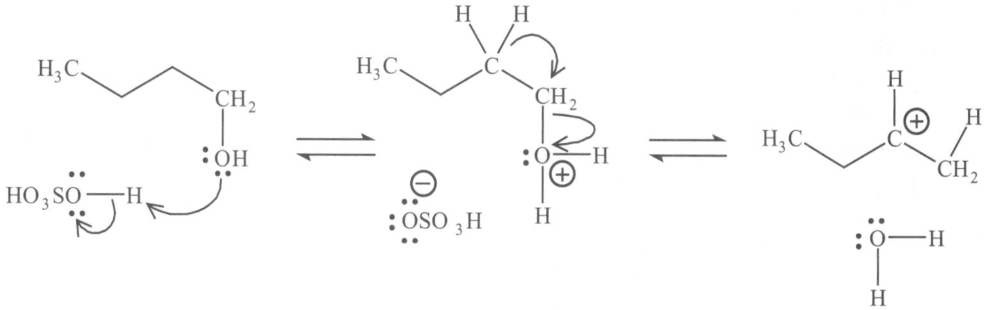

H.2. DEHYDROHALOGENATION OF ALKYL HALIDES

General Reaction:

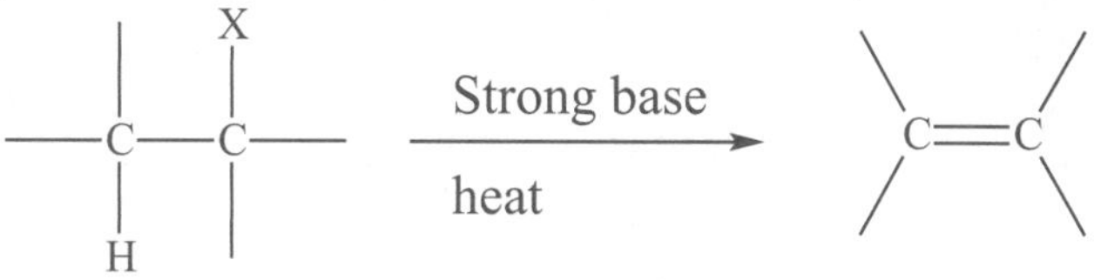

Specific Examples:

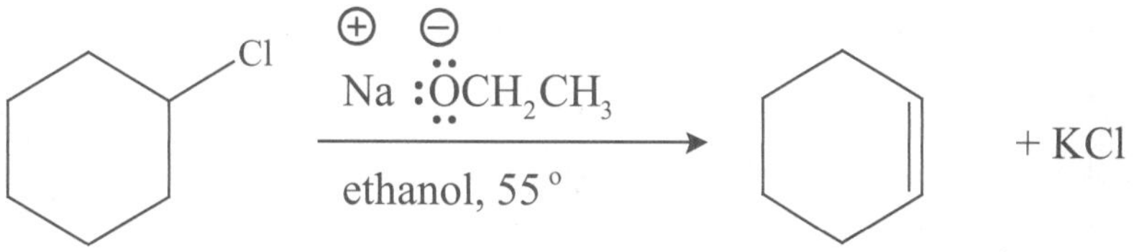

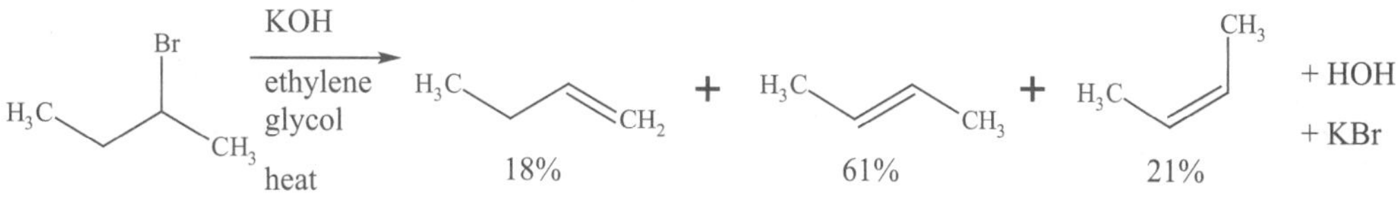

Treatment of an alkyl halide with a strong base is another way to produce an alkene. Because a hydrogen and the halogen are removed to form the C=C, the reaction is called **dehydrohalogenation**. The mechanism of this reaction has been shown to not involve cations and has no distinct intermediates. The general mechanism is as follows, using hydroxide ion as the base.

The base removes a hydrogen from the opposite side of the alkyl halide from the halogen. This is referred to as **anti-elimination**, and is the most common mechanism of elimination. The Newman projection on the right shows the H and the X are anti, with a dihedral angle of 180°.

There is much experimental evidence for this mechanism, including the following. The E2 reaction involving (1R,2S)-1-bromo-2-methylcyclohexane and sodium ethoxide (NaOEt) yields the two alkenes shown.

However, the E2 reaction involving (1S,2S)-1-bromo-2-methylcyclohexane and sodium ethoxide yields only one alkene.

To help explain these results, we need to look at the actual shapes of these compounds. As was discussed in Chapter 2, cyclohexane derivatives are usually in chair conformations, so we will first examine the chair conformation of the (1S,2S) compound with the bromine axial.

To obtain a 180° dihedral angle between the bromine and an adjacent hydrogen, both the bromine and the hydrogen have to be in axial positions. As the preceding chair shows, only one hydrogen on an adjacent carbon to the bromine is axial. Therefore, only one alkene product is formed. Now let's look at the chair form of the (1R,2S) compound with the bromine axial.

Both of these H's are axial.

Now there are two axial hydrogens adjacent to the bromine, so either can be removed by the base to form an alkene product.

If more than one alkene product can be formed, usually the more stable alkene is the major product.

REVIEW EXERCISES FOR CHAPTER 4

1. Give the major organic product for the reaction of **1-ethylcyclopentene** with each of the following reagents. Clearly show stereochemistry, if needed.
    a. H_2, Pd catalyst
    b. Br_2
    c. H_2O, H_2SO_4 catalyst
    d. (1) BH_3; (2) H_2O_2, NaOH, H_2O
    e. HCl
    f. Dilute $KMnO_4$ in water
    g. Hot, basic $KMnO_4$ in water
    h. (1) $Hg(OAc)_2$, H_2O; (2) $NaBH_4$
    i. (1) O_3; (2) $(CH_3)_2S$
    j. Peracetic acid
    k. Product of j + dilute aqueous acid
    l. $CHBr_3$ + KOH

2. Give the major organic product for the reaction of **3-ethylhex-3-ene** with each of the above reagents. Clearly show stereochemistry, if needed.

3. Give a reasonable arrow-pushing mechanism for the following reaction.

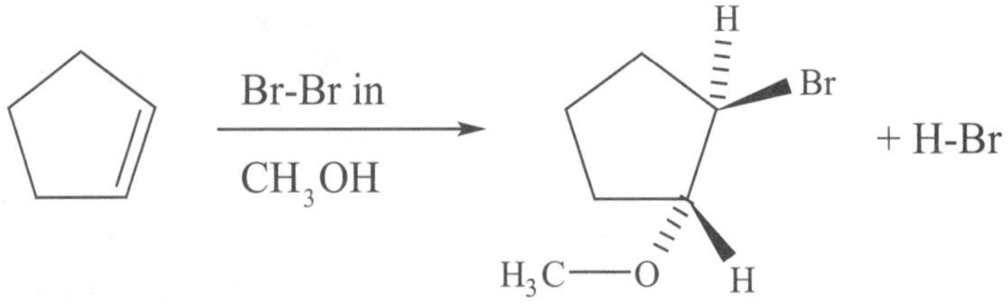

Br-Br in

CH_3OH

+ H-Br

4. When 1-methylcyclopentene reacts with D-Cl (D is ^2H, an isotope of hydrogen, and is called deuterium), there are two major products formed in approximately equal amounts. What are the structures of the products? Draw arrow-pushing mechanisms to show how they would be formed.

5. Give IUPAC names for the following structures. Use R, S, E, and Z when appropriate. You do not need to use E and Z in rings smaller than eight atoms.

a.

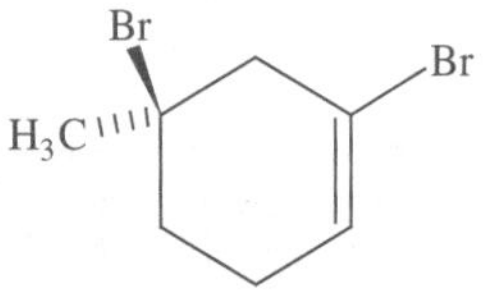

b.

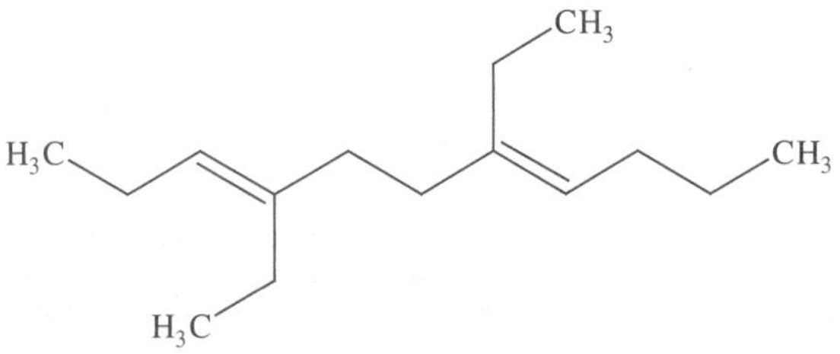

6. Draw structures that correspond to the following names.
   a. (Z)-4-ethyloct-3-ene
   b. (E)-5-cyclobutyl-2-fluoronon-3-ene
   c. 1,4,4-trimethylcyclohexene

7. Show how you would make the following products from the indicated starting materials, plus any other reagents.

   Starting Material   **Product**

   a. Bromocyclohexane   Cyclohexane
   b. Cycloheptanol   Cyclohepta-1,3-diene

Alkynes

WHAT YOU WILL LEARN

In this chapter, you will learn:

- what alkynes look like and how to name them;
- how alkynes react with a variety of reagents;
- how to draw the arrow-pushing mechanism of many alkyne reactions;
- how to synthesize alkynes from alkyl halides and from smaller alkynes.

Alkynes are those organic compounds that have a C≡C. There are fewer examples of C≡C containing compounds found in nature than of alkenes, but a few are shown below.

$$H_2C=CH-(CH_2)_2-C\equiv C-C\equiv C-C\equiv C-(CH_2)_7-COOH$$

Oropheic acid: Isolated from *Orophea enneandra*, it has antifungal activity versus *Cladosporium* species.

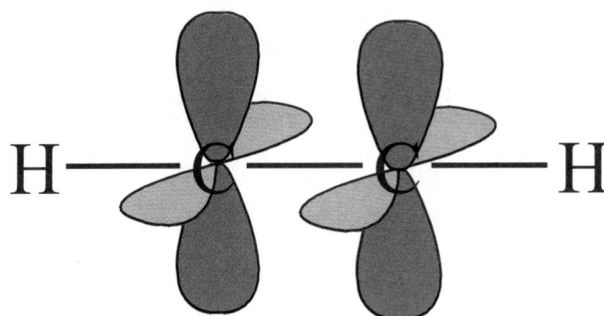

Cicutoxin: This poisonous compound, isolated from the water hemlock, has shown anti-leukemic activity.

A. Structure and Bonding

The H–C≡C bond angle of acetylene is 180°. To accomplish this bond angle, *sp* hybrid orbitals are used. Therefore, one C–C bond is formed by overlap of two *sp* orbitals. This leaves two unhybridized *p* orbitals on each carbon atom (shown below). The *p* orbitals are perpendicular to each other, as well as to the line defined by the H–C–C–H.

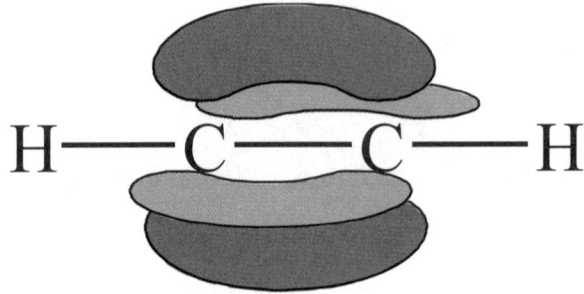

These *p* orbitals can overlap to form two pi bonds between the carbons, as shown below, to form the other two bonds.

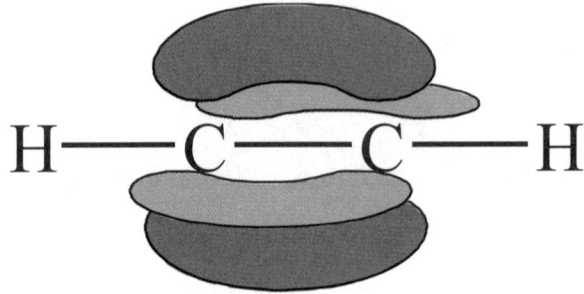

B. IUPAC Nomenclature

The IUPAC ending for an alkyne is -yne. Otherwise, alkynes are named just like alkenes. Some examples are given below.

H_3C ⎯ CH_2–CH_2–CH_2–C≡C ⎯ CH_3	**Hept-2-yne.** There are seven carbons in the chain, and we start numbering from the right end.
 1 H_3C 　　2　3　4　5/6　7　8 　　CH⎯C≡C⎯CH 　／ H_3C　　　　CH_3 　　　　　CH⎯CH_3 　　　　／ 　　H_3C (with CH_3 and CH_2–CH–CH_3 groups)	**5-Isopropyl-2,7-dimethyloct-3-yne.** You number from the left to give the alkyne the lower number. The groups are alphabetized (but not the *di-* prefix).
C≡C H_2C 8　1　2　3 CH ⎯ CH_2 H_3C　　　　　　　CH_3 　　CH 7　　　4 CH_2 　CH_2　6　5 　　CH_2–CH_2	**3,7-Diethylcyclooctyne.** Numbering is as shown so that the C≡C is numbered 1 and 2, and you run into a group on the next carbon.
H_3C 　　CH_2–C≡C ⎯ C≡C ⎯ CH_3	**Hepta-2,4-diyne.** The chain is numbered from the right, so that the C≡C's get the lowest numbers. IUPAC puts an *a* after *hept-*, so that the name flows better when saying it.

Alkynes are barely lower in IUPAC naming priority than alkenes. If there is both a C=C and a C≡C in a molecule, you number from the end that is closer to either the C=C or the C≡C. If it is the same distance to both, then the C=C gets the lower number. However, the *en* in the name comes before the *-yne* ending. I guess IUPAC thinks *en-yne* is easier to say than *yn-ene*. A couple of examples are given on the next page.

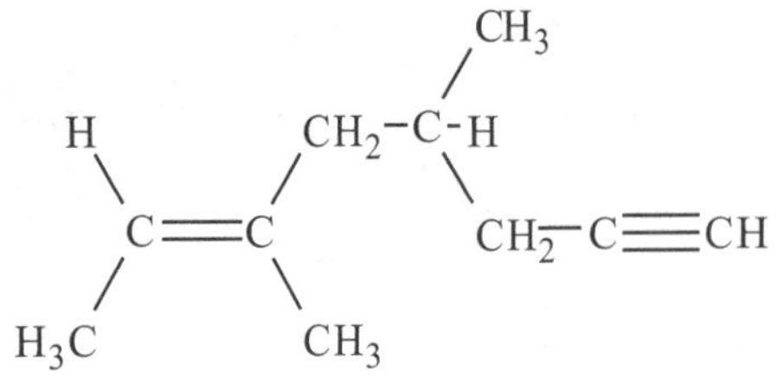	**(6E)-4,6-Dimethyloct-6-en-1-yne.** Because the C≡C is closer to one of the ends, we number from that end. The C=C is *E* because the higher priority groups on the C=C (the CH₃ on the left C and the five-carbon piece on the right C) are on opposite sides of the C=C.
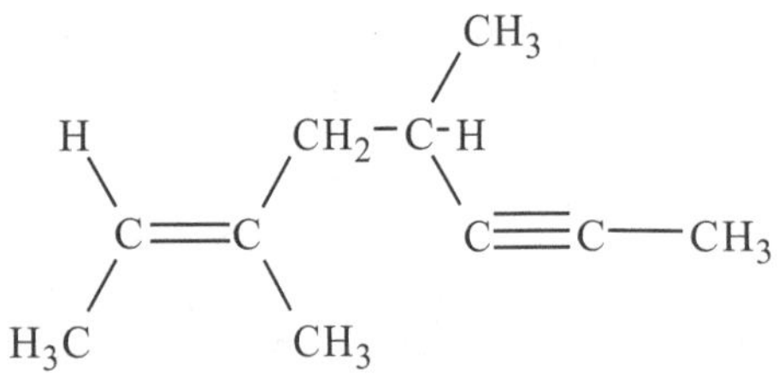	**(2E)-3,5-Dimethyloct-2-en-6-yne.** Now the C=C and the C≡C are the same distance from the ends, so the C=C is given the numbering priority.

C. The General Reaction of Alkynes: Addition to the C≡C

In Chapter 4, the general reaction of an alkene was addition to the pi bond of a C=C. Because a C≡C has two pi bonds, its general reaction is similar. In many cases, the general reaction of a C≡C is addition to both pi bonds. This is shown below: G–Z is a generic reagent.

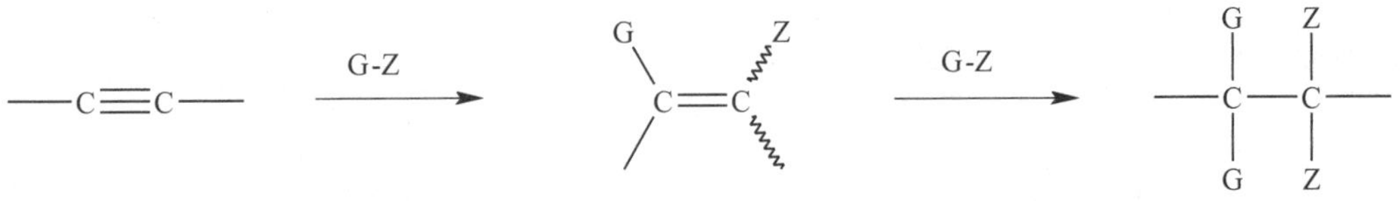

The wiggly lines in the middle structure mean that a mixture of *E* and *Z* isomers can be formed, depending on the exact reaction conditions.

D. Additions of H–X and Water

D.1. ADDITION OF H–X

Just like with an alkene, H–X adds to the pi bonds of an alkyne. The first equivalent adds to make a haloalkene, and the second equivalent adds to make a dihaloalkane. Markovnikov's rule is followed as well, so the major product of the reaction of a 1-alkyne is easy to predict.

General Reaction:

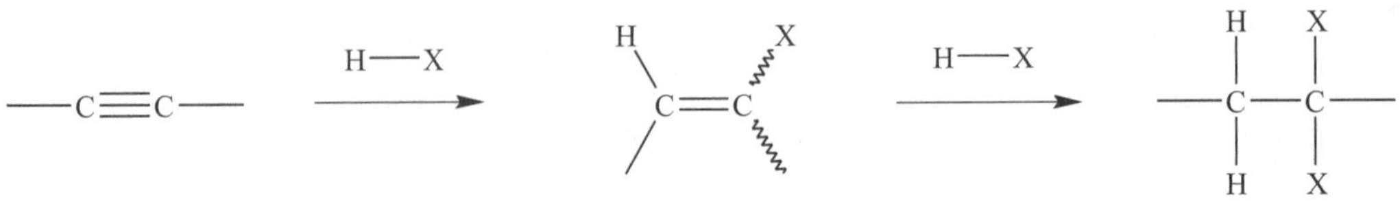

Specific Examples:

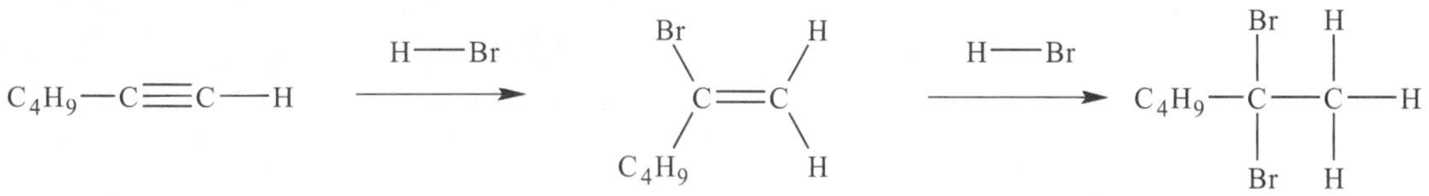

The mechanism is usually presented as analogous to addition of H–X to an alkene, as shown below.

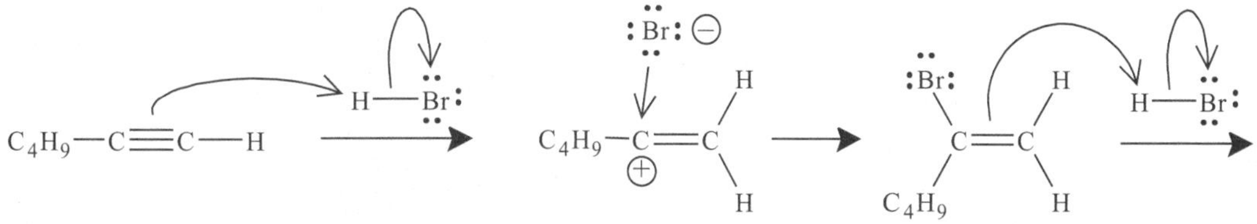

A "vinyl cation"

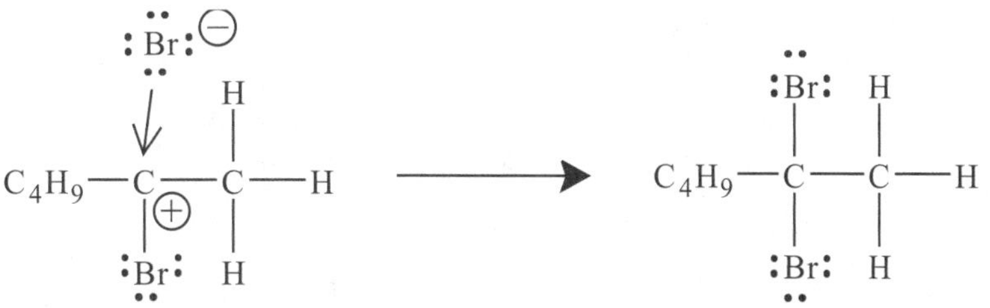

There is some substantial debate on whether a vinyl cation intermediate is really formed, or whether the addition of H and Br is concerted. For example, see the following articles in the *Journal of Chemical Education:* "The Electrophilic Addition to Alkynes." H. M. Weiss, *J. Chem. Educ.* **1993**, *70*, 873; "The Electrophilic Addition to Alkynes Revisited," T. T. Tidwell, *J. Chem. Educ.* **1996**, *73*, 1081; "Further Comments upon the Electrophilic Addition to Alkynes: A Response to Criticism from Professor Thomas T. Tidwell," H. M. Weiss, *J. Chem. Educ.* **1996**, *73*, 1082.

D.2. ADDITION OF WATER

General Reaction:

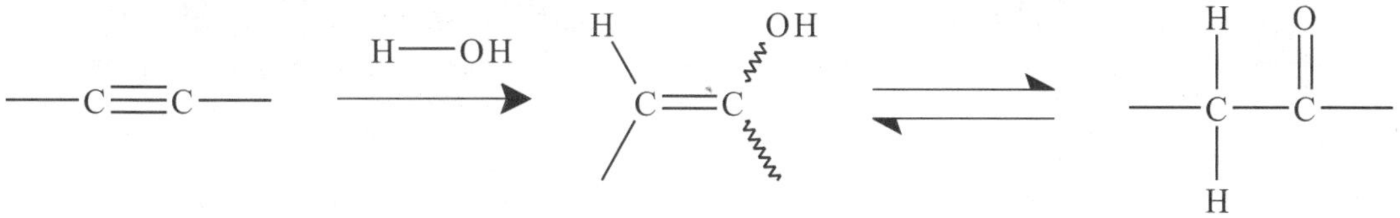

Specific Example:

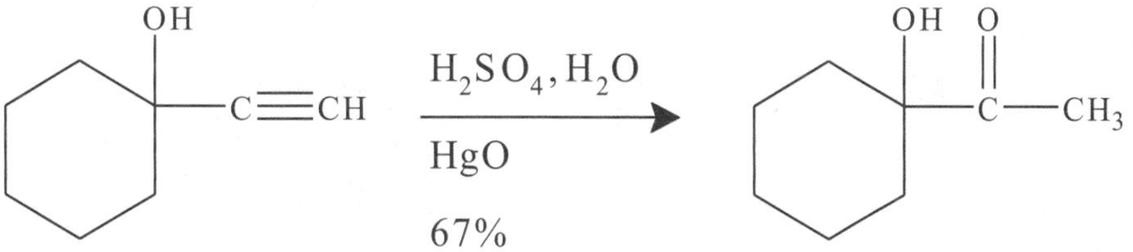

Water adds to an alkyne in a similar way it adds to an alkene. An acid catalyst is required. Usually, a mercury(II) salt is also added as a catalyst to speed things up, but it is not always required. Addition of water to a 1-alkyne follows Markovnikov's rule to make the ketone rather than the aldehyde. The addition of one equivalent of water results in the formation of an enol. Enols are less thermodynamically stable than the isomeric ketones, so they tautamerize to the ketones.

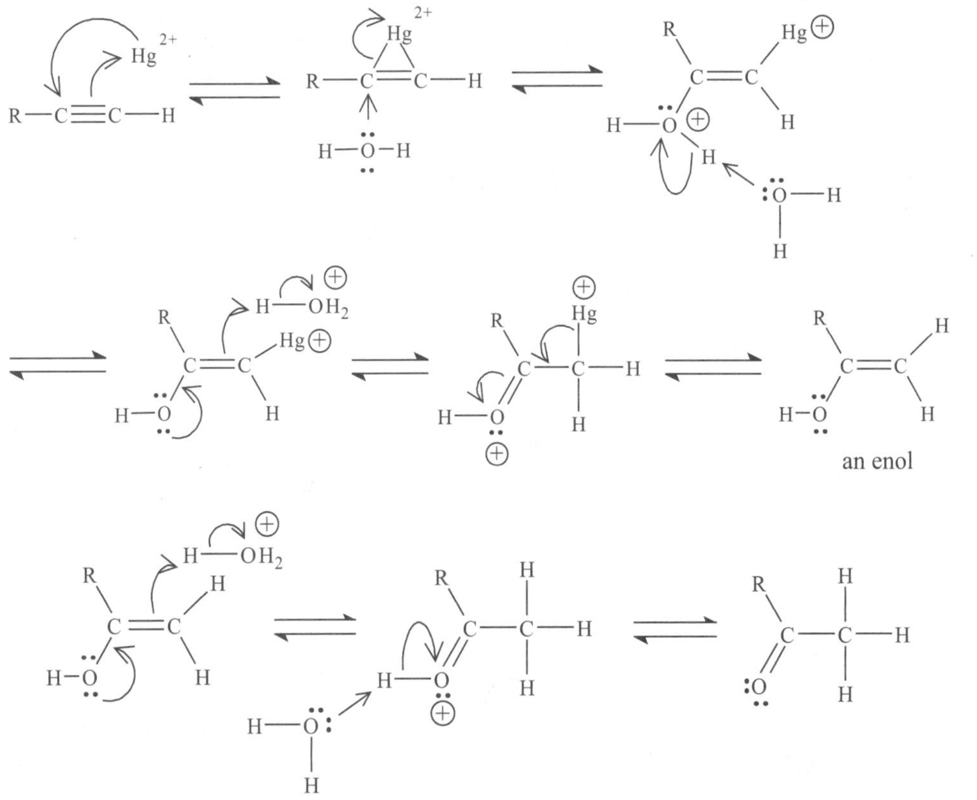

E. Addition of X₂

Halogens can be added to alkynes as well. One equivalent adds to form the dihaloalkene. Often, a mixture of *E*- and *Z*-isomers is formed, so this reaction is not often synthetically useful. Addition of another equivalent of halogen gives the tetrahaloalkane. An example is shown below.

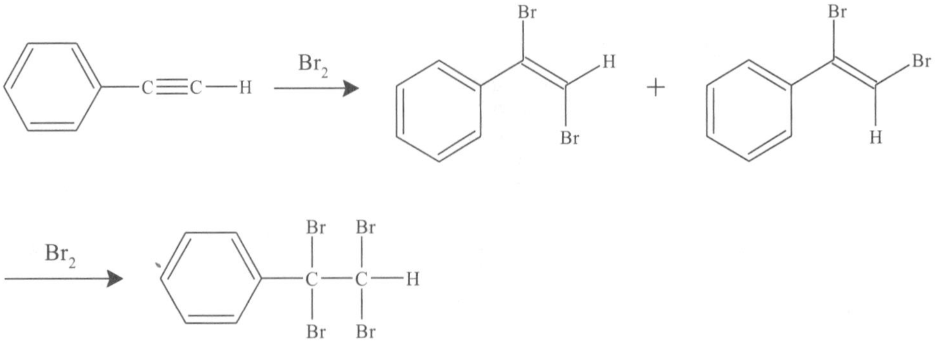

F. Hydroboration

Alkynes can be hydroborated in a similar manner to alkenes, and the resulting organoborane intermediates can be oxidized. In this case, C=O compounds are the isolated products. In the case of 1-alkynes, aldehydes are the major products. Usually, rather hindered dialkylboranes are used as the hydroborating agents to maximize the selectivity of the addition reaction. An example is shown below.

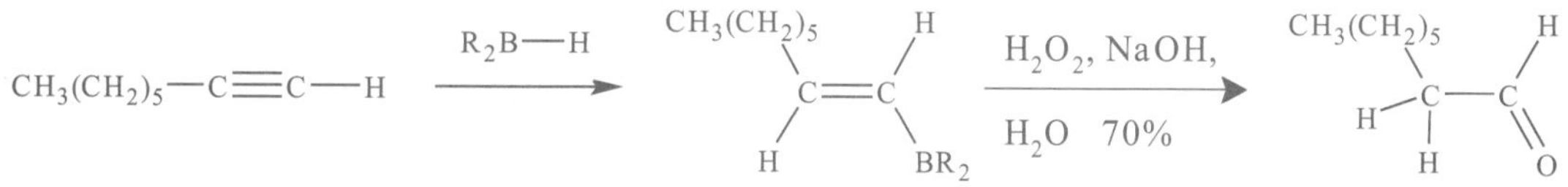

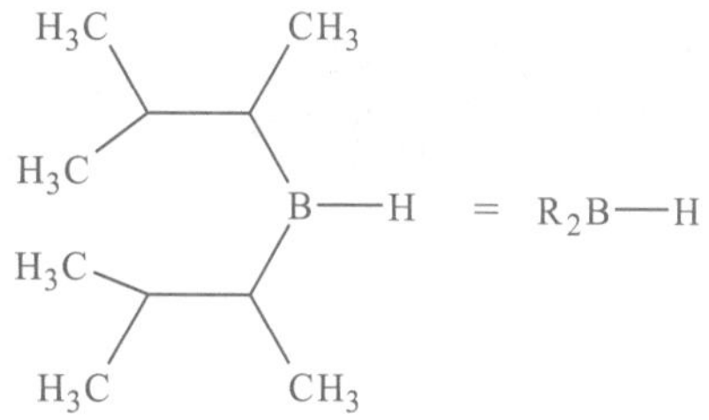

$$CH_3(CH_2)_5 - C \equiv C - H \xrightarrow{R_2B-H} \begin{array}{c} CH_3(CH_2)_5 \quad H \\ C=C \\ H \quad BR_2 \end{array} \xrightarrow[H_2O \ 70\%]{H_2O_2, \ NaOH,} \begin{array}{c} CH_3(CH_2)_5 \quad H \\ H-C-C \\ H \quad O \end{array}$$

$$\begin{array}{c} H_3C \quad CH_3 \\ H_3C \\ \quad B-H \quad = \quad R_2B-H \\ H_3C \\ H_3C \quad CH_3 \end{array}$$

G. Oxidation of Alkynes with KMnO₄ and Ozone

Internal alkynes can be oxidized at the C≡C by KMnO₄ under mild conditions to form a diketone. One example is shown below. PTA stands for phase-transfer agent, which is a compound that allows the permanganate ion to travel into the organic layer of the reaction to react with the alkyne.

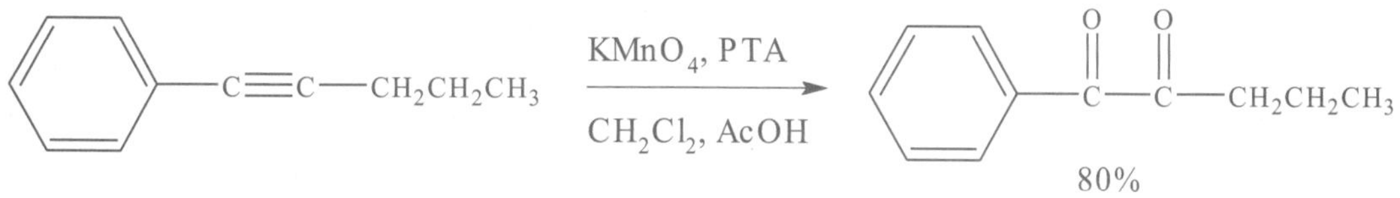

$$\begin{array}{c} \text{Ph}-C \equiv C-CH_2CH_2CH_3 \xrightarrow[CH_2Cl_2, \ AcOH]{KMnO_4, \ PTA} \text{Ph}-\overset{O}{\underset{\|}{C}}-\overset{O}{\underset{\|}{C}}-CH_2CH_2CH_3 \\ 80\% \end{array}$$

Ozone can do a similar oxidation, as long as a reductive work-up procedure is used (Zn in acetic acid, or dimethyl sulfide).

Oxidation of 1-alkynes usually forms the carboxylic acid with one carbon *fewer*, due to loss of the terminal carbon as CO_2. One example is shown below.

$$\xrightarrow[73\%]{O_3, \ CHCl_3}$$

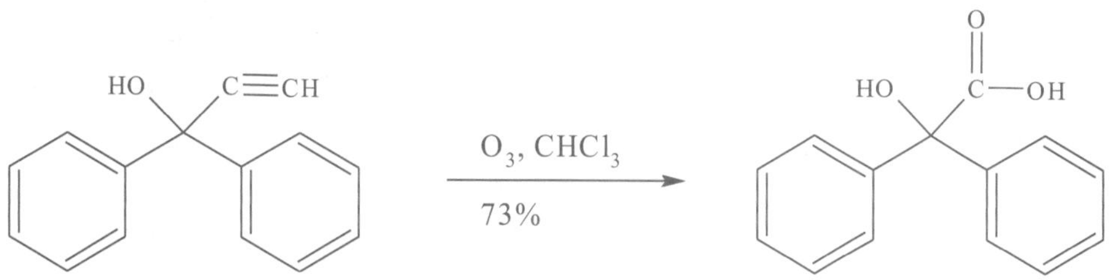

Ozone cleaves internal alkynes to carboxylic acids, when an oxidative workup is used (H_2O_2), as does vigorous oxidation with $KMnO_4$

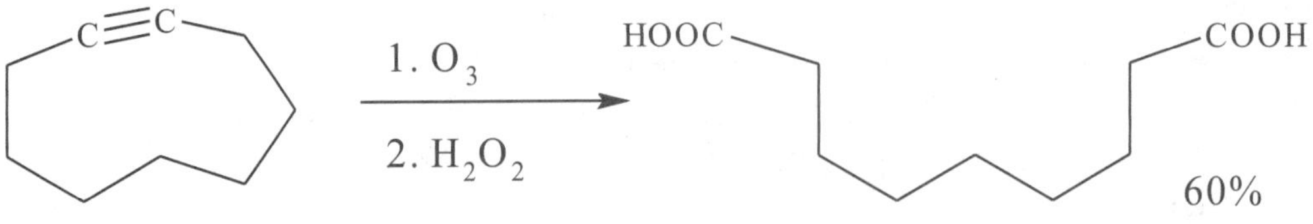

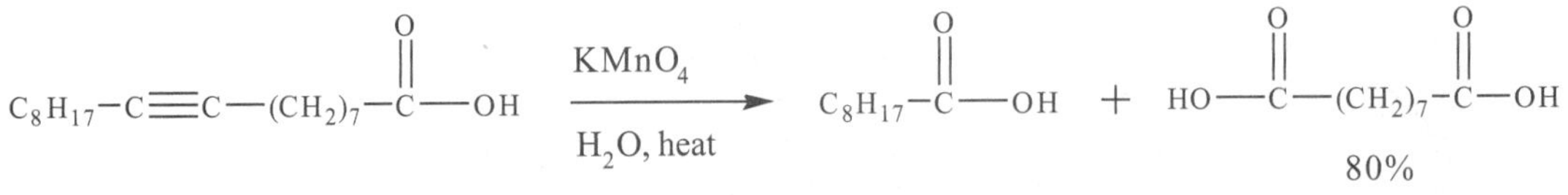

60%

80%

H. Reduction

Alkynes can be reduced all the way to alkanes by reaction with two equivalents of hydrogen in the presence of a platinum or palladium catalyst.

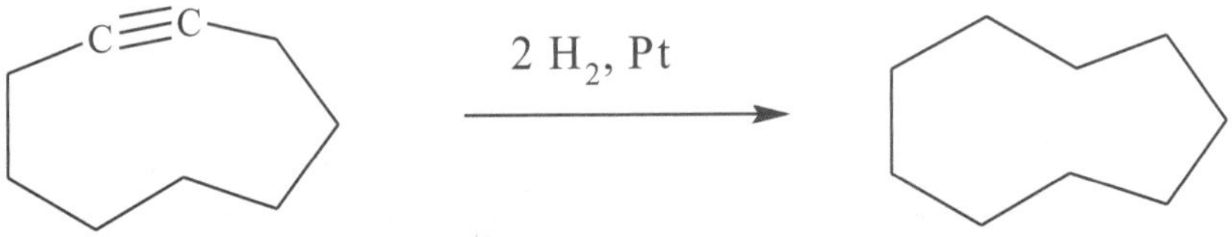

Cyclononyne → Cyclononane, 100%

The partial reduction of an alkyne to a *cis*-alkene can be accomplished by using hydrogen and a special catalyst. Two common catalysts are the Lindlar catalyst ($Pd/CaCO_3$ conditioned with $Pb(OAc)_4$, with added quinoline, or $Pd/BaSO_4$ with added quinoline) and P-2 nickel boride, which is easily prepared from sodium borohydride and nickel acetate. These catalysts react much faster with C≡C's than with C=C's, although they will reduce C=C's slowly. Usually, the amount of hydrogen reacted is monitored, and stopped when one equivalent of hydrogen has been used.

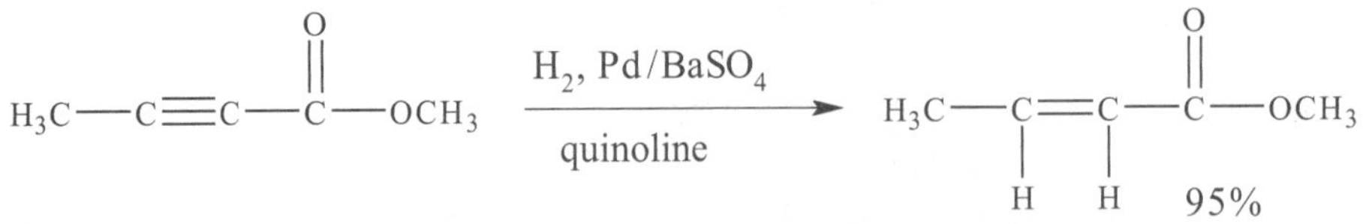

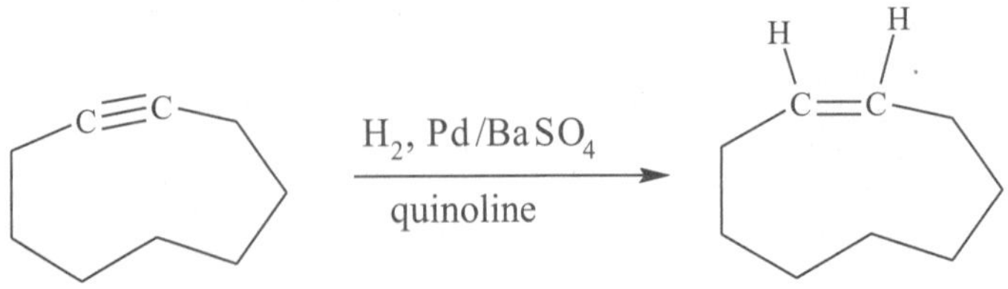

95%

Cyclononyne → *cis*-Cyclononene, 81%

Reduction of alkynes to *trans*-alkenes is usually accomplished by treating the alkyne with sodium in liquid ammonia.

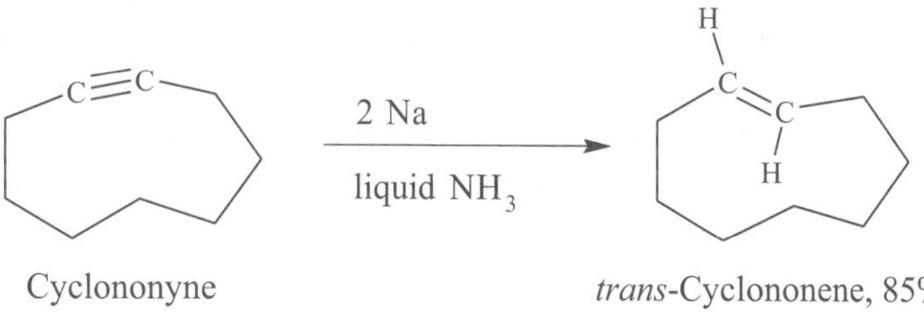

| Cyclononyne | *trans*-Cyclononene, 85% |

I. Acidity of 1-Alkynes

The H attached to the *sp* carbon of a 1-alkyne is much more acidic (p$Ka \approx 25$) than a H on an sp^3 carbon of an alkane (p$Ka > 40$). Therefore, it is possible to remove that hydrogen by reacting the 1-alkyne with a sufficiently strong base, such as NaH (sodium hydride), $NaNH_2$ (sodium amide), or a Grignard reagent (RMgX), to produce the salt of the alkyne. This is illustrated below.

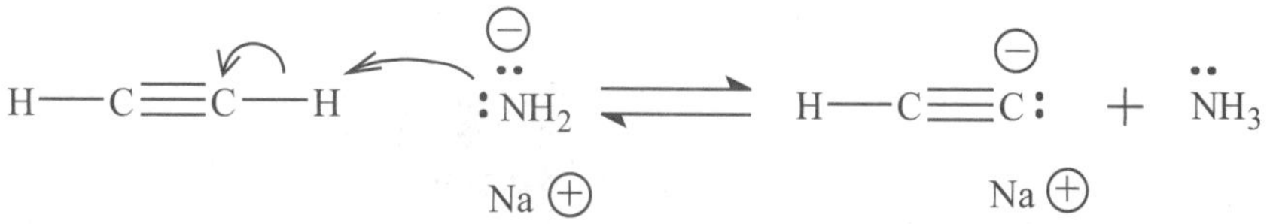

The salts of the alkynes react with primary alkyl halides ($R–CH_2–X$) to form a bigger alkyne. In this process, we have formed a new carbon–carbon bond. See below.

This reaction can be used to prepare 1-alkynes and internal alkynes. See the example below.

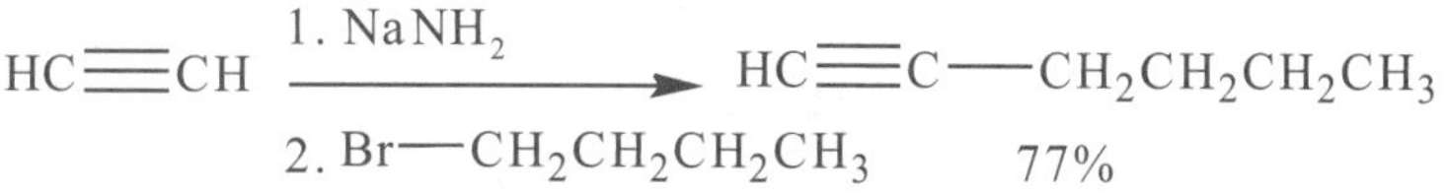

J. Preparations of Alkynes

In addition to the alkylation of salts of 1-alkynes discussed in Section I, there is only one common preparation of alkynes. This is the double dehydrohalogenation of dihalides. Conceivably, two types of dihalides could form an alkyne: a 1,1-dihalide (also referred to as a **geminal dihalide**) and a 1,2-dihalide (also called a **vicinal dihalide**). The general reactions of these dihalides with two equivalents of base are shown below.

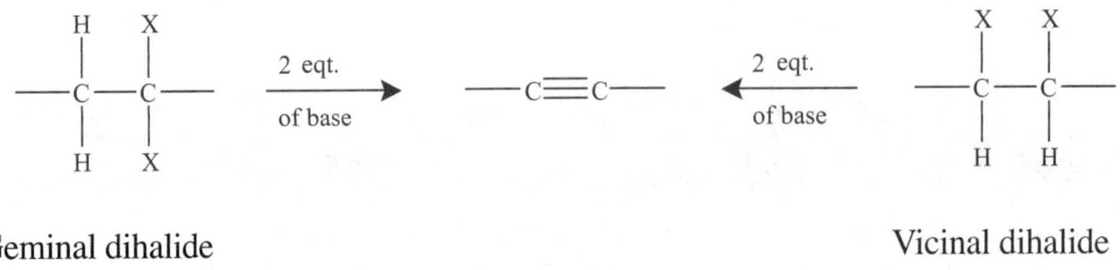

Geminal dihalide   Vicinal dihalide

Vicinal halides are easily obtained from addition of bromine to alkenes, so this reaction provides a convenient method of obtaining alkynes from alkenes in two steps. An example is shown below.

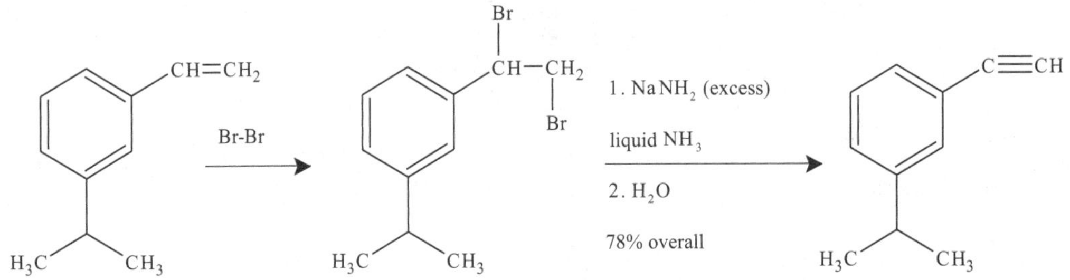

Sodium amide ($NaNH_2$) is a commonly used base, although hydroxides and other bases have been used. Sodium amide is a strong enough base to deprotonate a 1-alkyne, so the salt of a 1-alkyne is formed initially. Water is added to protonate the salt to give the 1-alkyne.

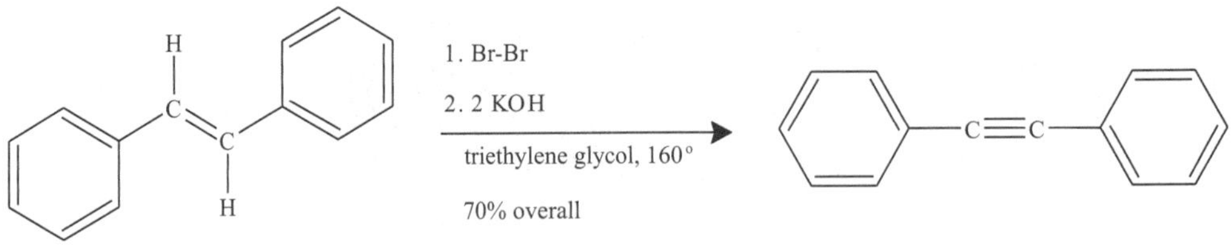

REVIEW EXERCISES FOR CHAPTER 5

1. Draw structures that correspond to the following names.
   a. 2-Cyclopentylhept-3-yne
   b. 1-Ethynylcyclohexene
   c. (Z)-4,5-Difluoro-2-octen-6-yne
   d. 1,5,9-Cyclododecatriyne (try to draw this the way it would actually look)

2. Give IUPAC names for the following compounds.

   a.

   b.

3. Give the major organic product for the reaction of **1-butyne** with each of the following reagents.
   a. 2 H_2, Pd catalyst
   b. 2 Br_2
   c. H_2O, H_2SO_4, $HgSO_4$ catalyst
   d. (1) NaNH2; (2) 1-iodopropane
   e. 2 HCl

4. Give the major organic products of the following reactions. You do not have to include the inorganic by–products.

   a.

   $$1.$$
   $$2.\ H_2O_2,\ NaOH$$

   b.

   $$H_2O$$
   $$H_2SO_4\ (cat.)$$

c.

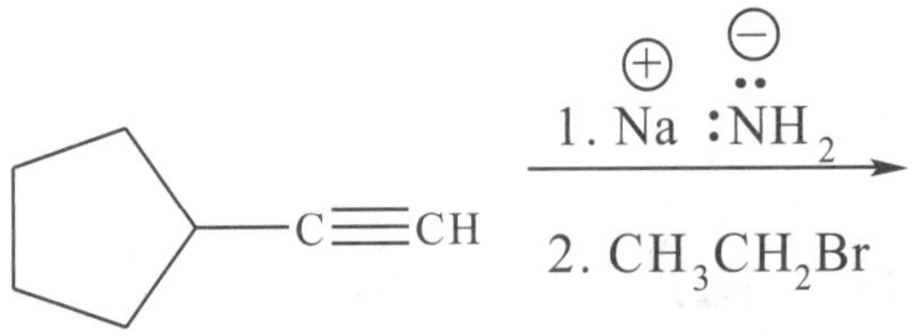

$$\xrightarrow{\begin{array}{c}\text{1. Na }\overset{\oplus}{}\ \overset{\ominus}{:}\!NH_2\\[4pt]\text{2. CH}_3\text{CH}_2\text{Br}\end{array}}$$

d.

$$\xrightarrow[\text{Lindlar Pd}]{H_2}$$

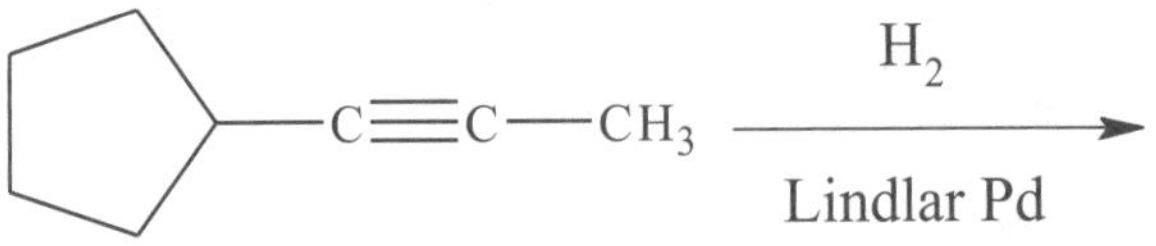

5. For the following reaction, give arrow-pushing mechanism showing how bonds are broken and formed to get the products.

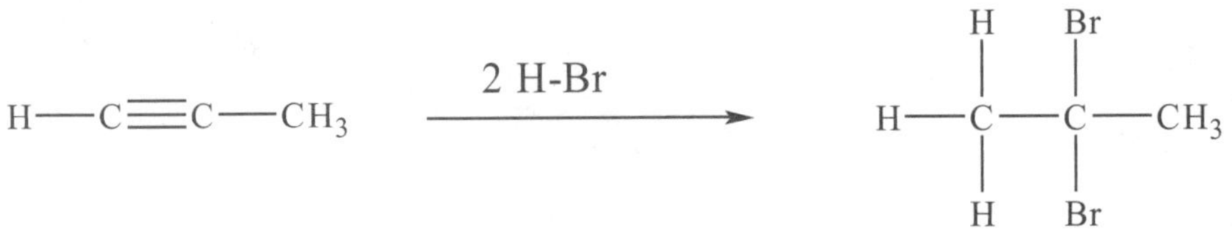

$$H-C\equiv C-CH_3 \xrightarrow{\text{2 H-Br}} $$

6. Explain why the following compound has never been prepared, and is unlikely ever to be.

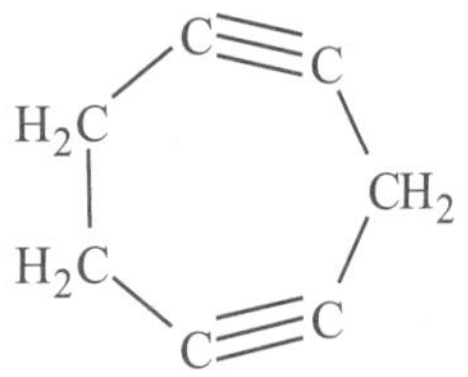

7. Provide a reasonable synthesis of the following products from the indicated starting material, plus any other reagents. More than one step is required in each case.

Starting Material   **Product**

a.

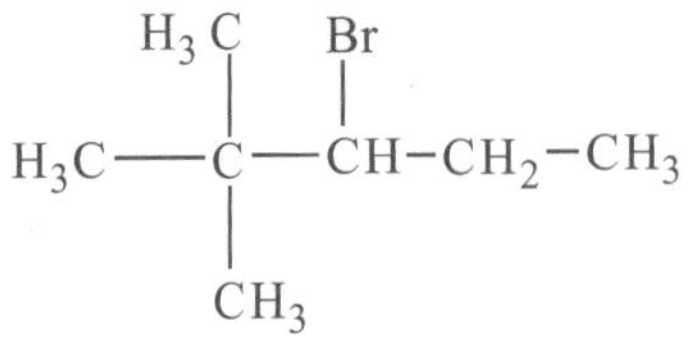

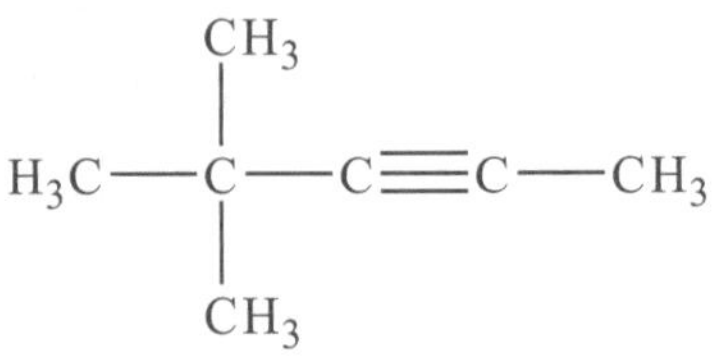

b.

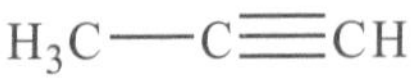

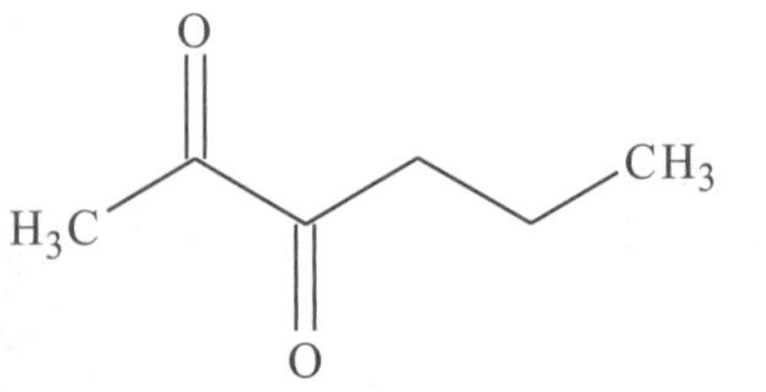

Nucleophilic Substitution and Elimination Reactions

WHAT YOU WILL LEARN

In this chapter, you will learn:

- what nucleophiles are, and how to classify them as strong or weak;
- how to classify alkyl halides;
- how and why different types of nucleophiles react with the different types of alkyl halides;
- how to draw the different types of arrow-pushing mechanisms for nucleophilic substitution and elimination mechanisms.

SECTIONS IN THIS CHAPTER

- Base Strength and Nucleophile Strength in Water and Hydrogen-Bonding Solvents
- Base Strength and Nucleophile Strength in Other Solvents
- Reactions of Methyl and Primary Alkyl Halides with Strong Nucleophiles/Bases
- Reactions of Tertiary Alkyl Halides with Strong Nucleophiles/Bases
- Reactions of Secondary Alkyl Halides with Strong Nucleophiles/Bases
- Reactions of Tertiary Alkyl Halides with Weak Nucleophiles/Bases
- Reactions of Secondary Alkyl Halides with Weak Nucleophiles/Bases
- Reactions of Methyl and Primary Alkyl Halides with Weak Nucleophiles/Bases
- Carbocation Rearrangements
- Summary of Nucleophilic Substitution and Elimination Reactions

T his chapter covers some fundamental reaction mechanisms central to organic chemistry. You will see these mechanisms, or ones similar to these, throughout your organic chemistry courses.

A few definitions are in order. A **nucleophile** is a molecule that can donate an electron pair to cause a substitution reaction. This is similar to the Lewis definition of a base, which is an electron pair donor. When we talk about a base, usually it is donating its electron pair to a partially positive or fully positive hydrogen. A nucleophile usually donates its electron pair to a partially positive or fully positive carbon. So in many ways, a base and a nucleophile are similar.

Another term we need to define is leaving group. A **leaving group** accepts a pair of electrons from the atom to which the nucleophile donated electrons. This is similar to the definition of a Lewis acid, which is an electron pair acceptor. See below.

The base donates a pair of electrons to the H of H–X, to form the Base–H bond. The H–X bond breaks, and X accepts the pair of electrons.

The nucleophile (Nu:) donates a pair of electrons to the partially positive carbon, to form the Nu–C bond. The C–X bond breaks, and the X (leaving group) accepts the pair of electrons.

Leaving groups usually contain electronegative atoms, or atoms to which a negative charge can be stabilized by resonance. The best leaving groups are very weak bases. Common leaving groups include the halogens Cl, Br, and I, as well as sulfonates, which have the general structure $O-SO_2R$. Most of the time, I will refer to the compound a nucleophile reacts with as an alkyl halide, even if the leaving group is a sulfonate rather than a halide.

Alkyl halides are classified according to the number of carbons that are bonded to the carbon bonded to the halogen, as shown on the next page.

Type of Alkyl Halide or Sulfonate	Generic Example	Real Example
Methyl	CH_3–X	CH_3–I
Primary (1°)		H_3C—CH_2—CH_2—Br
Secondary (2°)		In this case, the leaving group is a methanesulfonate, rather than a halide.
Tertiary (3°)		

A. Base Strength and Nucleophile Strength in Water and Hydrogen-Bonding Solvents

When you think of a base, you probably think of something like sodium hydroxide, NaOH. NaOH is a salt, and is soluble in water and in alcohol solvents. When investigators started looking at reactions of alkyl halides with bases and nucleophiles, they commonly used water, ethanol, or a mixture of the two as a solvent. Hence, much of the early work on determining nucleophile and base strength was done in solvents that could hydrogen-bond.

As was discussed earlier, nucleophiles and bases are similar in that both donate an electron pair to something. Therefore, many strong bases, such as hydroxide, are also strong nucleophiles. Another way to put this is that the conjugate base of a weak acid is a strong base and often a strong nucleophile. See the examples below.

Acid and	CH_4	NH_3	H_2O	HF
pKa	>40	33	15.7	3.45
Acid strength	Weakest	$\rightarrow$		Strongest
Conjugate base	$\ominus$:CH_3	$\ominus$:$\overset{..}{N}H_2$	$H\overset{..}{O}$:$\ominus$	:$\overset{..}{\underset{..}{F}}$:$\ominus$
Base strength	Strongest	$\rightarrow$		Weakest
Nucleophile strength	Strongest	$\rightarrow$		Weakest

One general trend from the periodic table is that acidity increases going from left to right across a period. This is shown in the hydrogen acids of period 2: HF is the strongest acid, while CH_4 is the weakest. This relates to the electronegativity of the atom: The more electronegative the atom, the more acidic the compound is. As such, base strength of the conjugate bases of these compounds decreases going from left to right. Nucleophile strength usually parallels base strength.

Acid and pKa	H_2O 15.7	(Phenol) 10	(Acetic acid) 4.76	(Methanesulfonic acid) −2.0
Acid strength	Weakest		→	Strongest
Conjugate base	$HO:^{\ominus}$	(Phenoxide)	(Acetate)	(Methanesulfonate)
Base strength	Strongest		→	Weakest
Nucleophile strength	Strongest		→	Weakest

In this case, we are comparing acid strengths of various O–H compounds, and the base strengths of the conjugate bases. The increased acidity has to do with inductive effects of atoms pulling electron density away from the O–H bond, as well as increasing stabilization of the negative charge in the anion by resonance. These effects will be discussed more later. The main point here is that the base strength and the nucleophile strength orders are the same.

The principal places where base strength and nucleophile strength are not parallel are when comparing compounds containing elements in the same row. The classic example is the following.

Acid and pKa	Acid strength	Conjugate base	Base strength	Nucleophile strength
HF 3.45	Weakest	$:\ddot{F}:^{\ominus}$	Strongest	Weakest
HCl −7		$:\ddot{Cl}:^{\ominus}$		
HBr −9		$:\ddot{Br}:^{\ominus}$		
HI −10	Strongest	$:\ddot{I}:^{\ominus}$	Weakest	Strongest

As you can see, fluoride is the strongest base, but the weakest nucleophile, in a hydrogen-bonding solvent. The reason for this again relates to electronegativity. Fluoride is the most electronegative. When water or an alcohol hydrogen-bonds to it, the hydrogen-bonding effectively ties up the electron pairs of fluoride, so they can't act as nucleophiles. This is shown below. Fluoride is also the smallest of these anions, so the negative charge is concentrated in a small space. In the case of iodide, the charge is more spread out over the larger atom, and waters don't hydrogen-bond as tightly to the less electronegative iodide anion.

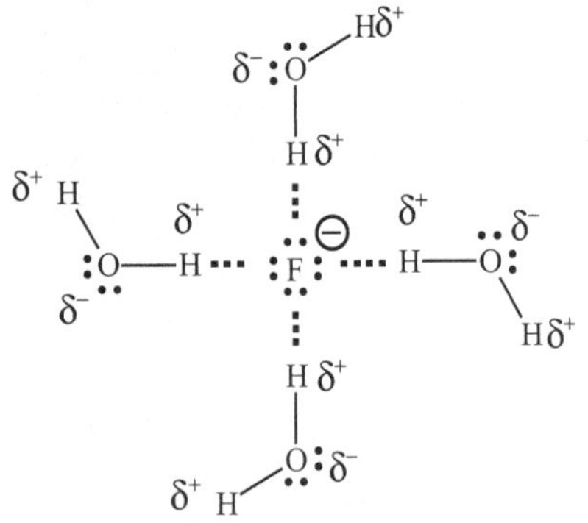

This shows water hydrogen-bonding to a fluoride ion. It really isn't flat: The waters would have a tetrahedral geometry around the fluoride.

There is a wide range of nucleophile strengths. For my students, I arbitrarily classify a nucleophile as either strong or weak. I then subdivide nucleophiles into those that are strong or weak bases. It is admittedly a big simplification, but for most examples, it works pretty well. These are shown below.

Nucleophile Strengths in Water or Another Hydrogen-Bonding Solvent		
Weak Nucleophiles and Weak Bases	**Strong Nucleophiles and Strong Bases**	**Strong Nucleophiles, but Weak Bases**
Water, alcohols (ROH), fluoride ion, chloride ion, nitrate ion, and carboxylic acids 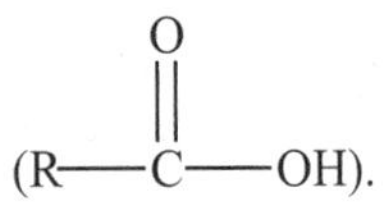 $$(R-\overset{\overset{\textstyle O}{\|}}{C}-OH).$$	$(-)OH$, $(-)OR$, R_3N, $RS(-)$, and almost anything else that is not the salt of a strong acid	Iodide ion, bromide ion, RSH, R_3P, and carboxylate ions 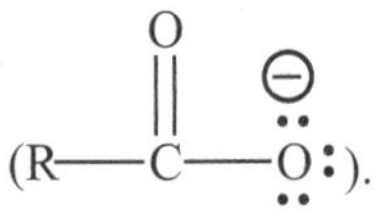 $$(R-\overset{\overset{\textstyle O}{\|}}{C}-\overset{\ominus}{O}:).$$

B. Base Strength and Nucleophile Strength in Other Solvents

There are many solvents (for example, hexane, toluene, and diethyl ether) that cannot hydrogen-bond as water and alcohols can. However, many nucleophiles, such as NaOH, have little solubility in these solvents, making these solvents useless for reactions involving nucleophiles. However, there are several polar solvents that cannot hydrogen-bond in which many nucleophiles will dissolve. Some of these are shown below.

Solvent	Structure and Any Important Resonance Forms
Dimethyl sulfoxide (DMSO)	
N,N-Dimethylformamide (DMF)	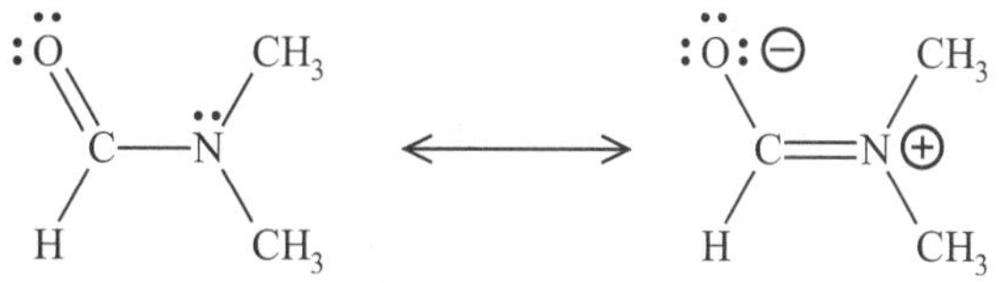
Hexamethylphosphorictriamide (HMPA)	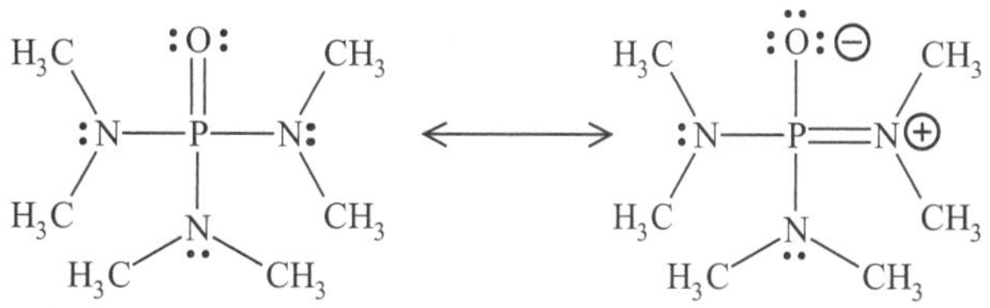 (plus two more equivalent forms involving the other nitrogens)

These and similar solvents are called polar, aprotic solvents. Why are these solvents important? In a polar, aprotic solvent, the stronger bases are the stronger nucleophiles. So when comparing the halide ions in DMSO, fluoride is the strongest nucleophile, and iodide is the weakest. This doesn't mean iodide becomes a weak nucleophile: It is still strong. Fluoride becomes a much stronger nucleophile in a polar aprotic solvent than in water.

C. Reactions of Methyl and Primary Alkyl Halides with Strong Nucleophiles/Bases

When a strong nucleophile reacts with a methyl halide, the only reaction that can take place is substitution. With a primary alkyl halide, the major reaction is substitution. See the examples below.

General Reaction and Mechanism of a Strong Nucleophile with a Methyl Halide:

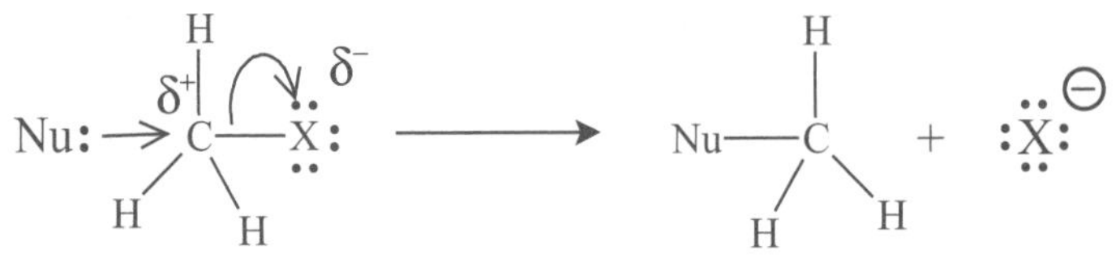

Specific Example:

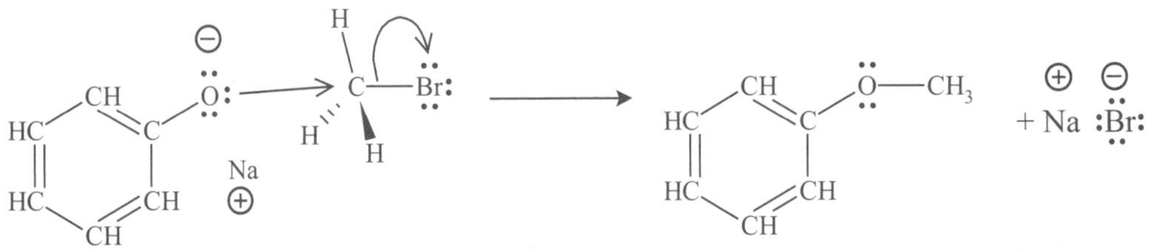

General Reaction and Mechanism of a Strong Nucleophile with a Primary Halide:

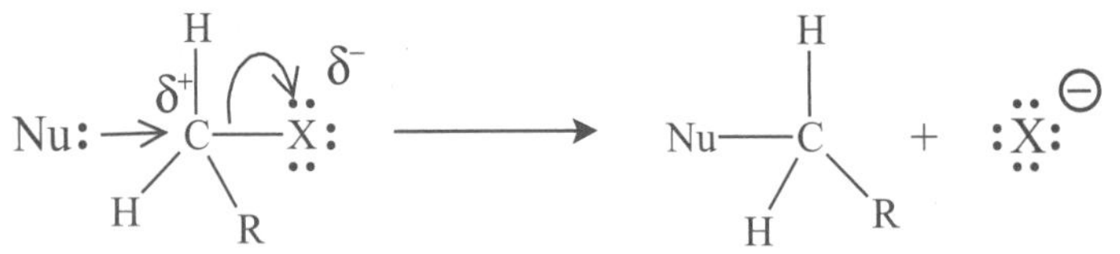

Specific Example:

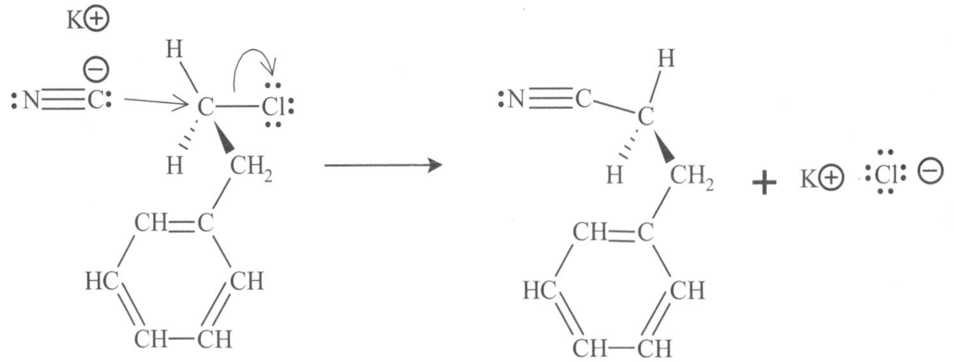

This reaction is referred to as nucleophilic substitution. People who have measured the rates of these reactions have determined that the rate of the reaction is dependent upon the concentration of the nucleophile and on the concentration of the alkyl halide. Therefore, these reactions are called S_N2 reactions—for Substitution, Nucleophilic, and 2 molecules involved in the rate-determining step (or bimolecular).

Evidence for the attack of the nucleophile from the backside of the carbon bearing the leaving group comes from several sources. One neat piece of evidence comes from using chiral alkyl halides, which is shown below.

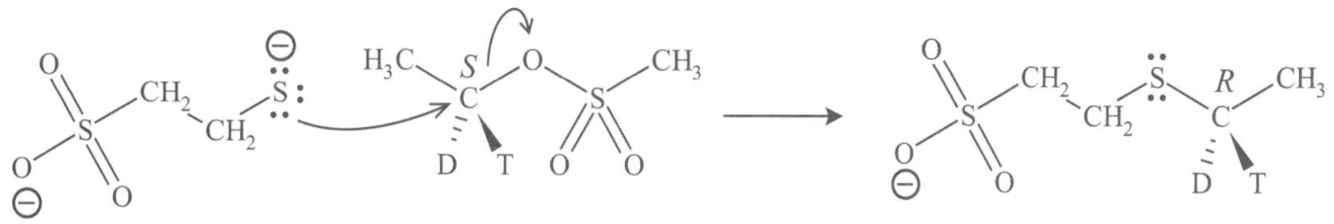

Nucleophile   (S)-Methanesulfonate   (R)-Product from substitution

By using two isotopes of hydrogen in the (S)-methanesulfonate (D = ^2H and T = ^3H), it is possible to have a chiral primary alkyl sulfonate. Since the product had the (R)-configuration, it provides good evidence for attack of the nucleophile as shown.

Although the major product of the reaction of a strong nucleophile with a primary alkyl halide is the substitution product, some amount of elimination product can also form. This is more likely if the nucleophile is also a strong Brønsted base, such as the salt of an alcohol, which is called an alkoxide. See the examples below. The yields given are the isolated yields, which sometimes don't add up to 100%. Real life isn't perfect, as you have undoubtedly found out in lab.

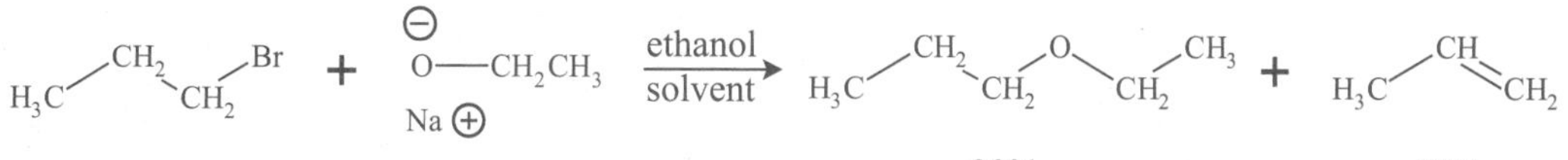

The substitution product is the major product. The alkene is formed by an E2 elimination reaction, which was discussed in Chapter 4.

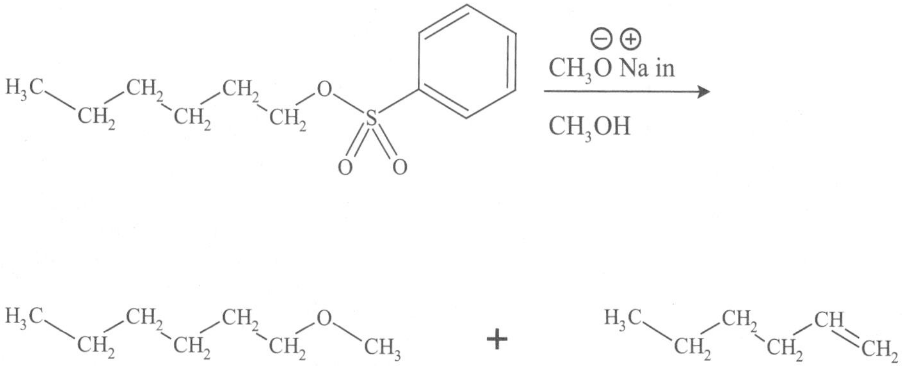

Again, the substitution product is the major product.

Substitution is still the major product, but now we are getting more elimination. This is because the size of the nucleophile/base is becoming larger. A larger nucleophile has a harder time getting to the backside of the alkyl halide to cause a substitution reaction. If the nucleophile can't do substitution, it can act as a base and do elimination. See below. This nucleophile is called potassium *t*-butoxide.

The *t*-butoxide ion is acting as a nucleophile, and attacking the backside of the carbon.	The *t*-butoxide ion is acting as a base, pulling off a hydrogen, and causing an elimination reaction.

This is a general trend: The harder it is for the nucleophile to attack the backside of the carbon, the less substitution product is formed. We will see more of this trend below.

D. Reactions of Tertiary Alkyl Halides with Strong Nucleophiles/Bases

In general, tertiary alkyl halides produce alkenes, when reacting with strong nucleophiles. The reaction mechanism is predominately E2. S_N2 reactions do not occur significantly because the nucleophile can't get to the backside of the carbon bonded to the halogen very easily. See the examples below.

63%   37%

No substitution occurs, only elimination. The major product is the more substituted alkene, as was discussed in Chapter 4. Sodium ethoxide is a fairly small base and can pull off either beta-hydrogen fairly easily.

27%   73%

No substitution, only elimination, occurs. This is the same alkyl halide as in the previous reaction, but the base is *t*-butoxide, which is bulkier. The bulky base has a more difficult time getting close to the hydrogens of the CH_2 next to the C–Br, so less of the more substituted alkene is formed in this case.

E. Reactions of Secondary Alkyl Halides with Strong Nucleophiles/Bases

Because primary alkyl halides give mostly substitution, and tertiary halides give essentially all elimination, you might guess that secondary halides plus a strong nucleophile/base would give a mixture of elimination and substitution. You are correct. The proportion of substitution and elimination varies with the specific alkyl halide and nucleophile/base, but, generally, you get more elimination. See the examples below.

79%   21%

Mostly elimination occurs, even though sodium ethoxide is a fairly small nucleophile/base.

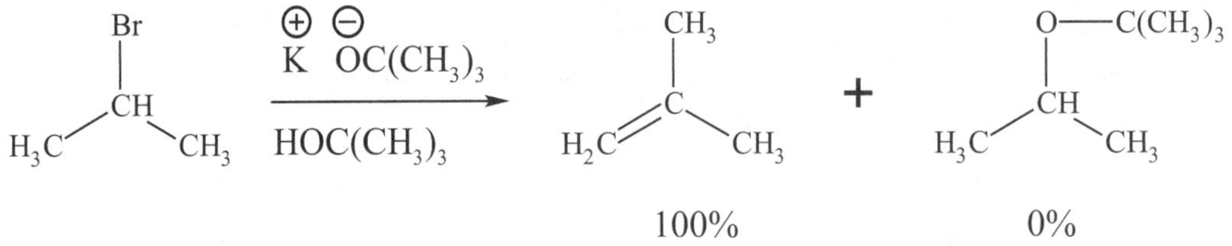

Potassium *t*-butoxide is a bulky base, and no substitution occurs in this case.

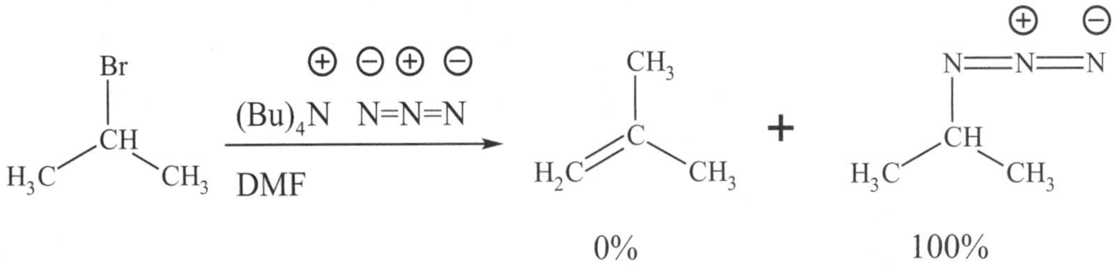

The N_3 (–) anion is called **azide**. Azide anion is a relatively weak base but a good nucleophile. Therefore, it does the S_N2 nucleophilic substitution solely, with no observed elimination.

F. Reactions of Tertiary Alkyl Halides with Weak Nucleophiles/Bases

The reaction products formed from the reaction of 2-bromo-2-methylpropane (*t*-butyl bromide) with ethanol are shown below.

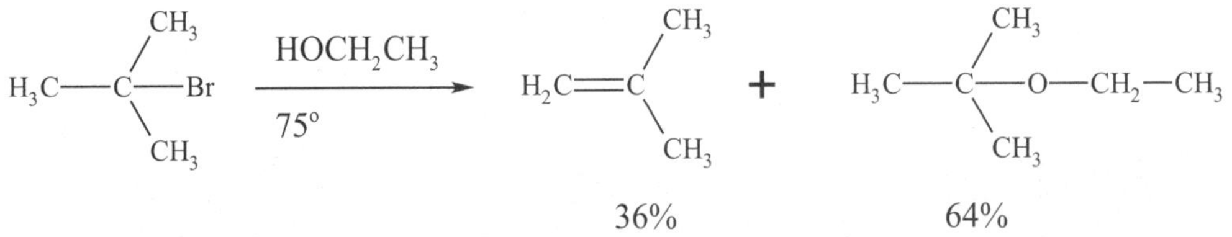

Two products are formed, but now the major product is the substitution product. With a tertiary halide, it is still unlikely that the nucleophile can attack the backside of the tertiary carbon and do an S_N2 substitution. Therefore, a different mechanism must be occurring. This mechanism for the substitution reaction, to form the ether product, is shown below.

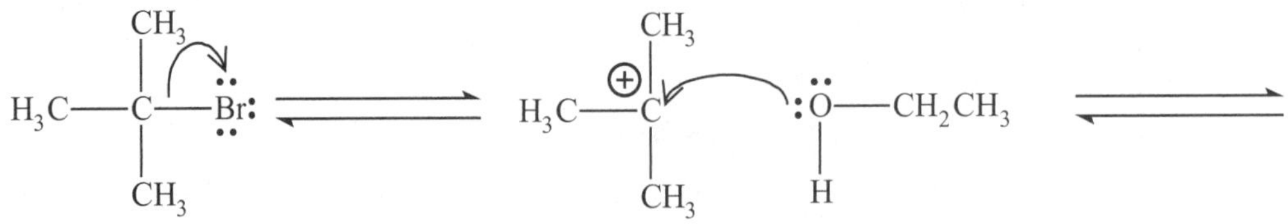

The carbon–bromine bond breaks, and both electrons go to the bromine because it is more electronegative than carbon. A carbocation is formed, which is attacked by the oxygen of the nucleophile, ethanol.

The intermediate formed is the ether product, with an extra hydrogen on the oxygen. Ethanol pulls off the extra hydrogen to give us the ether substitution product, plus a protonated ethanol.

To finish up, bromide pulls a hydrogen off the protonated ethanol molecule, to give H–Br and ethanol as the products. This mechanism is called an S_N1 mechanism—for substitution, nucleophilic, and unimolecular. The unimolecular comes from the studies of the reaction rates of these reactions, where the rate is dependent only upon the concentration of the alkyl halide.

The mechanism for the formation of the elimination product (the alkene) is similar to the substitution mechanism, up to a point. See below.

The carbon–bromine bond breaks, and both electrons go to the bromine because it is more electronegative. This forms a carbocation. To form the elimination product, the oxygen of the ethanol acts as a base and pulls a hydrogen from a carbon next to the carbocation. The electron pair flips in to form the C=C.

This mechanism is called E1, for elimination and unimolecular. It is unimolecular because the rate of the reaction is dependent only upon the concentration of the alkyl halide.

Some more examples of reactions of tertiary halides with weak nucleophiles are shown below.

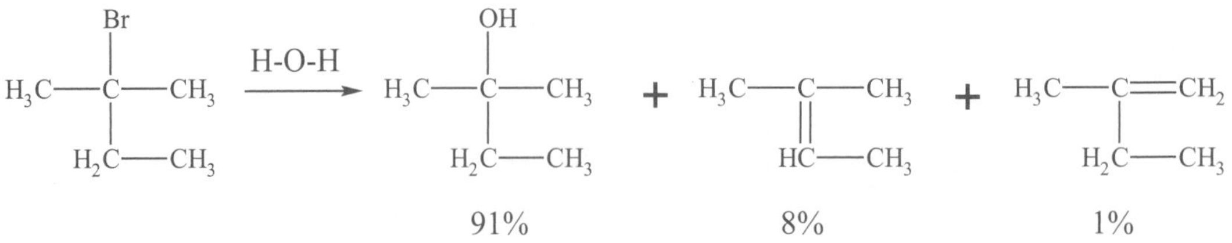

91%   8%   1%

Water is a weak nucleophile, so the major product is substitution. Of the alkene products, the more substituted one is formed in higher amounts. Water is very polar and very small and tends to produce more substitution product than ethanol does.

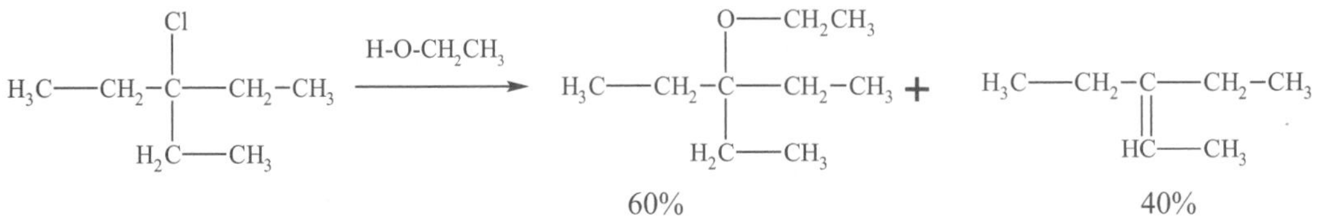

60%   40%

As the groups around the halide get bigger or bulkier, the percent substitution tends to decrease.

G. Reactions of Secondary Alkyl Halides with Weak Nucleophiles/Bases

In general, secondary alkyl halides undergo mostly S_N1 substitution, with competing E1 elimination. The reactions of secondary halides are considerably slower than those of tertiary halides because secondary carbocations are generally much less stable than tertiary cations. Therefore, these reactions are often not very useful practically to synthesize molecules: Usually, there are better ways. Some examples are shown below.

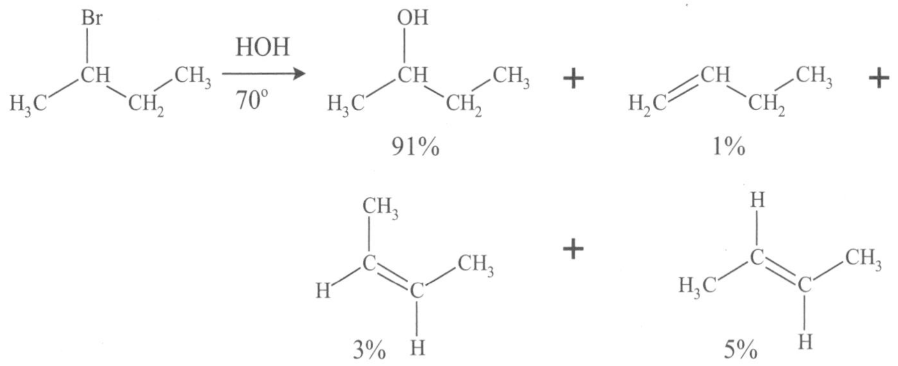

91%   1%

3%   5%

The major product is the substitution product, with minor amounts of the three possible alkene products. Of the alkene products, the more stable *trans*-but-2-ene is the major product. This reaction took 3 days to go to completion!

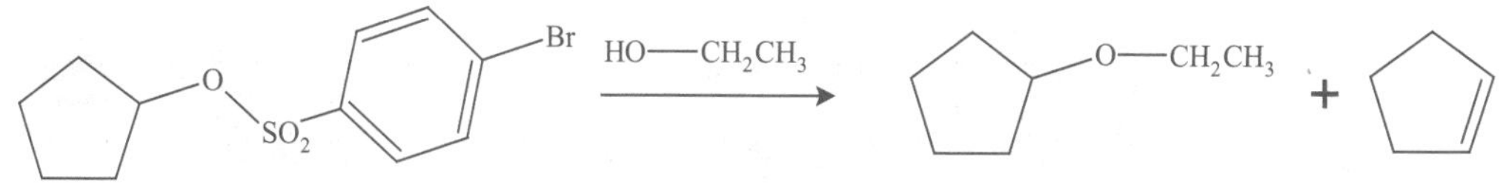

88% 12%

Again, the substitution product is the major product. The bromobenzenesulfonate (sometimes called a **brosylate**) is a very good leaving group.

H. Reactions of Methyl and Primary Alkyl Halides with Weak Nucleophiles/Bases

In general, primary and methyl halides do not undergo S_N1 reactions because primary and methyl cations are quite unstable thermodynamically. Therefore, no reaction occurs with these halides and weak nucleophiles. The only exception is primary cations that can be stabilized by resonance: These will react with weak nucleophiles at relatively slow rates. See the examples below.

$$H_2C{=}CH{-}CH_2{-}Cl \longrightarrow H_2C{=}CH{-}CH_2{-}OH + H{-}Cl$$

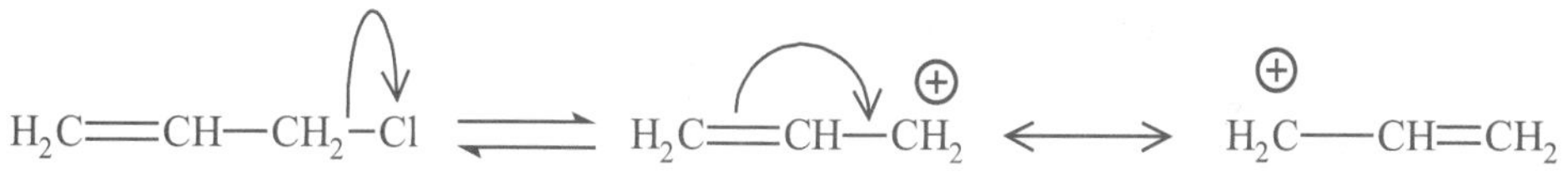

Resonance-stabilized cation

The reaction occurs at about the same rate as that of a secondary alkyl halide. The initial cation formed is stabilized by sharing the charge with the carbon on the other end of the chain, as shown above.

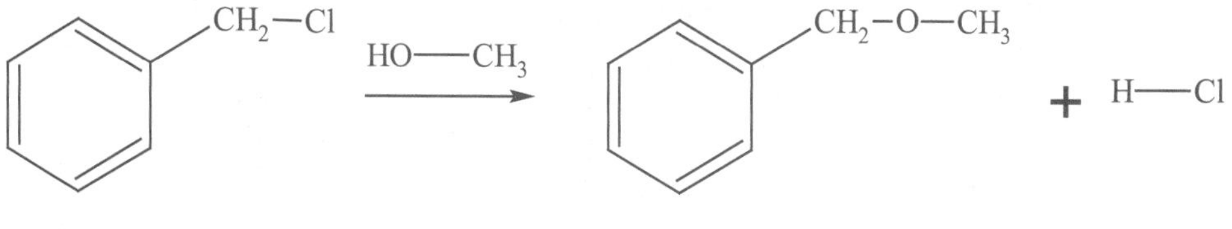

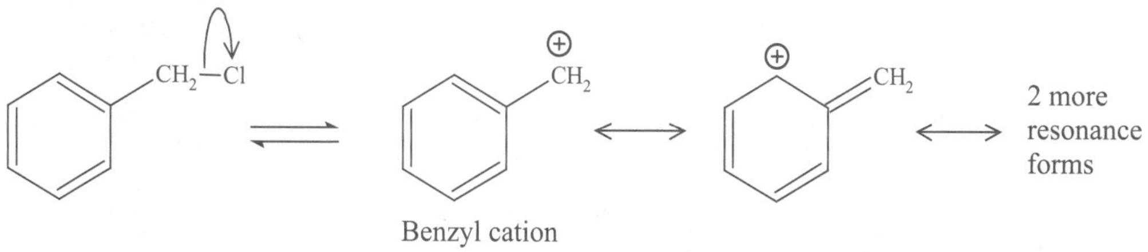

Benzyl cation

The starting material, whose common name is benzyl chloride, ionizes to form the benzyl cation, which has three additional resonance forms. Benzyl chloride reacts somewhat faster than a normal secondary alkyl halide, but it is still much slower than a tertiary alkyl halide.

I. Carbocation Rearrangements

Carbocations have the following order of stability: 3° >2° >1° > methyl. Therefore, when a cation forms in a reaction, it may not be the most stable possible one. If the initially formed cation can rearrange to a more stable one, it usually does. There are a large number of examples of carbocations rearranging to more stable ones from addition reactions of alkenes and from dehydration of alcohols. These reactions were covered in Chapter 4. An example is given below. This is the addition of HBr to 3,3-dimethyl-1-butene.

The first product is what you would expect from Markovnikov addition of H–Br to an alkene (the H adds to the carbon of the C=C with the most hydrogens, and the Br adds to the other carbon). It is however, the minor product. The major product has a different carbon skeleton. A methyl group has "moved" from carbon 3 to carbon 2. This movement is called a **rearrangement**, in chemical jargon. Let's look at the mechanism to see how and why this occurred.

Here is the standard mechanism for addition of H–Br to an alkene, just as it was presented in Chapter 4. So how is the other product formed? See below.

A secondary carbocation   A tertiary carbocation

A methyl group, with its pair of electrons, moves from carbon 3 to the secondary cation on carbon 2. This makes carbon 3 a cation, but in this case, the cation is tertiary, which is much more stable than the original secondary cation. Bromide attacks the new cation to form the final product.

The secondary cation most likely exists for some period of time before it rearranges to the tertiary cation; otherwise, we would not observe the first product, 3-bromo-2,2-dimethylbutane. However, rearrangement must occur shortly after the original secondary cation is formed because the major product is the second product, 2-bromo-2,3-dimethylbutane. Notice that when we name the products, we have to number from the left end of the carbon chain as shown.

Any reaction that goes through a carbocation mechanism has the potential for rearrangement. If a more stable carbocation can be formed from the original cation, rearrangement will occur, and you will see products from rearrangement, as well as unrearranged products.

J. Summary of Nucleophilic Substitution and Elimination Reactions

There is a lot of material in this chapter. When you are approaching a problem involving a nucleophile/base and an alkyl halide, I suggest approaching it in the following way.

1. Classify the alkyl halide as methyl, primary, secondary, or tertiary.
2. Determine if you are using a polar aprotic solvent. If no solvent is given, usually you assume it is water or an alcohol.
3. Classify the nucleophile as strong or weak. If it is strong, classify it as a strong or weak base.
4. Use the following table.

Nucleophile	Type of Alkyl Halide			
	Methyl	**Primary**	**Secondary**	**Tertiary**
Strong	S_N2 only.	Mostly S_N2. E2 is significant only if the base is strong and bulky.	Mixture of E2 and S_N2. E2 predominates with strong bases; S_N2, with weaker bases.	Mainly E2.
Weak	No reaction.	No reaction, unless cation formed can be stabilized by resonance.	Slow S_N1, with some E1. The weaker the nucleophile, the more E1.	S_N1, with some E1. The weaker the nucleophile, the more E1.

Remember, rearrangement is a possibility with any S_N1 or E1 reaction.

This table simplifies things somewhat, but it works for most cases. If you master the principles behind this table, you will do well on problems of this type. Just memorizing this table is not enough. I know my students do memorize this table because I find it written on exams. However, I can easily tell which students understand the principles behind this table when I grade their exams.

REVIEW EXERCISES FOR CHAPTER 6

1. For each of the following reactions, (i) classify the nucleophile as strong or weak; (ii) classify the alkyl halide as 1°, 2°, or 3°; (iii) draw the structure of the major organic product; and (iv) give the type of mechanism that was used to form the major organic product (S_N1, S_N2, E1, or E2).

   a.

   b.

   c.

   Sodium azide: found in airbags

   d.

2. Draw arrow-pushing mechanisms for the following reactions:

   a.

   + H-Cl

   b.

c.

3. What combinations of alkyl halide and nucleophile would you use to prepare each of the following compounds? Explain your reasoning.

   a.   b.

4. When 1-bromo-1-methylcyclohexane reacts with sodium ethoxide in ethanol, a mixture of two alkene products, A and B, is formed in a ratio of 2:1. When potassium *t*-butoxide is used as the base, the ratio of A:B is 1:20. What are the structures of products A and B? Explain how the base used influences the ratio of the products.

Free-Radical Reactions

WHAT YOU WILL LEARN

In this chapter, you will learn:

- what free-radicals are and how they are formed;
- the general mechanism of halogenation of alkanes with chlorine and bromine;
- why free-radical bromination of alkanes is more selective than chlorination.

In contrast to the polar types of mechanisms discussed in Chapter 6, there are other mechanisms that you will encounter as you go through organic chemistry. One of these is the mechanism of halogenation of alkanes. Because alkanes are nonpolar and have only C–H and C–C single bonds, there is nothing to which a nucleophile might be attracted. Alkanes are really weak acids, so there is no significant reaction with most of the bases we have encountered so far. However, alkanes react with most halogens in the presence of heat or ultraviolet (UV) light (even sunlight!).

A. Introduction to Free-Radicals

A **free-radical** is a molecule with an unpaired electron, such as the following.

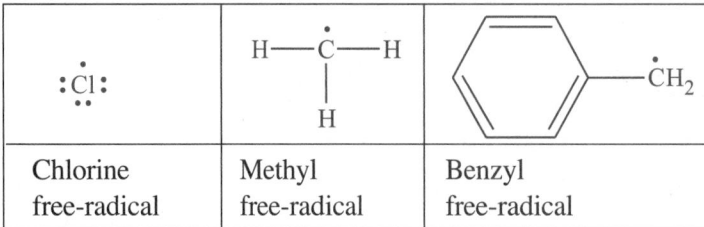

:Cl:	H—C—H with H below	benzyl CH₂
Chlorine free-radical	Methyl free-radical	Benzyl free-radical

The "free" part of free-radical is to distinguish them from the old use of the term "radical." Originally, a radical was a group. Therefore, in the molecule $CH_3CH_2CH_2Cl$, the CH_3 part of the molecule was referred to as a methyl radical. The CH_3 was bonded to the rest of the molecule, so you might think of it as a "bound" radical. To distinguish the species with an unpaired electron, •CH_3, from a bound radical, it was called a free-radical. Nowadays, if you call something a radical, everyone would assume you are referring to a free-radical.

Radicals are generally formed by heating a molecule, usually to a high temperature, or by irradiating it with UV light of the proper wavelength. When a molecule is heated, the bond that breaks first is usually the weakest bond. A table of Bond Dissociation Energies (BDE) for homolytic cleavage is given on the next page. Your text may provide a more extensive list, or you can find them in the *CRC Handbook of Chemistry and Physics*.

Bond	BDE (kcal/mol)	BDE (kJ/mol)
H–H, H–X, and X–X bonds		
H–H	104	435
F–F	38	159
Cl–Cl	58	243
Br–Br	46	192
I–I	36	151
H–F	136	569
H–Cl	103	431
H–Br	88	368
H–I	71	297
HO–H	119	498
HO–OH	51	213
HO–F	56	234
HO–Cl	65	272
HO–Br	56	234
HO–I	47	197
Methyl bonds		
CH_3–H	104	435
CH_3–F	108	452
CH_3–Cl	84	351
CH_3–Br	70	293
CH_3–I	56	234
Bonds to primary carbons		
$CH_3CH_2CH_2$–H	98	410
$CH_3CH_2CH_2$–F	106	444
$CH_3CH_2CH_2$–Cl	81	339
$CH_3CH_2CH_2$–Br	69	289
$CH_3CH_2CH_2$–I	53	222

Bond	BDE (kcal/mol)	BDE (kJ/mol)
Bonds to secondary carbons		
$(CH_3)_2CH$–H	95	397
$(CH_3)_2CH$–F	105	439
$(CH_3)_2CH$–Cl	81	339
$(CH_3)_2CH$–Br	68	285
$(CH_3)_2CH$–I	53	222
Bonds to tertiary carbons		
$(CH_3)_3C$–H	92	385
$(CH_3)_3C$–F	106	444
$(CH_3)_3C$–Cl	80	335
$(CH_3)_3C$–Br	64	268
$(CH_3)_3C$–I	51	213
Carbon–carbon bonds		
CH_3–CH_3	88	368
CH_3CH_2–CH_3	85	356
$(CH_3)_2CH$–CH_3	85	356
$(CH_3)_3C$–CH_3	80	335
CH_3CH_2–CH_3	84	351
Other bonds		
O=O	119	498
$H_2C=CHCH_2$–H (the C–H)	87	364
$C_6H_5CH_2$–H	85	356
C_6H_5–H (benzene C–H)	112	469
$H_2C=CH$–H (alkene C–H)	108	452
O=C=O (each C=O)	127	531

You can use the BDEs to predict the overall enthalpy change ($\Delta H°$) of a reaction. The amount of energy to break a bond is the BDE. When a bond is formed, the amount of energy released is the negative of the BDE. Let's look at an example.

HO—H + Cl—Cl → HO—Cl + H—Cl			
Bonds Broken	**$\Delta H°$ (per mole)**	**Bonds Formed**	**$\Delta H°$ (per mole)**
HO–H	+119 kcal (498 kJ)	HO–Cl	–65 kcal (–272 kJ)
Cl–Cl	+58 kcal (243 kJ)	H–Cl	–103 kcal (–431 kJ)
Total	+177 kcal (741 kJ)	Total	–168 kcal (–703 kJ)
$\Delta H° = +177$ kcal + (–168) kcal = +9 kcal/mol, or $\Delta H° = +741$ kJ + (–703) kJ = +38 kJ/mol			

The enthalpy change of this reaction under standard conditions is 9 kcal/mol, or 38 kJ/mol. When the change in enthalpy is positive, the reaction is said to be endothermic. When the change in enthalpy is negative, the reaction is exothermic. Therefore, this reaction is endothermic.

B. Free-Radical Chlorination of Methane

If you heat methane and chlorine together, or irradiate a mixture with UV light, you can produce chloromethane and HCl. The mechanism for this reaction takes place in a series of steps, which is shown below.

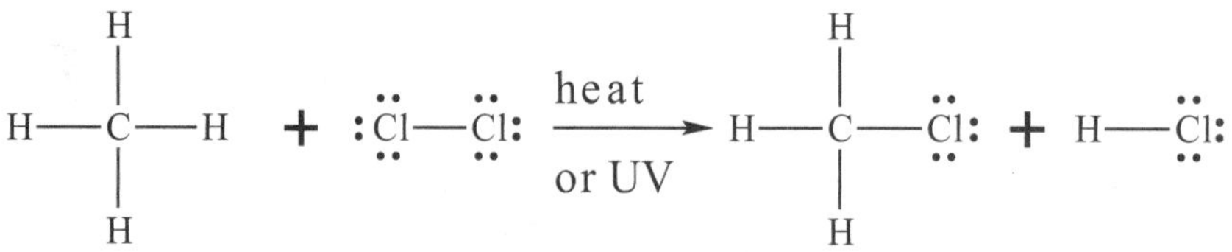

Here is the overall reaction. We can calculate $\Delta H°$ for this reaction, just like before.

$\Delta H° = 104$ kcal/mol + 58 kcal/mol + (–84) kcal/mol + (–103) kcal/mol = –25 kcal/mol, or

$\Delta H° = 435$ kJ/mol + 243 kJ/mol + (–351) kJ/mol + (–431) kJ/mol = –104 kJ/mole

Therefore, this reaction is exothermic.

Step 1.

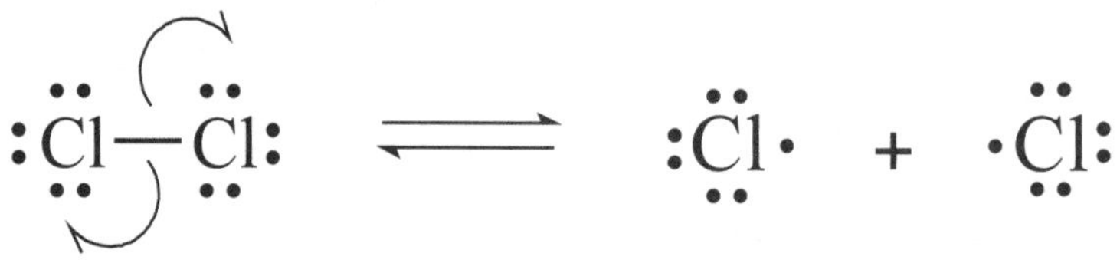

This is called the **initiation step**. Some radicals are formed to get the reaction started. Because this reaction is highly endothermic ($\Delta H° = 58$ kcal/mol), only a few chlorine radicals are formed. Notice that the curved arrows have only single prongs on them, like fishhooks. This is because each of these arrows only signifies the movement of one electron.

Step 2.

Once we have formed a few chlorine radicals, they can now react with something. If a chlorine radical bumps into a methane hydrogen, the C–H bond can break, and a new H–Cl bond forms. We also have formed a methyl radical. Because the overall number of radicals stays the same, this reaction is called a **propagation step**. We can calculate $\Delta H°$ for this step.

$$\Delta H° = 104 \text{ kcal/mol} + (-103) \text{ kcal/mol} = 1 \text{ kcal/mol, or}$$

$$\Delta H° = 435 \text{ kJ/mol} + (-431) \text{ kJ/mol} = 4 \text{ kJ/mol}$$

This step is endothermic by a very small amount.

Step 3.

Once the methyl radical is formed, it can react with another molecule. The most likely possibilities are another methane or a chlorine molecule. It is unlikely that a methyl radical will find a chlorine radical because we made only a few chlorine radicals to begin with. If the methyl radical reacts with a chlorine molecule, chloromethane is formed, along with another chlorine radical. Because the number of radicals hasn't changed, this is also called a **propagation step**. We can calculate $\Delta H°$ for this step.

$$\Delta H° = 58 \text{ kcal/mol} + (-84) \text{ kcal/mol} = -26 \text{ kcal/mol, or}$$

$$\Delta H° = 243 \text{ kJ/mol} + (-351) \text{ kJ/mol} = -108 \text{ kJ/mol}$$

This step is exothermic.

In step 3, a new chlorine radical is formed. That chlorine radical can react with another methane molecule, as shown in step 2, to form a new methyl radical. The new methyl radical can react with another chlorine molecule to form chloromethane, plus a new chlorine radical. Steps 2 and 3 can repeat over and over again, until all of the chlorine and methane are used up. This is shown on the next page.

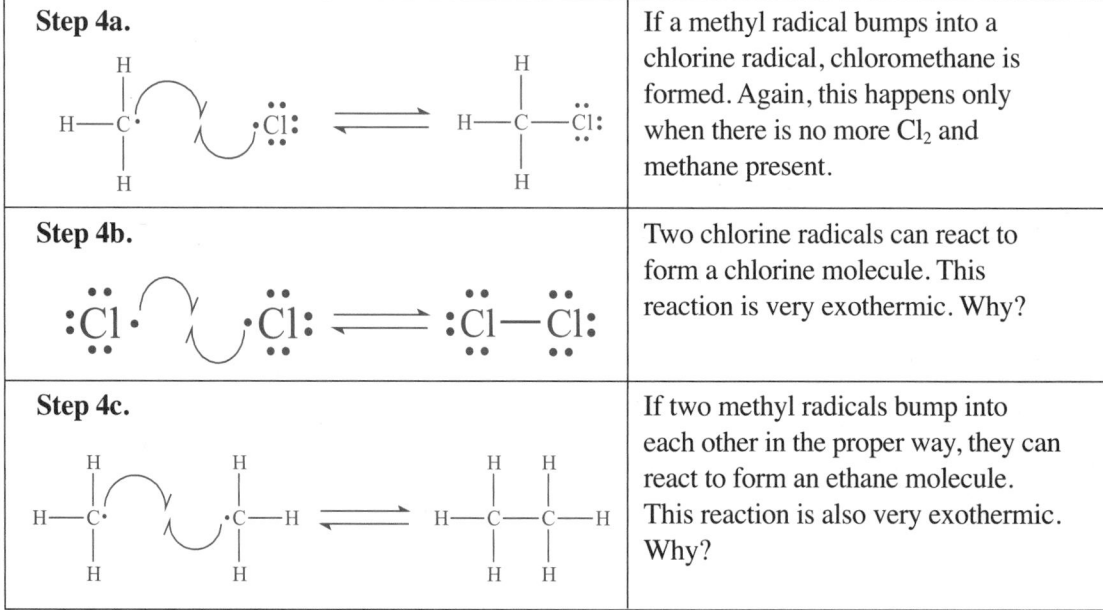

This process is a chain reaction. Eventually, as all the methane and chlorine molecules are used up, all we will have is radicals. If two radicals meet up, they can form products that are not radicals, as shown below. Steps that decrease the number of radicals are called **termination steps.** If the two radicals that reacted are the last two radicals, then the reaction would stop because there would be no more radicals present.

Step 4a.	If a methyl radical bumps into a chlorine radical, chloromethane is formed. Again, this happens only when there is no more Cl_2 and methane present.
Step 4b.	Two chlorine radicals can react to form a chlorine molecule. This reaction is very exothermic. Why?
Step 4c.	If two methyl radicals bump into each other in the proper way, they can react to form an ethane molecule. This reaction is also very exothermic. Why?

Step 4c produces ethane, rather than chloromethane. Ethane is an undesired by–product of the reaction. Because it forms only when two radicals collide, very little is actually formed. However, other by–products can and do form in significant amounts, unless care is taken to try and minimize their formation. One of these is dichloromethane, CH_2Cl_2. Why does dichloromethane form? Think about the processes going on in the propagation steps (steps 2 and 3) above. As the methane is used up, more and more chloromethane is forming. It is becoming more likely, therefore, that a chlorine radical will bump into a chloromethane, instead of a methane. If this occurs, the following propagation steps can occur.

Step 2'.

The chlorine radical pulls a hydrogen off a chloromethane, forming a chloromethyl radical and H–Cl.

Step 3'.

The chloromethyl radical reacts with a chlorine molecule to form dichloromethane and another chlorine radical, which can react with something else.

You can envision similar series of reactions to produce trichloromethane (chloroform, $CHCl_3$) and tetrachloromethane (carbon tetrachloride, CCl_4) as by–products as well.

So what do we get if we try to chlorinate methane? Potentially, we get a lot of organic products! See below.

Compounds	CH_4	Cl_2	$\rightarrow$	CH_3Cl	CH_2Cl_2	$CHCl_3$	CCl_4	HCl	Others
Boiling points	−164	−35		−24	40	62	77	−85	

Because methane and chlorine are gases, they are usually reacted in a special apparatus under pressure. Because the products all have different boiling points, they can be separated by distillation. Therefore, it is relatively easy to isolate pure chloromethane. However, what do you do with the by–products? If you have a market for them, you could potentially sell them to recoup the cost of producing them. If you don't, then you generally have to pay someone to take them off your hands and dispose of them somehow. It would be better if you didn't make them in the first place. So how can you avoid making the by–products?

One way is to use a large excess of methane. That way, when a chlorine radical bumps into another molecule, it is almost always a methane. Therefore, when all of the chlorine is used up, you will have chloromethane, HCl, and a lot of unreacted methane. The unreacted methane can be separated, and fed back into the reactor to make more chloromethane, so the extra methane is not wasted.

Let me make one other point before we go on. Generally, there is only one initiation step, and it involves breaking a rather weak bond in some molecule. In this case, the weakest bond is the Cl–Cl bond, so it is the bond that will break, to give us our initial radicals. The CH_3–H bonds are much stronger than the Cl–Cl bonds and are not broken spontaneously by normal heating. So when you write a mechanism, don't just break whatever bonds randomly: I see, for example, the following mechanism frequently on exams.

	The first reaction is fine because Cl–Cl has a weak bond. The second step does not occur to any significant degree under normal reaction conditions because the C–H bond is much stronger than the Cl–Cl.
	Now we have the massive radical recombination steps to form the products. Because relatively tiny amounts of radicals form, the chance of two radicals finding each other is very small. This mechanism gets a score of 2 out of 10 points from me. I assume you want a better score, so don't write this one down!

C. The Energetics of Halogenation of Methane with Other Halogens

What happens if we treat methane with other halogens in the presence of UV light? Potentially, we can make other halomethanes. Let's look at the values of $\Delta H°$ for the propagation steps for these reactions, as well as the overall reactions.

Overall reaction.
Step 1.

Step 2.

Step 3.

Halogen	$\Delta H°$ for Step 2 (per mole)	$\Delta H°$ for Step 3 (per mole)	$\Delta H°$ Overall (per mole)
Fluorine	−32 kcal (−117 kJ)	−70 kcal (−293 kJ)	−102 kcal (−420 kJ)
Chlorine	1 kcal (4 kJ)	−26 kcal (−108 kJ)	−25 kcal (−104 kJ)
Bromine	16 kcal (67 kJ)	−24 kcal (−101 kJ)	−8 kcal (−34 kJ)
Iodine	33 kcal (138 kJ)	−20 kcal (−83 kJ)	13 kcal (55 kJ)

Free-radical fluorination is extremely exothermic. As such, it is very difficult to control and is rarely done. On the other end of the scale, free-radical iodination is an overall endothermic process and is rarely done as well. The most common free-radical halogenations are done with chlorine and bromine. Both are overall exothermic, but they are not as exothermic as hologenations with fluorine. They can be rather easily controlled, and there are several examples of lab experiments involving free-radical halogenation.

D. Free-Radical Chlorination of Other Alkanes

Other alkanes can be chlorinated in a manner similar to that of methane. For example, the results with propane and 2-methyl propane are given below.

At first glance, chlorination does not look very selective because the two products in each case are formed in similar amounts. However, there is a certain amount of selectivity. Propane has two different types of hydrogens: the methyl hydrogens, which are primary, and the CH_2 hydrogens, which are secondary. There are two secondary hydrogens, and replacement of either one of them leads to 2-chloropropane. Therefore, replacement of either secondary hydrogen leads to 59/2, or 29.5% of the total products. Likewise, there are six methyl hydrogens, replacement of any one of those primary hydrogens leads to 41/6, or 6.8% of the total product. What this is telling us is that replacement of a secondary hydrogen leads to more of the product then replacement of a primary hydrogen. Another way of putting this is that a secondary hydrogen is more reactive than a primary hydrogen under these conditions. If we take the ratio of the percentages (29.5/6.8 = 4.3), we conclude that a secondary hydrogen is 4.3 times as reactive as a primary hydrogen in free-radical chlorination at 100°.

We can do the same analysis for the relative reactivities of a tertiary hydrogen and a primary hydrogen, using the yield data from the chlorination of 2-methylpropane. There is only one tertiary hydrogen, and it accounts for 44% of the total product. There are nine primary hydrogens, which account for 56% of the product, so each primary hydrogen accounts for 56/9 = 6.2% of the product. Therefore, a tertiary hydrogen is 44/6.2, or 7.1 times as reactive as a primary hydrogen under these conditions. So under these conditions, the relative reactivities of a tertiary hydrogen: secondary hydrogen: primary hydrogen are 7.1:4.3:1.

Your textbook may give slightly different relative reactivities than those given here. The reaction conditions can affect the relative yields of the products, which will change the ratios. For example, at 25° in the presence of UV light, the relative reactivities are 5.5:4.5:1. The bottom line is not what the actual numbers are, but that a tertiary hydrogen is more reactive than a secondary hydrogen, which is more reactive than a primary hydrogen. However, the differences in reactivity are not huge. And practically, if you chlorinate something with two or more different types of hydrogens, you will get a mixture of monochlorinated products. For a synthesis problem, a mixture of products is rarely good. Therefore, only use free-radical chlorination if there is only one possible monochlorinated product. Some examples are shown below.

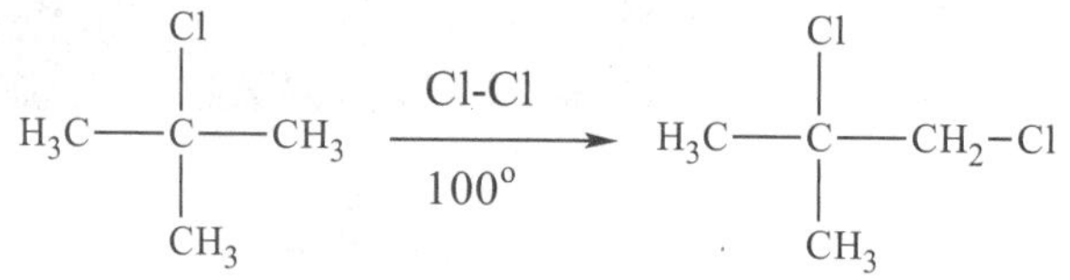

There are only methyl groups in the starting material, and they are all equivalent, so there is only one possible monochlorinated product.

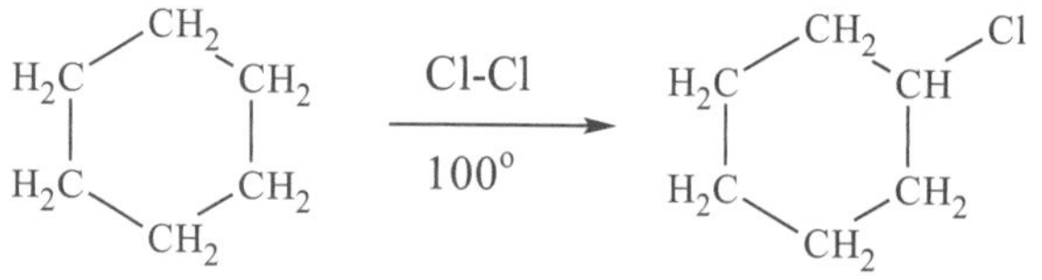

In cyclohexane, all the CH_2 groups are equivalent, so replacement of any hydrogen leads to the same product.

E. Free-Radical Bromination of Alkanes

In contrast to chlorination, free-radical bromination is much more selective. See the examples below.

$$H_3C-CH_2-CH_3 \xrightarrow[125°, \text{UV}]{\text{Br-Br}} H_3C-\overset{\displaystyle Br}{\underset{\displaystyle |}{CH}}-CH_3 \quad + \quad H_3C-CH_2-\overset{\displaystyle Br}{\underset{\displaystyle |}{CH_2}}$$

97%   3%

>99%   <1%

92%   4%   4%

From the first example, we can calculate the difference in reactivity between a secondary and a primary hydrogen in bromination under these conditions, which is 97:1. From the third example, the relative reactivities of tertiary and secondary hydrogens are 46:1, so the relative reactivities of a tertiary hydrogen and a primary hydrogen are 4462:1. The bottom line is that tertiary hydrogens are much more reactive than secondary hydrogens, which are much more reactive than primary hydrogens. The big question you may have is: Why is bromination much more selective than chlorination? That is our next topic.

F. The Selectivity of Free-Radical Bromination Versus Chlorination

Let's consider the energetics of the first propagation steps in the chlorination and bromination of propane. These are shown below.

Reaction	$\Delta H°$ (per mole)
H_3C—CH_2–CH_3 + $\cdot \ddot{C}l:$ $\rightleftharpoons$ H_3C—CH_2–$\dot{C}H_2$ + H—$\ddot{C}l:$	–5 kcal (–21 kJ)
H_3C—CH_2–CH_3 + $\cdot \ddot{C}l:$ $\rightleftharpoons$ H_3C—$\dot{C}H$—CH_3 + H—$\ddot{C}l:$	–8 kcal (–34 kJ)
Both of the above reactions are exothermic. The second is more exothermic because a secondary C–H bond is slightly weaker than a primary C–H bond.	
H_3C—CH_2–CH_3 + $\cdot \ddot{B}r:$ $\rightleftharpoons$ H_3C—CH_2–$\dot{C}H_2$ + H—$\ddot{B}r:$	10 kcal (42 kJ)
H_3C—CH_2–CH_3 + $\cdot \ddot{B}r:$ $\rightleftharpoons$ H_3C—$\dot{C}H$—CH_3 + H—$\ddot{B}r:$	7 kcal (29 kJ)
Both of the above reactions are endothermic. The second is less endothermic because a secondary C–H bond is slightly weaker than a primary C–H bond.	

The fact that chlorination is exothermic, while bromination is endothermic, leads to the difference in selectivity. This is shown in the following figures.

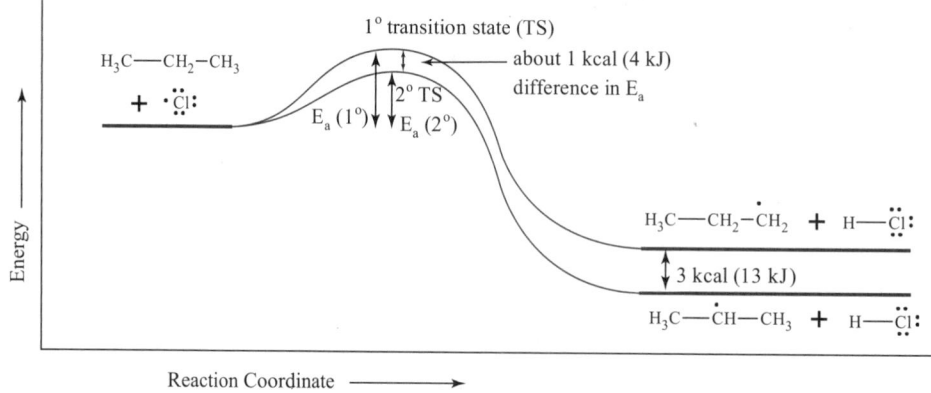

Figure 1

For an exothermic reaction, the starting materials are higher in energy than the products. The transition states are closer in energy to the starting materials than the products. Therefore, the transition state is more similar to the starting materials, which implies that the C–H bonds are just beginning to be broken. That implies that there is little radical character on the carbons, so the differences in stability of the two transition states are much less than in the two radicals themselves. A small difference in stability means the difference in activation energies to get to the two radicals is small, so a significant number of molecules will have enough energy to form the primary radical. This is shown in Figure 1.

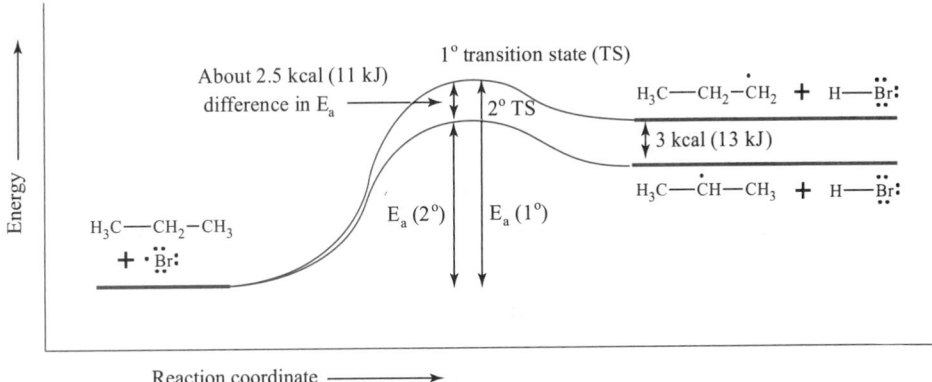

Figure 2

For an endothermic reaction, Figure 2, the starting materials are lower in energy than the products. The transition states are closer in energy to the products than the starting materials. Therefore, the transition state is more similar to the products, which implies that the C–H bonds are almost totally broken. That implies that there is much radical character on the carbons, so the differences in stability of the two transition states are similar to those of the two radicals themselves. A large difference in stability means the difference in activation energies to get to the two radicals is large, so a very few molecules will have enough energy to form the primary radical.

These reactions are an example of a general principle called the Hammond postulate. The Hammond postulate says: Related species that are similar in energy have similar structures. The structure of the transition state resembles the structure of the closest stable species. Free-radical halogenation is the main place in introductory organic chemistry that the Hammond postulate is discussed.

G. Other Halogenating Agents

Because chlorine is a corrosive and toxic gas, and bromine is a volatile, corrosive liquid, they are usually not used in most undergraduate organic laboratories. Other reagents can be used in place of these, although they are not without toxicity. Many of these reagents do not spontaneously form radicals upon heating and require another molecule to be present that can more easily form some radicals. These molecules are called **initiators**, not too surprisingly. Two common initiators are dibenzoylperoxide (DBP) and azobisisobutyronitrile (AIBN for short). They form radicals as shown on the next page.

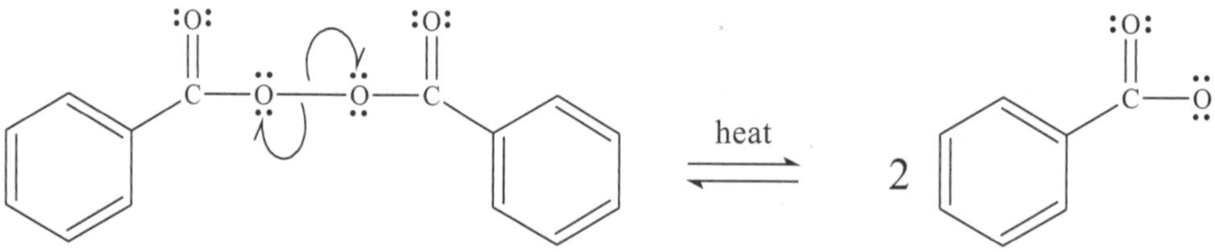

Dibenzoyl peroxide

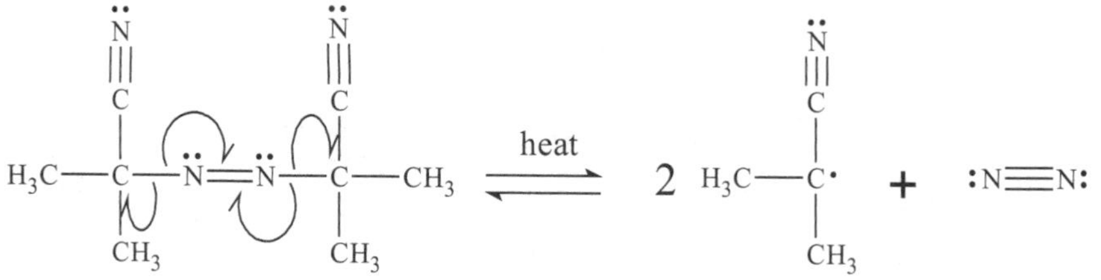

AIBN

Rather than drawing out the structure of the initiator radical, it is usually abbreviated as In•.

Sulfuryl chloride, SO_2Cl_2, is a dense, corrosive liquid that reacts with water to form sulfuric acid and HCl. It can be used in place of chlorine for free-radical reactions. The mechanism of chlorination of cyclohexane with sulfuryl chloride and an initiator is shown below.

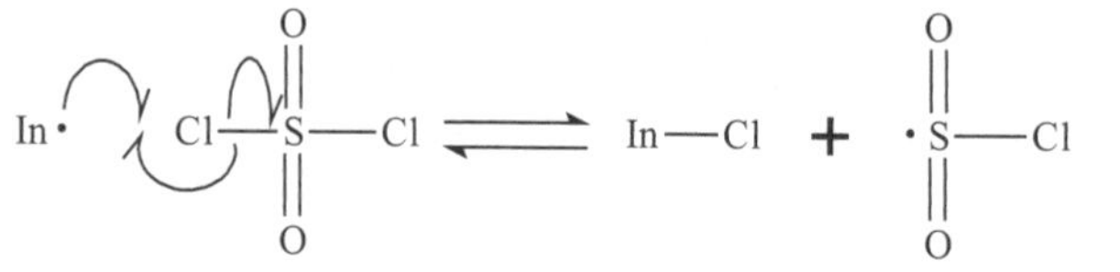	**Step 1.** The initiator pulls a chlorine off sulfuryl chloride.
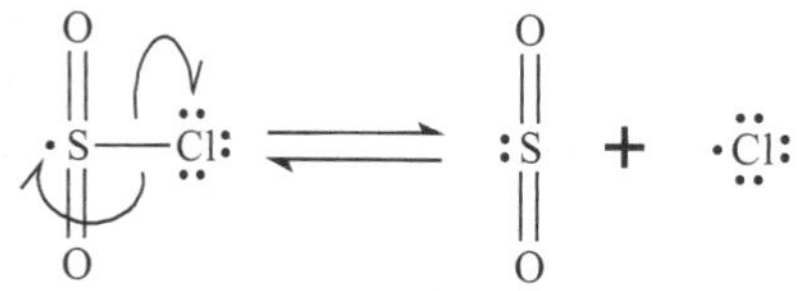	**Step 2.** The SO_2Cl radical loses the chlorine radical and forms SO_2, as a stable by–product.
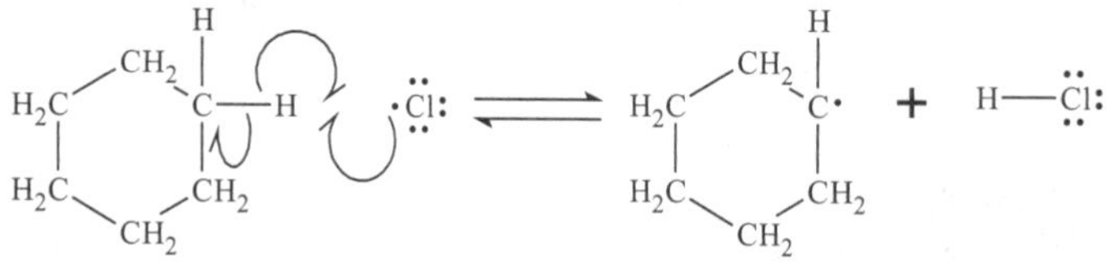	**Step 3.** The chlorine radical pulls a hydrogen off of cyclo-hexane, to form the cyclohexyl radical and H–Cl.
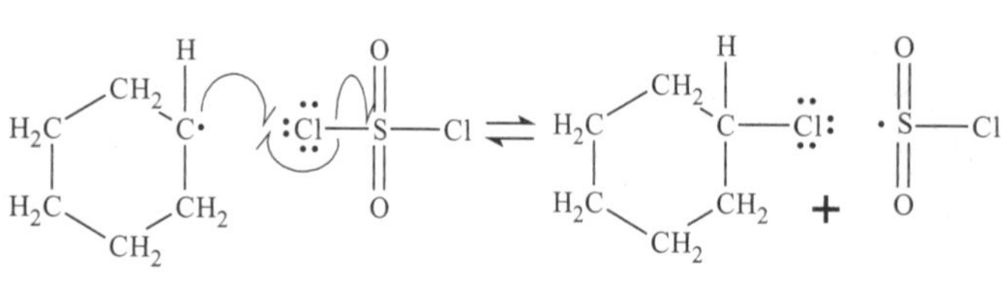	**Step 4.** Cyclohexyl radical pulls a chlorine off a sulfuryl chloride to make the product, plus another SO_2Cl radical, which can go back to step 2 and carry on the chain reaction.

N-Bromosuccinimide (NBS) is a solid that is easily handled and can be used for bromination. It is most often used in CCl_4, a solvent that it is only slightly soluble, and in the presence of an initiator. The mechanism of the reaction has been exhaustively studied. The reaction depends upon a trace of H–Br or Br_2 being present initially, probably as an impurity in the NBS. NBS reacts with H–Br in an ionic mechanism to produce a low concentration of Br_2, which is the actual brominating agent. The HBr by–product formed reacts with NBS to produce more Br_2. The by–product, succinimide, is insoluble in CCl_4 and is filtered off after the reaction is done. Removal of the CCl_4 yields the crude product, which can be purified as needed. The radical part mechanism is shown below, using cyclohexane.

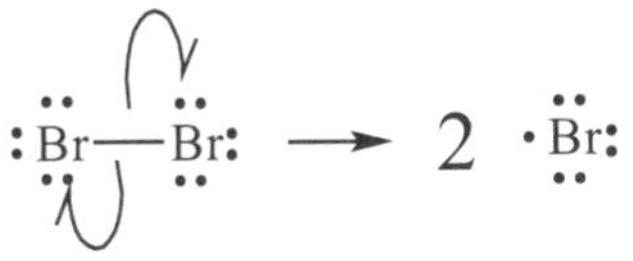 $:Br{-}Br: \longrightarrow 2 \ \cdot Br:$	Heat breaks a Br–Br bond, or the initiator present gets things started.
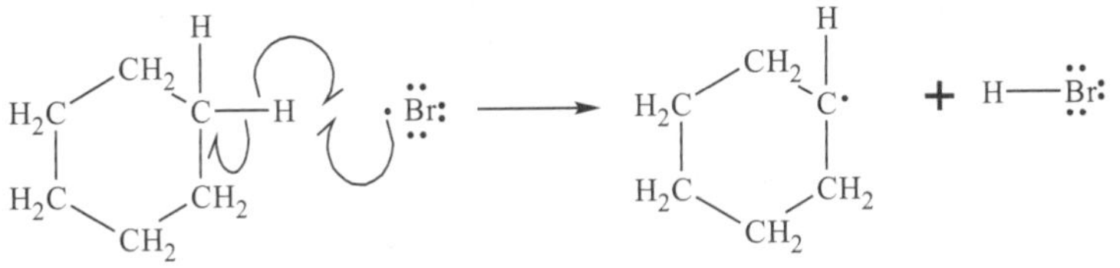 (cyclohexane + $\cdot Br:$ → cyclohexyl radical + $H{-}Br:$)	Bromine radical pulls a hydrogen off cyclohexane.
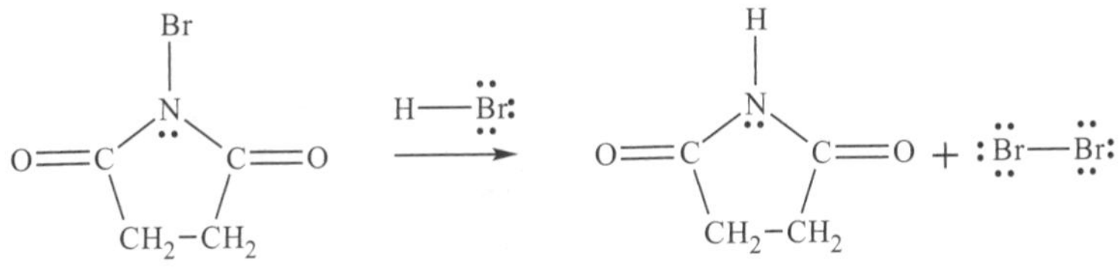 (NBS + $H{-}Br:$ → succinimide + $:Br{-}Br:$)	This is an ionic not a radical mechanism.
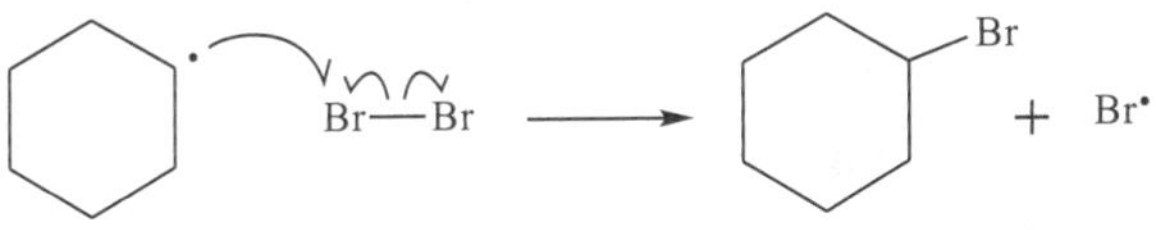 (cyclohexyl radical + $Br{-}Br$ → bromocyclohexane + $Br^\cdot$)	Cyclohexyl radical reacts with bromine to form the product and a now bromine radical.

H. Free-Radical Halogenation of Allylic and Benzylic Hydrogens

Here are another couple of organic chemistry jargon terms you need to know. An **allylic carbon** refers to the carbon next to a C=C. The hydrogens on an allylic carbon are called **allylic hydrogens**. Similarly, the carbon next to a benzene ring is called a **benzylic carbon**, and the hydrogens on a benzylic carbon are called **benzylic hydrogens**. These terms are illustrated below.

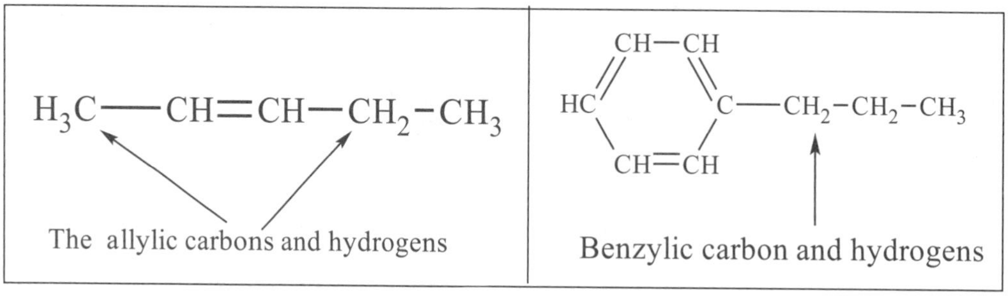

$$H_3C\!\!-\!\!CH\!\!=\!\!CH\!\!-\!\!CH_2\!\!-\!\!CH_3$$

The allylic carbons and hydrogens

Benzylic carbon and hydrogens

Allylic and benzylic positions are more reactive than other positions in free-radical halogenations, as the following examples show.

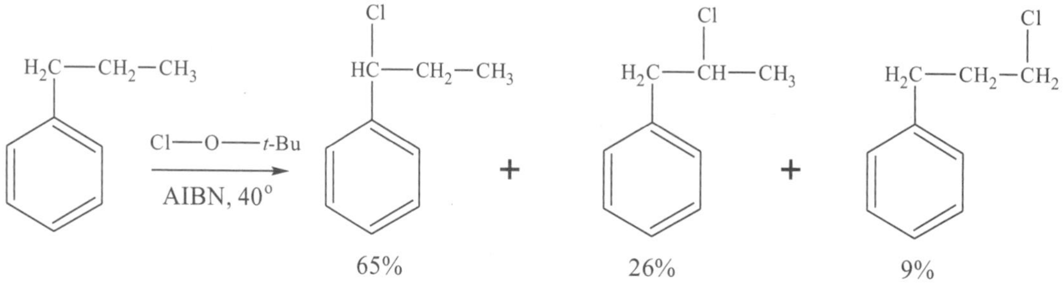

65% + 26% + 9%

Chlorination of the 2° benzylic position is favored by more than 2:1 over chlorination of the other 2° position. A small amount of chlorination occurs at the methyl.

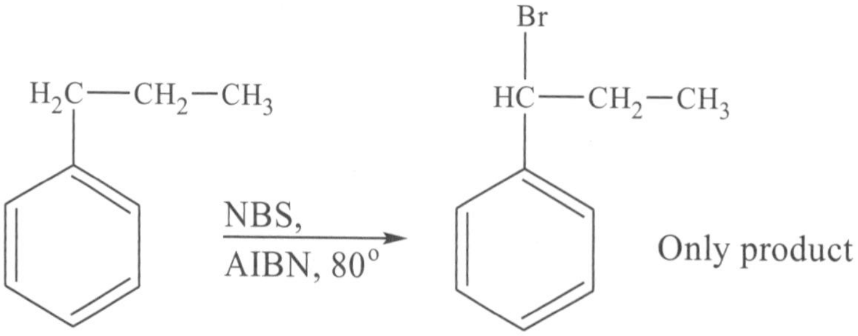

NBS, AIBN, 80°   Only product

Bromination is much more selective than chlorination, so only benzylic bromination is observed.

The radicals formed in benzylic and allylic halogenation are resonance-stabilized. That lowers the energies of those intermediates; therefore, the products derived from them are the major products.

I. Oxygen as a Free-Radical

If I asked you to write a Lewis structure for O_2, dioxygen, or molecular oxygen, you would probably write the structure on the left, below. Unfortunately, even though this structure looks great, it is not what molecular oxygen looks like. It is really the structure on the right.

Incorrect structure for dioxygen   Correct structure for dioxygen

A molecule with unpaired electrons is attracted to a magnetic field and is called **paramagnetic**. A nice demonstration of the paramagnetic properties of liquid oxygen is available on the web at *http://www.chem.uiuc.edu/clcwebsite/liquido2.html*. Notice that the person handling the liquid gases is not wearing gloves, even though liquid oxygen boils at −183°C, and liquid nitrogen boils at −196°C!

REVIEW EXERCISES FOR CHAPTER 7

1. Give the major organic product of each of the following reactions. Assume only one halogen will be incorporated into the product.

   a.

   Br_2, heat

   b.

   NBS,
   DBP, 80°

   c.

   Cl_2, UV light

d.

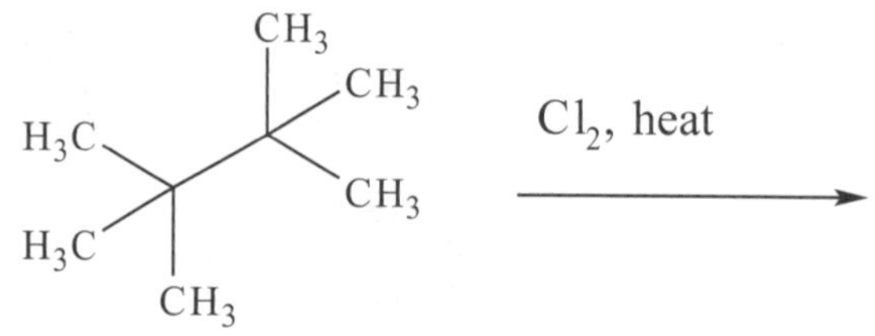

Cl_2, heat

e.

NBS,
DBP, $80°$

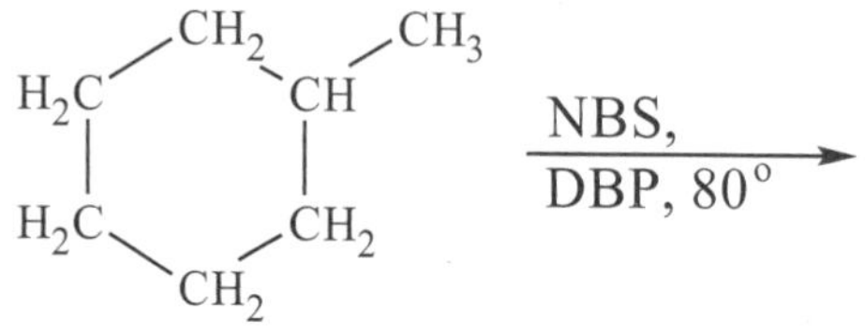

2. Write arrow-pushing mechanisms for Problems 1a and 1b.

3. Explain why the following mechanism for free-radical halogenation of ethane is probably not correct. Think about $\Delta H°$ for each step, as well as how bonds would have to break and form.

$$Cl-Cl \rightleftharpoons 2\ Cl\bullet$$

$$Cl\bullet + CH_3CH_3 \rightleftharpoons H\bullet + CH_3CH_2-Cl$$

$$H\bullet + Cl-Cl \rightleftharpoons H-Cl + Cl\bullet$$

4. In the discussion of NBS in the text, I mentioned that it forms a low concentration of Br_2. Why would a low concentration of Br_2 be a good thing when performing allylic bromination of an alkene? What reaction would likely compete with allylic bromination of an alkene?

5. Calculate $\Delta H°$ for each of the following reactions. Which ones are exothermic?
   a. $Cl_2 + CH_3CH_3 \rightarrow 2\ CH_3Cl$
   b. $Br_2 + CH_3CH_3 \rightarrow 2\ CH_3Br$
   c. $(CH_3)_3C-H + I_2 \rightarrow (CH_3)_3C-I + H-I$
   d. $CH_4 + 2O_2 \rightarrow CO_2 + H_2O$

Alcohols and Thiols

WHAT YOU WILL LEARN

In this chapter, you will learn:

- how to name alcohols using IUPAC rules;
- how the structure of an alcohol influences its physical properties and acidity;
- how alcohols can be converted into alkyl halides and other functional groups;
- how alcohols can be prepared from alkenes and carbonyl compounds;
- how to name, prepare, and oxidize thiols.

SECTIONS IN THIS CHAPTER

- IUPAC Nomenclature of Alcohols
- Physical Properties of Alcohols
- Reactions That Cleave the O–H Bond
- Reactions That Cleave the C–O Bond
- Oxidation of Alcohols
- Preparations of Alcohols
- Thiol Naming and Properties
- Thiol Reactions and Preparations

Alcohols (R–O–H) and thiols (R–S–H) are functional groups found in lots of naturally-occurring molecules. Some examples are shown below.

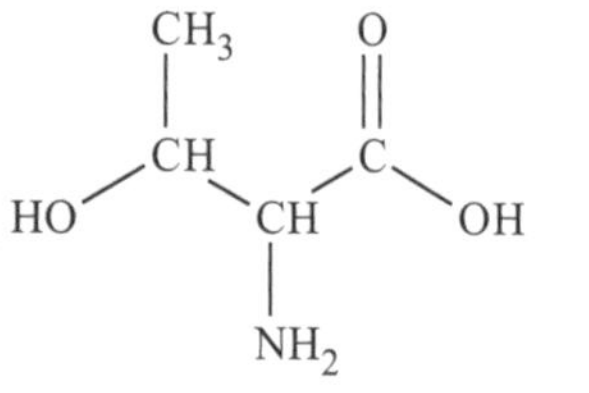 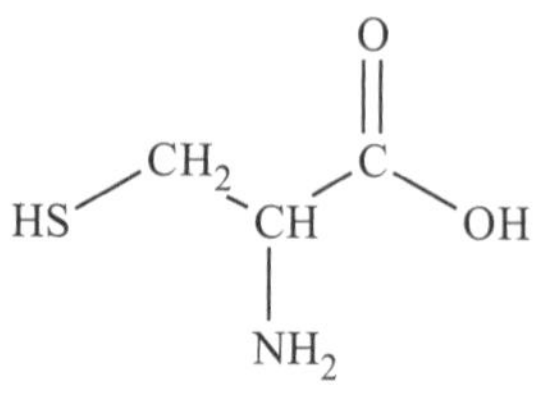	Here are the amino acids threonine and cysteine. Threonine contains an alcohol, and cysteine contains a thiol.
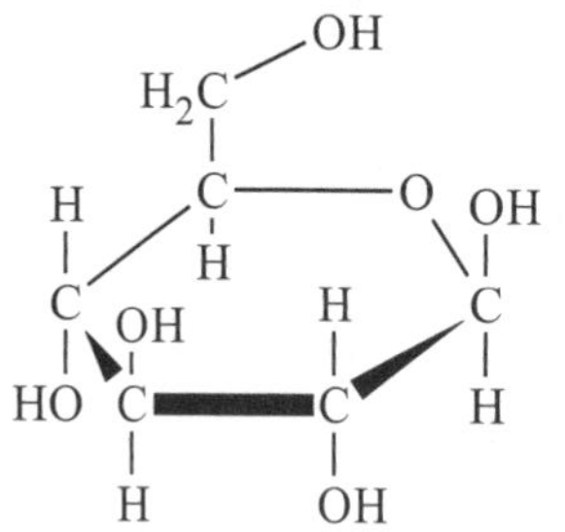	This is one representation of glucose (a Haworth projection), a common sugar. Glucose contains several alcohol groups.

Alcohols come in three varieties, primary, secondary, and tertiary, just like alkyl halides. The key is how many carbons are bonded to the carbon connected to the OH. These are illustrated below.

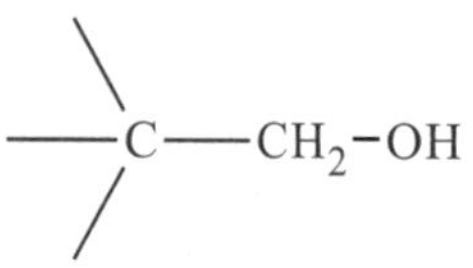

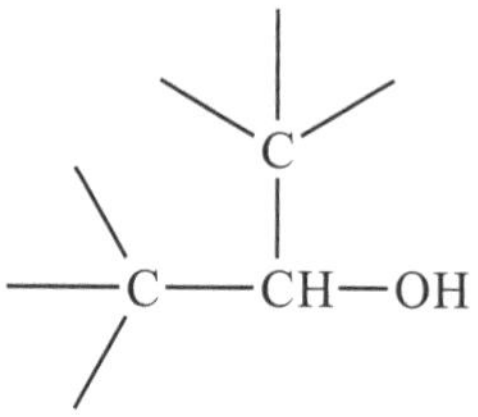

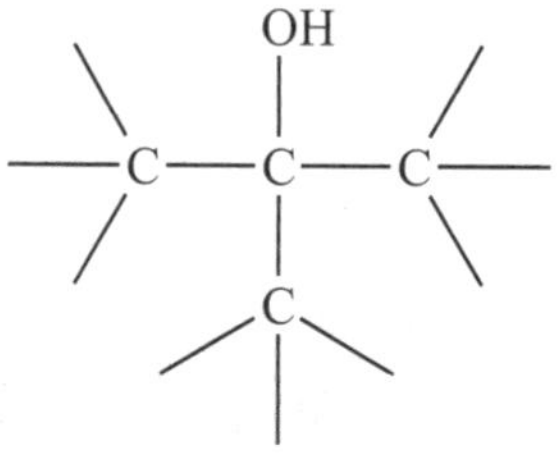

Primary (1°)   **Secondary (2°)**   **Tertiary (3°)**

A. IUPAC Nomenclature of Alcohols

Alcohols are higher in naming priority than alkenes, alkynes, amines, thiols, benzene rings, alkanes, ethers, and halides. The IUPAC ending for an alcohol is -*ol*.

OH on second carbon of pentane chain	This is a five-carbon chain, so the base part of the name is pent. There are no C=C's or C≡C's, so it is a pentanol. The OH group is on the second carbon from the right, so it is **pentan-2-ol**. Some texts use the name **2-pentanol**, but the most recent IUPAC rules put the 2 before the -ol suffix.
$H_3C-CH_2-CH_2-CH-C≡C-C-CH_3$ with OH and two Cl	This is an eight-carbon longest chain, so it is oct. There is an alkyne and an alcohol, so it is an octynol. The OH has the higher priority, so it is an oct-5-yn-4-ol. Numbering from the left puts the chlorines on carbon 7, so it is **7,7-dichlorooct-5-yn-4-ol**.
cyclohexane ring with OH (two representations) or	There are six carbons in the ring, with an alcohol attached. The correct name is **cyclohexanol**. No number is needed for the OH, which is defined as being at carbon 1, because it is the highest priority functional group.
seven-membered ring with HO and Br stereocenters	This is a seven-membered ring with an alcohol, so it is a cycloheptanol. There is a bromine on the third carbon from the OH, so it is a 3-bromocyclo-heptanol. Because both carbons 1 and 3 are stereocenters, we need to classify them as R or S. Therefore, the entirely correct name is **(1R,3S)-3-bromocycloheptanol**.
HO−CH$_2$−CH(OH)−CH$_3$	This is commonly called propylene glycol, and is used as an antifreeze. To name something with two alcohols, we have to put the di before the ol. Therefore, this is a propanediol. The e is retained to make the name sound better when pronounced. Adding the numbers 1 and 2 gives us **propane-1,2-diol**.

B. Physical Properties of Alcohols

Three physical properties of alcohols are fairly important in organic chemistry: water-solubility, boiling point, and acidity. We will deal with these below.

B.1. WATER-SOLUBILITY AND BOILING POINT

Because alcohols have the polar O–H bond, they can participate in hydrogen-bonding with water. This attractive force helps solubilize alcohol molecules in water. The alcohol molecules can also hydrogen-bond with themselves. These attractions are illustrated below with propan-1-ol.

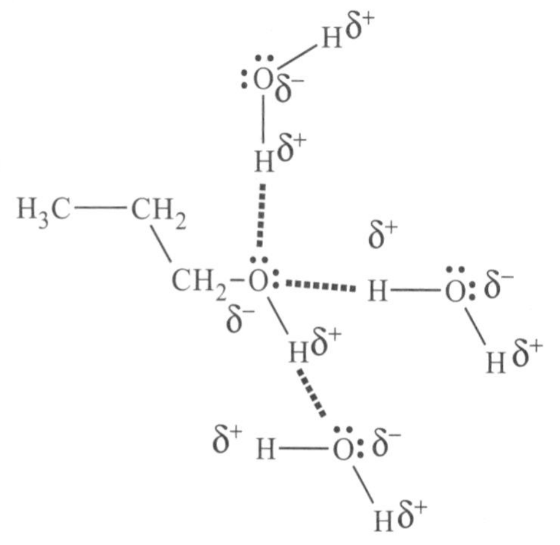 The partially positive hydrogens of water molecules are attracted to the partially negative oxygen of propan-1-ol. The partially negative oxygen of another water molecule is attracted to the partially positive hydrogen of propan-1-ol. Because C–H bonds are not very polar, water is not attracted significantly to them.	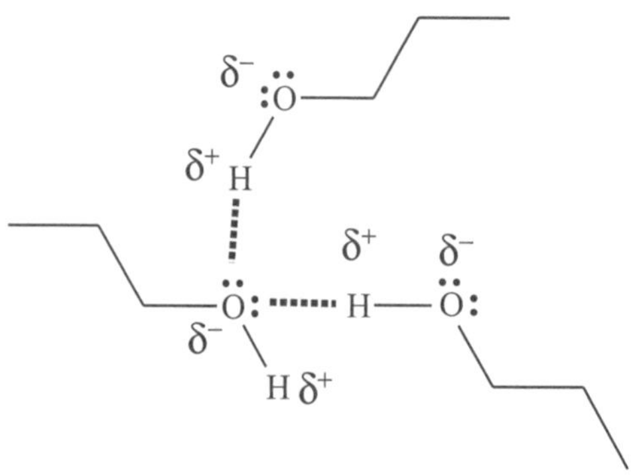 The partially positive hydrogens of the alcohol molecules are attracted to the partially negative oxygens of other alcohol molecules. This relatively strong intermolecular force must be overcome in order to separate alcohol molecules, so they can enter the vapor phase. Therefore, alcohols have significantly higher boiling points than similar molecules that cannot hydrogen-bond.

The following table lists the water-solubilities and boiling points for the first ten "straight-chain" alcohols.

Name	Structure	Water-Solubility (g/100 mL of water)	Boiling Point (°C)
Methanol	CH_3OH	Miscible	65
Ethanol	CH_3CH_2OH	Miscible	78
Propan-1-ol	$CH_3(CH_2)_2OH$	Miscible	97
Butan-1-ol	$CH_3(CH_2)_3OH$	7.4	118
Pentan-1-ol	$CH_3(CH_2)_4OH$	2.3	138
Hexan-1-ol	$CH_3(CH_2)_5OH$	0.60	156
Heptan-1-ol	$CH_3(CH_2)_6OH$	0.17	176
Octan-1-ol	$CH_3(CH_2)_7OH$	0.049	194
Nonan-1-ol	$CH_3(CH_2)_8OH$	0.014	214
Decan-1-ol	$CH_3(CH_2)_9OH$	0.0037	233

The first three alcohols are listed as miscible, which means soluble in all proportions. Some handbooks list them as infinite. After that, the longer the carbon chain gets, the lower the water-solubility becomes. This is because the nonpolar part is getting longer, and, hence, a smaller proportion of the molecule is interacting with the water. Once we get to decan-1-ol, the molecule is essentially insoluble in water, and very similar in water-solubility to an alkane.

As the length of the carbon chain increases, the boiling point increases by approximately 20°C. This is due to the increased van der Waals attractive forces between molecules.

Shape also influences water-solubility and boiling point, as shown by the table below.

Name	Three-Dimensional Shape from Chem3D®	Water-Solubility (g/100 mL of water)	Boiling Point (°C)
Pentan-1-ol		2.3	138
Pentan-2-ol		4.6	118
2-methyl butan-2-ol		12.4	103

In the drawings in this table, the white balls are hydrogens, the light gray balls are carbons, and the darkest ball is the oxygen. These molecules are constitutional isomers and differ in shape. The more linear pentan-1-ol is the least soluble, while the more spherically shaped 2-methylbutan-2-ol is the most water-soluble. The more spherical the molecule is, the less surface area it has. Therefore, the nonpolar surface area is minimized in 2-methylbutan-2-ol, so it is the most soluble.

Because 2-methylbutan-2-ol has less surface area, it has fewer van der Waals attractive forces between molecules of itself and has a lower boiling point than the other isomeric alcohols. Pentan-1-ol has the largest surface area and the highest boiling point.

B.2. ACIDITY

In general, alcohols are slightly less acidic than water, with pK_a's in the range of 15–19. This makes sense for two reasons. First, they are similar in structure to water, so their pK_a's should be similar. Second, as the size of the alkyl group increases, solvation of the conjugate base of the alcohol decreases, leading to a decrease in acid strength as the size of the alkyl group increases. See the table on the next page.

$$R-\overset{\cdot\cdot}{\underset{\cdot\cdot}{O}}-H \ + \ H-\overset{\cdot\cdot}{\underset{\cdot\cdot}{O}}-H \ \rightleftharpoons \ R-\overset{\cdot\cdot}{\underset{\cdot\cdot}{O}}{:}^{\ominus} \ + \ H-\overset{\oplus}{\underset{\underset{H}{|}}{\overset{\cdot\cdot}{O}}}-H$$

Alcohol	pK_a
CH_3OH, methanol	15.5
CH_3CH_2OH, ethanol	15.9
$(CH_3)_3COH$, 2-methylpropan-2-ol	18.0

C. Reactions That Cleave the O–H Bond

C.1. PREPARATION OF ESTERS

General Example:

$$R_1-OH \ + \ HO-\overset{\overset{O}{\|}}{C}-R_2 \ \underset{}{\overset{H^+ \text{ cat.}}{\rightleftharpoons}} \ R_1-O-\overset{\overset{O}{\|}}{C}-R_2 \ + \ H-O-H$$

Specific Examples:

$$H_3C-OH \ + \ HO-\overset{\overset{O}{\|}}{C}-\bigcirc \ \underset{\text{reflux 2 hours}}{\overset{H^+ \text{ cat.}}{\rightleftharpoons}} \ H_3C-O-\overset{\overset{O}{\|}}{C}-\bigcirc \ + \ H-O-H$$

excess   75%

$$H_3C-\overset{\overset{CH_3}{|}}{\underset{\underset{CH_3}{|}}{C}}-OH \ + \ Cl-\overset{\overset{O}{\|}}{C}-CH_3 \ \underset{\text{reflux 2 hours}}{\overset{R_3N}{\longrightarrow}} \ H_3C-\overset{\overset{CH_3}{|}}{\underset{\underset{CH_3}{|}}{C}}-O-\overset{\overset{O}{\|}}{C}-CH_3$$

68%

Esters can be made from alcohols by treatment with a carboxylic acid and an acid catalyst. This reaction is an equilibrium reaction, so an excess of one reactant is usually needed to obtain a good yield of the ester product. Esters can also be prepared by treating an alcohol with an acid chloride or an acid anhydride. In this case, the reaction is not an equilibrium reaction. The mechanisms of these reactions are discussed in Chapter 13.

C.2. PREPARATIONS OF SULFONATE ESTERS

General Example:

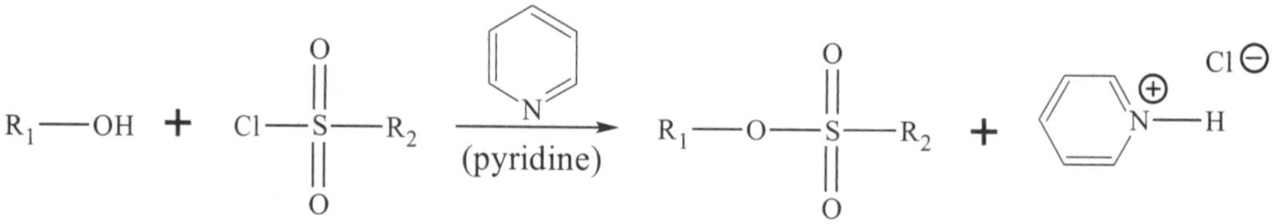

Specific Examples:

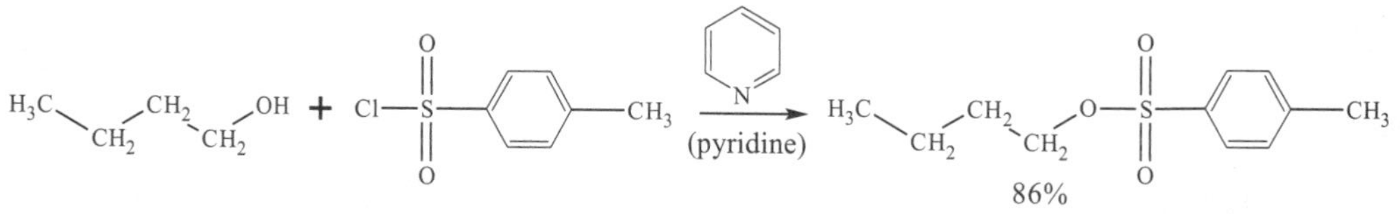

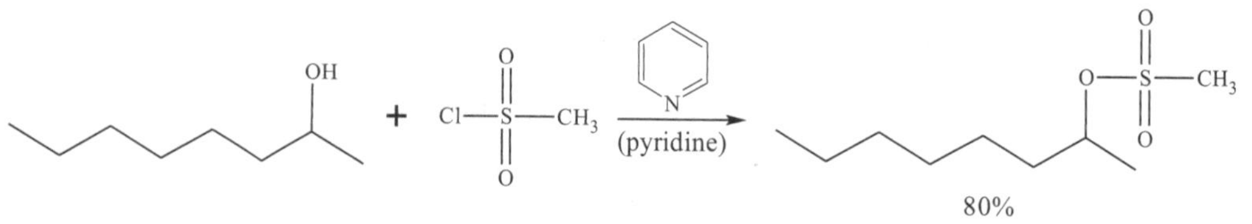

86%

80%

This is the reaction to convert alcohols into sulfonates, which are good leaving groups. The reaction mechanism is similar to that of the reaction of alcohols with acid chlorides. The pyridine serves as a solvent and a base, and reacts with the HCl produced as a by–product. The top example gives butyl *p*-toluenesulfonate (sometimes called butyl tosylate) as the product. Tosylates are often crystalline and relatively easy to purify. The bottom reaction produces (1-methylheptyl) methanesulfonate (or 1-methylheptyl mesylate).

C.3. REACTIONS WITH VERY STRONG BASES

General Reaction:

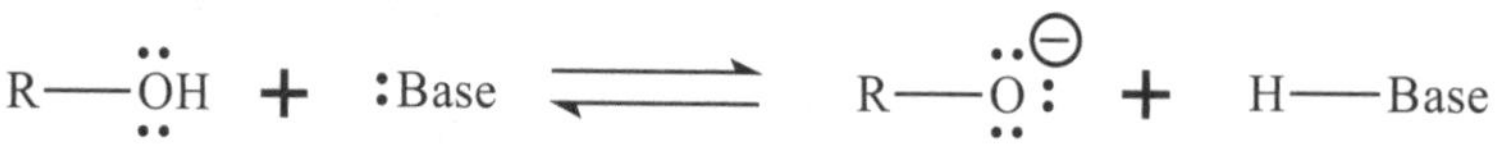

This is a general way to form alkoxides, which are strong nucleophiles and bases, from alcohols, which are weak nucleophiles and bases. Hydroxide is generally not a strong enough base to completely remove the hydrogen from an alcohol. Stronger bases such as sodium amide ($NaNH_2$) and sodium hydride (NaH) are often used to completely remove the alcohol hydrogen. These produce ammonia and hydrogen gas as by–products, respectively, along with the sodium alkoxide.

C.4. REACTIONS WITH ALKALI METALS (LITHIUM, SODIUM, AND POTASSIUM)

General Reaction:

$$2 \ R\!-\!\overset{..}{\underset{..}{O}}H \ + \ 2\,M \ \longrightarrow \ 2 \ R\!-\!\overset{..}{\underset{..}{O}}\!\!:^{\ominus} \ M^{\oplus} \ + \ H_2$$

M = Li, Na, or K

This reaction is another way to produce alkoxides from alcohols. The metal is added slowly to a large excess of the alcohol, and the hydrogen produced is vented away. Sodium and potassium react very exothermically, while lithium reacts more slowly. This gives a solution of the alkoxide in the excess alcohol as a solvent, which can be used for whatever reaction is desired.

D. Reactions That Cleave the C–O Bond

D.1. DEHYDRATION OF ALCOHOLS

General Reaction:

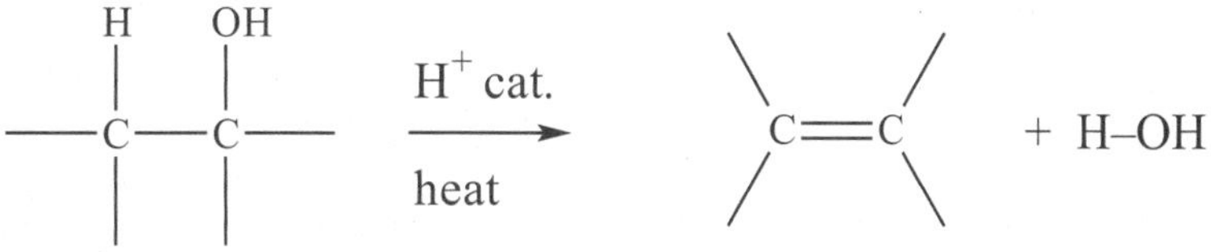

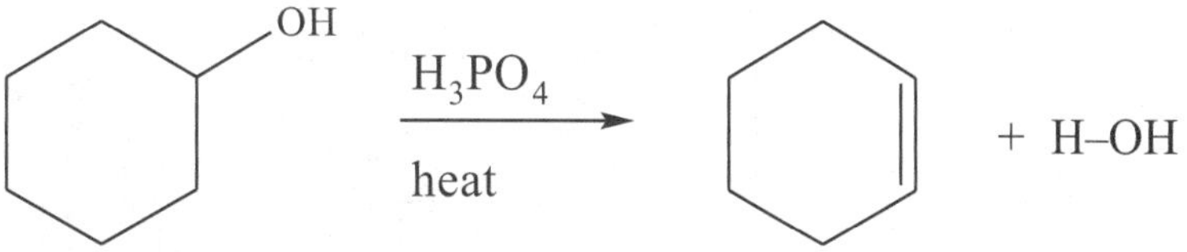

This reaction is an E1 elimination of water, and is discussed in detail in Chapter 4, Section I.

D.2. REACTIONS OF ALCOHOLS WITH CONCENTRATED H–X ACIDS

General Reaction:

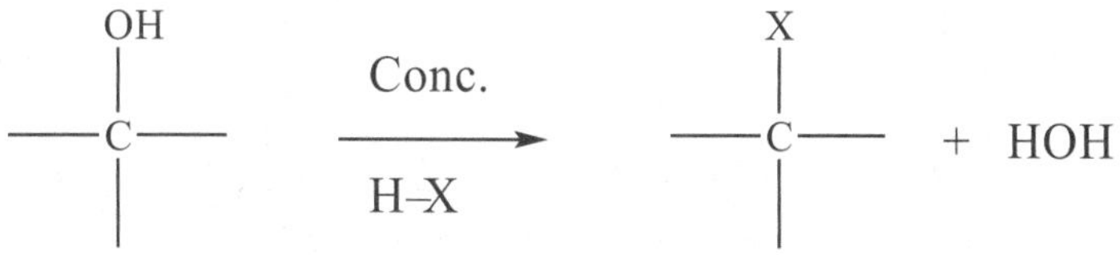

Specific Examples:

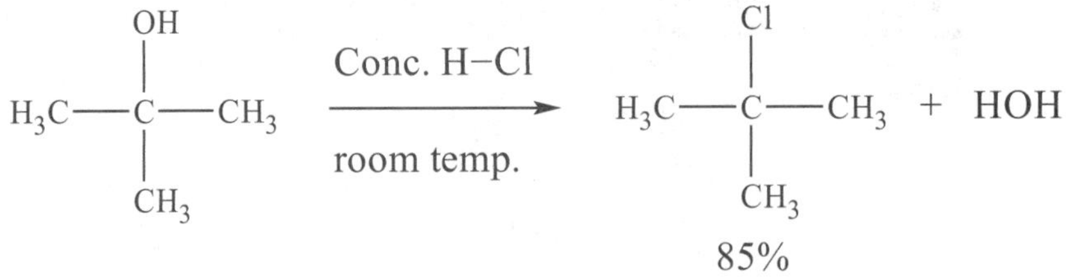

$$H_3C{-}(CH_2)_6{-}CH_2{-}OH \xrightarrow[\text{reflux 5 hours}]{\text{Conc. H–Br}} H_3C{-}(CH_2)_6{-}CH_2{-}Br + HOH$$

91%

This relatively simple-looking reaction has been much studied and is a bit more complicated than it looks. Concentrated H–F doesn't react well with alcohols, so we won't consider it. If you can identify the type of alcohol (primary, secondary, or tertiary), then you can make some generalizations about the type of mechanism by which the reaction proceeds.

D.2.a. TERTIARY ALCOHOLS: S_N1 SUBSTITUTION

The S_N1 mechanism involves three steps: protonation of the alcohol, loss of water to form a carbocation, and attack of the halide ion on the carbocation. This is illustrated below.

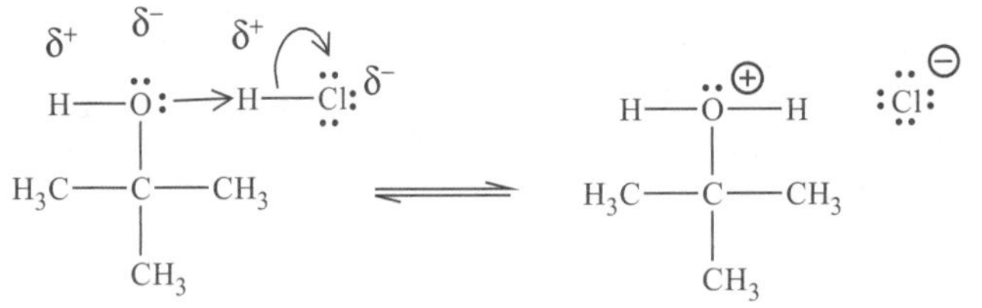	The slightly negative oxygen of the alcohol is attracted to the slightly positive H of H–Cl. The chlorine pulls the electrons in the H–Cl bond toward itself.
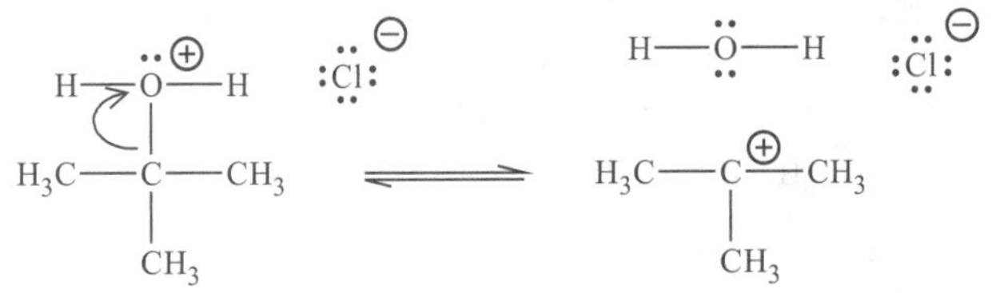	The C–O bond breaks to form water and the tertiary carbocation. This is the slow step in the mechanism.
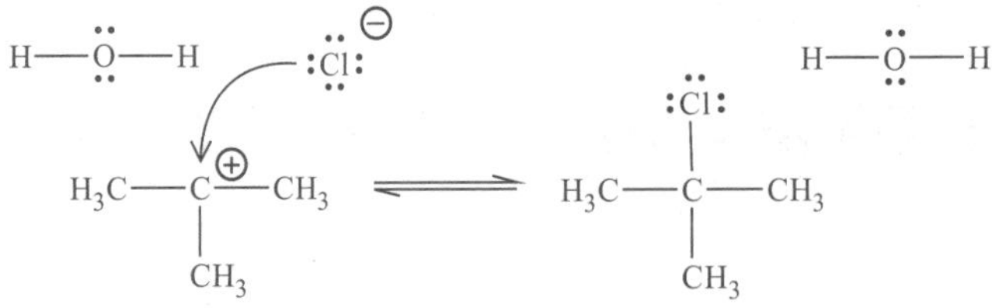	Chloride ion attacks the carbocation to form the product.

Generally, a small amount of E1 elimination also occurs, but it is usually minor because halide ions are very weak bases.

D.2.b. PRIMARY ALCOHOLS: S_N2 SUBSTITUTION

Because simple primary carbocations are much less stable than tertiary carbocations, they are virtually never formed under the normal reaction conditions used in these reactions. Therefore, primary alcohols react via an S_N2 mechanism (discussed in Chapter 6), which involves protonation of the alcohol, followed by backside attack of halide to form the product. This is shown below.

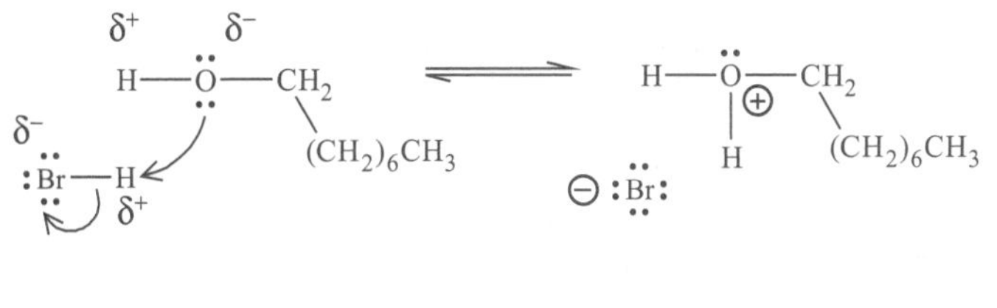	The slightly negative oxygen of the alcohol is attracted to the slightly positive H of H–Br. The bromine pulls the electrons in the H–Br bond toward itself.
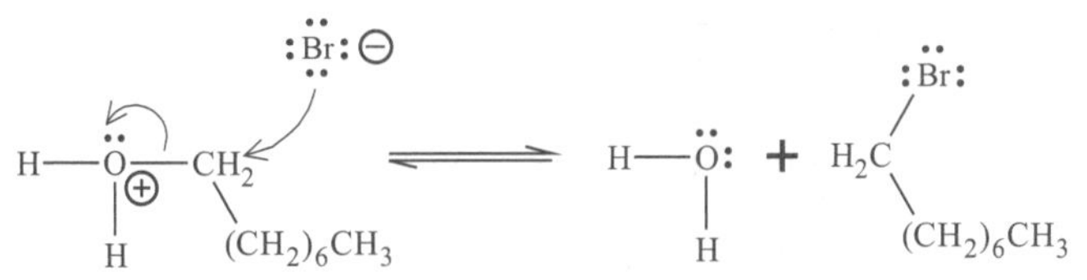	The bromide attacks the backside of the carbon and pushes the protonated alcohol group out to form water and the alkyl halide.

Concentrated HCl reacts very slowly with primary alcohols because chloride is a very weak nucleophile. To make HCl work, a catalyst such as $ZnCl_2$ must be used. The $ZnCl_2$ complexes the OH group to make it a better leaving group, so even a weak nucleophile such as chloride can displace it.

D.2.c. SECONDARY ALCOHOLS: S_N2 AND S_N1 SUBSTITUTION

Secondary alcohols react with hydrogen halides via a mixture of S_N2 and S_N1 substitution. Carbocation rearrangements can occur; this means that more than one product can be formed. Some examples of this are shown below.

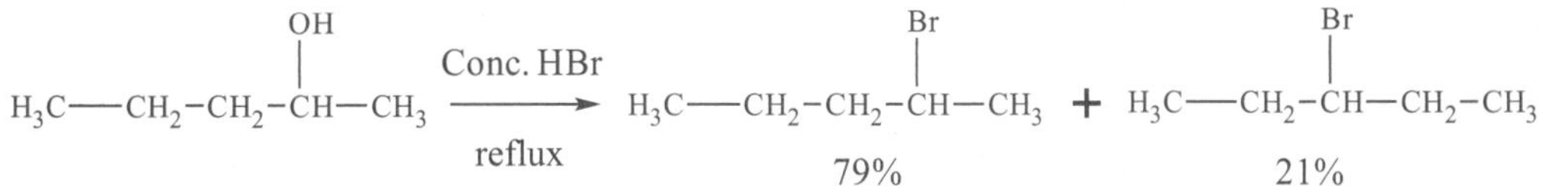

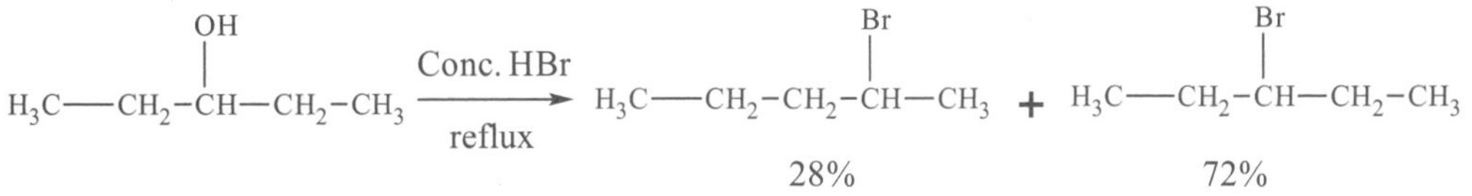

The fact that there is rearrangement occurring in both reactions shows that some S_N1 reaction is occurring. It is probably not all S_N1, because the major product in both cases is the unrearranged product. If the mechanism were purely S_N1, with an equilibrium distribution of secondary cations, then the ratio of 2-bromopentane to 3-bromopentane should be 2:1 for both reactions. I have left this as a problem for you to work out. As an aside, elimination to form an alkene, followed by addition of H–X to form the alkyl halide, has been shown not to occur.

Here is another example. How do you account for the two products?

~50%   ~50%

Let's reason it out. The first product is not rearranged, while the second product is rearranged. The first product could be formed by either an S_N1 mechanism or an S_N2 mechanism, while the second product most likely formed by an S_N1 mechanism. See below.

	H–Cl protonates the alcohol.
	To form the substituton product, chloride attacks the backside of the carbon and pushes the proto-nated alcohol group out to form water and the alkyl halide.
	In the S_N1 mechanism, water leaves to form the secondary car-bocation.
	The unrearranged product can also be formed when chloride attacks the carbocation.
	The secondary carbocation can rearrange to a more stable tertiary carbocation by a hydride shift.

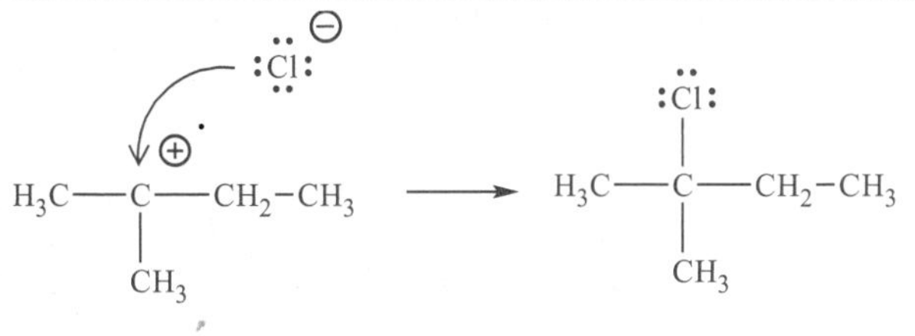

Chloride attacks the tertiary carbocation to form the rearranged product.

The propensity of rearrangements of secondary cations prompted chemists to try to find reagents that can do this reaction without rearrangement.

D.3. REACTIONS OF ALCOHOLS WITH THIONYL CHLORIDE, PHOSPHORUS TRIBROMIDE, AND SIMILAR REAGENTS

Specific Examples:

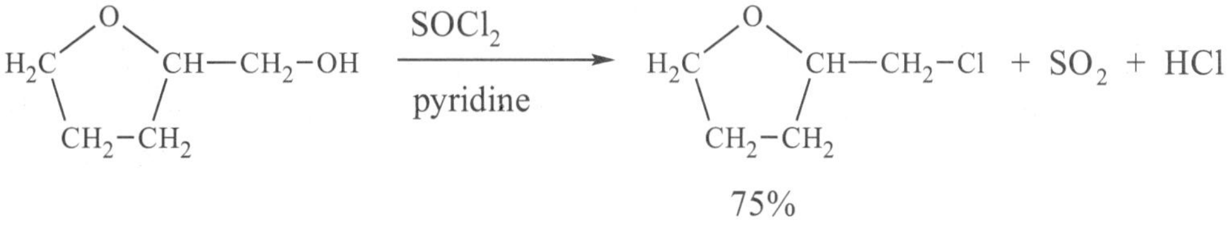

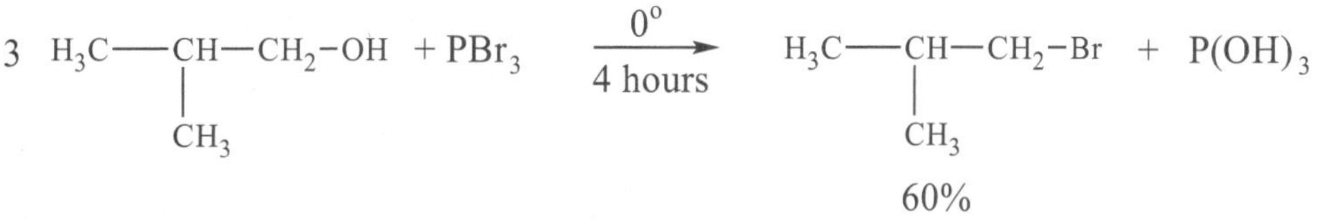

For primary alcohols, thionyl chloride ($SOCl_2$) and phosphorus tribromide (PBr_3) are good reagents for production of primary alkyl halides. They have the advantage of not being strong acids, although thionyl chloride produces HCl as a by–product. By running the reaction in pyridine as a solvent, the HCl is neutralized as it is formed. A number of texts claim or infer that PBr_3 gives alkyl halides without rearrangement, but there are numerous examples of rearrangements involving PBr_3 with alcohols. Two are shown below.

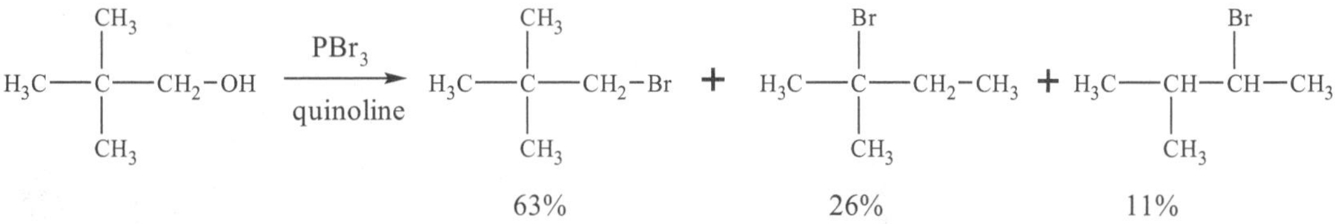

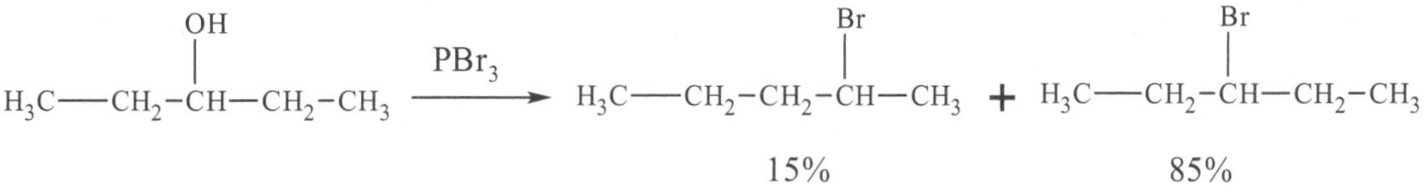

Thionyl chloride does not give rearranged products, in general. If the reaction of S-(+)-octan-2-ol with thionyl chloride were run in a nonpolar solvent, or no solvent, the major product would be R-(–)-2-chlorooctane. The stereogenic center had been inverted, as shown below.

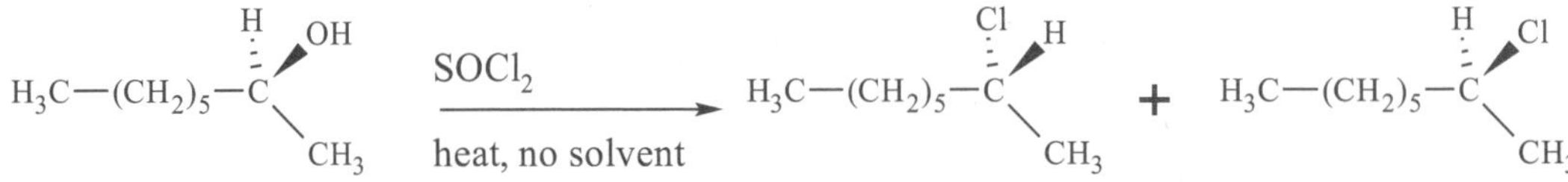

S-(+)-Octan-2-ol

R-(–)-2-Chlorooctane   S-(+)-2-Chlorooctane

The reaction appears to take place by the following mechanism.

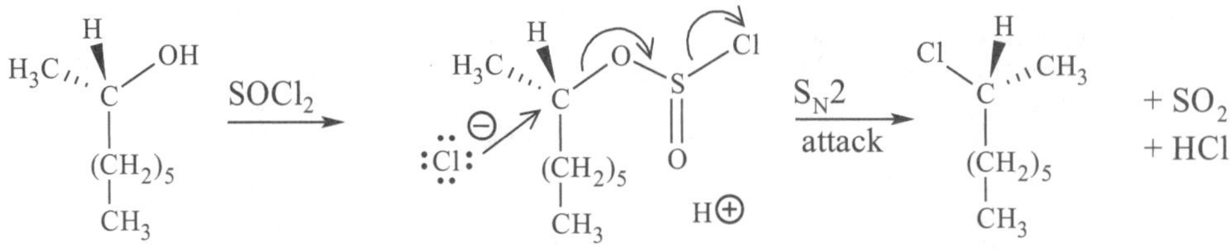

The "chlorosulfite"   R-(–)-2-Chlorooctane

Thionyl chloride initially reacts with the alcohol to form the chlorosulfite intermediate. Chloride ion does an S_N2 attack on the backside of the carbon to produce the alkyl halide with the inverted configuration.

A number of other reagents have been developed to overcome the problem of rearrangement. Among these are the triphenylphosphine-halogen adducts. When primary and secondary alcohols are treated with these reagents, only the corresponding alkyl halides are formed, with no traces of isomeric products. See the examples below.

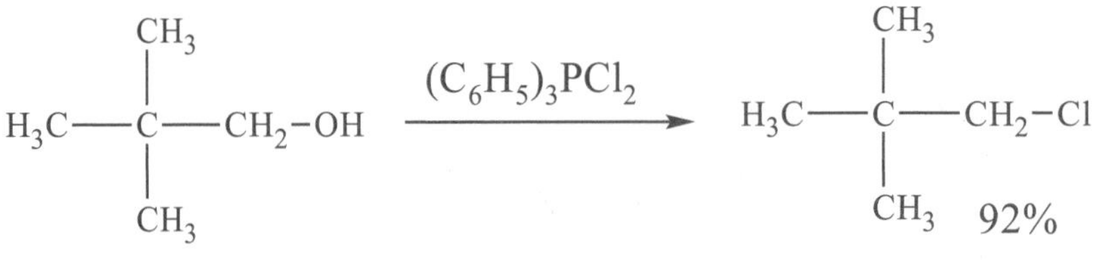

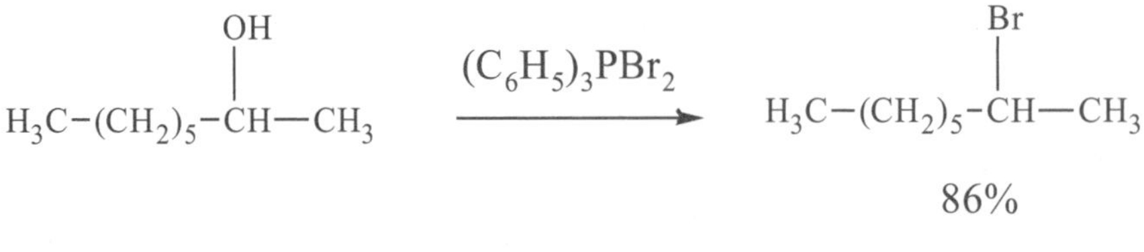

E. Oxidation of Alcohols

You can recognize oxidation of an organic compound by comparing the product with the starting material. If the product has more oxygens and/or fewer hydrogens, then an oxidation has occurred. Oxidation of an alcohol generally involves removal of the hydrogen on the oxygen and a hydrogen on the carbon bonded to the oxygen. This is shown below. The symbol [O] means oxidation, or a generic oxidizing agent.

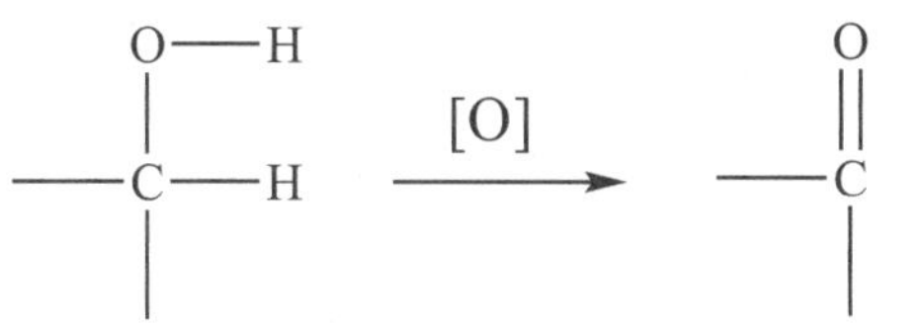

In this generic oxidation reaction, two hydrogens have been removed. The product has fewer hydrogens than the starting material, so an oxidation has taken place.

Because tertiary alcohols do not have a hydrogen on the carbon bonded to the oxygen, they are generally not oxidized by normal oxidizing agents. So if you are asked what the product of oxidation of a tertiary alcohol is, you should write, "No reaction."

Secondary alcohols are oxidized to ketones. Common reagents used are a variety of chromium oxides in acidic solution, such as potassium dichromate ($K_2Cr_2O_7$) in sulfuric acid and water, or chromium trioxide (CrO_3) in sulfuric acid and acetone, which is commonly called **Jones' Reagent**. Bleach, which is a solution of sodium hypochlorite (NaOCl) in water, is a mild oxidizing agent. Some examples are shown below.

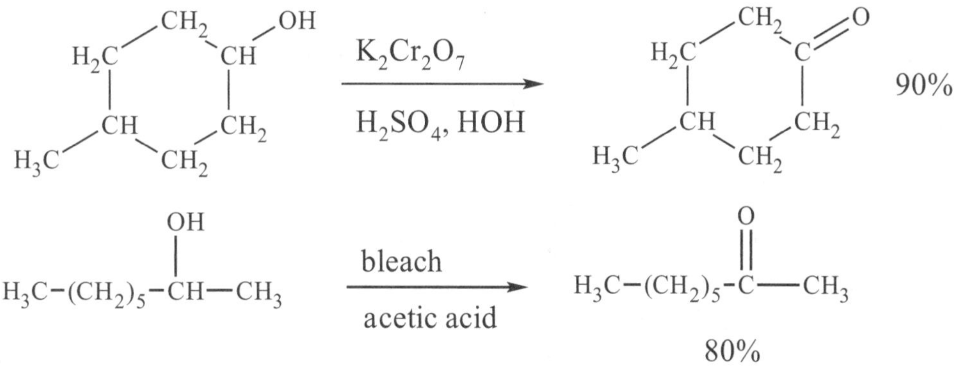

Chromium reagents are toxic, and many chromium(VI) salts are carcinogenic, so these need to be used cautiously. The by–product Cr(III) salts can be difficult to work with and can make product purification more difficult. Bleach (NaOCl) has the advantage that the by–product is NaCl, but the alcohol must be somewhat soluble in water for the reaction to work well.

In the presence of water, primary alcohols can be oxidized in two stages, as shown below.

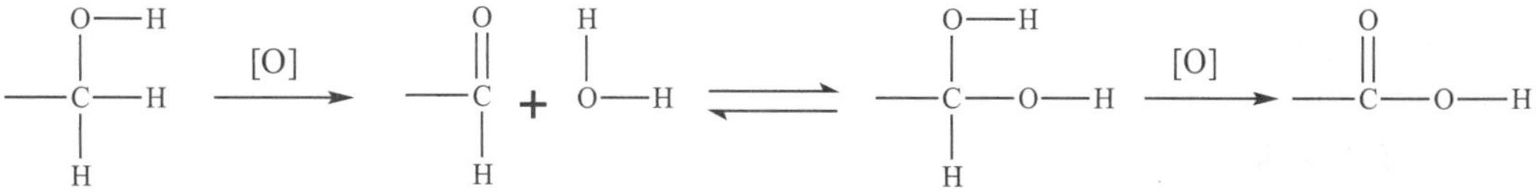

The first oxidation forms an aldehyde. Water can add to the aldehyde to form a hydrate (we will cover this reaction in Chapter 12). The hydrate can then be oxidized to form a carboxylic acid.

All of the oxidizing agents we have mentioned so far contain water, so all of them will oxidize a primary alcohol to a carboxylic acid. Attempts to limit the amount of oxidizing agent to stop the oxidation at the aldehyde stage are usually unsuccessful. However, if you had an oxidizing agent that did not contain water, the oxidation could be stopped at the aldehyde stage. Once it was discovered that CrO_3 was soluble in pyridine, it opened the door for development of a variety of CrO_3-pyridine derived reagents that could oxidize primary alcohols to aldehydes. These include CrO_3-pyridine in CH_2Cl_2, pyridinium chlorochromate (PCC), and pyridinium dichromate (PDC). Some examples are shown below.

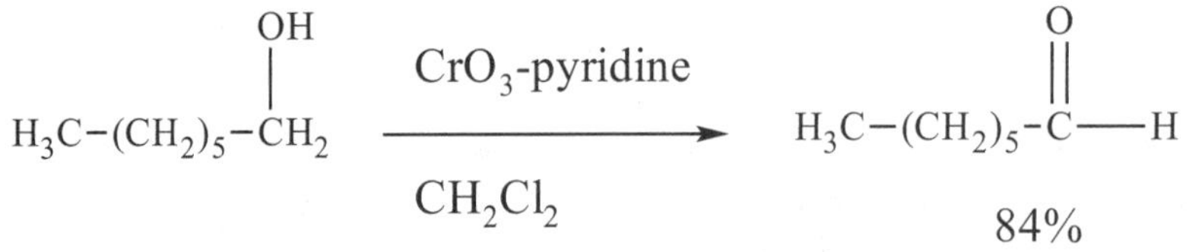

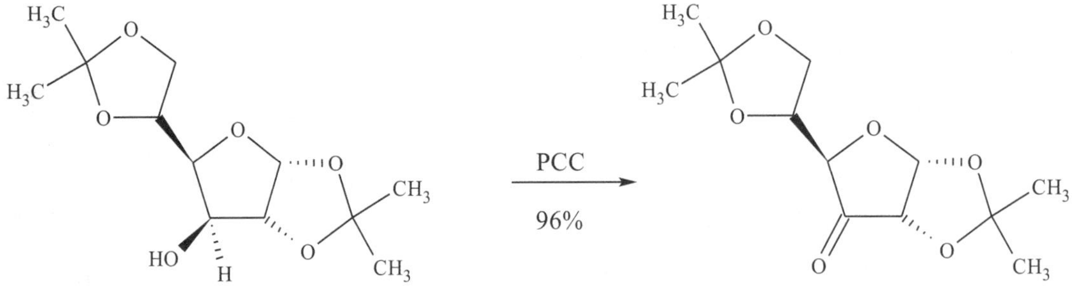

The last example shows that these reagents also oxidize secondary alcohols to ketones in good yields. For the purposes of an exam, my students use PCC to oxidize primary alcohols to aldehydes and secondary alcohols to ketones. Make sure your instructor will give you credit if you use the acronym PCC: I tell my students that they can use it, but if they give me the wrong acronym, such as PPC, I will count it totally wrong.

F. Preparations of Alcohols

We can divide the reactions commonly used to make alcohols into two general classes: those that do not involve making a new carbon–carbon bond and those that do involve making a new carbon–carbon bond. We have already encountered a few of the former type, so we will begin with those.

F.1. REACTIONS THAT DO NOT INVOLVE MAKING A NEW CARBON–CARBON BOND

F.1.a. HYDRATION OF AN ALKENE

General Reaction:

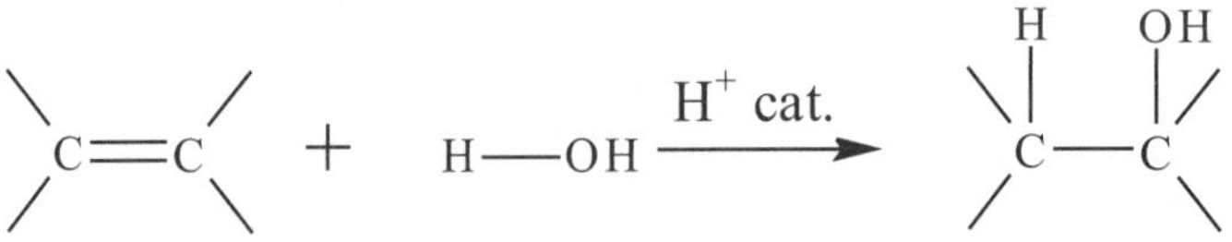

Specific Example:

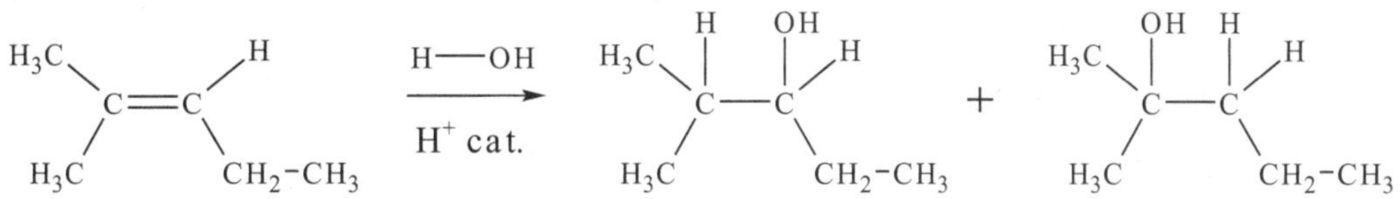

None observed Only observed product

This reaction was discussed in detail in Chapter 4. Water adds to the C=C in the Markovnikov fashion.

F.1.b. OXYMERCURATION

General Reaction:

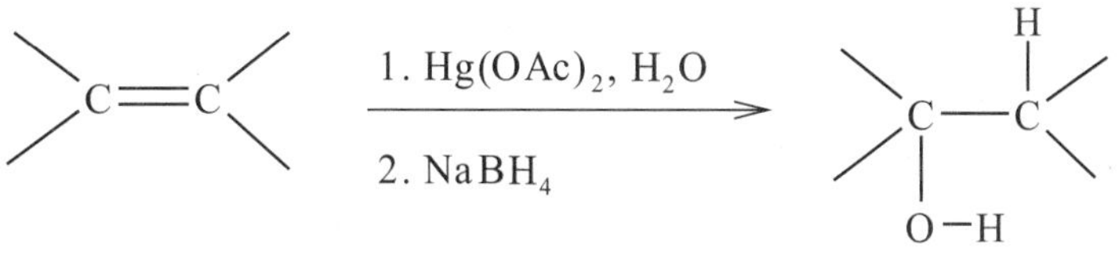

Specific Example:

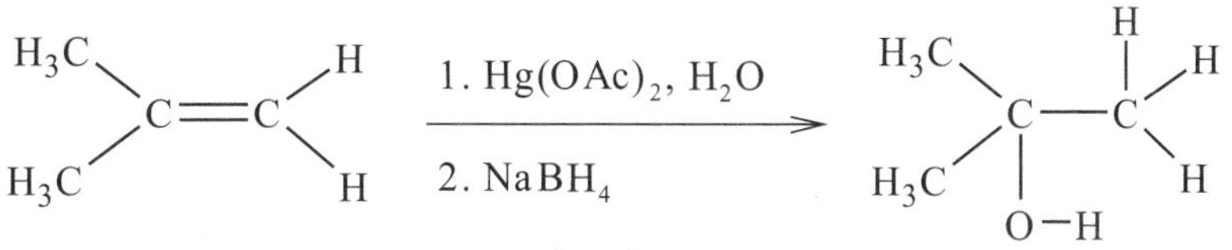

The oxymercuration/demercuration procedure is a two-step sequence that adds water to a C=C in a Markovnikov fashion. The yields are often higher than direct acid-catalyzed addition of water. Also, this procedure is less prone to carbocation rearrangements. This reaction is discussed in detail in Chapter 4.

F.1.c. HYDROBORATION/OXIDATION

General Reaction:

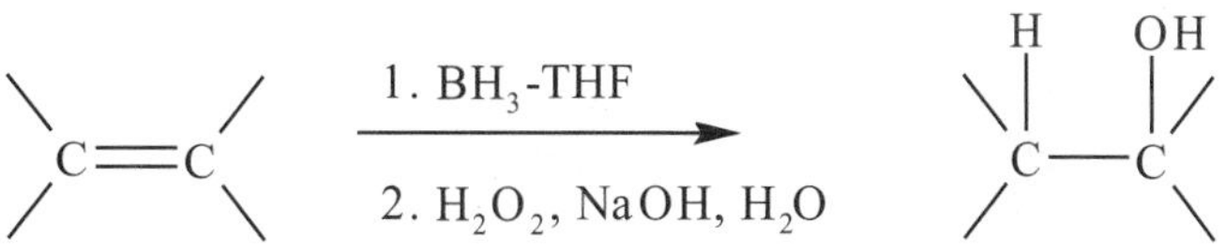

Specific Example:

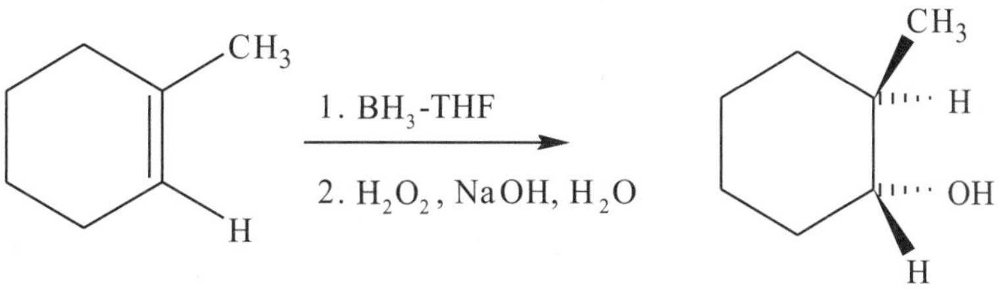

This two-step procedure results in a net "anti-Markovnikov" addition of water to a C=C: The H ends up bonded to the more substituted C or the C=C, and the OH ends up bonded to the least substituted carbon. In addition, the H and the OH are added to the same side of the C=C, which is referred to as **syn addition**, as illustrated in the reaction of 1-methylcyclohexene shown above. This reaction is discussed in detail in Chapter 4.

F.1.d. REDUCTIONS OF ALDEHYDES AND KETONES

General Reaction:

[H] stands for reduction, or a generic reducing agent.

Specific Examples:

Although hydrogen and a catalyst such as Pd or Pt does reduce the C=O of an aldehyde or a ketone, the reaction is somewhat slow. A C=C is reduced faster under these conditions than the C=O. In the 1940s and 1950s, two reducing agents that reduced aldehydes and ketones rapidly were developed. These reagents, sodium borohydride (NaBH₄) and lithium aluminum hydride (LiAlH₄), revolutionized the reductions of various types of C=O compounds, and we will discuss these reagents in more detail in Chapters 12 and 13.

Both of these reagents have the four hydrogens bonded to the central B or Al, with the Na or Li serving mainly to balance the charge. These reagents are shown below.

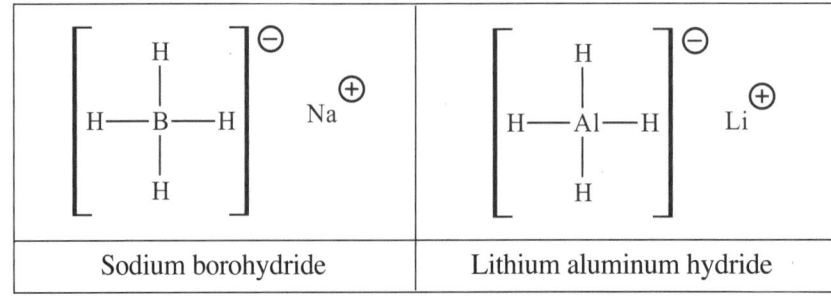

Hydrogen is slightly more electronegative than boron and somewhat more electronegative than aluminum. Therefore, each of these reagents function as if they are sources of H(–), which is hydride. Because there is a greater electronegativity difference between H and Al, lithium aluminum hydride functions as a more powerful hydride donor than sodium borohydride. As we will discover in Chapter 13, $LiAlH_4$ reduces all kinds of C=O, C≡N, and C=N containing compounds. Sodium borohydride is mainly used to reduce aldehydes and ketones, along with certain C=N compounds. Another difference is their reactivity toward water and alcohols. Sodium borohydride is soluble in water and smaller alcohols and only reacts with them very slowly to produce hydrogen gas. If there is some acid present, sodium borohydride reacts rapidly with water to produce four moles of hydrogen per mole of borohydride.

The simplified version of the mechanism of the reduction of a C=O with $LiAlH_4$ is shown below. The same mechanism applies to $NaBH_4$.

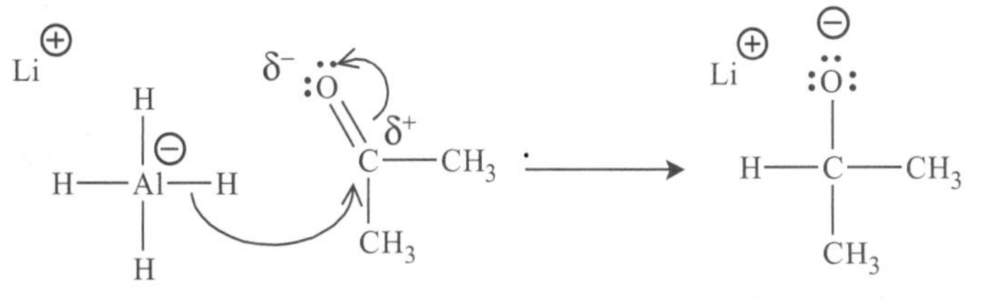	The hydrogen with its pair of electrons attacks the partially positive carbon of the ketone. One pair of electrons from the C=O moves up on to the O, to form an alkoxide.
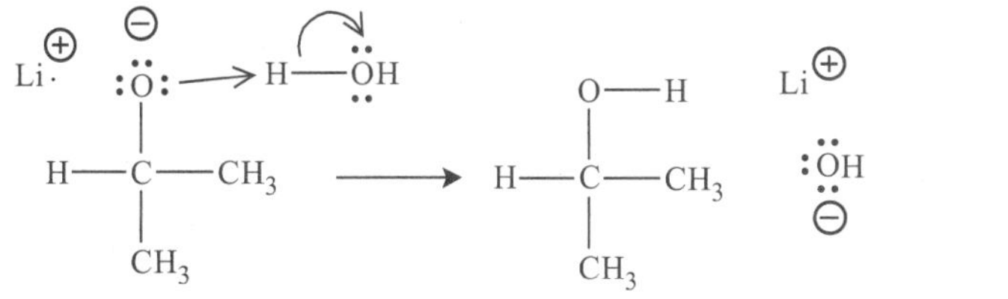	Because an alkoxide is a strong base, it can pull a hydrogen off water. This forms the alcohol product, plus hydroxide ion.

I mentioned that this is the "simplified" mechanism. It is actually somewhat more complex. Each mole of borohydride or aluminum hydride can reduce four moles of aldehyde or ketone. See below.

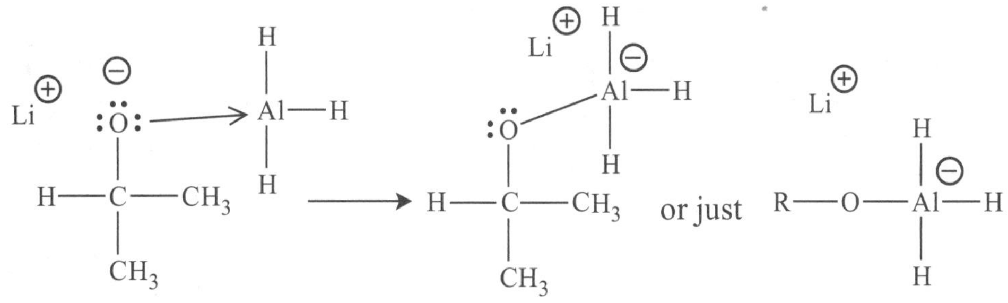

After the aluminum hydride ion donated a hydride, AlH_3 was left over. It is a Lewis acid, so the alkoxide could react with it as shown. This forms an alkoxy aluminum hydride ion.

$$
\underset{\displaystyle \text{Li}^{\oplus}}{} \quad
\begin{array}{c}
 \quad \text{H} \\
 \quad | \quad \ominus \\
\text{R} - \text{O} - \text{Al} - \text{H} \\
 \quad | \\
 \quad \text{H}
\end{array}
$$

1. React with a ketone and form an alkoxide.

2. Alkoxide reacts with aluminum by-product.

3. Repeat 1 and 2 two more times.

$$
\begin{array}{c}
 \quad \text{O} - \text{R} \quad \text{Li}^{\oplus} \\
 \quad | \quad \ominus \\
\text{R} - \text{O} - \text{Al} - \text{O} - \text{R} \\
 \quad | \\
 \quad \text{O} - \text{R}
\end{array}
$$

Tetraalkoxy aluminum ion

The alkoxy aluminum hydride ion can react with three more ketone molecules, until a tetraalkoxy aluminum ion is formed.

$$
\begin{array}{c}
 \quad \text{O} - \text{R} \quad \text{Li}^{\oplus} \\
 \quad | \quad \ominus \\
\text{R} - \text{O} - \text{Al} - \text{O} - \text{R} \\
 \quad | \\
 \quad \text{O} - \text{R}
\end{array}
\quad \xrightarrow{\text{4 HOH}} \quad 4\ \text{ROH} + \text{LiOH} + \text{Al(OH)}_3
$$

Tetraalkoxy aluminum ion

The tetraalkoxy aluminum ion reacts with water to form lithium and aluminum hydroxides, plus the four alcohol molecules.

Normally I ask my students for the simplified mechanism only. Ask your instructor what you need to know.

F.2. REACTIONS THAT DO INVOLVE MAKING A NEW CARBON-CARBON BOND

We noted earlier that the C=O of an aldehyde or ketone is polar, that is, the carbon is partially positive and the oxygen is partially negative. Therefore, if we had a reagent where the carbon was negative (a carbanion, a strong nucleophile), it would be attracted to the partially positive carbon of the C=O, and would form a new carbon–carbon bond. I am arbitrarily using lithium as the metal counterion: It could be sodium, potassium, or a variety of other metals. See the next page.

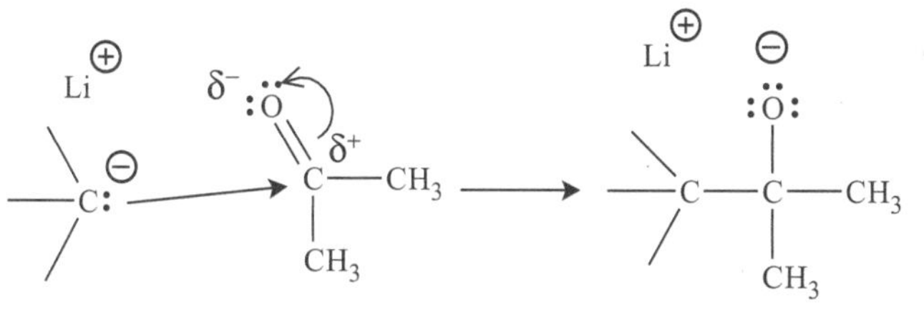	The lithium salt of the carbanion attacks the carbon of the C=O, pushing an electron pair up onto the oxygen. This forms the lithium salt of an alkoxide.
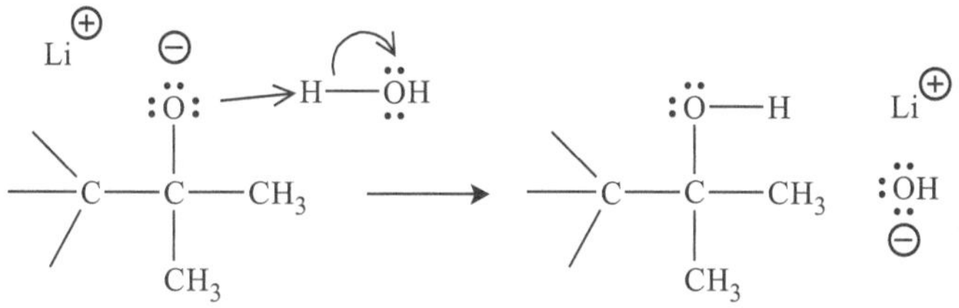	In a second step, the alkoxide, being a strong base, deprotonates water. This forms the alcohol product, plus hydroxide ion.

Carbanion-type compounds that are associated with a metal counterion are called organometallic compounds. Virtually every metal has been investigated, and varying the metal significantly changes the reactivity of the carbanion. In this chapter, we will consider only three kinds of organometallic compounds, postponing the others for later chapters.

F.2.a. SALTS OF ACETYLENE AND 1-ALKYNES

In Chapter 5, we discussed how the terminal hydrogen of acetylene or a 1-alkyne could be pulled off with a very strong base and how the resulting carbanion could react in an S_N2 fashion with a primary alkyl halide. This is shown below.

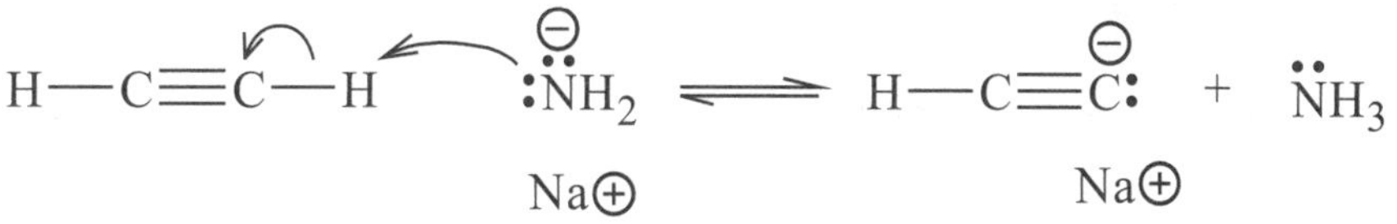

Sodium amide, $NaNH_2$, is a very strong base. The amide anion pulls the terminal hydrogen off acetylene to form sodium acetylide and ammonia.

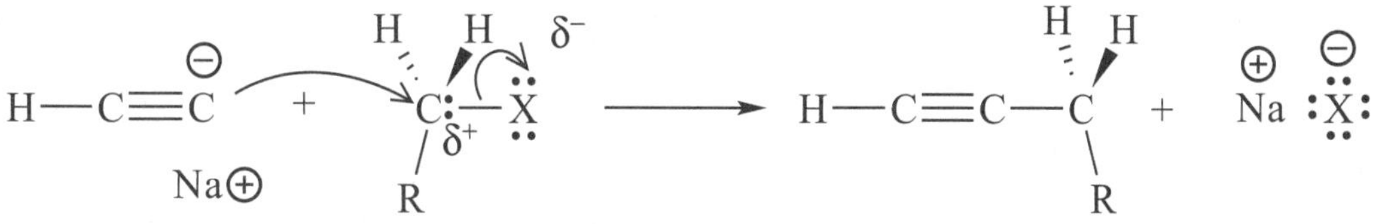

The acetylide anion does an S_N2 attack on the backside of the partially positive carbon of a primary alkyl halide to form a larger alkyne.

Because an aldehyde or a ketone also has a partially positive carbon, the acetylide ion can attack it and form a new carbon–carbon bond. See below.

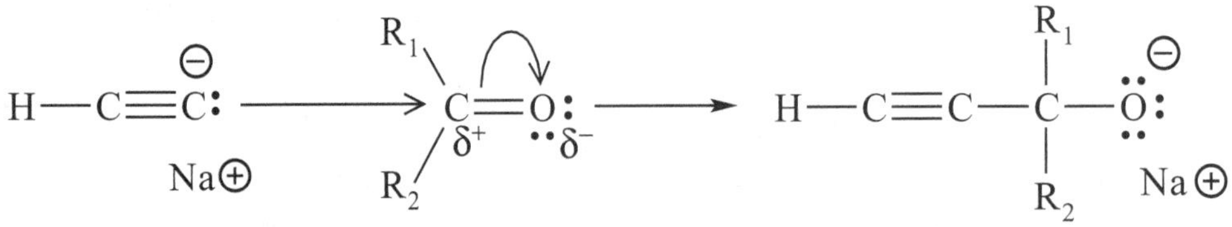

The acetylide ion attacks the partially positive carbon of the C=O and pushes an electron pair onto the oxygen. This forms an alkoxide. Notice that we have made a new carbon–carbon bond in the process.

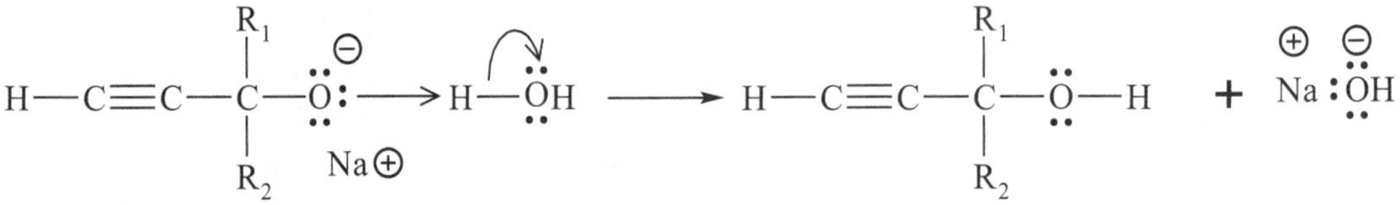

In a second step, the alkoxide, being a strong base, pulls a hydrogen off water. This forms the alcohol product, plus hydroxide ion.

A couple of examples are shown below.

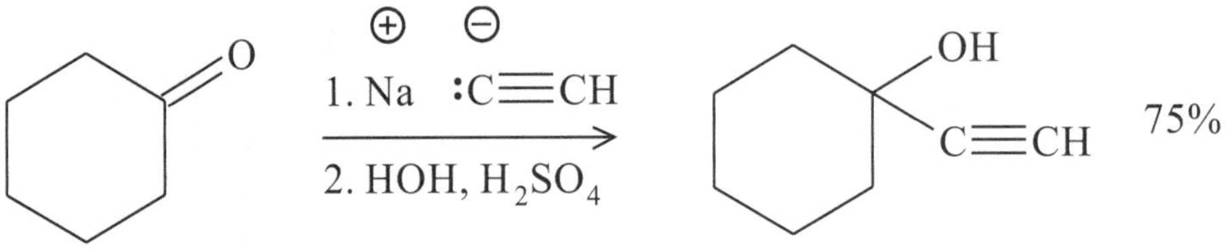

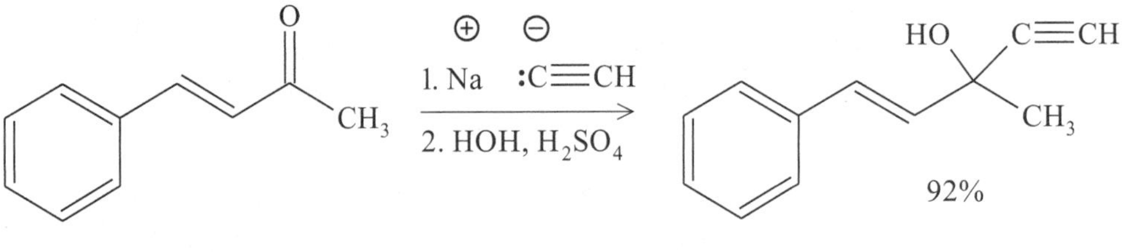

F.2.b. GRIGNARD REAGENTS AND ORGANOLITHIUM REAGENTS

When an alkyl halide is treated with magnesium in ether, the magnesium inserts between the carbon and the halogen to form an organomagnesium halide, which is called a Grignard reagent. This process was developed by a French chemist, Victor Grignard. Grignard's development of the applications of Grignard reagents in organic synthesis resulted in his being awarded the Nobel Prize in chemistry in 1912. The formation of the Grignard reagent is shown below.

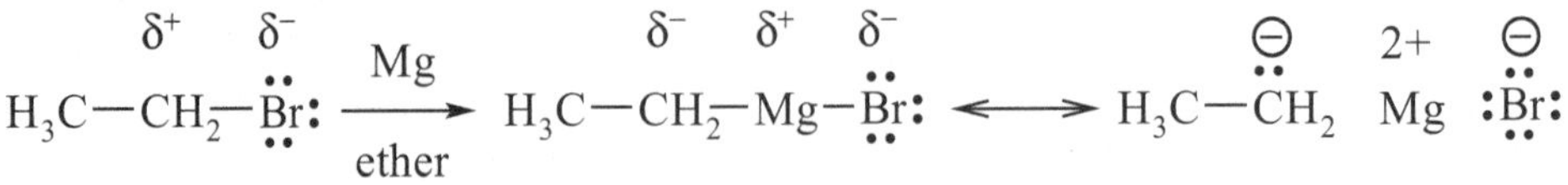

The ether solvent helps dissolve the Grignard reagent because the ether oxygens are coordinating to the magnesium, as shown below.

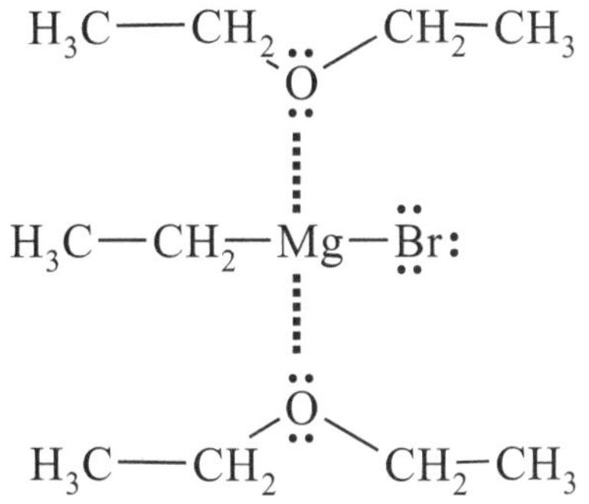

$$H_3C—CH_2 \ddot{O} CH_2—CH_3$$

$$H_3C—CH_2—Mg—\ddot{B}r:$$

$$H_3C—CH_2 \ddot{O} CH_2—CH_3$$

By choosing the appropriate aldehyde or ketone, you can make a primary, secondary, or tertiary alcohol. See the examples below.

Type of Alcohol	C=O Compound	Example
Primary	**H₂C=O (formaldehyde)**	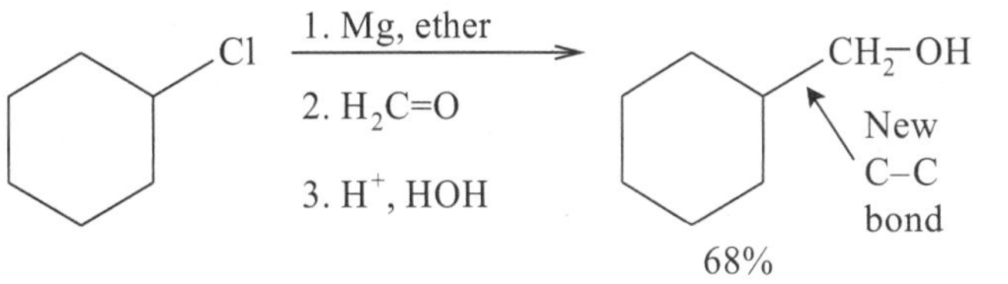 1. Mg, ether 2. H₂C=O 3. H⁺, HOH → CH₂—OH (New C–C bond) 68%
Secondary	**Any other aldehyde**	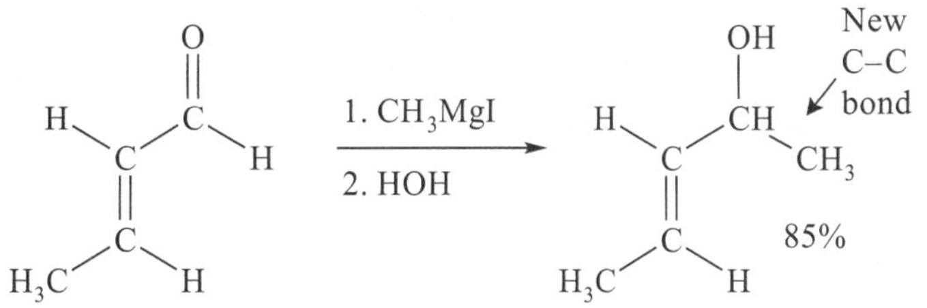 1. CH₃MgI 2. HOH → OH (New C–C bond) CH, CH₃ 85%
Tertiary	**Ketone**	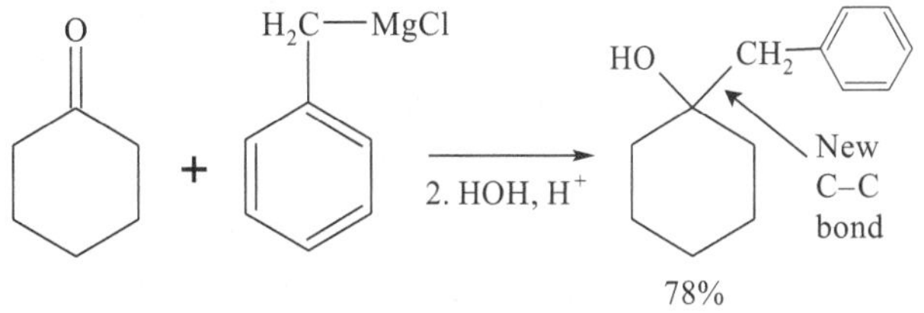 + H₂C—MgCl 2. HOH, H⁺ → HO CH₂ (New C–C bond) 78%

There are some other important aspects of the Grignard reaction that you should know.

1. In addition to alkyl halides, aryl halides (halogen attached to a benzene ring) and vinyl halides (halogen attached to an alkene) also will react with Mg to form Grignard reagents.
2. Iodine is the most reactive halogen, followed by bromine and then chlorine. Fluorine is quite unreactive.

3. Because Grignard reagents are strong nucleophiles, and hence strong bases, they react readily with any acidic compound, even weakly acidic ones such as water and alcohols. Therefore, all solvents and reactants must be dry, as should the glassware being used. Care needs to be taken to keep moisture out of the apparatus. Whenever I have students do a Grignard reaction in lab, it always seems to rain, or at least be very humid outside. In addition, the alkyl halide and the aldehyde or ketone cannot contain other acidic functional groups, such as carboxylic acids, alcohols, thiols, and primary or secondary amines and amides.

Grignard reactions are often used to synthesize alcohols. When you want to make a particular alcohol, you need to be able to figure out what Grignard reagent you want to react with what aldehyde or ketone. This means you need to be able to look at the alcohol product and identify what carbon–carbon bond could be made with a Grignard reaction. One helpful hint is that the carbon that the alcohol was bonded to in the product was the C=O carbon in the starting material. Therefore, any carbon–carbon bond to that carbon could have been potentially made by a Grignard reaction. See the example below.

OH $\|$ $H_3C\!-\!\!-\!CH\!-\!CH_2\!-\!CH_2\!-\!CH_3$	Here is our target compound. It is a secondary alcohol, so we could make it by reacting a Grignard reagent with an aldehyde.
OH $\|$ $H_3C\!-\!\!-\!CH\!-\!CH_2\!-\!CH_2\!-\!CH_3$ a b	The bonds labeled a and b are the two carbon–carbon bonds to the carbon bonded to the OH group. Therefore, either of them could possibly have been made with a Grignard reaction.
OH $\|$ $H_3C\!\!\nmid\!\!CH\!-\!CH_2\!-\!CH_2\!-\!CH_3$	To make bond a, we need two pieces. I have drawn a wiggly line through the bond to show what the pieces are. The piece on the right-hand side contains the OH, so it must have originally been the aldehyde. Therefore, I need the four-carbon aldehyde as a starting material. The left-hand piece must have come from the Grignard reagent. Because it contains only one carbon, I need a methyl Grignard reagent.

$\underset{OH}{\underset{|}{H_3C}} \!\!\!\sim\!\!\!\underset{}{CH}\!-\!CH_2\!-\!CH_2\!-\!CH_3$ 2. HOH $H_3C\!-\!MgI \;\;+\;\; \underset{\underset{}{HC}}{\overset{O}{\overset{\|}{}}}\!-\!CH_2\!-\!CH_2\!-\!CH_3$	Here is the combination of Grignard reagent and aldehyde that we need to make the product alcohol. Water is added in the second step to protonate the alkoxide formed after the initial reaction.
$\underset{HO}{\underset{|}{H_3C\!-\!CH}}\!\!\!\sim\!\!\!CH_2\!-\!CH_2\!-\!CH_3$	I've drawn the wiggly line through bond b to help us see the two pieces that we need. The alcohol is on the two-carbon piece, so we need the two-carbon aldehyde. The other piece we need must be a three-carbon Grignard reagent.
$\underset{HO}{\underset{|}{H_3C\!-\!CH}}\!\!\!\sim\!\!\!CH_2\!-\!CH_2\!-\!CH_3$ 2. HOH $\underset{\underset{}{H_3C\!-\!CH}}{\overset{O}{\overset{\|}{}}} \;\;+\;\; \underset{MgBr}{\underset{|}{H_2C}}\!-\!CH_2\!-\!CH_3$	Here is the combination of aldehyde and Grignard reagent we need to make bond b.

If an alkyl halide is treated with two equivalents of lithium metal, an alkyllithium reagent is produced plus an equivalent of lithium halide. Alkyllithium reagents do the same types of reactions with aldehydes and ketones to form alcohols. We will encounter some differences in the reactivity of alkyllithium reagents and Grignard reagents as we look at reactions of other functional groups. See below.

$H_3C\!-\!CH_2\!-\!CH_2\!-\!CH_2\!-\!Br \xrightarrow{\;2\,Li\;} H_3C\!-\!CH_2\!-\!CH_2\!-\!CH_2\!-\!Li \;\;+\;LiBr$ $(n\text{-}C_4H_9Li)$	Formation of the alkyllithium reagent, butyl lithium.
1. n-C$_4$H$_9$Li 2. HOH HO—CH$_2$—CH$_2$—CH$_2$—CH$_3$ 89%	Reaction of butyl lithium with a ketone to make a tertiary alcohol.

G. Thiol Naming and Properties

Thiols are named by adding *thiol* to the name of the hydrocarbon. Thiols are just below alcohols in naming priority. If a thiol is a substituent, it is called **mercapto**. This comes from the old name for a thiol, which was mercaptan. Thiols bind tightly to mercury and have been used to remove mercury from solutions, so that's were the *merc* part of the name comes from. See the following examples.

CH_3SH	This is **methanethiol**. Without the sulfur, it would be methane. Adding thiol to that makes methanethiol.
H_3C—CH_2—CH_2—SH	This is **propane-1-thiol**. There is a three-carbon chain, with the thiol on the end carbon.
	This is **2-cyclopentenethiol**. Because the thiol is the highest priority group, we start numbering at the carbon it is attached to. Numbering counterclockwise gets us to the C=C the quickest.

Thiols have characteristically strong and pungent odors. If you have smelled natural gas, you have really been smelling methanethiol. Natural gas has no odor, so methanethiol is added to it at a parts per million level so natural gas leaks can be detected. Skunks use a variety of thiols as part of their defensive arsenals, including 3-methylbutane-1-thiol and 2-butene-1-thiol.

Sulfur is less electronegative than oxygen, so an S–H bond is less polar than an O–H bond. Therefore, thiols are less water-soluble than alcohols and have lower boiling points than alcohols of similar molecular weight. See the table below.

Name	Structure	Water Solubility (g/100 mL of water)	Boiling Point (°C)
Methanethiol	CH_3SH	2.4	6
Ethanethiol	CH_3CH_2SH	0.68	36
Propane-1-thiol	$CH_3CH_2CH_2SH$	Slightly soluble	68
Butane-1-thiol	$CH_3CH_2CH_2CH_2SH$	0.06	99

Thiols are stronger acids than alcohols. For example, ethanethiol has a pK_a of 10.5, whereas ethanol has a pK_a of 15.9.

H. Thiol Reactions and Preparations

H.1. REACTIONS WITH BASES

Because thiols are stronger acids than alcohol, they react with bases such as sodium hydroxide and sodium carbonate to form salts, which are called **thiolates**. The thiolates are very good nucleophiles and can participate effectively in S_N2 reactions. See the example below.

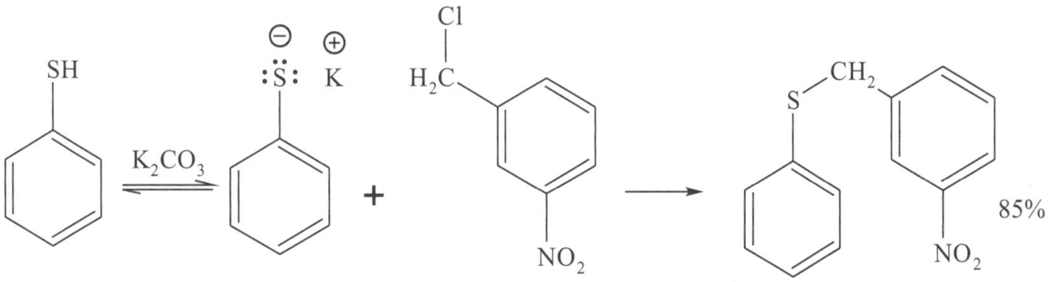

H.2. MILD OXIDATION

On treatment with mild oxidizing agents, thiols can be oxidized to disulfides. Hydrogen peroxide is a common reagent, but sometimes it does not give good yields. Lead tetraacetate [$Pb(OAc)_4$ or LTA] oxidizes a variety of thiols to disulfides in good yields. Some examples are shown below.

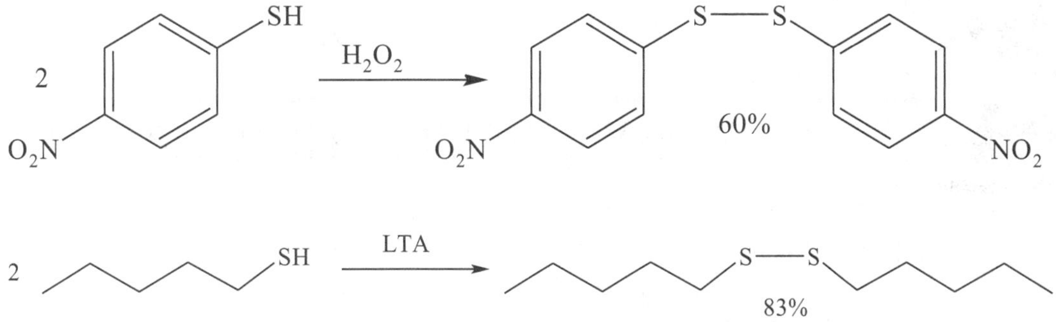

The by–products are lead(II) acetate and 2 molecules of acetic acid.

H.3. VIGOROUS OXIDATION

Thiols, as well as the corresponding disulfides, can be further oxidized to sulfonic acids and related compounds. Nitric acid is a common reagent, as is bromine and water. Chlorine and nitric acid can convert a thiol or a disulfide into a sulfonyl chloride. The sulfonyl chloride can be converted into the sulfonic acid by heating with water. See the examples below.

$$\text{HO}-\overset{\overset{\displaystyle O}{\|}}{\text{C}}-\underset{\underset{\displaystyle NH_2}{|}}{\text{CH}}-\text{CH}_2-\text{S}-\text{S}-\text{CH}_2-\underset{\underset{\displaystyle NH_2}{|}}{\text{CH}}-\overset{\overset{\displaystyle O}{\|}}{\text{C}}-\text{OH} \quad\xrightarrow[\text{6 HOH}]{\text{5 Br}_2}\quad 2\ \text{HO}-\overset{\overset{\displaystyle O}{\|}}{\text{C}}-\underset{\underset{\displaystyle NH_2}{|}}{\text{CH}}-\text{CH}_2-\overset{\overset{\displaystyle O}{\|}}{\underset{\underset{\displaystyle O}{\|}}{\text{S}}}-\text{OH} \ +\ 10\ \text{HBr}$$

Cystine   Cysteic acid   90%

[structure of bis(2-nitrophenyl) disulfide] $\xrightarrow[\text{HNO}_3]{\text{Cl}_2}$ [benzenesulfonyl chloride structure]   84%

Thiols can be prepared by the S_N2 substitution reaction of NaSH (formed from H_2S and NaOH) on primary or relatively unhindered secondary alkyl halides. A large excess of NaSH is used to ensure that the alkyl halide reacts with NaSH, and not the product thiol, which is also nucleophilic.

REVIEW EXERCISES FOR CHAPTER 8

1. If the reactions of 2-pentanol and 3-pentanol with HBr went solely by an S_N1 mechanism, what would the ratios of 2-bromopentane to 3-bromopentane be in each case? Explain your reasoning.

2. Give IUPAC names for the following structures.

a.	b.

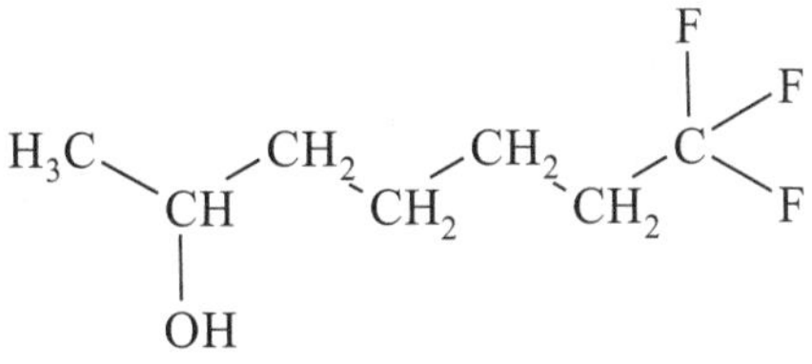

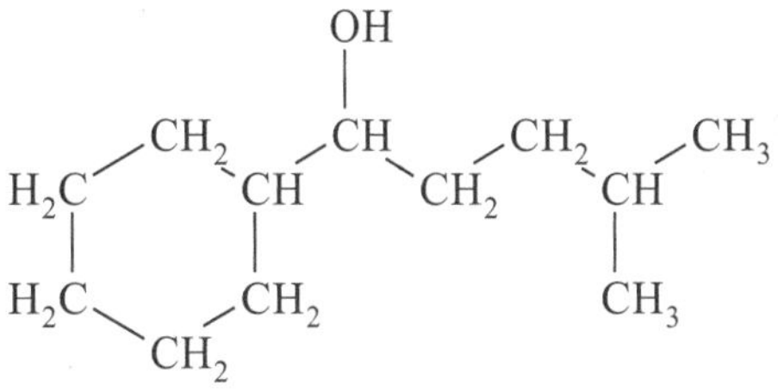

c.	d.

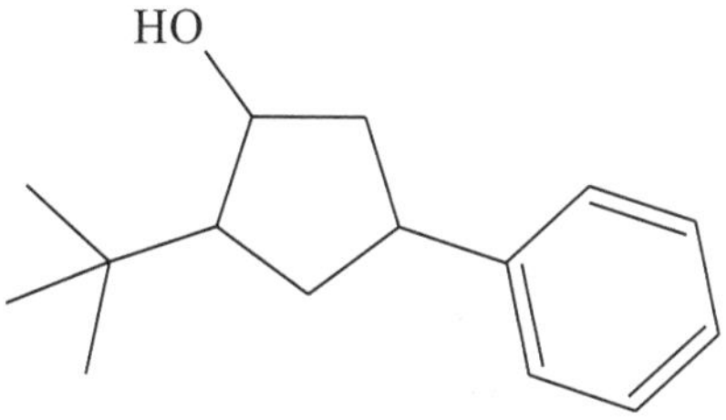

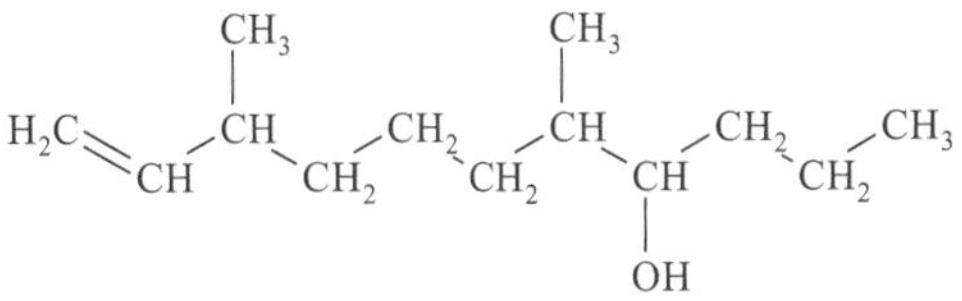

e.	f.

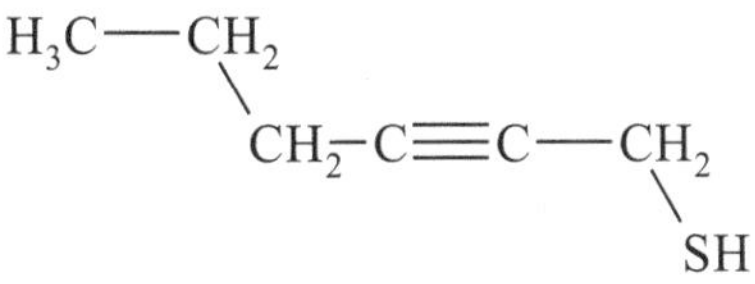

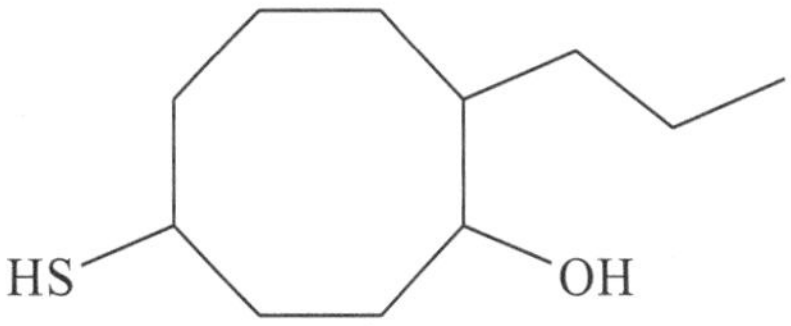

3. Give the major organic products from the following reaction sequences. Show the intermediates formed after each step. You do *not* have to draw the mechanisms.

a.

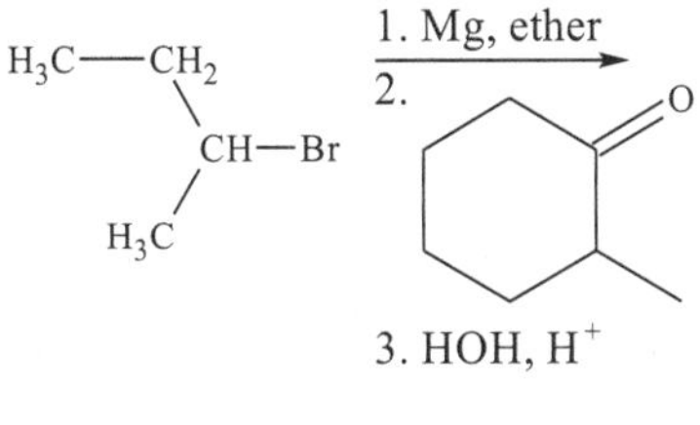

1. Mg, ether
2.

3. HOH, H$^+$

b.

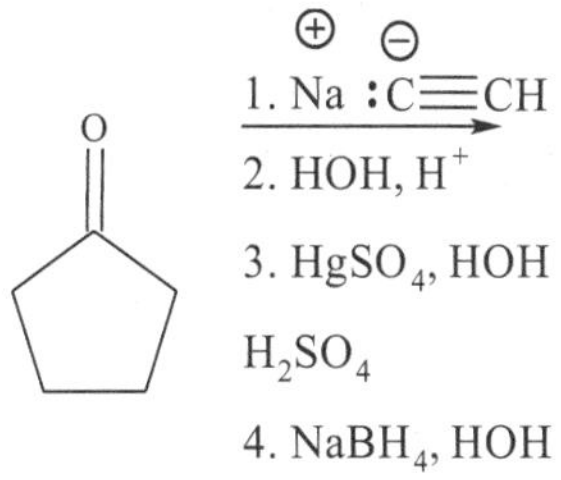

1. Na :C≡CH (⊕ ⊖)
2. HOH, H$^+$
3. HgSO$_4$, HOH
   H$_2$SO$_4$
4. NaBH$_4$, HOH

c.

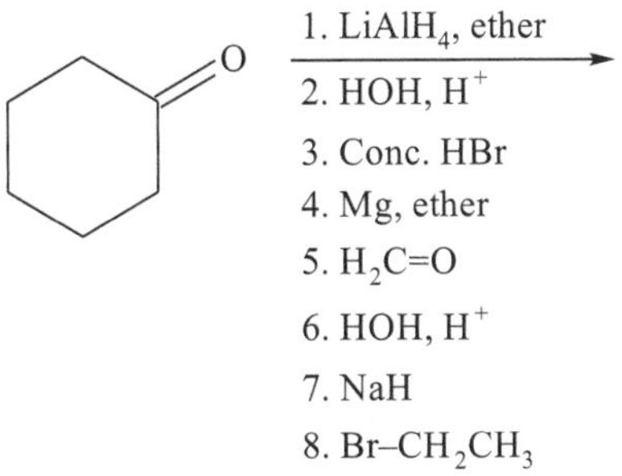

1. LiAlH$_4$, ether
2. HOH, H$^+$
3. Conc. HBr
4. Mg, ether
5. H$_2$C=O
6. HOH, H$^+$
7. NaH
8. Br–CH$_2$CH$_3$

4. What are all the possible combinations of Grignard reagent and aldehyde or ketone needed to make each of the following alcohols?

a.

b.

5. How could you make the following alcohol:

a. Using a Grignard reaction?

b. Using a reduction of a C=O compound?

c. From an alkene?

d. Starting with 3-methyl-1-butyne?

Ethers, Thioethers, and Epoxides

WHAT YOU WILL LEARN

In this chapter, you will learn:

- how to name ethers, thioethers, and epoxides using IUPAC rules;
- how to prepare ethers and thioethers from alkyl halides;
- how ethers and epoxides react with nucleophiles and acids;
- how thioethers can be oxidized to sulfoxides and sulfones.

SECTIONS IN THIS CHAPTER

- Structure and Nomenclature
- Physical Properties
- Preparations
- Reactions

A. Structure and Nomenclature

Ethers have the general structure R–O–R', where R and R' are carbon-containing groups that do not have a C=O bonded to the O. Thioethers are the sulfur analogs of ethers, with the general formula R–S–R'. Epoxides are three-membered ring ethers. Epoxides react differently from most other ethers, so they are considered separately.

When naming an ether or a thioether, you first must decide whether R or R' has the higher priority. You must determine which part has the higher priority functional group in it. If they have the same priority functional groups, then you go with the longest carbon chain. See the examples below.

$H_3C\!-\!CH_2\!-\!CH_2\!-\!O\!-\!CH_3$ **A methoxy** group	Both groups attached to the O are alkanes, so we use the larger as the higher priority functional group. Therefore, this is a propane. The other carbon part has only one carbon, which is meth. It is not methyl because there is an oxygen. The ending for an oxygen is oxy, so the group is methoxy. The methoxy group is on an end carbon, so it is located at carbon 1. Therefore, the name is **1-methoxypropane**.
  3 CH_2 4 $O\!-\!CH_2CH_3$ 2 HC CH 1 HC CH CH_2 5 $O\!-\!CH_2CH_3$ 6	When you have two ether groups, the part of the molecule they are both attached to is generally the "longest chain." In this case, it is cyclohexene. There are two ethoxy groups attached, so it is a diethoxycyclohexene. You have to number the carbons of the C=C as 1 and 2, so the complete name is **4,5-diethoxycyclohexene**.
$H_3C\diagdown S\diagdown CH_2\diagdown CH_2\diagdown CH_2\diagdown CH_2\diagdown CH\diagup OH \diagup CH_3$ ... CH_3 A(methylthio) group	Here is an alcohol with a thioether group. The longest carbon chain containing the alcohol is seven carbons, with the OH on the third carbon, so it is a heptane-3-ol. There is a methyl group on carbon 2. The thioether group is on carbon 7. Thioethers are named as *alkyl + thio*, so this is a (*methylthio*) group. Listing the groups alphabetically gives us **2-methyl-7-(methylthio)heptan-3-ol**.

Simple ethers and thioethers also have common names that you will often encounter, especially in the lab. The common system names the R and R' groups as substituent groups, followed by the word "ether" for ethers, or by "sulfide" for thioethers. This system only works for R groups that can be easily named as substituents. See the examples below.

Compound	Common Name
CH_3—CH_2—O—CH_2—CH_3	**Diethyl ether**, or just **ethyl ether**. This is the "ether" that has been used as a general anesthetic.
(phenyl ring)—S—CH_2—CH_2—CH_2—CH_2—CH_3	**Pentyl phenyl sulfide**.

The simplest epoxide is named systematically as oxirane. *Ox* means oxygen, -*ir*- means three-membered ring, and -*ane* means no C=C or C≡C. When numbering an epoxide, the oxygen gets the number 1. See the examples below.

Structure	Name
H_2C—CH_2 with O bridge	**Oxirane**, the simplest epoxide.
H_2C—CH—CH_2-CH_3 with O bridge, position 1	**2-Ethyloxirane**, or just **ethyloxirane**.
H_3C—C—CH—O—CH_3 with O bridge (position 1), H_3C below at 2, position 3	**3-Methoxy-2,2-dimethyloxirane**: *methoxy* comes before *methyl* in the alphabet, and *di* is not used in alphabetizing.

B. Physical Properties

Because ethers have an oxygen, they can hydrogen-bond with water. The oxygen makes ethers more water-soluble than alkanes of similar molecular weight. They are about as soluble as alcohols of similar molecular weight. Because thioethers have a less polar sulfur atom, they are not as water-soluble as ethers. See the table below.

Compound	Structure	Molecular Weight	Water-Solubility (g/100 mL)	Boiling Point (°C)
Methoxymethane	CH_3OCH_3	46	3.5	–25
Ethoxyethane	$CH_3CH_2OCH_2CH_3$	74	6.1	35
1-Methoxypropane	$CH_3OCH_2CH_2CH_3$	74	3.5	39
1-Propoxypropane	(structure)	102	0.4	90

Compound	Structure	Molecular Weight	Water-Solubility (g/100 mL)	Boiling Point (°C)
Methylthiomethane	CH_3SCH_3	62	1.9	37
Ethylthioethane	$CH_3CH_2SCH_2CH_3$	90	"Negligible"	92
Butan-1-ol	$CH_3CH_2CH_2CH_2OH$	74	7.4	118
Pentane	$CH_3CH_2CH_2CH_2CH_3$	72	0.036	36

Boiling points of ethers are similar to those of alkanes of similar molecular weight. This is because ether molecules cannot hydrogen-bond with each other as alcohols can.

C. Preparations

Ethers and thioethers are typically prepared by nucleophilic substitution reactions, either S_N2 (strong nucleophile + primary alkyl halide) or S_N1 reactions (weak nucleophile + tertiary alkyl halide). Symmetrical thioethers can be prepared by the reaction of sodium sulfide with two equivalents of a primary alkyl halide. See the examples below.

90%

A primary alkyl halide plus a strong nucleophile forms an ether by an S_N2 mechanism.

64%   36%

A tertiary alkyl halide plus a weak nucleophile forms an ether by an S_N1 mechanism.

80%

A primary alkyl halide plus a strong nucleophile forms a thioether by an S_N2 mechanism.

79%

Preparation of a thioether by reacting two equivalents of an alkyl halide with sodium sulfide.

Epoxides are typically prepared by treating an alkene with a peracid (RCO_3H), or by a two-step procedure in which the alkene is reacted with a halogen and water to form a 2-halo alcohol, which is then treated with a base to form the epoxide. These reactions are discussed in more detail in Chapter 4. Some examples are shown below.

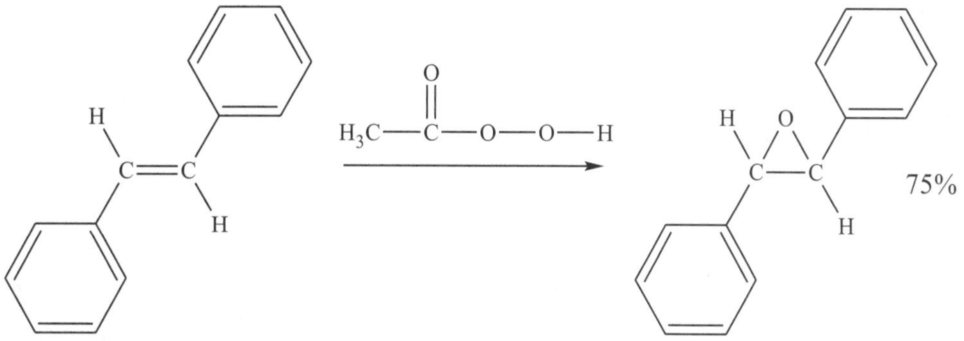

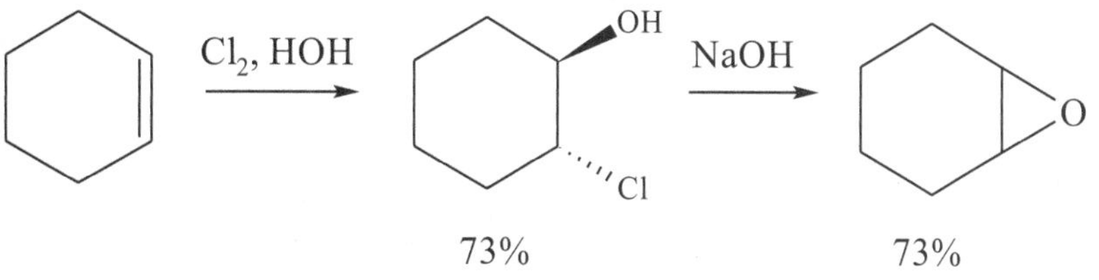

D. Reactions

D.1. ETHERS

Ethers are rather unreactive, which makes them good, general organic solvents. The main reaction of dialkyl ethers is their reaction with concentrated HI or HBr, to form the alkyl halides and water. If an aryl alkyl ether is used, the products are the phenol and the alkyl halide. Some examples are shown below.

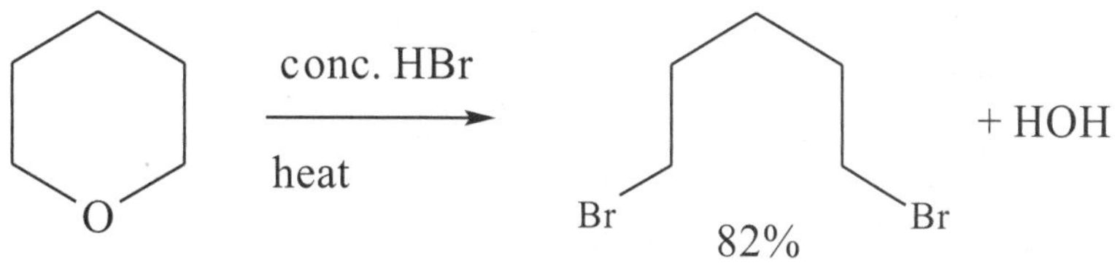

Excess HBr breaks both C–O bonds and replaces them with bromines.

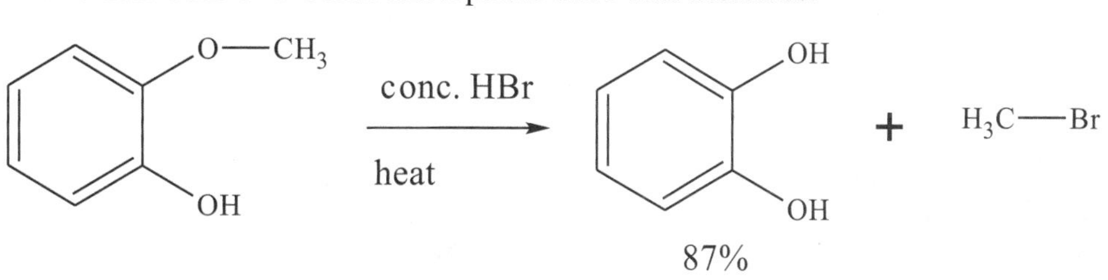

Benzene C–O bonds are not cleaved by HBr, so the product is the aromatic alcohol (a phenol).

The mechanism of the preceding reactions is similar to the reaction of an alcohol with H-Br. The mechanism of the first reaction is shown below.

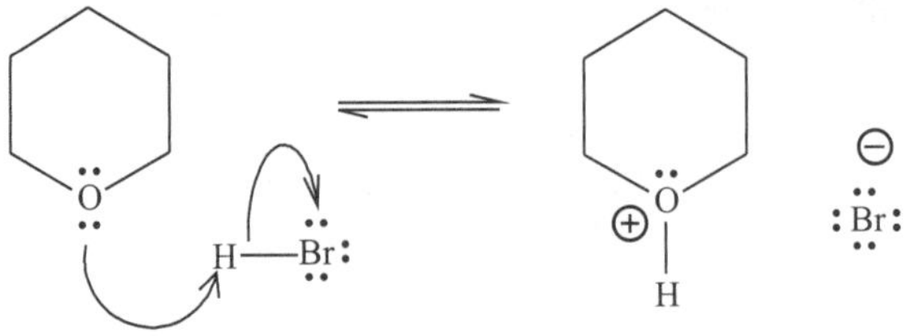

The oxygen pulls off the hydrogen of H–Br, to form a protonated ether and bromide ion.

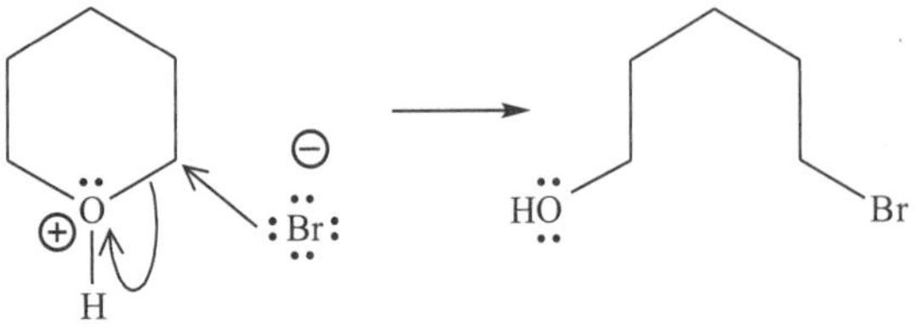

Bromide does an S_N2 attack on the primary carbon and pushes out the protonated oxygen. We are halfway through the mechanism, and we will repeat these two steps over again.

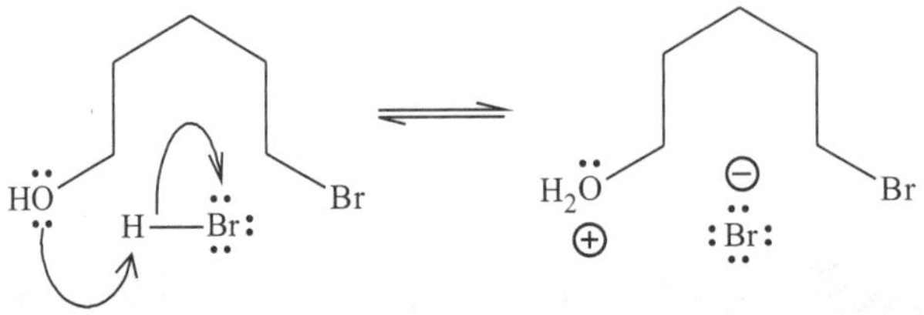

The oxygen pulls off the hydrogen of H–Br, to form a protonated alcohol and bromide ion.

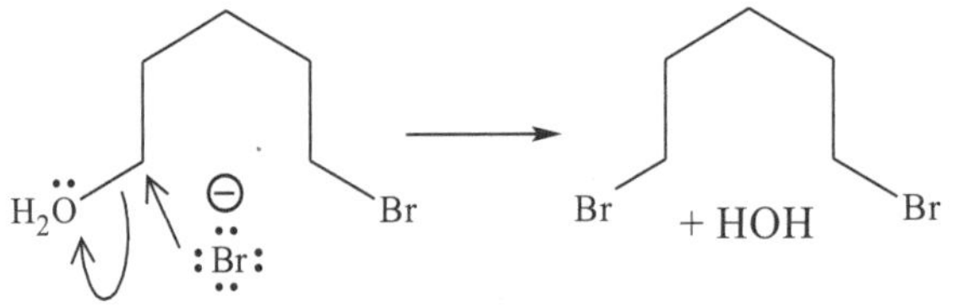

Bromide does an S_N2 attack on the primary carbon and pushes out the protonated oxygen. The dibromo product is formed.

If the carbon bonded to the ether oxygen is tertiary, that C–O bond is probably cleaved in a S_N1 process (protonation of the oxygen, cleavage of the C–O to form a tertiary cation, and attack by the halide ion on the cation). This is shown below.

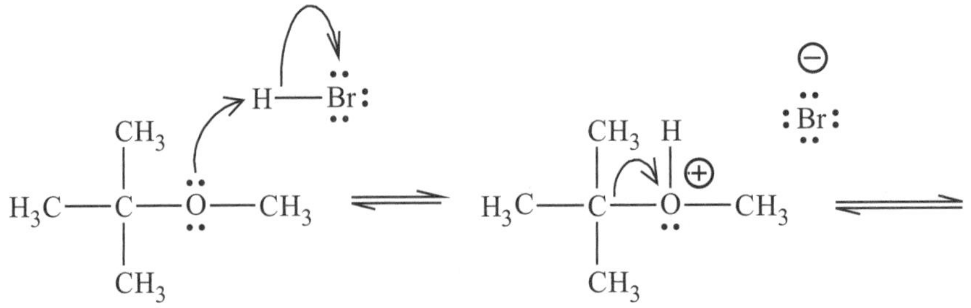

The oxygen pulls off the hydrogen of H–Br, to form a protonated ether and bromide ion. The C–O bond breaks to form a tertiary cation and methanol.

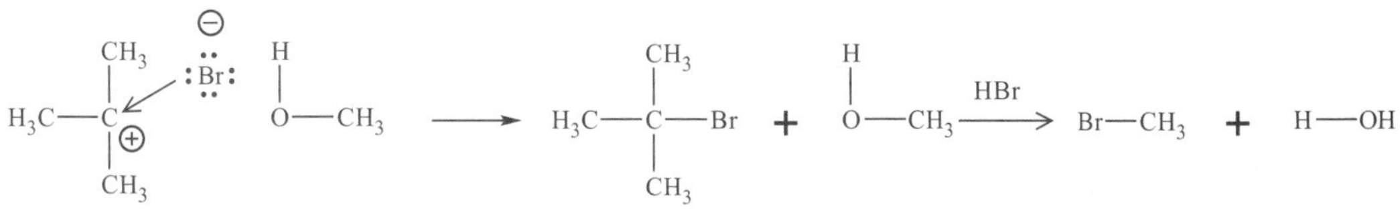

Bromide attacks the tertiary cation to form the alkyl halide. Methanol reacts with more HBr in an S_N2 process, like the previous example, to form bromomethane.

D.2. THIOETHERS

Thioethers react much more slowly with concentrated HI and HBr. Sulfur is not as basic as oxygen and is not protonated very effectively by these acids to form a good leaving group. The main reaction of sulfides is oxidation to either a sulfoxide or a sulfone. This is shown below.

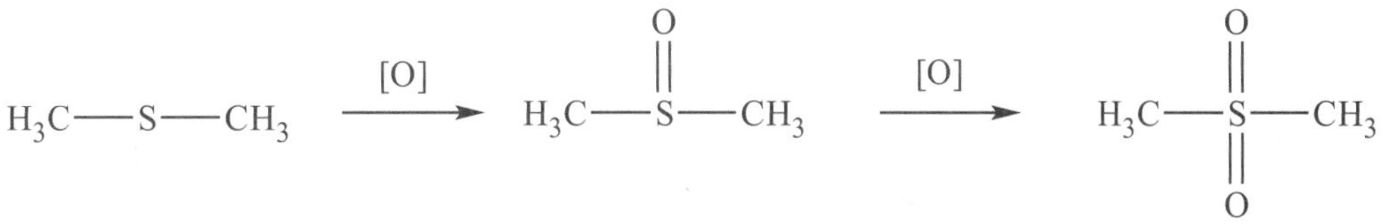

"Dimethyl sulfide"   "Dimethyl sulfoxide"   "Dimethyl sulfone"
                                  (DMSO)

Oxidation to the sulfoxide requires milder conditions, such as working at low temperatures and with one equivalent of the oxidizing agent. Sodium periodate ($NaIO_4$) and meta-chloroperbenzoic acid (MCPBA) are useful reagents for preparing sulfoxides. A variety of reagents, including potassium permanganate and hydrogen peroxide, oxidize sulfides to sulfones. See the examples below.

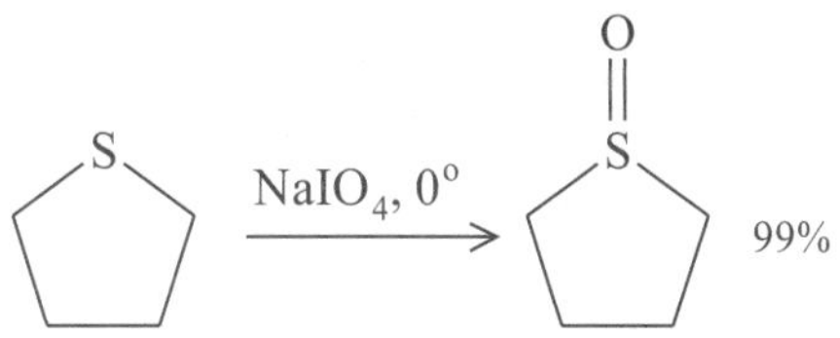

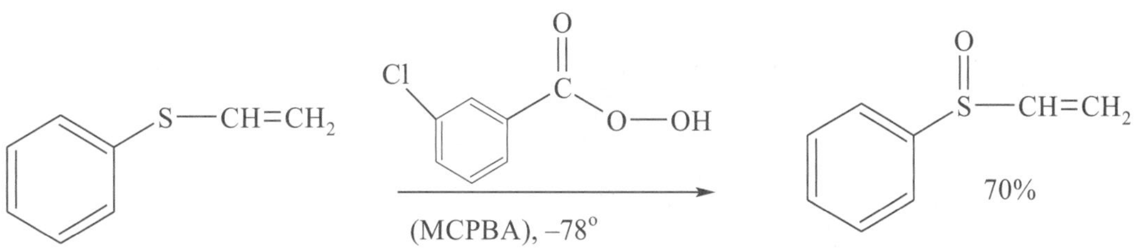

At low temperature, peracids selectively oxidize thioethers in preference to epoxidizing alkenes.

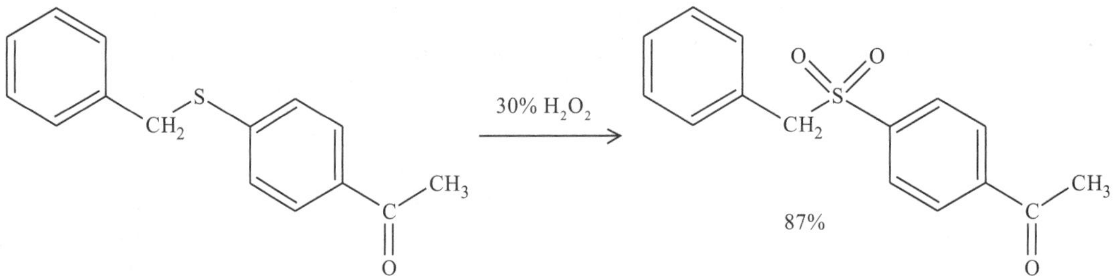

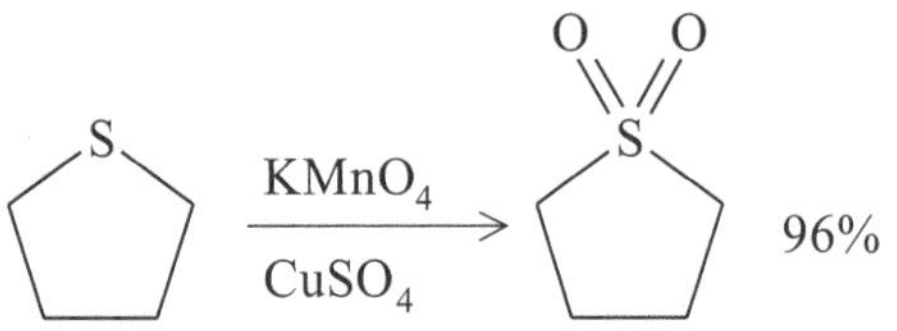

Sulfoxides are sometimes written as a resonance form, as shown below, to emphasize the polarity of the S–O bond.

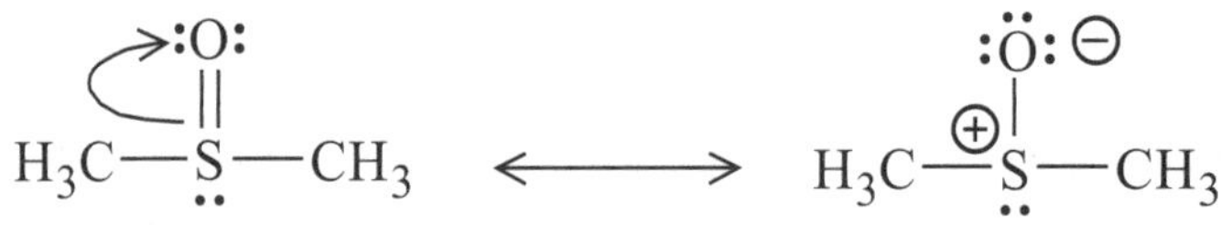

The sulfur has an unshared pair of electrons on it in each form. The form on the left seems to violate the octet rule, but as I mentioned in Chapter 1, some elements, such as sulfur and phosphorus, can do so. The structure on the right satisfies the octet rule, but has formal charges on the sulfur and the oxygen.

D.3. EPOXIDES

Because epoxides have a three-membered ring, there is considerable bond-angle strain, as well as eclipsing strain from interactions of the hydrogens on the neighboring carbon atoms. If a strong nucleophile attacks one of the partially positive carbons, the ring could open to relieve these strains. This is shown below.

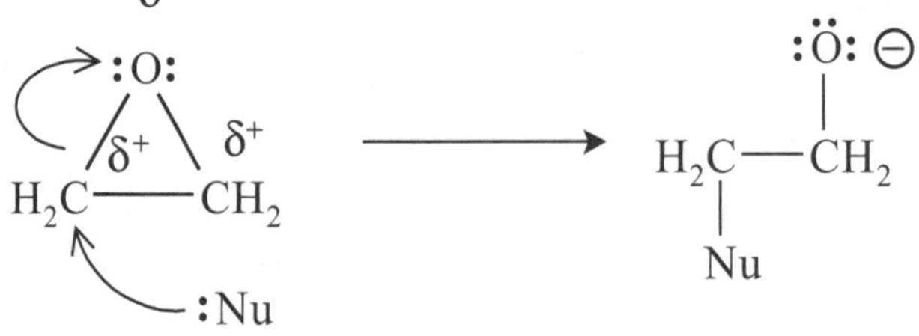

If the epoxide is not symmetrical, then potentially the nucleophile could attack at either carbon to produce isomeric products. Strong nucleophiles generally attack the less substituted carbon, as we saw in S_N2 reactions. Weak nucleophiles generally attack the more substituted carbon, as is more typical with S_N1 reactions. See the examples below.

The strong nucleophile attacks at the less substituted carbon. No product from the attack at the tertiary carbon was observed. See the mechanism below.

The nucleophile attacks the less substituted carbon, which causes the C–O bond to break and pushes the electron pair onto the oxygen.

The alkoxide ion pulls a hydrogen off methanol to form the alcohol product.

The product is derived from the attack of the weak nucleophile at the more substituted carbon. No product from the attack at the secondary carbon was observed. See the mechanism below.

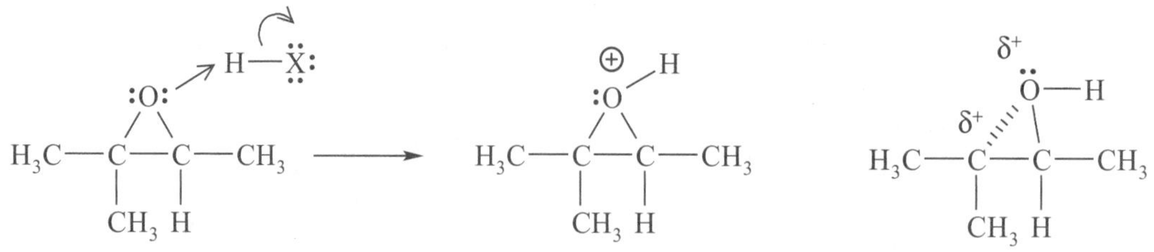

The epoxide oxygen attacks the acid hydrogen to form a protonated epoxide. This protonated epoxide has a significant amount of positive charge on the more substituted carbon (as shown by the structure at the right), but it does not become a free carbocation: The oxygen remains partially bonded to it.

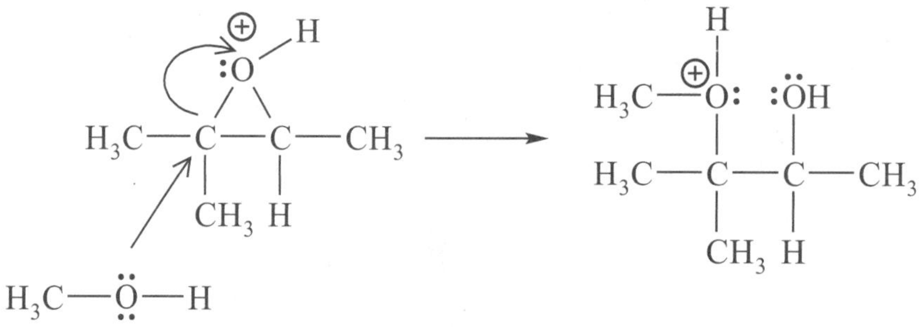

Methanol attacks the tertiary carbon and pushes the electron pair up onto the oxygen.

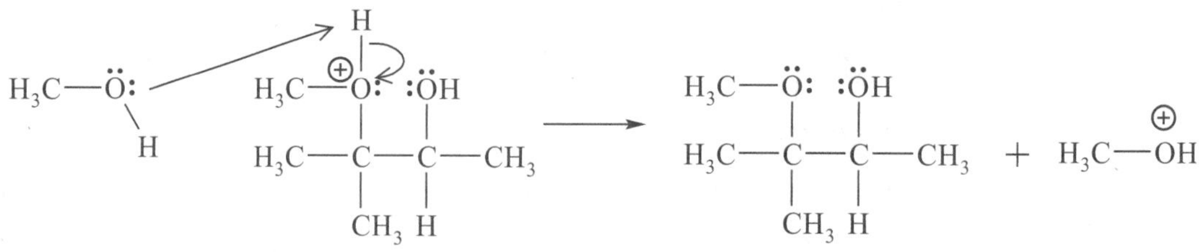

Methanol deprotonates the intermediate to form the product.

The following example shows that weak nucleophiles attack from the opposite side of the molecule from the epoxide oxygen, as indicated by the above mechanism.

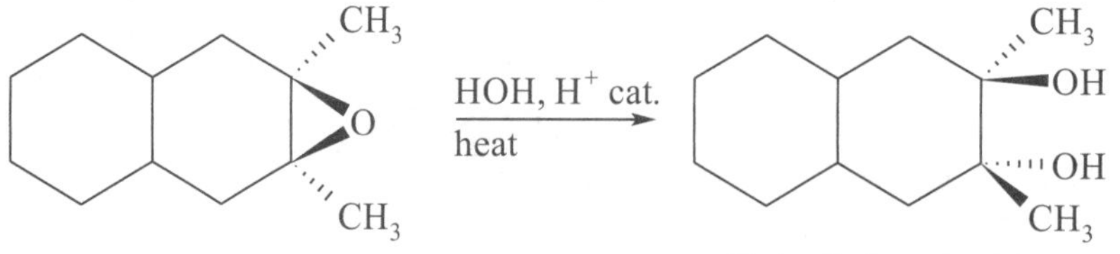

Only diol product

Because the two OH groups are on opposite sides of the ring in the product, water must have attacked from the opposite side of the molecule.

The following are some examples of a strong nucleophilic reaction to epoxides.

$$83\%$$

The strong nucleophile attacks at the less substituted carbon.

$$62\%$$

Grignard reagents react with oxirane to give primary alcohols with two more carbons.

$$99\%$$

Lithium aluminum hydride reduces epoxides, with the hydride adding to the less substituted carbon.

REVIEW EXERCISES FOR CHAPTER 9

1. Give IUPAC names for the following structures.

   a.

   $$H_3C—CH_2—CH—CH—C\equiv C—CH_3$$

   with OH on the CH, and a cyclopentyl substituent (CH with H$_2$C, CH$_2$, CH$_2$—CH$_2$)

   b.

   (structure: ethoxy chain with SH and terminal alkene)

   c.

   (epoxide attached to cyclopropane)

   d. (Methoxyflurane, an anesthetic)

   $$H_3C—O—\underset{\underset{F}{|}}{\overset{\overset{F}{|}}{C}}—\underset{\underset{Cl}{|}}{\overset{\overset{Cl}{|}}{C}}—H$$

   e.

   $$H_3C=CH—CH_2—S—C\equiv C—CH_3$$

   with CH$_3$ on the CH

f.

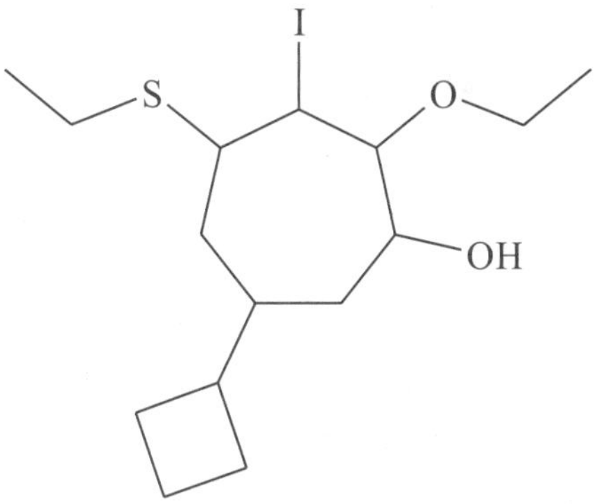

2. Give the major organic product of each of the following reactions:

a.

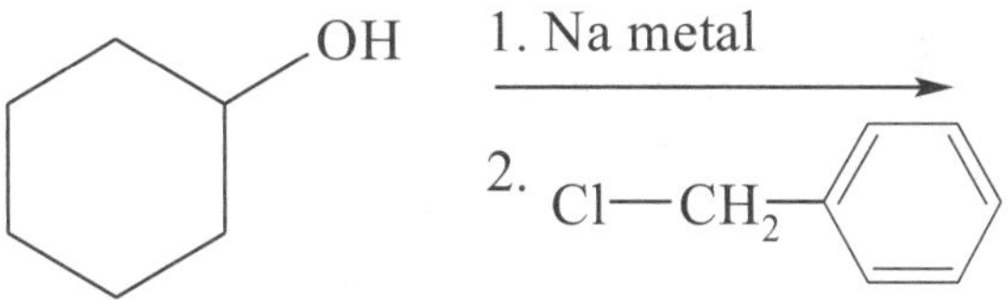

1. Na metal

2. $Cl-CH_2-$ (phenyl)

b.

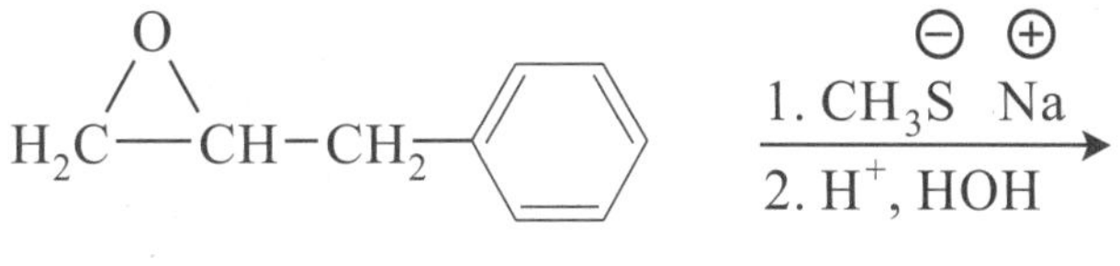

1. $CH_3\overset{\ominus}{S}$ $\overset{\oplus}{Na}$

2. H^+, HOH

c.

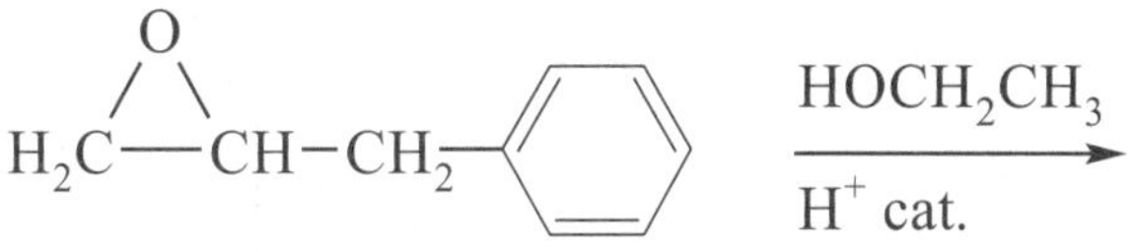

$HOCH_2CH_3$

H^+ cat.

d.

OH

1. Conc. HBr

2. $CH_3\overset{\ominus}{S}$ $\overset{\oplus}{Na}$

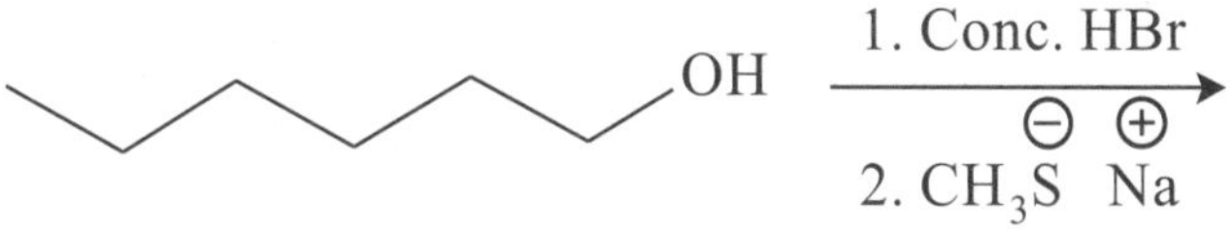

e.

Product of part d, plus MCPBA.

f.

Product of part d, plus H_2O_2.

g.

$$\xrightarrow[\text{(excess), heat}]{\text{conc. HI}}$$

3. Give arrow-pushing mechanisms for the reactions in Problems 2c and 2d.

4. Each of the following products can be synthesized in two steps from the indicated starting materials. Give the reagents needed for the two steps.

Starting Material	Product
a.	
b.	
c.	

5. Phenols are compounds with an OH group bonded to a benzene ring. When treated with a strong oxidizing agent, phenols can be oxidized to tarry by–products. Therefore, phenols are sometimes protected as ethers, the oxidation reaction is then done, and finally, the ether group is removed. Provide the reagents needed to accomplish the following sequence of reactions.

Aromatic Compounds

WHAT YOU WILL LEARN

In this chapter, you will learn:

- how to recognize aromatic, antiaromatic and non-aromatic compounds;
- how to name benzene derivatives and other aromatic compounds;
- how benzene reacts with electrophiles, and how to draw the mechanisms;
- how to design a synthesis of a substituted benzene using the directing abilities of the attached groups.

SECTIONS IN THIS CHAPTER

- Structure of Benzene
- Unexpected Stability of Benzene
- Hückel's Rules
- Aromatic and Anti-aromatic Compounds
- Nomenclature of Benzene Derivatives
- Electrophilic Aromatic Substitution of Benzene
- Electrophilic Aromatic Substitution of Substituted Benzenes: Directing Abilities
- Electrophilic Aromatic Substitution of Substituted Benzenes: Activating or Deactivating Effects
- Synthesis of Disubstituted and Trisubstituted Benzenes
- Phenol Naming and Properties
- Other Aromatic Compounds

A. Structure of Benzene

The chemical we now call benzene was first isolated from the gas used to light street lamps in 1825 by the famous English chemist, Michael Faraday. Faraday discovered this compound had the empirical formula, CH, and called it "carbureted hydrogen." Other scientists determined the molecular formula to be C_6H_6. Various other names were proposed, including pheno, from the Greek, *phainein*, which means to shine. Pheno never caught on, but we use a variation of this, *phenyl*, to describe the C_6H_5 group. Other names included benzin and benzol, both of which are still found in some literature. French and English chemists preferred and promoted benzene.

Once the molecular formula was determined, chemists proposed a variety of possible structures. Four are shown here.

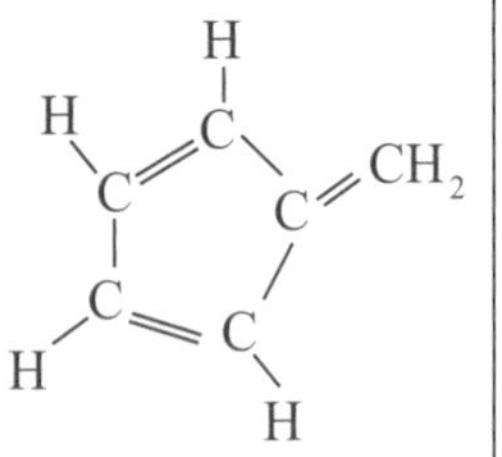

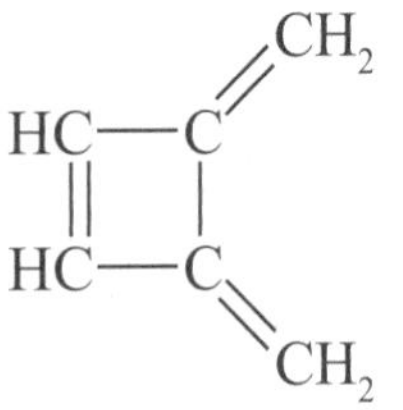

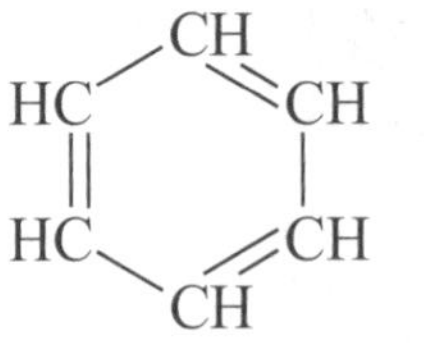

Proposed by August Kekulé in 1865

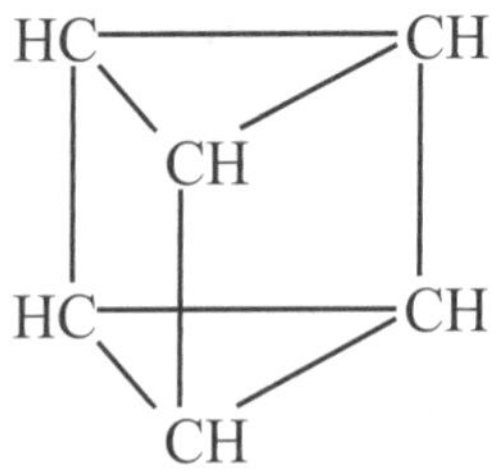

Proposed by Albert Ladenburg in 1869

They did reactions with benzene to try to eliminate some of the possibilities. One reaction was with Cl_2 in the presence of $AlCl_3$ as a catalyst. They found benzene produced only one product in which one chlorine had substituted for a hydrogen. Another way of saying this is that there is only one monochlorobenzene. They also found that benzene formed three products in which two chlorines had substituted for two hydrogens (three dichlorobenzenes).

The first structure could have three different monochlorobenzenes and many more than three dichlorobenzenes. The second structure has two monochlorobenzenes and four dichlorobenzenes, if you ignore stereoisomers. Both Kekulé's and Ladenburg's benzene structures fit with these data: I have left these for you as a problem. As chemists tried to prepare Kekulé's structure, they found that they made benzene, so Kekulé's structure was thought to be the correct one.

Kekulé's structure did have one problem: It predicted that there would be two different 1,2-dichlorobenzenes, as shown below.

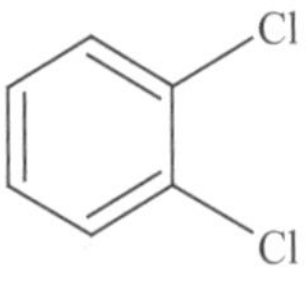

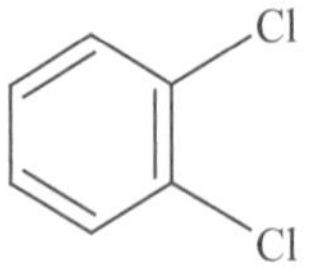

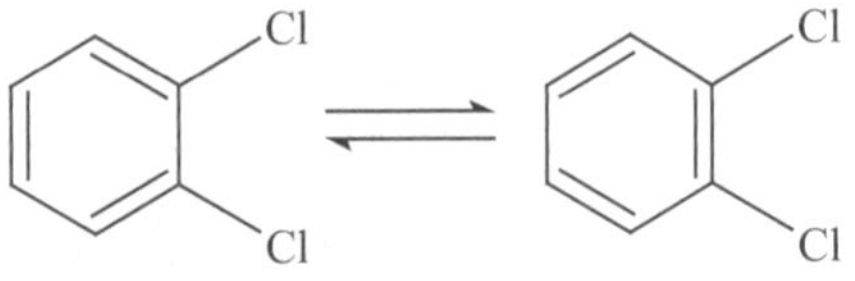

However, only one 1,2-dichlorobenzene is known. To account for this, Kekulé proposed that the two structures are in equilibrium, as shown on the right-hand side. However, as more work was done, it appears that there are not two structures in rapid equilibrium. Instead, these structures appear to be resonance structures; the real structure of benzene has each carbon–carbon bond with a partial double-bond character. That is why benzene is often written as a hexagon with a circle inside it: to emphasize that there are not "normal" C=C's in a benzene ring. See below.

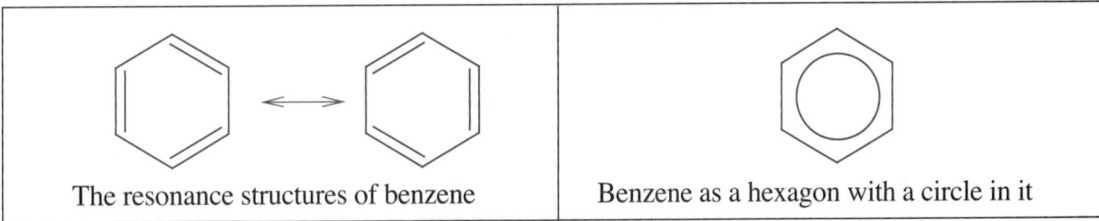

The resonance structures of benzene	Benzene as a hexagon with a circle in it

Depending on what I want to emphasize, I may write benzene either way, but no matter how I write it, it is still benzene, and not a cyclohexatriene.

B. Unexpected Stability of Benzene

Back in Chapter 4, we looked at stability of alkenes. We can also look at the stability of benzene versus the stability of cyclohexene, cyclohexadienes, and a hypothetical cyclohexatriene.

Reaction	$\Delta H°$, kcal/mol (kJ/mol)
Cyclohexene, our standard cyclic C=C $\xrightarrow{\text{H}_2,\ \text{Pd}}$	−28.6 (−119.8)
1,4-Cyclohexadiene, a nonconjugated diene $\xrightarrow{2\text{H}_2,\ \text{Pd}}$	−57.4 (−240.5)
1,3-Cyclohexadiene, a conjugated diene $\xrightarrow{2\text{H}_2,\ \text{Pd}}$	−55.4 (−232.1)
The conjugated diene is about 2 kcal/mol (8 kJ/mol) more stable than the nonconjugated diene. This is due to the extended overlap of the p orbitals in the conjugated system.	

Reaction	$\Delta H°$, kcal/mol (kJ/mol)
A hypothetical cyclohexatriene $\xrightarrow{3H_2,\ Pd}$	−85.8 (−359.5)
Benzene $\xrightarrow{3H_2,\ Pd}$	−49.8 (−208.7)

Benzene is about 36 kcal/mol more stable than a hypothetical cyclohexatriene with three cyclohexene C=C's. This extra stabilization reflects the overlap of all six p orbitals, rather than having three isolated C=C's, as well as the special stabilization that benzene has. These numbers are illustrated in Figure 1.

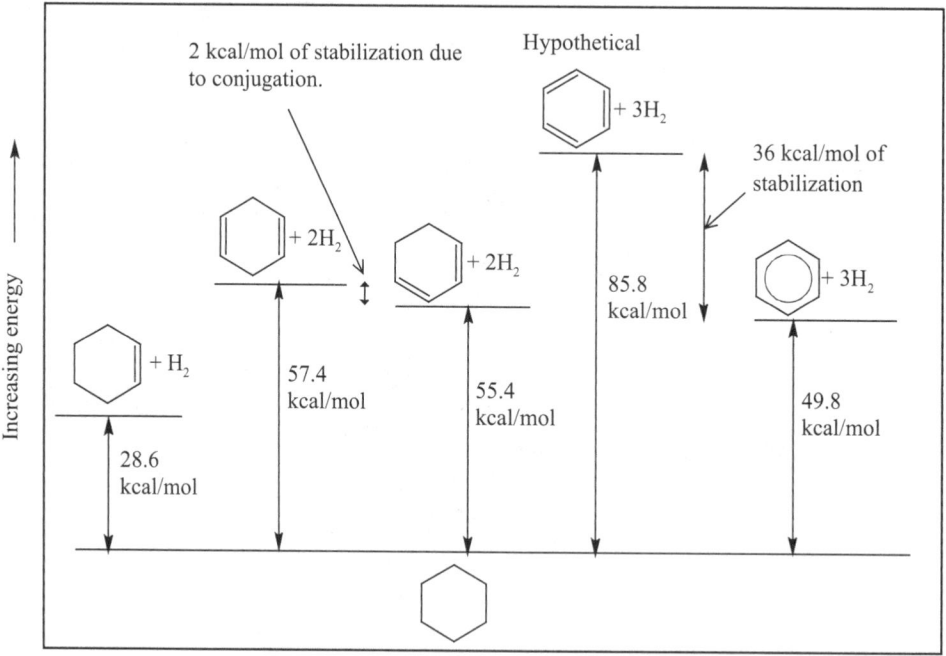

Figure 1

Not all conjugated cyclic alkenes exhibit this stabilization. 1,3-Cyclobutadiene is incredibly unstable and can only be isolated at temperatures around 30 K. 1,3,5,7-Cyclooctatetraene reacts just like a normal alkene and exhibits no special stability.

Because of this stability of benzene, benzene does not undergo the typical addition reactions of normal alkenes readily. Cyclohexene readily adds molecular bromine to give 1,2-dibromocyclohexane, whereas benzene is inert under these conditions. Benzene reacts only with bromine in the presence of a Lewis acid catalyst, and then gives substitution products, rather than addition products, and forms bromobenzene plus HBr. See below.

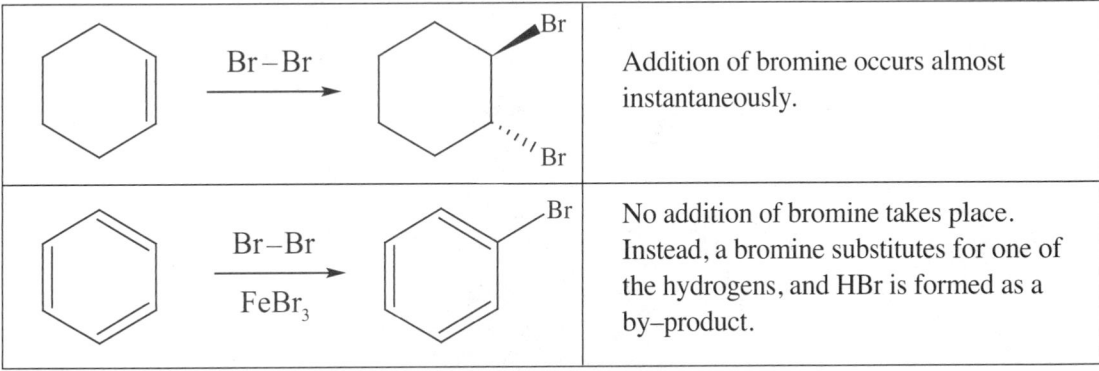

	Addition of bromine occurs almost instantaneously.
	No addition of bromine takes place. Instead, a bromine substitutes for one of the hydrogens, and HBr is formed as a by–product.

C. Hückel's Rules

In the 1930s, German scientist Erich Hückel formulated a set of rules that could be used to predict which compounds would exhibit the special stability that benzene has. A compound that exhibits this special stability was called an **aromatic compound**. These rules follow:

1. The aromatic part of the compound must contain a ring.
2. Each ring atom must have a p (or d) orbital.
3. The ring of p (or d) orbitals must be planar (flat), or nearly so.
4. There must be ($4n + 2$, $n = 0, 1, 2, 3, \ldots$) electrons in the system of p (or d) orbitals.

A compound must obey all the rules to be aromatic. Compounds that obey rules 1, 2, and 3, but have $4n$ electrons in the system of p (or d) orbitals are said to be **antiaromatic**. Antiaromatic compounds are especially unstable. Compounds that fail to obey these rules in any other way are said to be **nonaromatic**.

D. Aromatic and Antiaromatic Compounds

Let's consider benzene, 1,3-cyclobutadiene, and 1,3,5,7-cyclooctatetraene, as far as Hückel's rules are concerned. See below.

Rule			
Ring?	Yes	Yes	Yes
p or d orbital on each ring atom?	Yes	Yes	Yes
Planar?	Yes	Yes	No, tub-shaped
$4n + 2$ *p* (or *d*) electrons?	Yes ($n = 1$)	No ($n = 1/2$)	No ($n = 3/2$)
Classification	Aromatic	Antiaromatic	Nonaromatic

Of these rules, the hardest to determine without using models or chemical structure software is whether a ring is planar. Most rings with conjugated double bonds with fewer than eight members are usually planar. Larger rings can be planar, but sometimes they would need to have very large bond angles, or have atoms to bump into each other, which is not good. If so, they adopt a nonplanar geometry.

In addition to special stability, aromatic compounds have another property in common: the so-called **aromatic ring current**. When aromatic compounds are placed in a strong magnetic field, the pi electrons behave as if they were circulating in the system of *p* orbitals, and this induces a small magnetic field. The effects of this can be observed by the technique of nuclear magnetic resonance, which is discussed in Chapter 15. The net result of this is that the positions of the peaks for the hydrogens change due to the induced magnetic field. Normal alkene hydrogens appear at 5–6 ppm in the hydrogen NMR spectrum, whereas hydrogens on the **outside** of aromatic rings appear at **7–9 ppm**, and hydrogens on the **inside** of aromatic rings appear at **negative ppm** values. So NMR is a valuable technique for determining if a compound is aromatic. As different chemists made compounds to attempt to verify experimentally the predictions of Hückel's rules, NMR spectroscopy was used to see if the compounds exhibited a ring current.

Compounds containing elements other than carbon can also be aromatic. See the examples below.

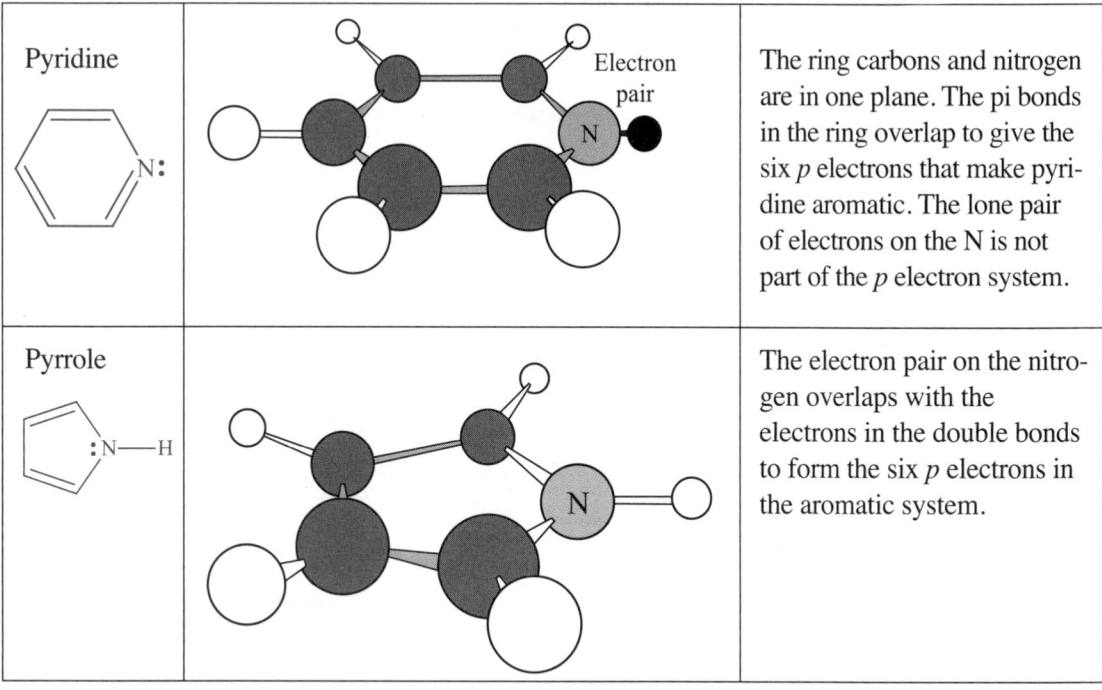

| Pyridine | | The ring carbons and nitrogen are in one plane. The pi bonds in the ring overlap to give the six p electrons that make pyridine aromatic. The lone pair of electrons on the N is not part of the p electron system. |
| Pyrrole | | The electron pair on the nitrogen overlaps with the electrons in the double bonds to form the six p electrons in the aromatic system. |

Certain carbocations and carbanions can also be aromatic. See the examples below.

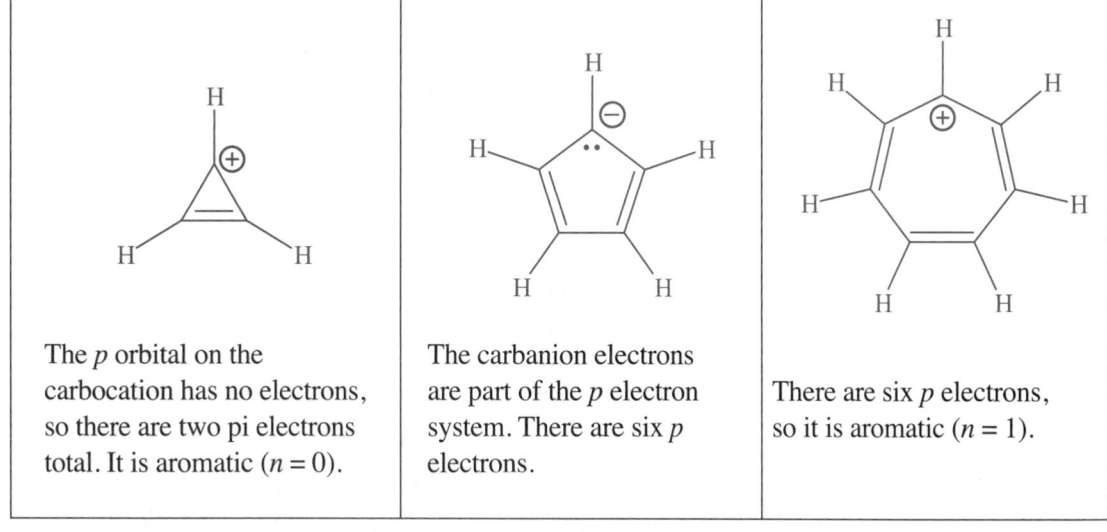

The p orbital on the carbocation has no electrons, so there are two pi electrons total. It is aromatic ($n = 0$).	The carbanion electrons are part of the p electron system. There are six p electrons.	There are six p electrons, so it is aromatic ($n = 1$).

E. Nomenclature of Benzene Derivatives

Benzene is just above alkanes in naming priority. Therefore, benzene rings with alkyl groups, halogens, ethers, and nitro groups attached are names as substituted benzenes. If the benzene ring is substituted onto an alkene, alkyne, alcohol, or thiol, it is named as a **phenyl group**. Let's name some monosubstituted benzenes.

$CH_2-CH_2-CH_2-CH_3$ attached to benzene ring	There is a standard four-carbon alkyl group attached to the benzene ring, so this is **butylbenzene**.
benzene ring with Br	This is **bromobenzene**.
H_3C-CH_2-O attached to benzene ring	This is an ether with a two carbon chain attached to a benzene, so it is **ethoxybenzene**.
benzene ring with Mercedes symbol	This is something I draw on the board with a straight face as I lecture. Some students recognize I am making a joke; others try to figure out what kind of a group I have stuck on the benzene ring. This is the hood ornament of a Mercedes, so this is . . . Mercedes Benzene!
alkene chain with phenyl, OH, and $O-CH_3$	The highest priority functional group is the alcohol. The longest chain containing the OH is nine, so it is a *nonenol* of some sort. Numbering from the right to give the OH the lower number gives us **1-methoxy-5-phenylnon-8-en-2-ol**.

There are some monosubstituted benzenes that have "special" names that you need to know. Many of these names have been used for over 100 years, and IUPAC "grandfathered them in," so to speak. These are shown below. Sometimes naming gets tricky with these, but I will show you how it works as we go along.

CH$_3$	NH$_2$	OH	Benzaldehyde	Benzoic acid	SO$_3$H
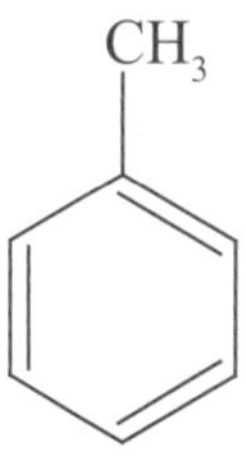 Toluene	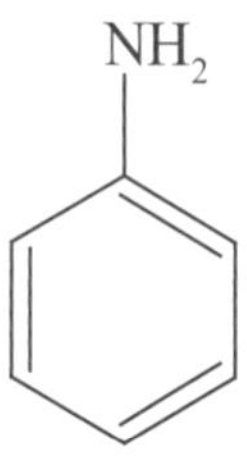 Aniline	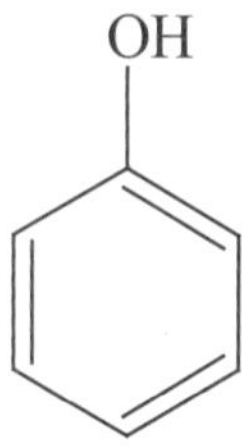 Phenol	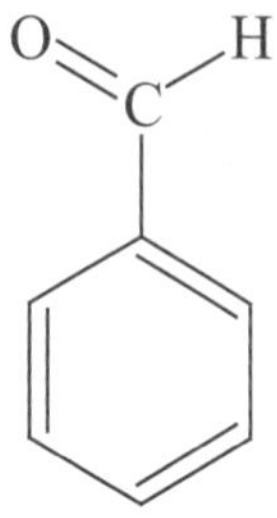 Benzaldehyde	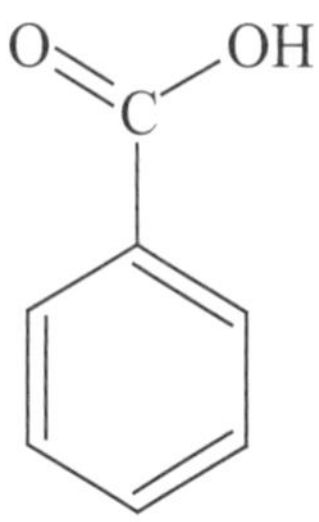 Benzoic acid	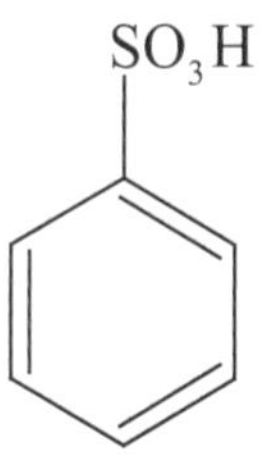 Benzene sulfonic acid

The functional groups on these molecules increase in naming priority from left to right.

As we discovered earlier, there are three different ways that benzene can be disubstituted, as shown below.

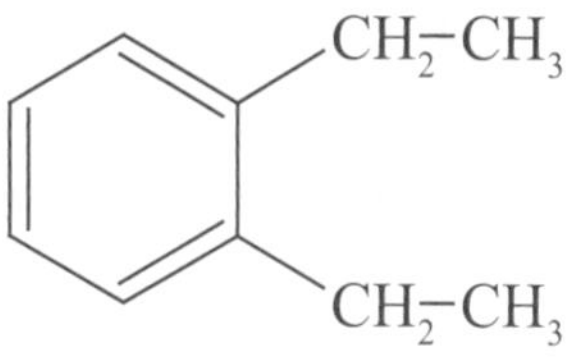 CH$_2$-CH$_3$ / CH$_2$-CH$_3$	This has two ethyl groups on adjacent carbons on the benzene ring, so it is named **1,2-diethylbenzene**. There is a common way to say that there are two groups in a 1,2-relationship on a benzene ring, and this is to put *ortho* or *o* in front of the words instead of numbers. So, a common name for this is *ortho*-diethylbenzene, or *o*-diethylbenzene. The *ortho* or *o* is italicized.
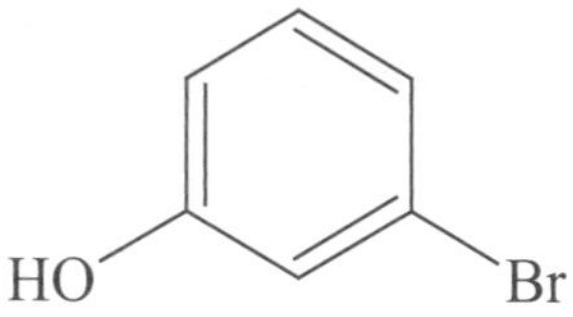 HO — Br	This has an OH on the benzene ring, so it is a *phenol*. The Br is on the third carbon from the OH, so this is **3-bromophenol**. Two groups in a 1,3 relationship are called *meta*, or *m*, in the common system, so a common name for this is *meta*-bromophenol.
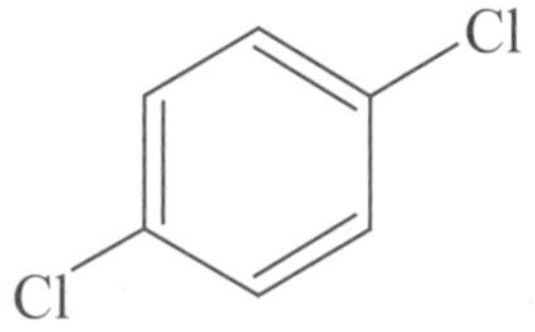 Cl / Cl	The two chlorines are on opposite sides of the ring, so this is **1,4-dichlorobenzene**. In the common system, two groups in a 1,4 relationship are called *para*, or *p*, so a common name is *para*-dichlorobenzene. This is commonly used in mothballs.

If there are more than two groups on a benzene ring, you find the special named group, if present, and start numbering there. If no special named group is present, then number as you would for any other ring. See the examples on the next page.

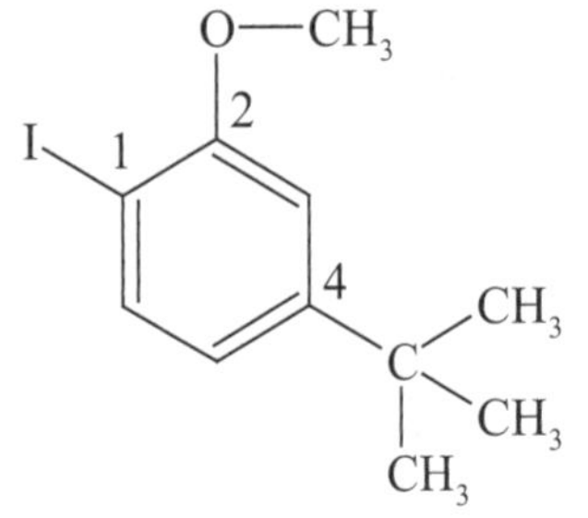	In this case, there are no special named groups, so this is a simple trisubstituted benzene. It is numbered as shown to give the lowest set of numbers. Putting the groups in alphabetical order gives us **4-*t*-butyl-1-iodo-2-methoxybenzene**.
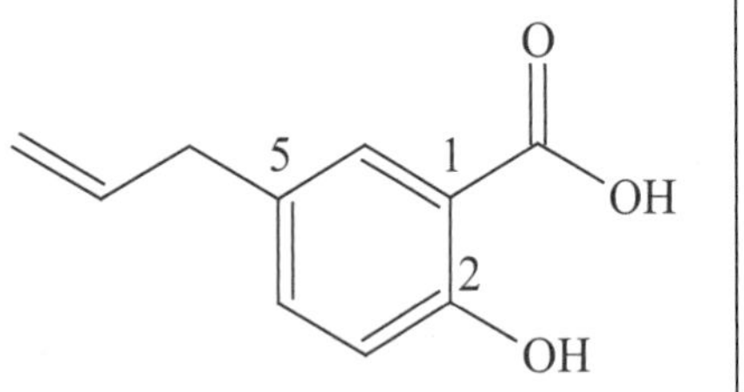	This has a carboxylic acid group on the benzene ring, so this is a benzoic acid. The carboxylic acid group is higher in priority than the OH. Carbon 1 is the carbon to which the carboxylic acid group is attached. We number clockwise because we come to the next group quicker taking this route. OH as a substituent is hydroxy. The other group is either allyl (accepted by IUPAC) or (2-propenyl). So the possible names are **5-allyl-2-hydroxybenzoic acid** or **2-hydroxy-5-(2-propenyl)benzoic acid**.
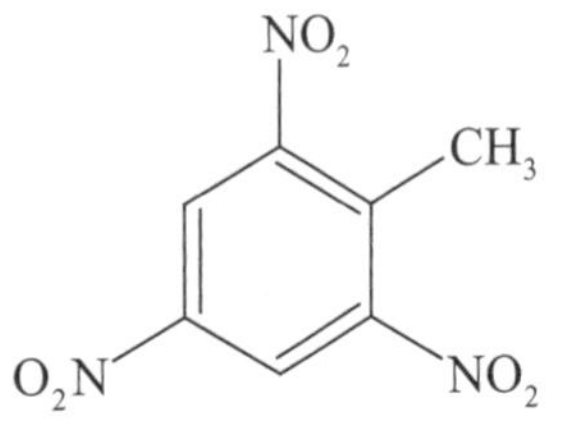	A methylbenzene is called toluene. There are three nitro groups attached, so this is a trinitrotoluene. We start numbering from the carbon where the methyl is attached, so this is **2,4,6-trinitrotoluene**. This is otherwise known as TNT, the explosive.

F. Electrophilic Aromatic Substitution of Benzene

The most common reaction of benzene is electrophilic substitution, which is, as it sounds, the substitution of a hydrogen by an electrophile. An electrophile is similar to a Lewis acid: It can accept a pair of electrons. The general mechanism is shown below: E^+ is a generic electrophile.

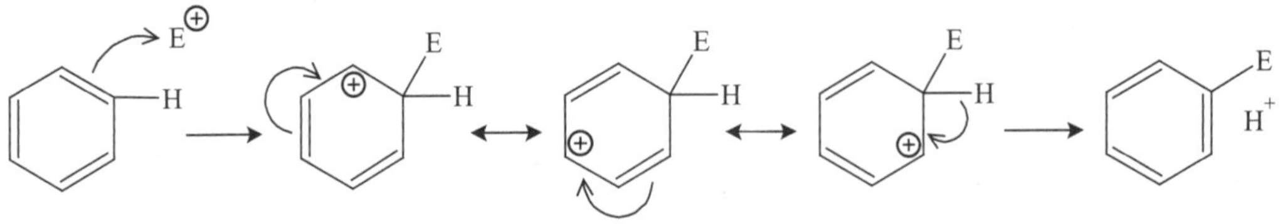

The electrophile is attacked by one of the electron pairs in the benzene ring. This forms the cation-containing structure shown, which is one of three resonance structures. Loss of H^+ reforms the aromatic system of *p* orbitals.

A common question on almost every organic chemistry exam involves drawing and understanding this mechanism. When you are drawing this mechanism, make sure that you know how to push electrons, that you don't draw any carbons making more than four bonds, and that the positive charge is not on a C=C.

There are five common electrophilic substitution reactions that occur by this general mechanism. They differ only in the structure of the electrophile and in how the electrophile is formed.

F.1. HALOGENATION

Halogenation of benzene generally occurs when benzene is treated with a halogen (usually bromine or chlorine) plus a Lewis acid catalyst, which is often the Fe(III) halide. Iodination does not occur under these conditions, but it does occur in the presence of an oxidizing agent or a silver(I) salt. See the examples below.

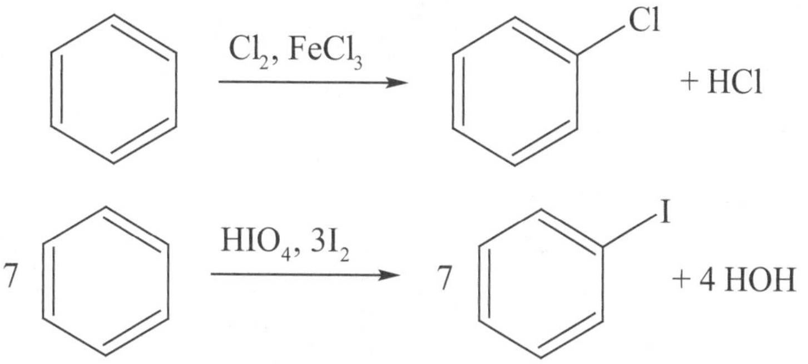

In chlorination or bromination, the electrophile is formed when the halogen complexes the iron halide, as shown below.

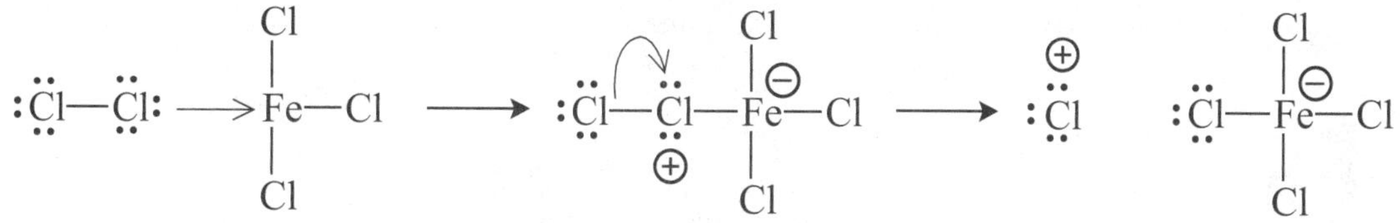

There is some evidence that the middle species is the real electrophile. Cl^+ is easier to write when drawing out a mechanism. For iodination, I^+ is thought to be the actual electrophile, formed in the oxidation-reduction reaction of I_2 and HIO_4.

F.2. NITRATION

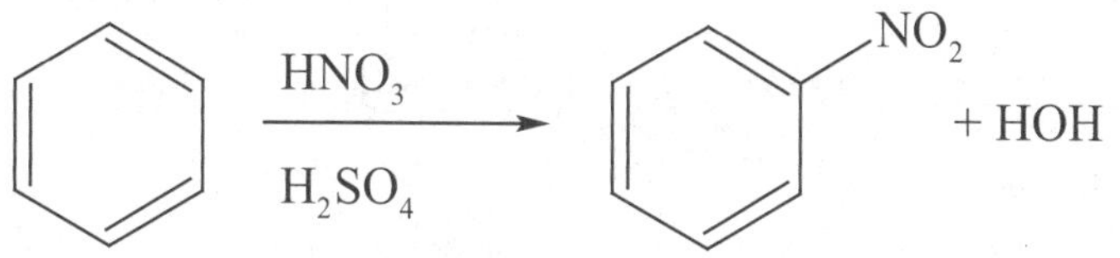

This is the most common way to introduce a nitro group onto a benzene ring. Nitric acid is the reagent, and sulfuric acid serves as a catalyst and solvent. The electrophile, NO_2^+, is formed as follows.

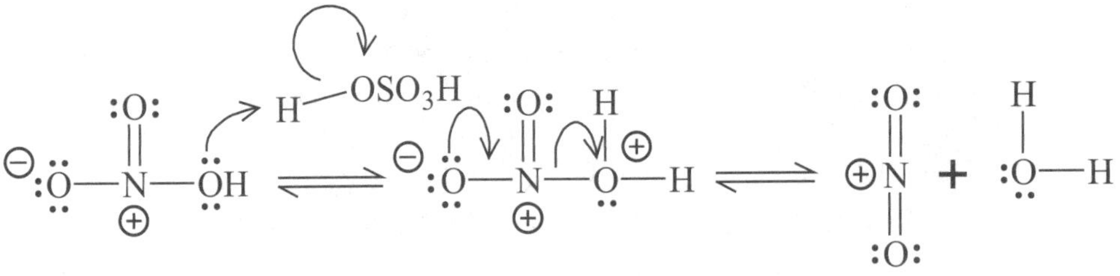

The mechanism of nitration is shown below. Because the nitrogen is positive, it is attacked by the electron pair from benzene. I didn't show all of the resonance structures.

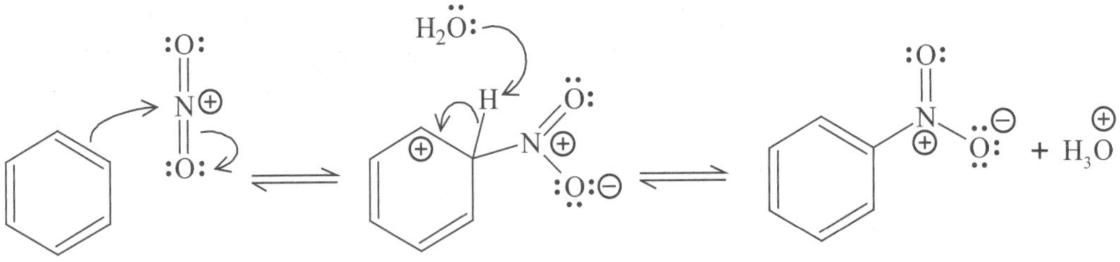

F.3. SULFONATION

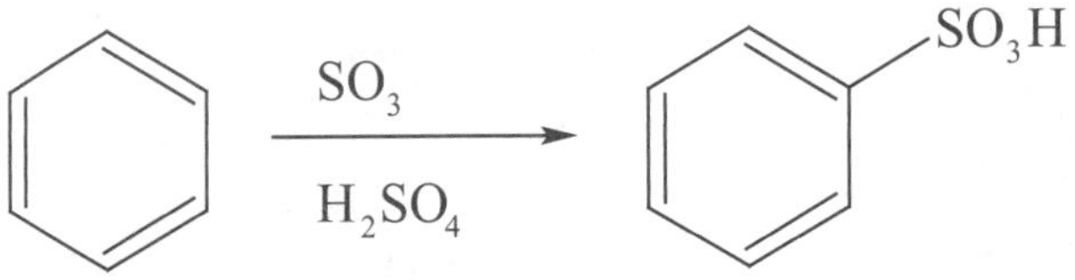

This is one way benzenesulfonic acid is made. Sulfur trioxide in sulfuric acid is called **oleum**, and is a very corrosive material. The electrophile is SO_3 itself. The mechanism of the substitution reaction is shown below.

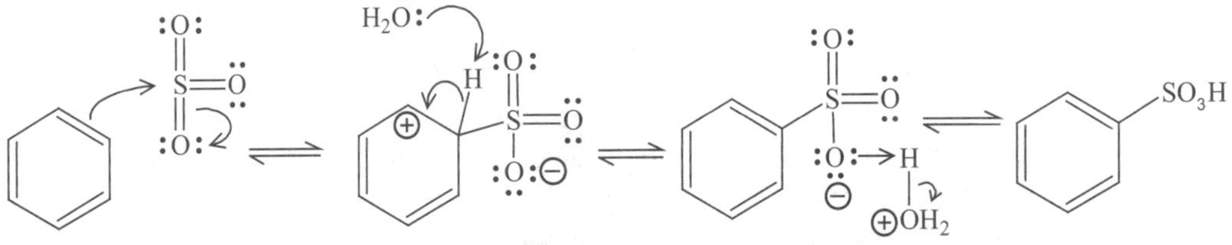

The sulfonation reaction is reversible. If benzenesulfonic acid is heated in dilute sulfuric acid, benzene is produced, and the SO_3 reacts with water to become sulfuric acid. We will see later why you would want to do this.

F.4. FRIEDEL–CRAFTS ALKYLATION

General Reaction:

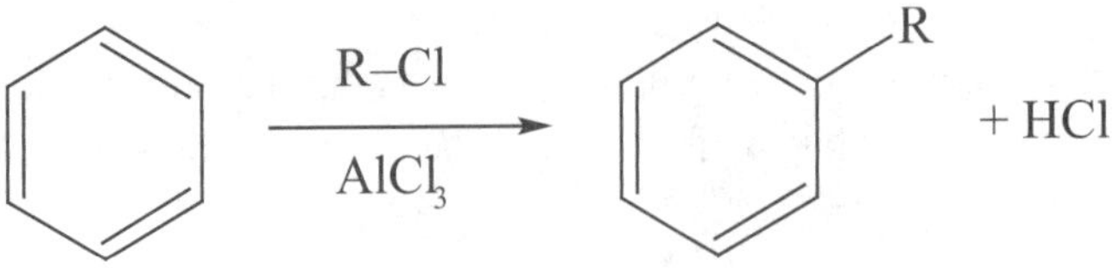

Specific Examples:

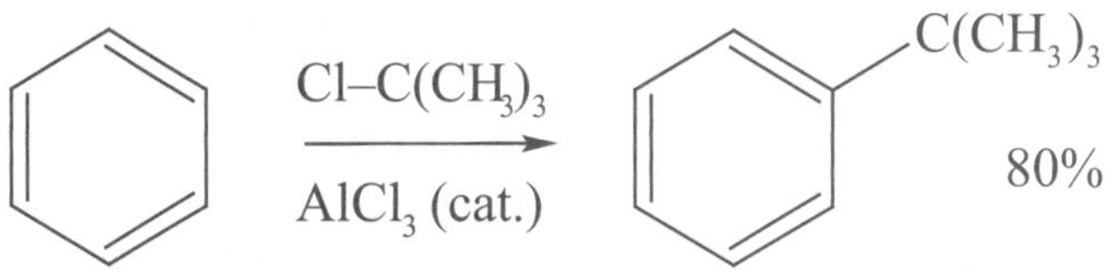

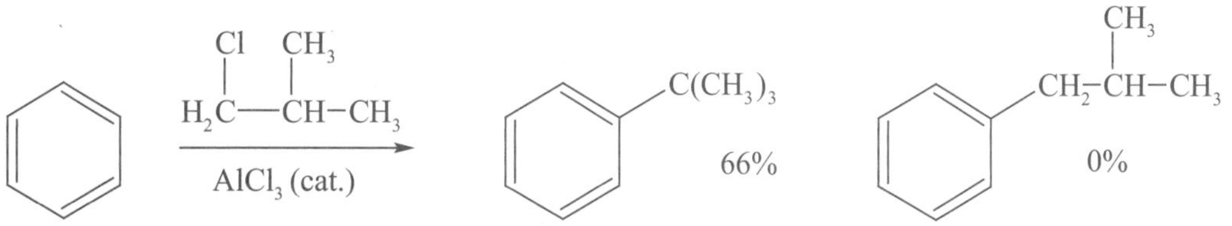

Friedel–Crafts alkylation (named after French chemist Charles Friedel and American chemist James Crafts) is a common way to introduce an alkyl group onto a benzene ring. Aluminum chloride is a common catalyst, although others work as well. As the second example shows, however, rearrangements can occur. This should not be surprising if the reaction proceeds through a cation mechanism. Therefore, this reaction works most predictably for alkyl groups that don't rearrange. The mechanism of the second reaction is shown below.

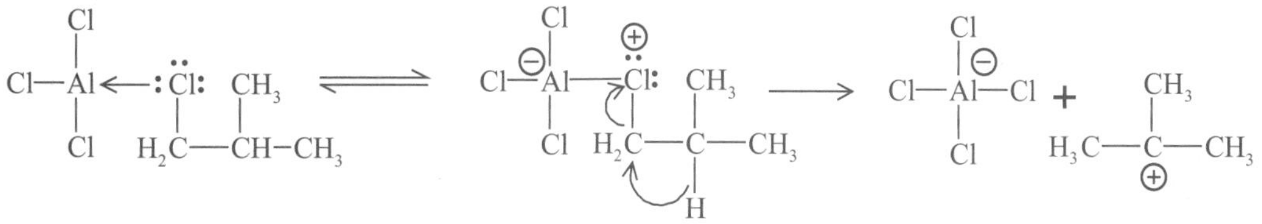

The $AlCl_3$ complexes the chloride, after which the carbon–chlorine bond breaks, with simultaneous hydride shift, to form the tertiary cation, and $AlCl_4^-$.

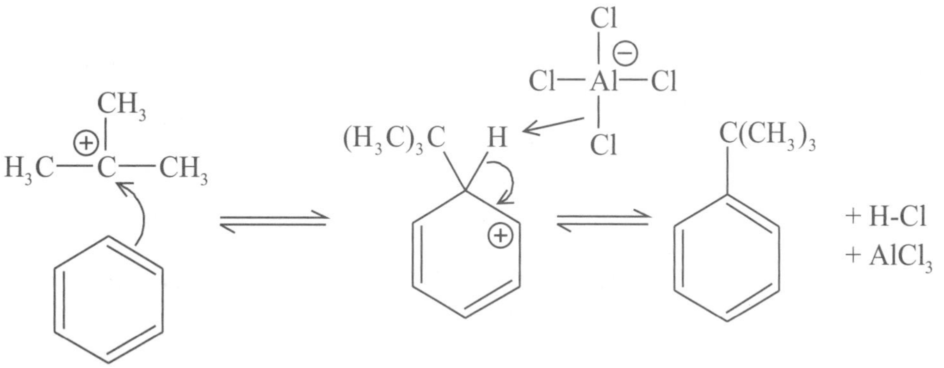

The benzene ring attacks the cation, then H^+ is lost as H–Cl to form the product.

Because carbocations are intermediates, other reactions that generate carbocations can lend themselves to Friedel–Crafts alkylations. See the examples on the next page.

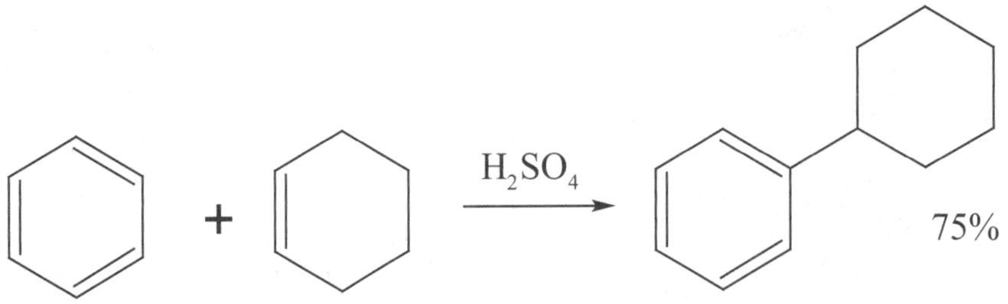

75%

Sulfuric acid protonates the C=C to generate a cyclohexyl cation.

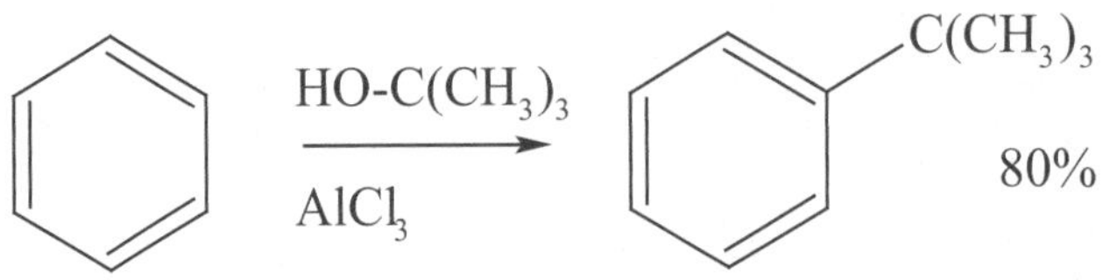

80%

In this case, the AlCl₃ is not catalytic; an equimolar amount as the alcohol is used.

Because cation rearrangements can occur with Friedel–Crafts alkylation, it cannot be universally used to make all kinds of alkylbenzenes. We will discover a way around this as we go on.

F.5. FRIEDEL–CRAFTS ACYLATION

General Reaction:

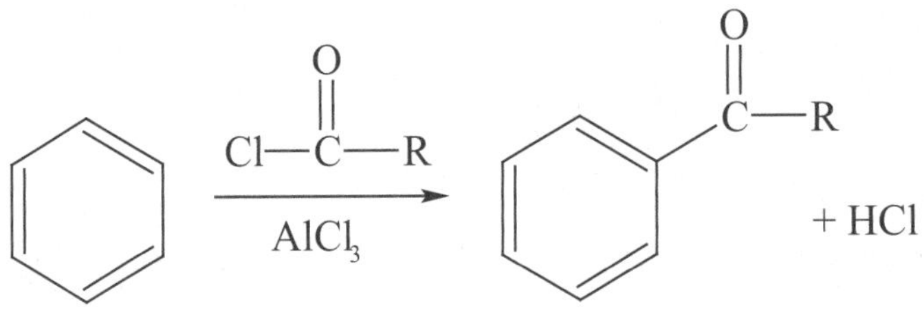

+ HCl

Specific Examples:

70%

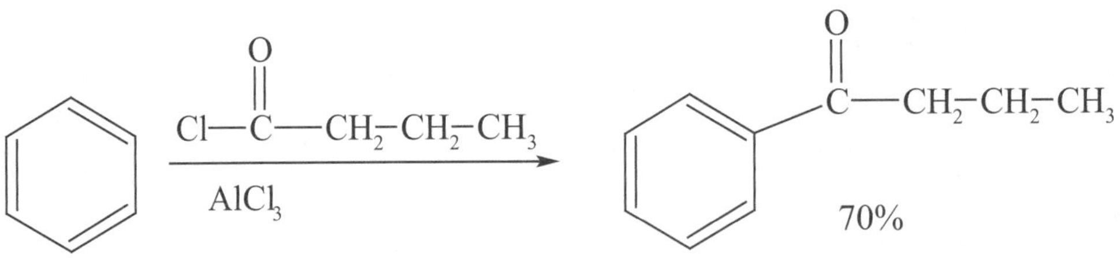

Notice that there was no rearrangement! We will look at why a little later.

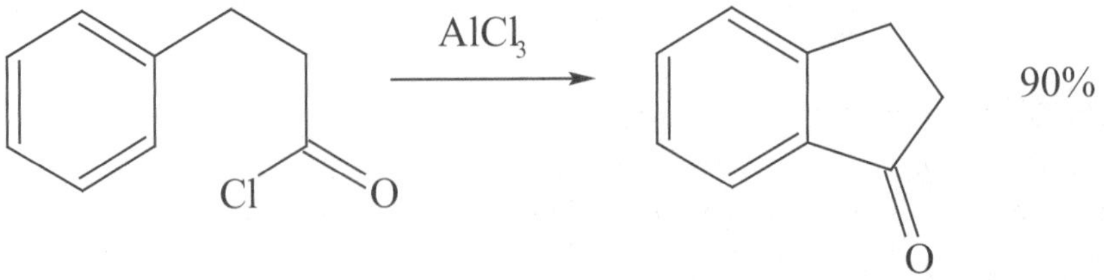

90%

If the benzene ring and the acid chloride are in the same molecule, the acid chloride can attack the benzene ring and form another ring. This works best to form five- and six-membered rings.

The mechanism of this reaction is similar to that of alkylation, except that there is no rearrangement. This is because the intermediate in acylation is resonance stabilized and is much more stable than a simple alkyl cation. See below.

The mechanism starts out similarly to halogenation, in that the Lewis acid complexes the chlorine to form the intermediate in the middle. The carbon–chlorine bond breaks to form the acyl cation.

The acyl cation is resonance stabilized. The resonance form on the right is especially stable because both the carbon and the oxygen have an octet of electrons.

By the hopefully now familiar mechanism, the benzene attacks the cation and then loses a hydrogen as H–Cl to reform the benzene ring.

One common variation of this reaction that students try on exams is the preparation of benzaldehyde by the following Friedel–Crafts reaction.

However, this acid chloride is not stable and cannot be isolated, so *DON'T DO THIS!* There are several indirect ways to introduce an aldehyde onto an aromatic ring. One of these is the Gatterman reaction, which involves treating the aromatic ring with zinc cyanide, HCl, and $AlCl_3$. HCN is formed and then protonated; it serves as the electrophile. An example is shown below.

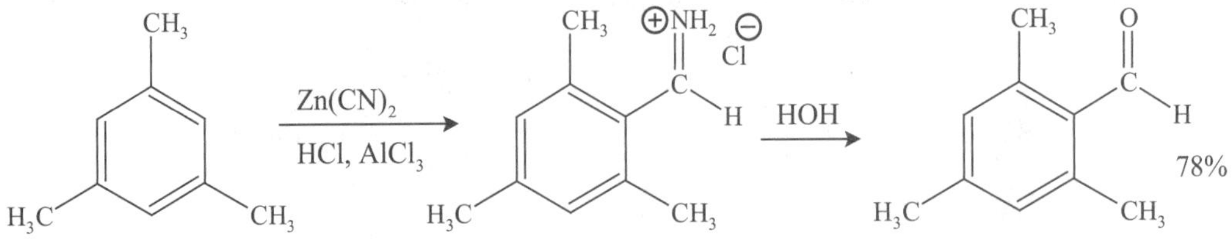

G. Electrophilic Aromatic Substitution of Substituted Benzenes: Directing Abilities

If there is already a group on the benzene ring, there are three possible places that the electrophile can be substituted for a hydrogen on the benzene ring: *ortho, meta,* or *para* to the original group.

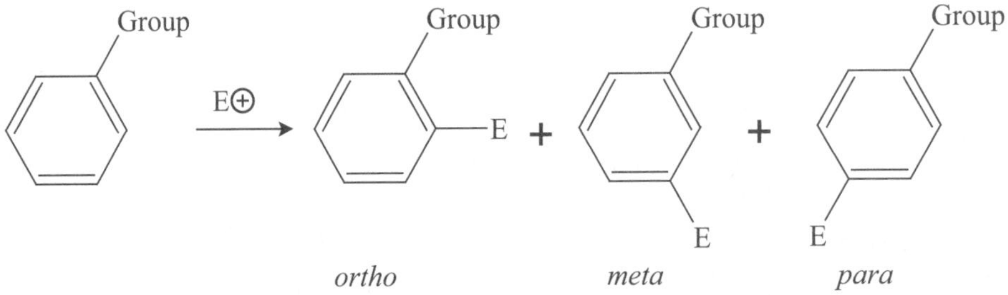

ortho   *meta*   *para*

Because there are two *ortho* positions, two *meta* positions, and one *para* position, one might expect that the ratio of the products to be in a ratio of 2:2:1. However, this ratio is never seen. Some groups give predominantly the *meta* product, and some groups give predominantly the *ortho* and *para* products. We will examine why below.

G.1. ORTHO-PARA DIRECTORS

Here is the mechanism for attack of an electrophile, E^+, on the ortho position of phenol. The positive charge is shared over three carbons in the benzene ring, as well as on the oxygen. This is because the O can donate a pair of electrons to the ring when the positive charge is on the adjacent C.

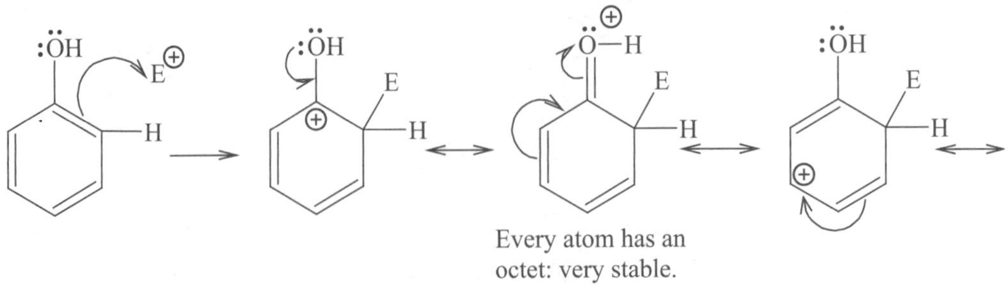

Every atom has an octet: very stable.

Here is the mechanism for attack of an electrophile, E^+, on the para position of phenol. Again, the positive charge is shared over three carbons in the benzene ring, as well as on the oxygen.

Every atom has an
octet: very stable.

Here is the mechanism for attack of an electrophile, E^+, on the meta position of phenol. The positive charge can be shared only over three carbons in the benzene ring. The positive charge is never on the carbon adjacent to the O, so the charge cannot be shared by the O.

There is nothing wrong with meta attack of the electrophile: Both ortho and para attacks allow the positive charge to be shared by the O as well. The more atoms the charge is shared over, the more stable the intermediate cation and the transition state leading to it are. Any group that has an unshared pair of electrons on the atom attached to the benzene ring will be an ortho-para directing group: OH, NH_2, OR, O(C=O)R, NH(C=O)R, halogens, and so on. Alkyl groups are also ortho-para directors, even if they don't have an unshared pair of electrons. Why? See the example on the next page.

This is a tertiary cation.

Here is the mechanism for ortho attack on toluene. The third resonance form is a tertiary cation, which is more stable than a secondary cation. Meta attack only leads to secondary cations, so ortho and para attack on toluene lead to a more stable cation intermediate and the transition state leading to it.

G.2. META DIRECTORS

Here is the mechanism for attack of an electrophile, E^+, on the ortho position of nitrobenzene. The positive charge is shared only over two carbons in the benzene ring because one resonance form has two (+) charges adjacent to each other. This resonance form is very high in energy and, thus, contributes little to the stabilization of the cation.

Two (+) charges next to each other: not very stable.

Here is the mechanism for attack of an electrophile, E^+, on the para position of nitrobenzene. Again, the positive charge is shared only over two carbons in the benzene ring, since one resonance form has two (+) charges adjacent to each other. This resonance form is very high in energy and, thus, contributes little to the stabilization of the cation.

Two (+) charges next to each other: not very stable.

Here is the mechanism for attack of an electrophile, E^+, on the meta position of nitrobenzene. The positive charge is shared over three carbons in the benzene ring because no resonance has two (+) charges adjacent to each other.

Meta substitution is favored because there is something bad about ortho and para substitution, not because there is anything especially good about meta substitution. Any group with a positive atom attached to the benzene ring, or a partially positive atom attached to the benzene ring is a meta-directing group: NO_2, CF_3, NH_3 (+), any group with a C=O attached to the benzene ring, etc.

H. Electrophilic Aromatic Substitution of Substituted Benzenes: Activating or Deactivating Effects

In addition to affecting the position of substitution of the electrophile on a benzene ring, the group also affects the rate of reaction of the benzene ring. The following table gives the relative rates of nitration of substituted benzenes, as well as the qualitative description of the effect of the group on reaction rate, and the group's directing ability.

Group on Benzene Ring	Relative Rate of Nitration	Qualitative Description of the Group's Effect on Reactivity	Ortho-Para or Meta Director
OH	1000	Strongly activating	Ortho-para
CH_3	25	Mildly activating	Ortho-para
H	1	The standard	
I	0.18	Mildly deactivating	Ortho-para
Cl	0.033	Mildly deactivating	Ortho-para
$COOCH_2CH_3$ (ester)	0.0037	Strongly deactivating	Meta
CF_3	2×10^{-5}	Strongly deactivating	Meta
NO_2	6×10^{-8}	Strongly deactivating	Meta

Notice a few points.

1. All meta directors are strongly deactivating. This means that they react very slowly with electrophiles because a meta-directing group is attracting electron density from the ring, either by an inductive effect, a resonance effect, or both.
2. All activating groups are ortho-para directors. This is due to the group's pushing electron density into the ring, either by resonance or induction.

3. Halogens are the only exceptions. They are deactivating because the halogen is pulling electron density from the ring by an inductive effect because halogens are more electronegative than carbon. They are ortho-para directors because electrons from the halogen can be donated to the ring by resonance. The two effects work against each other, and the net effect is that halogens are mildly deactivating.

By knowing and being able to apply a group's directing ability and activating or deactivating properties, you can design syntheses of substituted benzenes, which is the topic of the next section. One implication of strongly deactivating groups is the following: Friedel–Crafts reactions do not work on strongly deactivated benzenes (benzenes with meta-directing groups attached). So don't try it: It won't work!

I. Synthesis of Disubstituted and Trisubstituted Benzenes

Let's do a simple synthesis. Suppose you wanted to make 1-isopropyl-4-nitrobenzene, starting from benzene. How would you do it? First, you need to know how to put an isopropyl group and a nitro group on a benzene ring. These reactions were covered in Section F. Then you need to know their directing abilities. An isopropyl group is an ortho-para director, and a nitro group is a meta director. Finally, you need to know whether the groups are activating or deactivating. Isopropyl is mildly activating, and nitro is strongly deactivating. Now you have to apply what you know to decide what order to put the groups on the benzene ring. Because the two groups are para, you need to put the ortho-para directing group on first, so it will direct the nitro group into the para position. The synthesis looks like this.

The para product is usually the major product with an ortho-para director. Usually the ortho and para products have different enough physical properties that they can be easily separated.

What would happen if you reversed the order of the steps? See below.

Even if a reaction had occurred, the incoming group would have ended up meta to the nitro group because the nitro group is a meta director.

Let's try another. Suppose you wanted to make the following compound: How would you do it?

A bromine is an ortho-para director, and a ketone is a meta director. A bromine is mildly deactivating, and a ketone is strongly deactivating. Because the two groups are meta to each other, we need to put the ketone on first, so it will direct the bromine to the meta position. See below.

"Propiophenone"

I wrote this synthesis "backwards," that is, going from right to left. I did it intentionally because the best way to do syntheses is to work backwards. You ask yourself, "How do I make the product from anything?" In this case, brominating the benzene ring of propiophenone is one obvious way. Then you ask yourself, "Can I make propiophenone from benzene?" In this case, the answer is yes. If the answer were no, or if you forgot the Friedel–Crafts acylation, then you would ask yourself, "How do I make propiophenone from anything?" You would hopefully come up with a reaction that would work, and eventually get back to benzene. One possible route would be the following.

"Propiophenone"

Propiophenone could be made by oxidizing the secondary alcohol. The secondary alcohol could be made with a Grignard reaction. Now we have to make the Grignard reagent from benzene.

Adding magnesium to bromobenzene gives us the Grignard reagent. Bromobenzene is made by brominating benzene.

I presented this alternate synthesis to show you there is more than one way to make the product. You should be acquiring a collection of reactions, so if you forget one, you may be able to work around it. Now some reactions, like nitration, are the only way you are going to learn in this course to put a nitro group on a benzene ring. If you forget it, you are sunk.

Let's do another example: 1-*t*-butyl-2-iodobenzene. Because the two groups are ortho, and both groups are ortho-para directors, you might be tempted to do this in two steps, such as:

This approach would not work because most of the iodo compound produced in the second step would be para, not ortho. If you could somehow block the para position, then the iodine would have to go ortho to the *t*-butyl group. Removal of the blocking group would give the product. We can accomplish this by sulfonating *t*-butylbenzene, iodinating, and then removing the sulfonic acid group. See below.

There are three oxidation or reduction reactions which are commonly presented in the chapter on synthesis of aromatic compounds. These are the following.

I.1. REDUCTION OF A NITRO GROUP

Examples:

Aromatic nitro groups are reduced to amines by treatment with hydrogen and a metal catalyst (typically Pd, Pt, or Ni). Nitro groups reduce very quickly and can be reduced in the presence of C=O containing functional groups. Selective reduction of a nitro group in the presence of C=C's and C≡C's cannot easily be done. Fe, Sn, or Zn in hydrochloric acid also reduces nitro groups. These metals do not reduce C=C's or C≡C's. Fe and Sn do not reduce C=O's either. Nitro groups are *not* reduced by NaBH$_4$. The importance of this in synthesis is that a strongly deactivating, meta-directing nitro group is converted into a strongly activating, ortho-para-directing amino group.

I.2. THE CLEMMENSEN REDUCTION

Example:

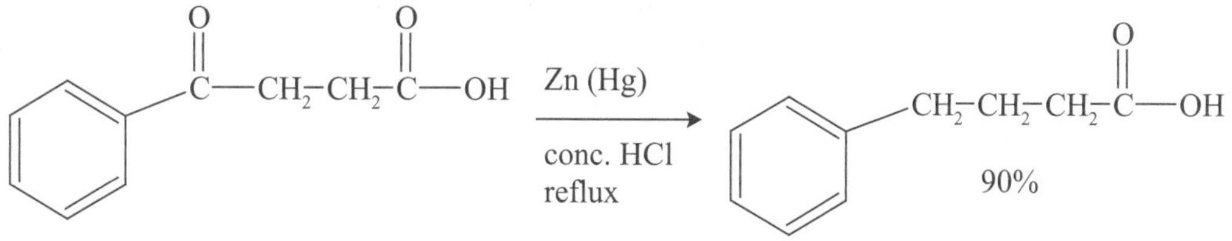

Aldehyde and ketone C=O's are reduced to CH$_2$'s by zinc amalgam (zinc metal that has been treated with mercuric chloride in the presence of HCl: It is commonly written as Zn(Hg)). Chlorines and bromines on a benzene ring are not affected (iodines *are* reduced), but halogens on alkyl groups and nitro groups are reduced. Other types of C=O compounds are generally not reduced. The importance of this in synthesis is that a strongly deactivating, meta-directing C=O group is converted into a mildly activating, ortho-para-directing alkyl group.

There are other ways to do this reduction that will be covered in Chapter 12.

I.3. OXIDATION OF ALKYLBENZENES TO CARBOXYLIC ACIDS

Examples:

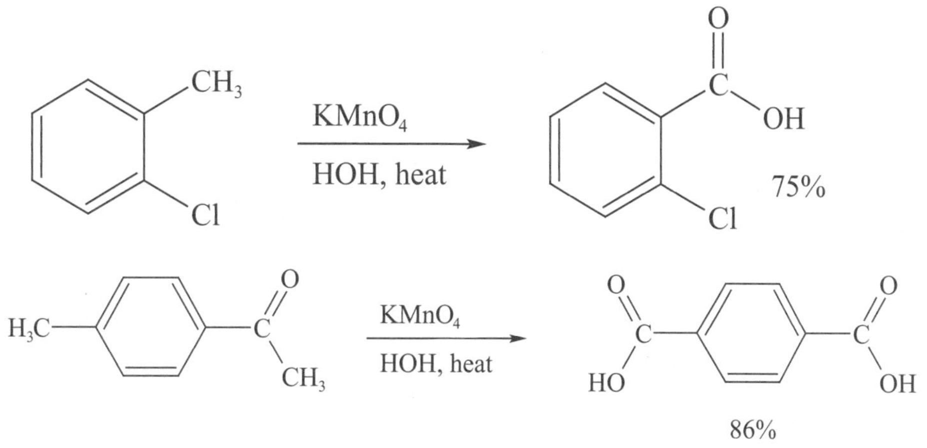

If a methyl group, a primary alkyl group, or a secondary alkyl group is attached to a benzene ring, it can be oxidized to a benzoic acid by a hot concentrated solution of potassium permanganate in water. Aldehydes and ketones attached directly to the benzene ring are also oxidized. Tertiary alkyl groups attached to the benzene ring are not oxidized. Halogens and nitro groups are unaffected, but amino groups are oxidized to mixtures of compounds. This converts an ortho-para-directing alkyl group into a meta-directing carboxylic acid.

Now you know these reactions, try the following synthesis problems, starting from benzene. First, how would you make 1-chloro-3-ethylbenzene?

Both of the groups on the ring are ortho-para directors, but they are meta to each other. That must mean there is a way to turn a meta director into one of these groups. The Clemmensen reduction will allow us to convert a ketone into an alkyl group, so this is the last step. Now all we have to do is make the ketone.

Friedel–Crafts acylation gives us the ketone. Chlorination puts the chlorine meta.

The problem here is the opposite of the preceding problem: Both groups are meta directors, and they are para to each other. Therefore, we must have to convert some ortho-para-directing group into one of these meta-directing groups. Oxidation of an alkylbenzene produces a benzoic acid, so this is the approach we will take. See below.

Friedel–Crafts alkylation gives us the toluene: I could have made any alkylbenzene, except something like *t*-butylbenzene. Nitration of toluene gives us 4-nitrotoluene, which we can separate from any ortho product. Oxidation of 4-nitrotoluene gives us the product.

J. Phenol Naming and Properties

Phenols are compounds with an OH group bonded to the benzene ring. Phenols are commonly found in a variety of naturally occurring compounds, such as the ones below.

	Vanillin, the major component of vanilla flavoring.
	Epinephrine, otherwise known as **adrenaline.** This is an adrenal hormone that is released when you need to "fight or flight."

The parent compound is named phenol. As a substituted benzene with a special name, numbering starts at the carbon where the OH is attached. See the examples below.

	This phenol has a methoxy and an allyl group attached. Because *a* comes before *m*, this is **4-allyl-2-methoxyphenol**. This is also know as **eugenol** and has the odor of cloves.
	When the OH group is not the highest priority group, it is called hydroxy as a substituent. This is named **2-hydroxybenzoic acid**, according to IUPAC.

Phenols are much more acidic than normal alcohols. See the table below.

Compound	pK_a	
Cyclohexanol	18	Cyclohexanol is a typical alcohol, with a pK_a somewhat higher than water (15.7). Phenol's pKa is almost 6 units lower than water, which means it is almost a million times more acidic than water and 100 million times more acidic than cyclohexanol. Why? When cyclohexanol ionizes, the negative charge is located on the oxygen only. When phenol ionizes, the negative charge can be shared by three carbons in the ring. This lessens the amount of negative charge on any one atom and makes the anion more stable. A more stable anion helps shift the equilibrium more toward products, which means more acid is produced.
Phenol	9.9	
4-methylphenol	10.2	
2-nitrophenol	7.2	
3-nitrophenol	8.3	
2-nitrophenol	7.2	
2,4-dinitrophenol	4.0	
2,4,6-trinitrophenol	0.4	

When cyclohexanol ionizes, the negative charge is only on the oxygen.

The negative charge can be shared by three carbons in the ring.

Three resonance structures with the negative charge on carbons in the ring

In 4-nitrophenol, the negative charge is also shared by an oxygen in the nitro group, which helps further stabilize the anion. A similar resonance form can be drawn for ionization of 2-nitrophenol, but it cannot be drawn for 3-nitrophenol. Therefore, 3-nitrophenol is the least acidic nitrophenol. A nitro group can inductively pull electron-density toward itself, so it is more acidic than phenol.

As more nitro groups are substituted *ortho* or *para* to the phenol, the acidity of the phenol increases. 2,4,6-Trinitrophenol is quite a strong acid for an organic compound, and is more acidic than most carboxylic acids.

K. Other Aromatic Compounds

A large number of aromatic compounds are either found in nature, or have been synthesized. Many of these compounds have two or more benzene rings fused together to make a large system of rings. One of the ring systems below is not completely aromatic: Which one is it?

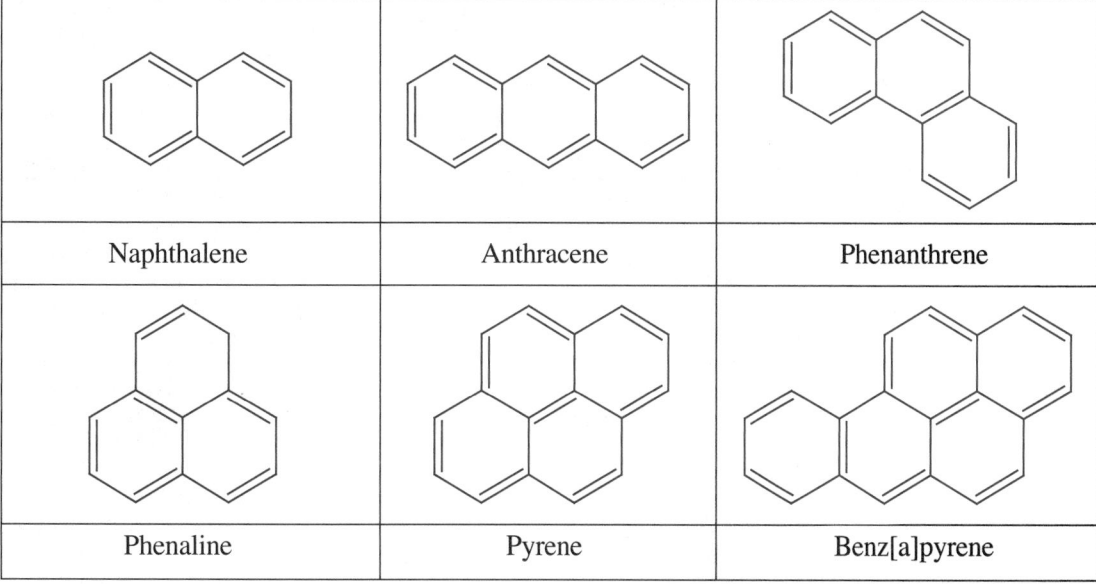

Naphthalene	Anthracene	Phenanthrene
Phenaline	Pyrene	Benz[a]pyrene

Phenaline is the one that is not completely aromatic: the top ring has a carbon that does not have a *p* orbital.

REVIEW EXERCISES FOR CHAPTER 10

1. Draw all of the monochloro and dichloro derivatives of the following two compounds. Ignore stereoisomers.

a.	b.	c.	d.

2. Give IUPAC names for the following structures.

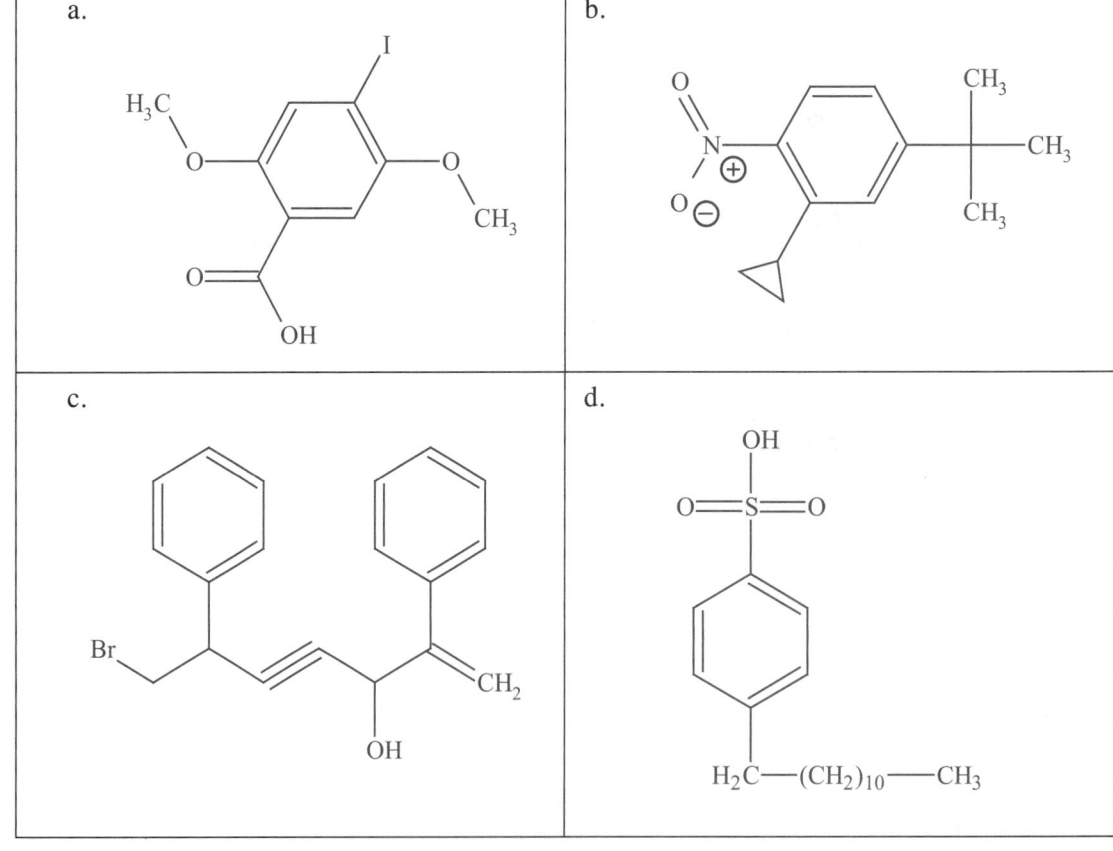

a.

b.

c.

d.

3. Give the major organic product of each of the following reactions. If both ortho and para products should form, draw only the para. If no reaction will occur, write "No Reaction." Briefly explain your reasoning.

a.

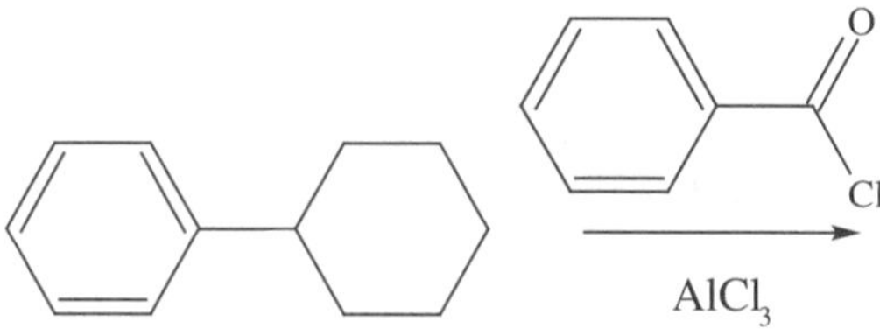

b.

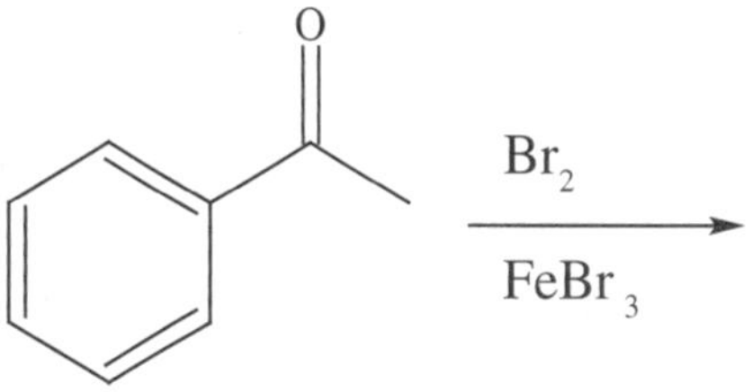

c.

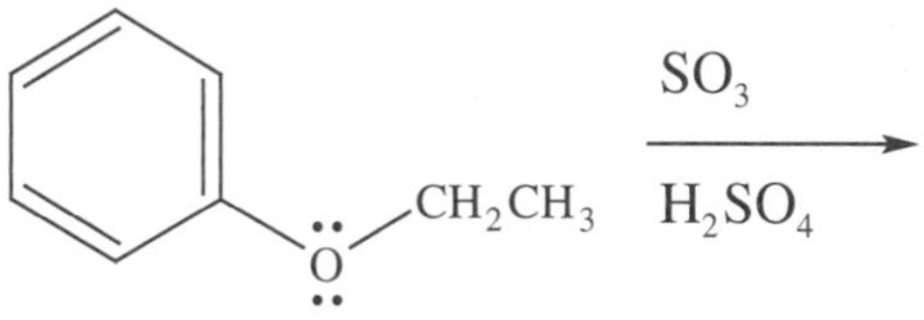

d.

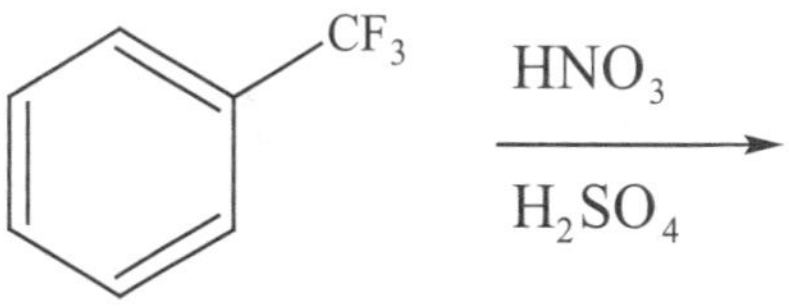

4. Propose syntheses of the following compounds, starting from benzene, plus any other reagents.

a.

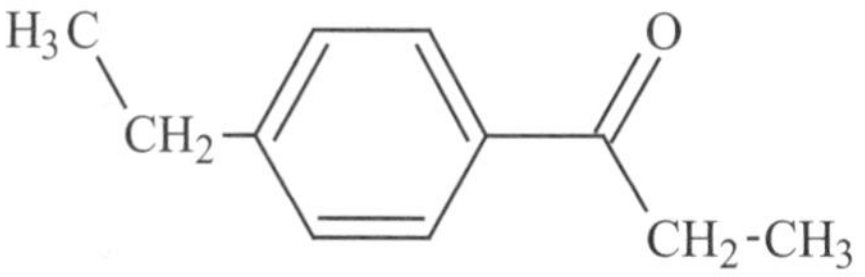

b.

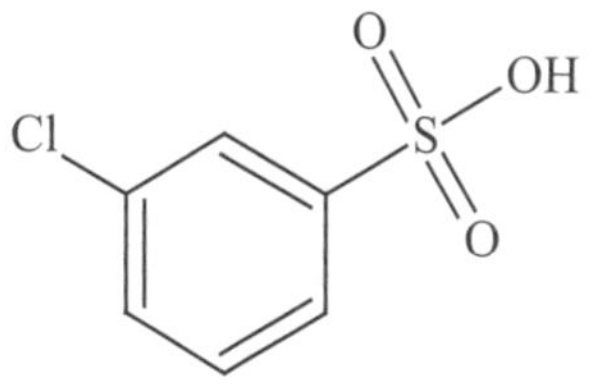

c.

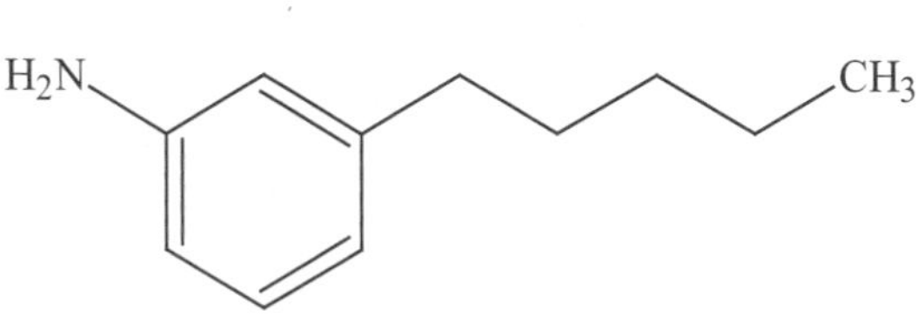

d.

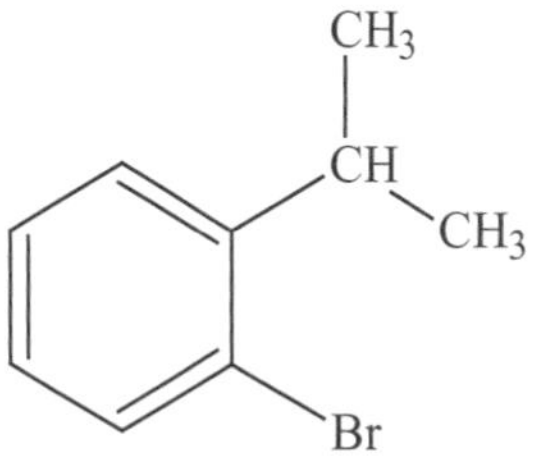

5. Classify the following structures as either aromatic, antiaromatic, or nonaromatic. Explain your reasoning. Consider only the atoms on the perimeter of the molecule.

   a.   b.   c.   d.

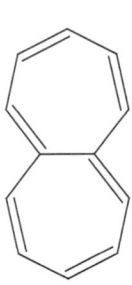 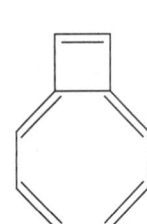 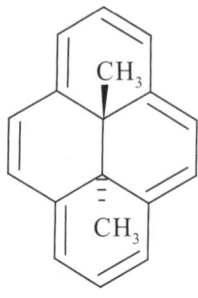

6. When the proton NMR spectrum of the structure in Problem 5c was taken, peaks showed up in two regions of the spectrum. Several peaks, which integrated for 10 H's, showed up between 8 and 9 ppm. A single peak, which integrated for 6 H's, showed up at –4.25 ppm (4.25 ppm to the *right* of the TMS standard peak at 0 ppm)! Are these data consistent with the structure in Problem 5c's being aromatic, antiaromatic, or nonaromatic? Explain your reasoning.

Amines

WHAT YOU WILL LEARN

In this chapter, you will learn:

- how to recognize different types of amines;
- how to name amines using IUPAC rules;
- how the structure of an amine influences its physical properties and basicity;
- how amines react with various electrophiles and with nitrous acid;
- how to prepare amines by nucleophilic substitutions, reductions, and rearrangements.

SECTIONS IN THIS CHAPTER

- Nomenclature

- Physical Properties

- Basicity

- Reactions as Nucleophiles

- Reactions of Amines with Nitrous Acid

- Preparations

A mines are those compounds in which the nitrogen is bonded to one or more carbon-containing groups that are not C=O's. There are three types of amines: primary, secondary, and tertiary. Some examples are shown below. Notice that for amines, primary, secondary, and tertiary have different meanings than for a tertiary alcohol or alkyl halide.

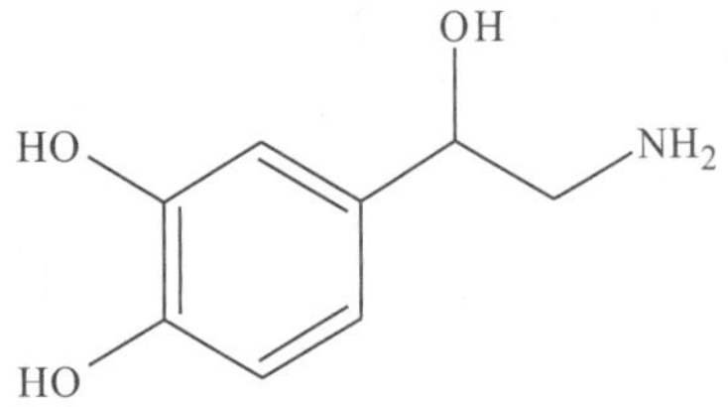	Norepinephrine, a neurotransmitter	**Primary amine**: the nitrogen is bonded to *one* carbon.
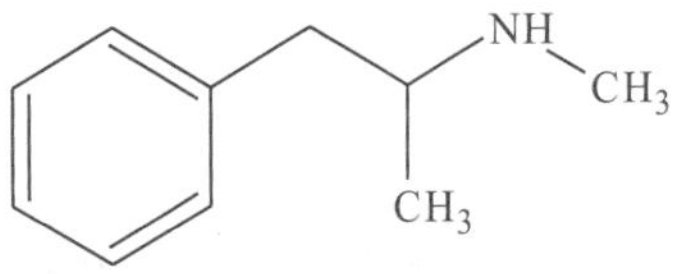	Methamphetamine, a synthetic drug of abuse	**Secondary amine**: the nitrogen is bonded to *two* carbons
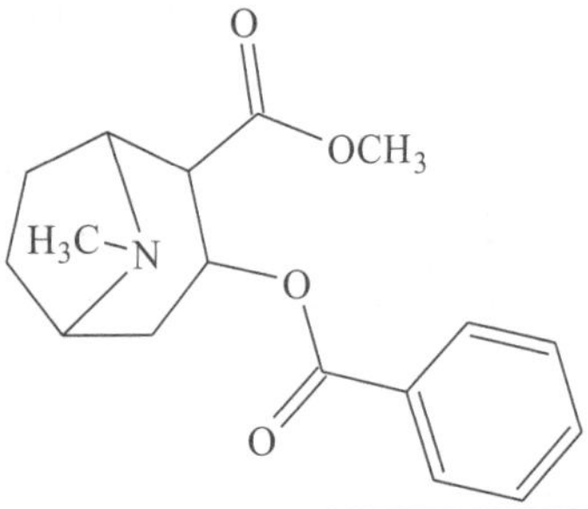	Cocaine, a naturally occurring drug of abuse	**Tertiary amine**: the nitrogen is bonded to *three* carbons.

A. Nomenclature

Amines are just below thiols and just above alkenes in functional group priority. See below.

Priority So Far	Functional Group	Ending If the Functional Group Is the Highest Priority	Substituent Name of the Functional Group
1.	Alcohol (–OH)	*-ol*	Hydroxy
2.	Thiol (–SH)	*-thiol*	Mercapto
3.	Amine (–NH$_2$)	*-amine*	Amino
4.	Alkene (C=C)	*-ene*	*-en-**
5.	Alkyne (C≡C)	*-yne*	*-yn-**
6.	Benzene	*benzene*	Phenyl
7.	Alkane (–R)	*-ane*	Alkyl
8.	Ether (–OR)		Alkoxy
	Halogens (–X)		Halo
	Nitro (–NO$_2$)		Nitro

* The *-en-* or *-yn-* replaces the *-an-* in the name.

We will look at naming primary amines first. Consider the following structures and try to name them.

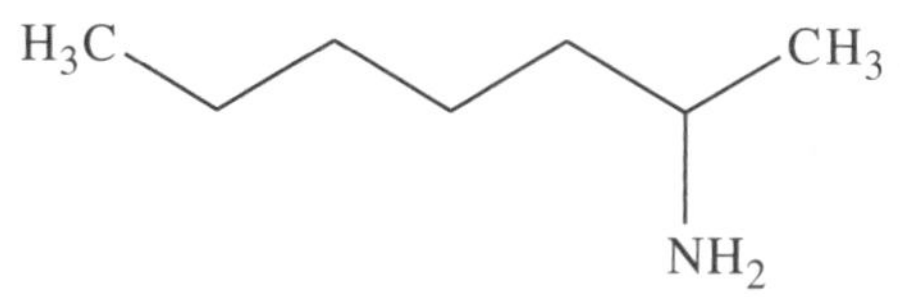	The longest carbon chain is seven. The amine is the highest priority functional group, so it is a heptanamine. The amine is attached to the second carbon, so it is a **heptan-2-amine**.
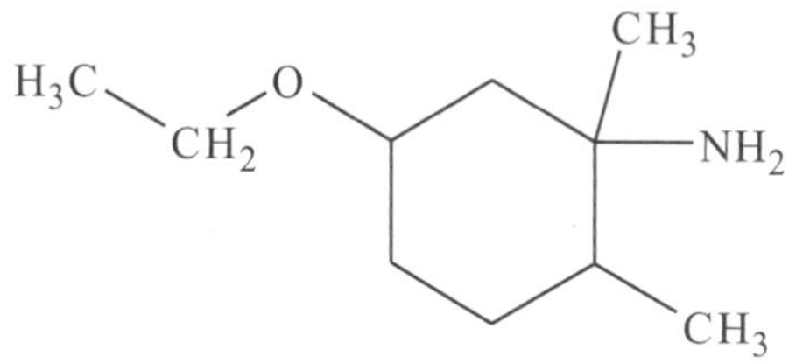	The amine is the highest priority functional group, and it is attached to a six-membered ring, so it is a cyclohexanamine. There are two methyl groups and an ethoxy group. Ethoxy comes first, but the methyls are closer to the amine, so we number clockwise. The name is **5-ethoxy-1,2-dimethylcyclohexanamine**.
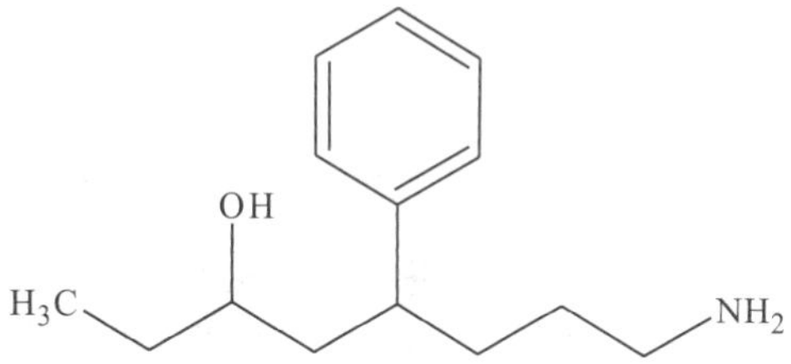	The alcohol is the highest priority functional group, so it is an octan-3-ol. It has an amino and a phenyl group attached, so it is **8-amino-5-phenyloctan-3-ol**.
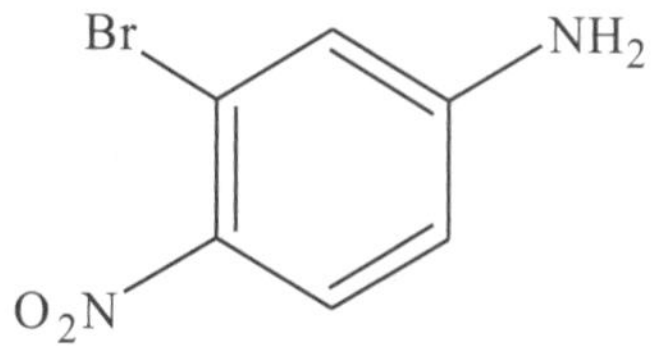	This example is just to keep you sharp with your "special-named" benzenes. A benzene with an amine is aniline, so this is **3-bromo-4-nitroaniline**.

Secondary and tertiary amines have a couple of twists, depending on whether the amine is the highest priority functional group. See the examples below.

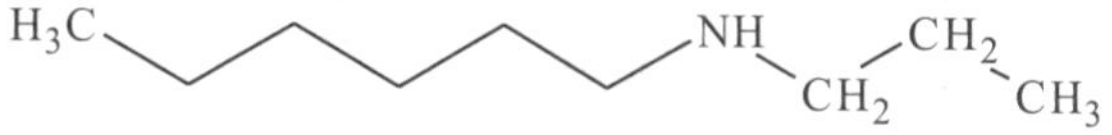	The longest carbon chain is six, with the amine on the end of it, so it is a hexan-1-amine. There is a propyl group attached to the nitrogen, so we use an *N-* to tell where it is. Therefore, this is *N*-**propylhexan-1-amine**.

	This aniline has four methyls attached, so it is a tetramethylaniline. Two methyls are on the nitrogen, and the others are on 2 and 5. IUPAC puts *N*'s before numbers, so this is *N*,*N*,2,5-**tetramethylaniline**.
	This is the same as the compound above, except with an added thiol, so this is a benzenethiol. There are two methyls, and a "dimethylamino" group attached. This is **4-(dimethylamino)-2,5-dimethylbenzenethiol**.

When naming amines such as pyrrole and pyridine, the nitrogen in the ring is numbered "1." See the example below.

	The ring is pyrrole. We start numbering at the N and go clockwise because we run into another group first going in that direction. The name of this compound is **1,2-dimethyl-4-nitropyrrole**.

B. Physical Properties

The most noticeable property of low-molecular-weight amines is their odors, which can be described as pungent. Some are fishy, while others are pretty obnoxious to me!

The boiling points of primary and secondary amines tend to be between those of comparable alkanes and alcohols. The boiling points of these amines are higher than those of alkanes because the amines can hydrogen-bond to each other, while alkanes cannot. The amines have lower boiling points than the alcohols, since nitrogen is less electronegative than oxygen, so there is less negative charge on the nitrogen than on the oxygen of an alcohol. Tertiary amines cannot hydrogen-bond with each other, so their boiling points are very similar to those of the corresponding alkanes. See the table on the next page.

Compound	Boiling Point (°C)	Water-Solubility (g/100 mL H_2O)
$CH_3CH_2CH_2CH_2OH$	118	7.4
$CH_3CH_2CH_2CH_2CH_3$	36	0.036
$CH_3CH_2OCH_2CH_3$	35	6.1
$CH_3CH_2CH_2CH_2NH_2$	77	Miscible
$CH_3CH_2NHCH_2CH_3$	56	Miscible
$CH_3CH_2N(CH_3)_2$	37	Miscible

Primary and secondary amines are more water-soluble than most alcohols and ethers of similar molecular weight. The reason for this is not obvious, but it is true in most cases concerning low-molecular-weight amines. It may have something to do with the increased basicity of the amines. See the table below as well. The aromatic compounds have higher boiling points than the nonaromatic compounds. This is probably due to the attraction of the benzene rings in each molecule for one another, the so-called **pi stacking interaction**. Aniline has a lower water-solubility than phenol, which may be due to aniline's reduced basicity versus a nonaromatic amine.

Compound	Boiling point (°C)	Water-Solubility (g/100 mL H_2O)	Compound	Boiling point (°C)	Water-Solubility (g/100 mL H_2O)
OH	161	3.6	OH	182	6.7
NH_2	134	Miscible	NH_2	185	3.4
CH_3	101	0.0014	CH_3	111	0.54

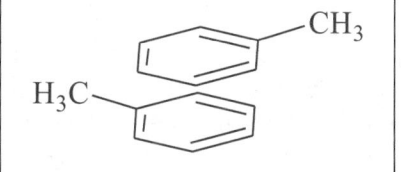 This shows two toluenes with the benzene rings pi-stacked.

C. Basicity

Amines are the strongest bases of any functional group covered in this book. In previous chapters, we have looked at acidity of a functional group using pK_a as a measure of acidity. Some texts use pK_b as a measure of basicity. I prefer to use the pK_a of the conjugate acid of an amine as a measure of basicity because the weaker the conjugate acid is, the stronger the base is, and the stronger the conjugate acid is, the weaker the conjugate base is. See the table below.

Compound	pK_a	Conjugate Base	Comments
NH_4^+ (ammonium)	9.26	NH_3 (ammonia)	This is our standard. We will compare other compounds with it.
$CH_3NH_3^+$	10.64	CH_3NH_2 methanamine	Adding an alkyl group increased the pK_a, which means the protonated amine is a weaker acid than ammonium. Therefore, methanamine is a stronger base than ammonia.
(Anilinium)	4.60	Aniline	Anilinium is a stronger acid than ammonia, so aniline is a weaker base than ammonia. The electron pair on the nitrogen can be delocalized into the benzene ring by resonance, making it less available to act as a base.
	5.25	Pyridine	Pyridine is a weaker base than ammonia, which is partially due to the hybridization state of the nitrogen.
	–1	Pyrrole	Pyrrole is a very weak base. If the electron pair on nitrogen accepts a proton, the ring is no longer aromatic. Therefore, it is not favorable for pyrrole to function as a base.
	11.3	Pyrrolidiene	Pyrrolidine is normal secondary amine and is not aromatic. Secondary amines are stronger bases than primary amines.
	1		4-Nitroaniline is a weaker base than aniline. The electron pair on the nitrogen can be delocalized into the benzene ring *and into the nitro group* by resonance, making it less available to act as a base.

D. Reactions as Nucleophiles

Because amines are good bases, they can also act as nucleophiles. Amines react with a variety of functional groups that have a partially positive carbon. Some common reactions are shown below.

D.1. REACTIONS WITH ALKYL HALIDES

Ammonia and amines react with primary alkyl halides to give more substituted amines. Secondary and tertiary alkyl halides give lower yields, as might be expected with an S_N2 reaction. Usually, excess ammonia or amine is used to minimize the formation of higher alkylated amines, although significant amounts can form. The products can usually be separated by distillation. See the examples below.

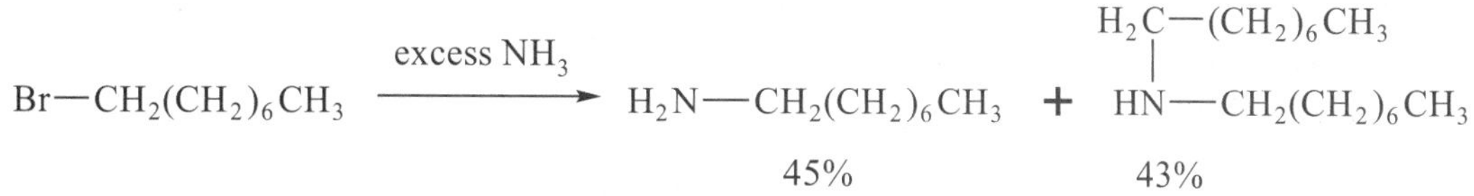

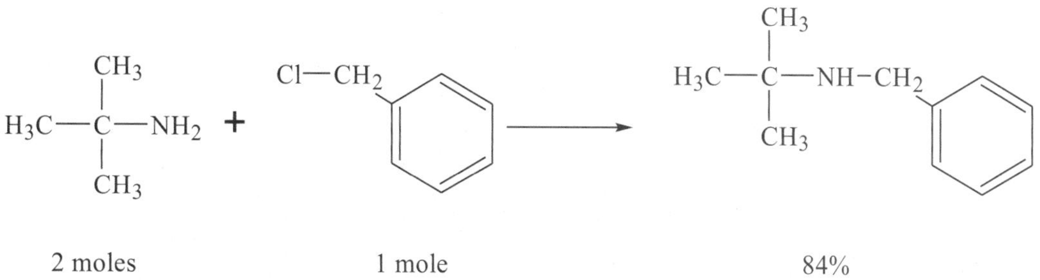

If a tertiary amine is treated with an alkyl halide, a quaternary ammonium salt is formed. Some quaternary ammonium salts are used in shampoos as cleansing agents, and some have antibacterial properties, such as Zephiran. See the examples below.

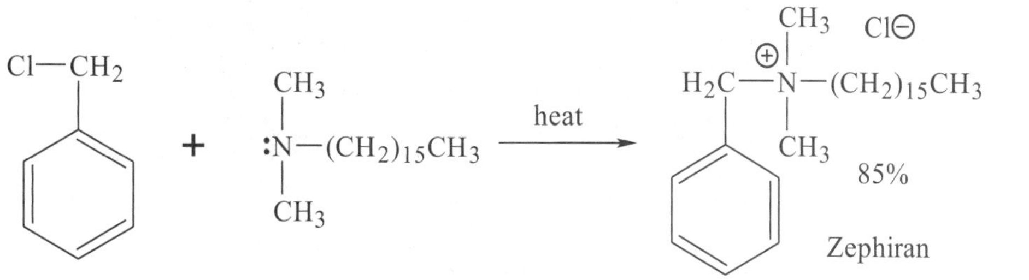

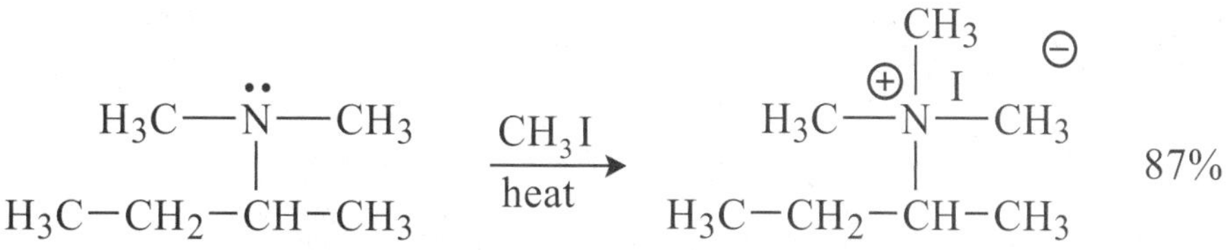

N,N,N-Trimethylbutan-2-aminium iodide

D.2. REACTIONS WITH ALDEHYDES AND KETONES

Ammonia and primary amines react with aldehydes and ketones to produce C=N compounds, which are called imines or Schiff bases. Some examples are shown below.

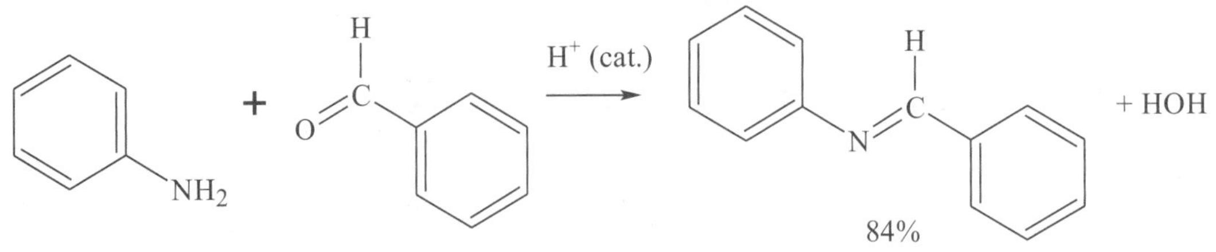

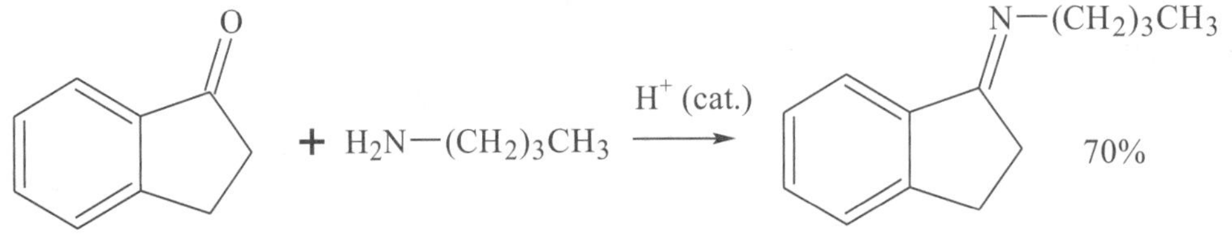

The mechanism of the first reaction is shown below. The R's represent phenyl groups.

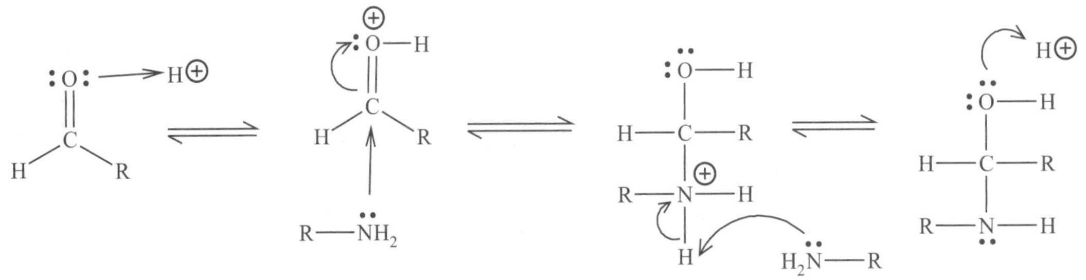

H^+ protonates the oxygen. The nitrogen attacks the partially positive carbon, and pushes an electron pair up on the oxygen. Nitrogen loses H^+, and H^+ protonates the oxygen again.

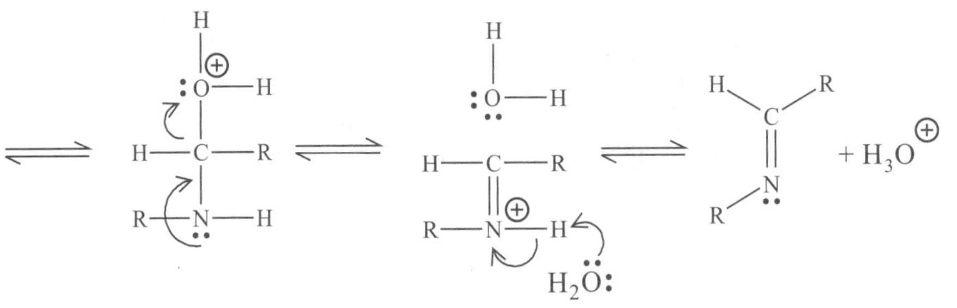

Water leaves, and the nitrogen donates a pair of electrons to form the C=N. Loss of H^+ forms the imine.

D.3. REACTION WITH ACID CHLORIDES

Ammonia, primary amines, and secondary amines react with acid chlorides and anhydrides to make amides. Usually excess amine or dilute sodium hydroxide is used to react with the acid that is produced as a by–product. See the examples below.

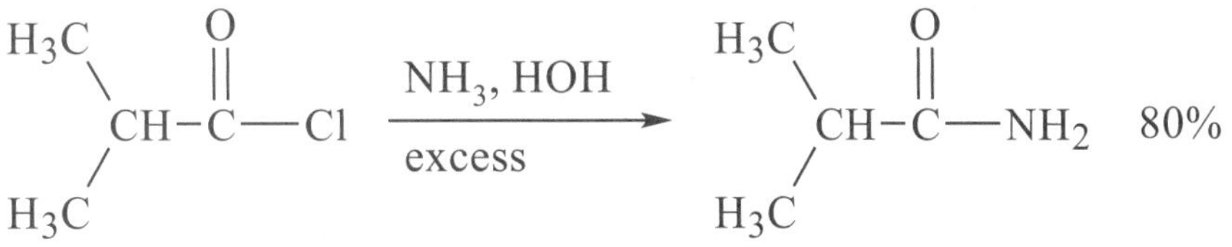

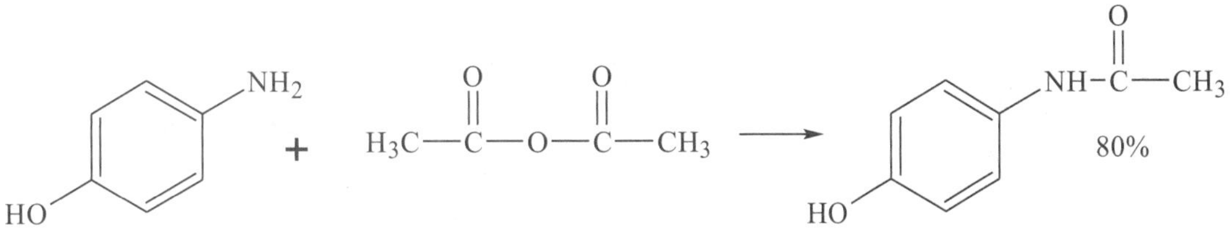

The mechanism of this reaction will be discussed in Chapter 13.

E. Reactions of Amines with Nitrous Acid

Nitrous acid (HNO_2) is a mild oxidizing agent. It is unstable as the pure acid and is prepared in solution from sodium nitrite ($NaNO_2$) and hydrochloric acid (or some other strong acid).

Primary amines react with sodium nitrite and HCl in water to form diazonium salts, as shown in the following equation. The mechanism for this reaction is complex and will not be shown.

$$R\!-\!NH_2 + NaNO_2 + 2\,HCl \longrightarrow \overset{\oplus}{R\!-\!N\!\equiv\!N} + NaCl + 2\,HOH$$
$$Cl^{\ominus}$$

Diazonium salts are relatively unstable, and are often difficult to isolate in the pure state. Aliphatic diazonium salts (R = an alkyl group) decompose by losing nitrogen and are generally too reactive to use in reactions. Aromatic diazonium salts are sufficiently stable in solution to be used as intermediates in reactions. Some of these reactions are shown below.

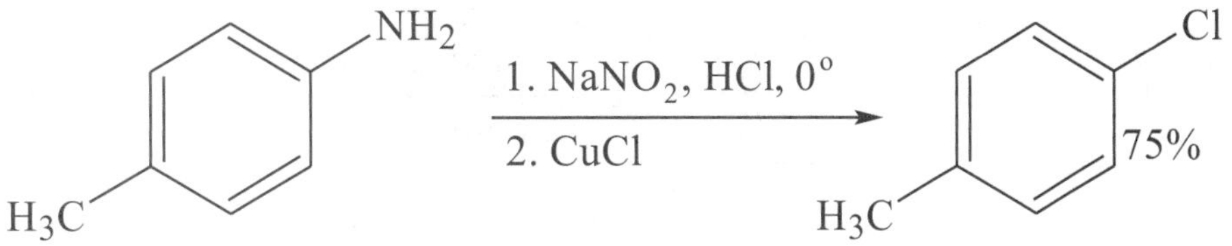

This is an example of the Sandmeyer reaction. An amino group can be replaced by a chloride or a bromide, by using the appropriate Cu(I) salt.

An amino group can be substituted by a nitrile in this variation of the Sandmeyer reaction.

After diazotization, an amine can be replaced by an iodine with KI.

After diazotization, treatment with HBF_4 (fluoboric acid) or HPF_6 (hexafluorophosphoric acid) replaces the amino group with a fluorine. Sometimes the diazonium BF_4 or PF_6 salt is isolated and heated to make the fluoro compound.

The aqueous solution of the diazonium salt is boiled, and the amine is replaced by an OH from water.

Diazonium salts can be reduced by sodium borohydride, hypophosphorus acid (H_3PO_2), or ethanol. In this case, the diazonium group is replaced by a hydrogen. This can be useful to make substituted benzenes not available by direct substitution. See the example below.

Reaction of aniline with three equivalents of bromine brominates all the ortho and para positions. No catalyst is needed because aniline is a very reactive benzene: The amine is a strong activating group. Diazotization, followed by heating with ethanol, reduces the amine to a hydrogen. This is a convenient synthesis of 1,3,5-tribromobenzene. Because bromine is an ortho-para director, 1,3,5-tribromobenzene cannot be made by three successive brominations of benzene. What would you make?

F. Preparations

In Section D.1., we saw that amines could be made by alkylation of ammonia and other amines. To avoid multiple alkylation, an excess of the amine starting material had to be used. When the product is isolated, the excess unreacted amine must be removed and either repurified for further use, or thrown away. In this section, we will look at more efficient preparations of amines.

F.1. REDUCTIONS

Amines can be prepared by reductions of a wide variety of nitrogen-containing functional groups. One common difficulty students have is remembering what reducing agent to use to reduce which functional group. I will discuss this in detail in each of the following sections and summarize this in a table at the end. In general, hydrogen and a catalyst reduces C=N (imine) and C≡N (nitrile), as well as NO_2 (nitro). These bonds are not as polar as carbonyls. Sodium borohydride reduces imines much faster than aldehyde or ketone carbonyls. Lithium aluminum hydride reduces all carbonyls, imines, and nitiles, but it is not generally useful for nitro groups.

F.1.a. REDUCTIVE AMINATION: REDUCTION OF C=N

In Section 11.D.2., we saw that amines and aldehydes or ketones react to form imines. Imines can be reduced to amines by sodium borohydride ($NaBH_4$), or hydrogen and a catalyst, such as Pd or Pt. See the example below.

The mechanism is shown below. One of the hydrides of borohydride attacks the partially positive carbon of the C=N, pushing an electron pair onto the nitrogen. The negative nitrogen pulls a hydrogen off the methanol solvent to form the product.

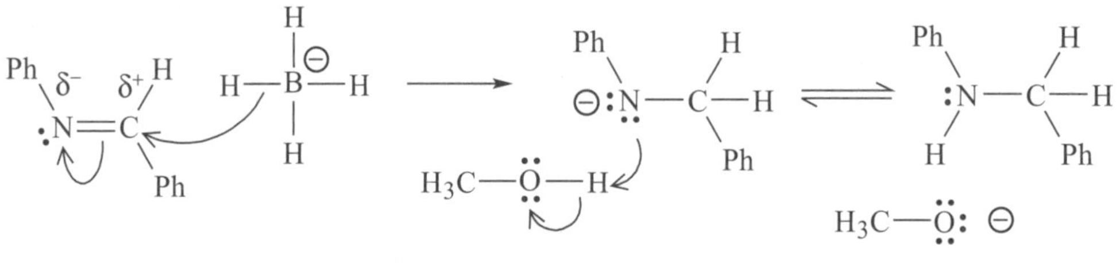

The two steps can be combined into a single step, if NaCNBH$_3$ (sodium cyanoborohydride) is used as a reducing agent. At a pH of 5–6, NaCNBH$_3$ reduces imines much faster than aldehydes or ketones. Therefore, if an amine is reacted with an aldehyde or a ketone in the presence of NaCNBH$_3$, the imine is reduced as it is formed. Hydrogen and Pd or Pt can also be used because imines are reduced much faster than aldehydes or ketones by this reagent. See the examples below.

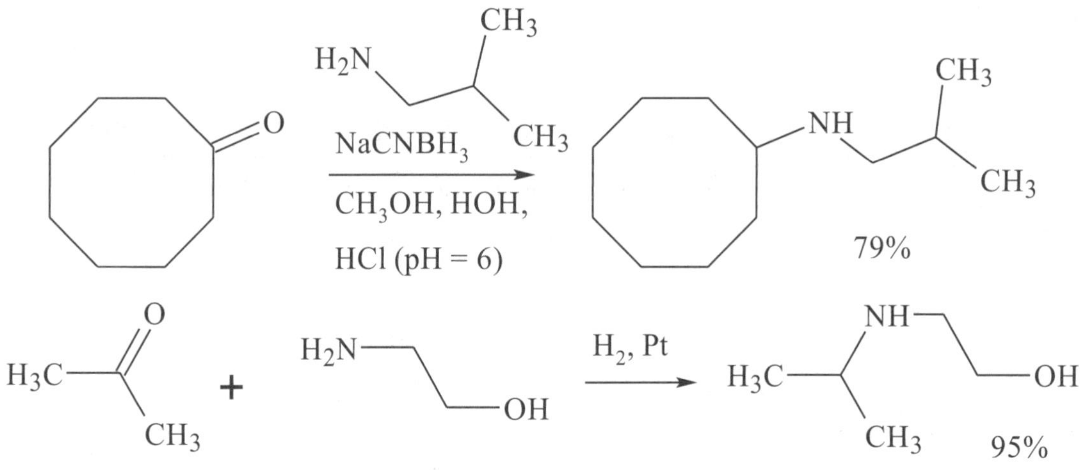

If formaldehyde is used as the aldehyde, tertiary amines containing two methyl groups can be prepared from primary amines. NaCNBH$_3$ can be used as the reducing agent. Alternatively, the amine, formaldehyde, and formic acid can be heated together to accomplish the same reaction in what is known as the Eschweiler–Clarke methylation. Secondary amines can also be methylated under these conditions. See the examples below.

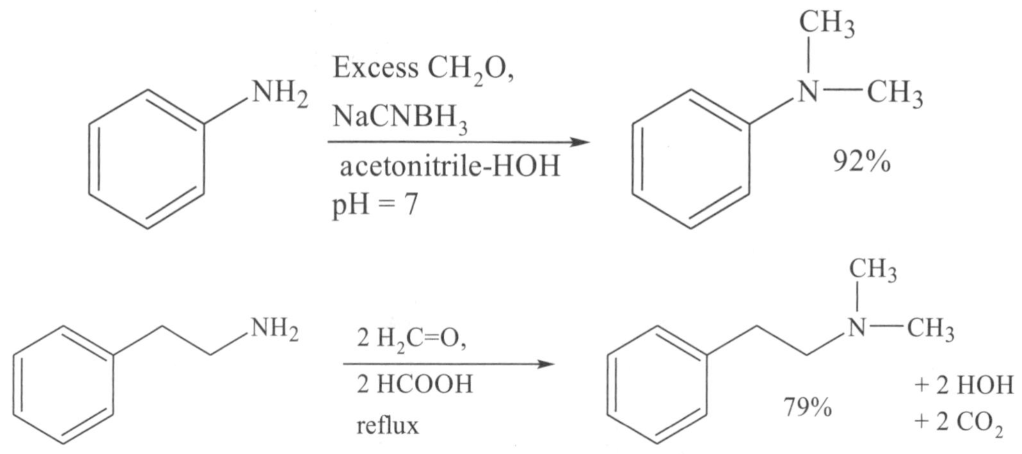

F.1.b. REDUCTIONS OF AMIDES

The preparation of amides was covered in Section D.3. Amides can be reduced by $LiAlH_4$ to amines. See the examples below.

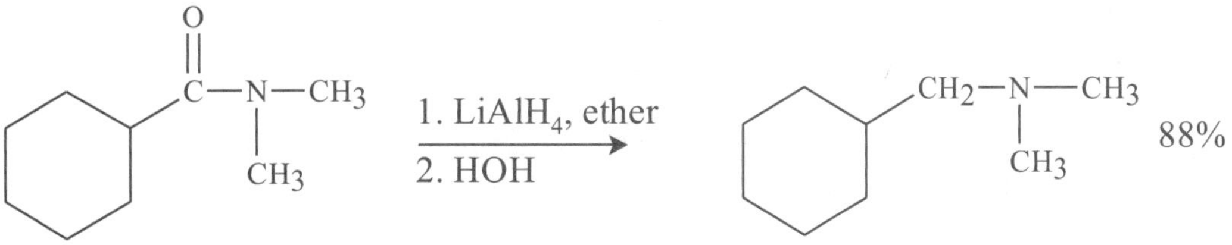

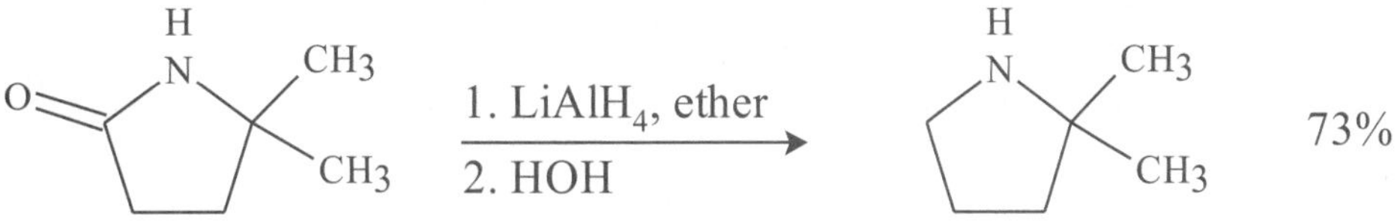

F.1.c. REDUCTION OF NITRILES

Nitriles can be reduced to amines by hydrogen and a catalyst, or by lithium aluminum hydride. Some examples are shown below.

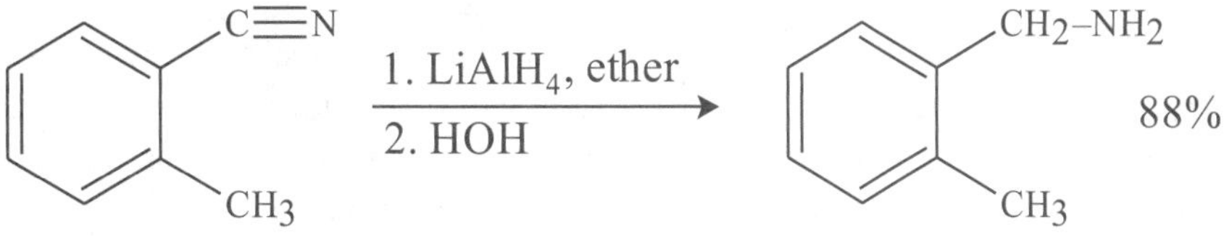

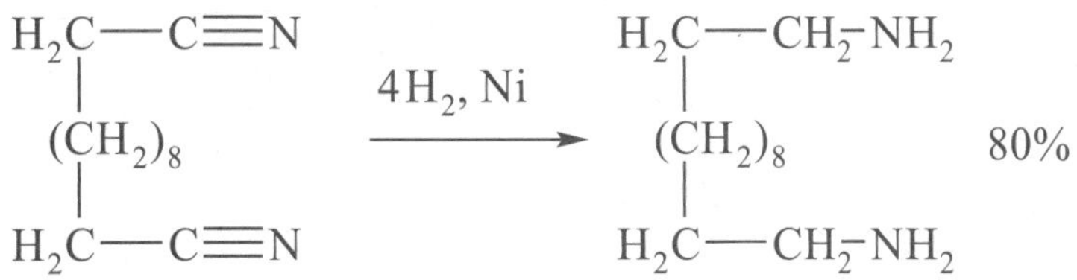

Because nitriles are usually made by substitution of cyanide for a leaving group, the combination of this substitution reaction, followed by reduction, allows for the synthesis of an amine with more carbons than the original starting material. For example, consider how you would make the nitrile starting materials in the preceding examples. See below.

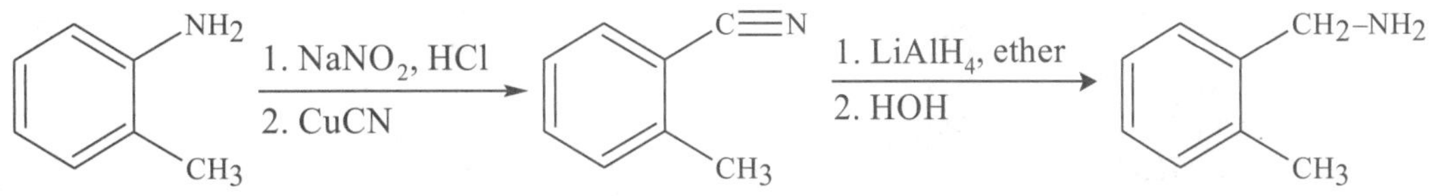

Diazotization of the amine gives the diazonium salt. CuCN replaces the diazonium salt with a nitrile. Reduction of the nitrile gives a new amine with one carbon more than the original starting material.

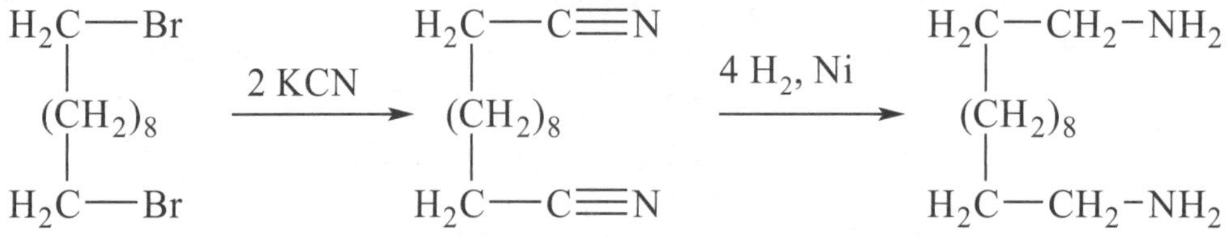

Reaction of the 1,10-dibromodecane with two equivalents of potassium cyanide gives the twelve-carbon dinitrile. Reduction of the dinitrile produces the twelve-carbon diamine.

F.1.d. REDUCTION OF NITRO GROUPS

As discussed in Chapter 10, aromatic nitro groups can be reduced to amines either by hydrogen and a catalyst or by tin or iron metal in the presence of acid. Some examples are shown below.

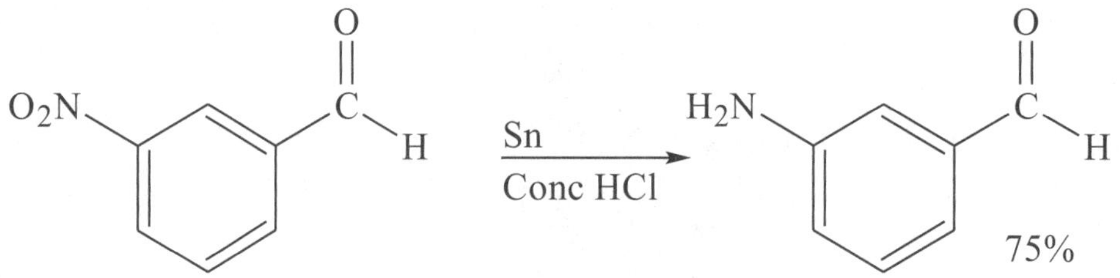

Tin and HCl is a relatively mild reducing agent and doesn't affect most other functional groups. Hydrogen and a catalyst can reduce a variety of other functional groups. Reduction of a nitro group is much faster than reduction of an aldehyde or ketone, so the reduction can be stopped before the aldehyde or ketone is reduced, as shown by the example below.

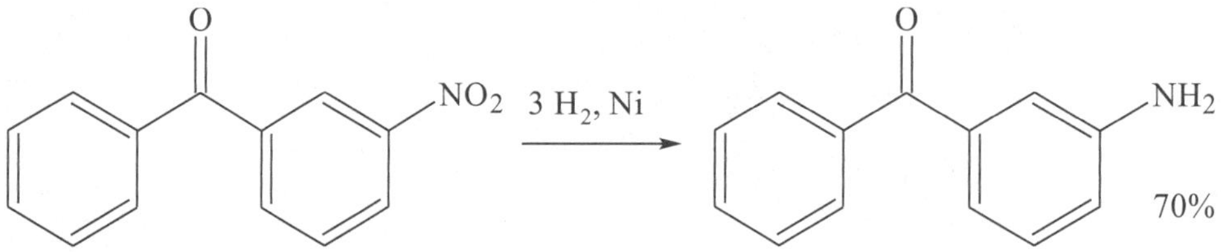

F.1.e. REDUCTION OF OXIMES

Oximes are prepared by reacting an aldehyde or a ketone with hydroxylamine, NH_2OH. This is similar to the reaction of amines with aldehydes and ketones discussed in Section D.2. Oximes can be reduced to primary amines either by hydrogen and a catalyst or by lithium aluminum hydride. Some examples are shown below.

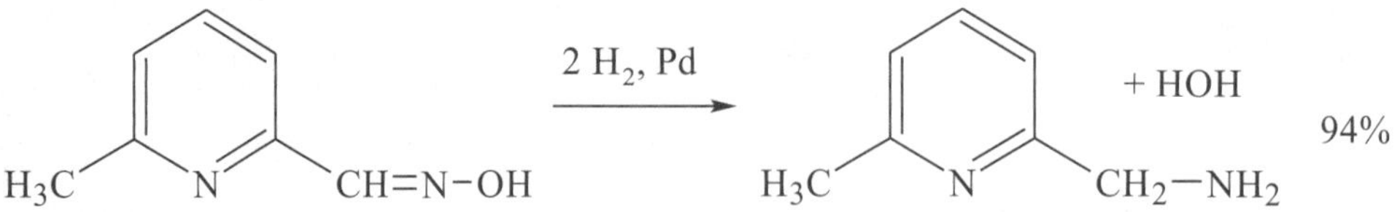

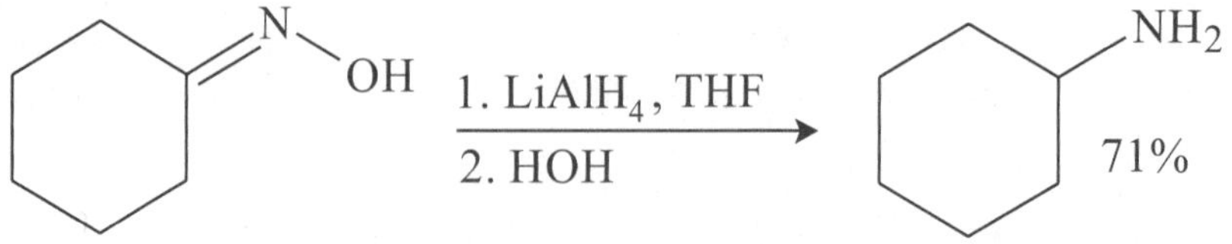

F.1.f. REDUCTION OF AZIDES

Sodium azide is a strong nucleophile. Alkyl azides are prepared by the reaction of a primary alkyl halide with sodium azide. Azides can be reduced by lithium aluminum hydride or with hydrogen and a catalyst. The net result is that an alkyl halide can be converted into a primary amine, without the problems of multiple alkylation that are observed when alkyl halides are treated with ammonia. See the example below.

The following table summarizes which functional groups are reduced by which reducing agents to form amines.

Functional Group	Reduced by H$_2$/Catalyst?	Reduced by NaBH$_4$ or NaCNBH$_3$?	Reduced by LiAlH$_4$?	Other Reducing Agents
Imine C=N	Yes	Yes	Yes	
Nitrile C≡N	Yes	No	Yes	BH$_3$
Amide $-\overset{\overset{\displaystyle O}{\|\|}}{C}-N\big\langle$	No	No	Yes	BH$_3$
Nitro NO$_2$	Yes	No*	No*	Sn + HCl
Oxime C=NOH	Yes	No	Yes	Na + ethanol
Azide N$_3$	Yes	No	Yes	

* Complex mixtures of products are formed.

F.2. PREPARATIONS OF AMINES BY REARRANGEMENT REACTIONS

We have previously encountered rearrangement reactions when we looked at the S$_N$1 reactions and other reactions that involved carbocations. In this section, we will look at rearrangements involving groups bonded to nitrogen. Both rearrangements, the Hofmann rearrangement and the Curtius rearrangement, allow us to convert carboxylic acids, or their derivatives, into amines containing one carbon *fewer*.

F.2.a. THE HOFMANN REARRANGEMENT

General Reaction:

$$R-\overset{\overset{\displaystyle O}{\|}}{C}-NH_2 \xrightarrow{\text{Br}_2,\ 2\ \text{NaOH}} R-NH_2 + 2\ NaBr + CO_2 + HOH$$

Specific Examples:

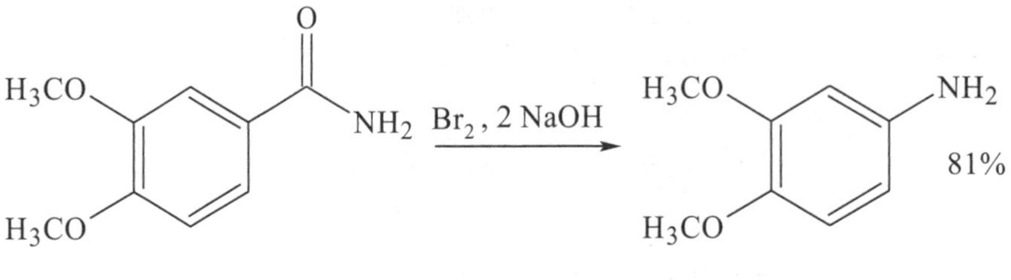

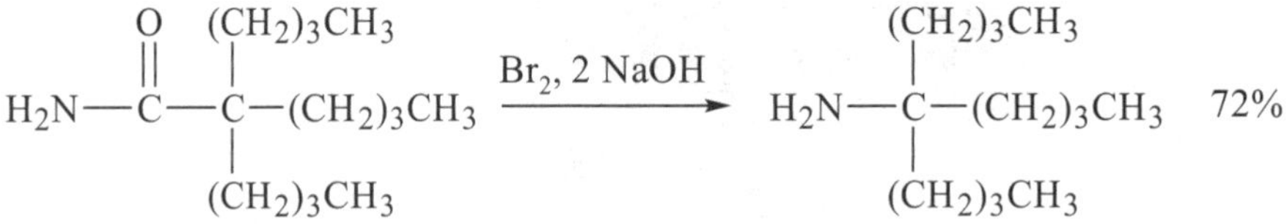

The mechanism of the Hofmann rearrangement has been extensively studied and is shown below.

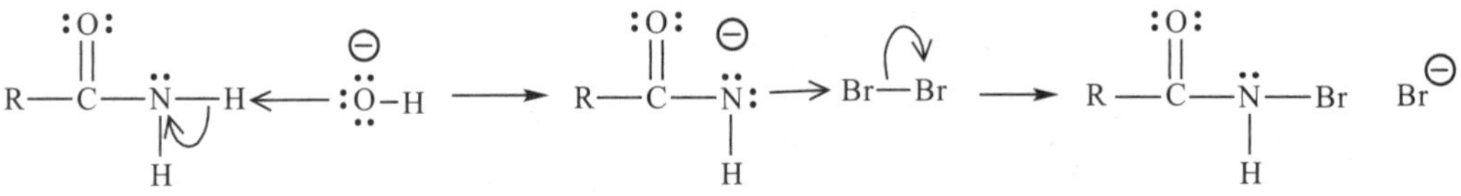

Hydroxide pulls off one of the partially positive hydrogens on the amide. The resulting anion attacks bromine to make the *N*-bromo amide. These two intermediates can be isolated under carefully controlled conditions.

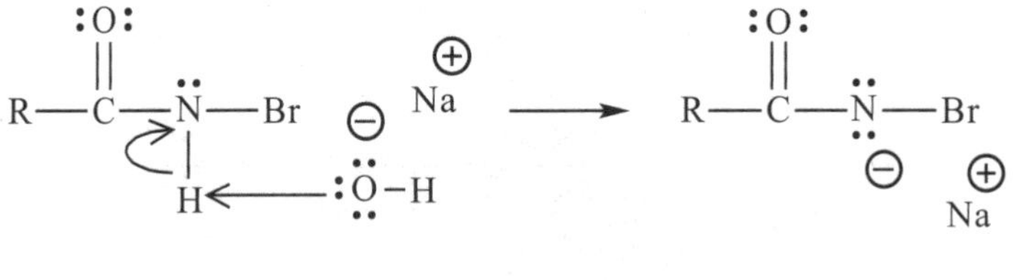

Hydroxide pulls off the other hydrogen from the amide. This intermediate has also been isolated.

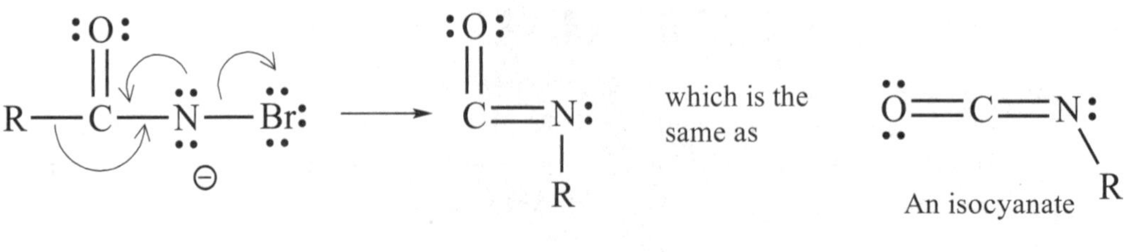

which is the same as

An isocyanate

Here is the critical rearrangement step. Bromine pulls the electron pair toward itself. This would create a positive nitrogen, so the R-group shifts to the nitrogen. This would create a positive charge on the carbon, so an unshared pair of electrons on the nitrogen forms a double bond with the carbon. All these "electron movements" occur simultaneously.

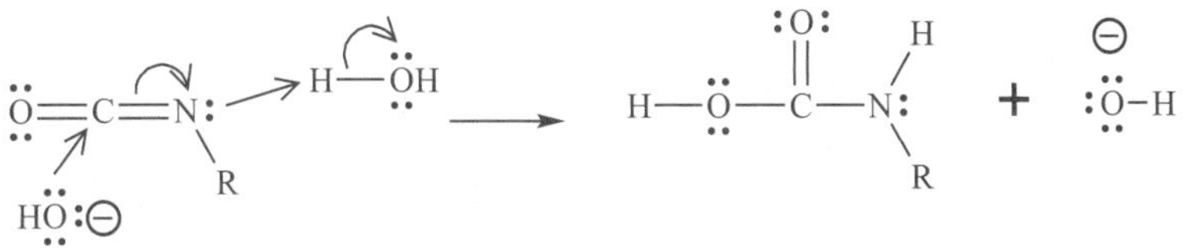

Hydroxide attacks the partially positive carbon and pushes an electron pair on the nitrogen. The negative nitrogen deprotonates water.

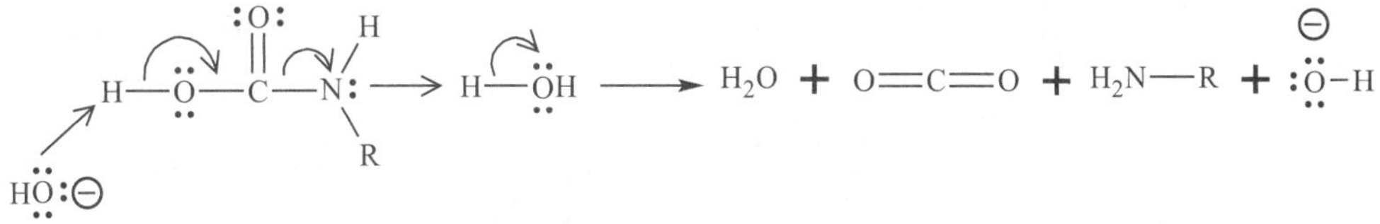

Hydroxide pulls the hydrogen off the oxygen. An electron pair moves in to form a C=O, which causes the bond to the nitrogen to break. The negative nitrogen pulls a hydrogen off water, to form the amine product.

F.2.b. THE CURTIUS REARRANGEMENT

General Reaction:

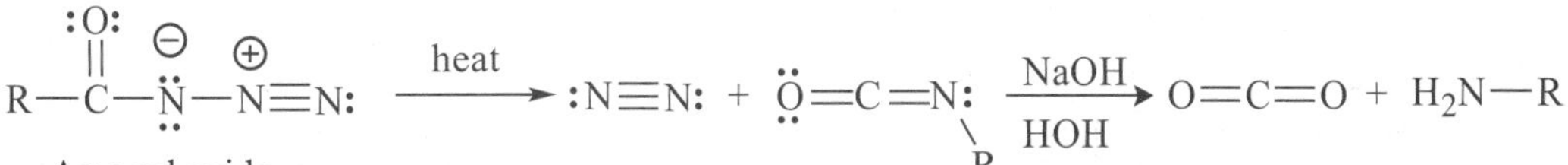

When an acyl azide is heated, nitrogen is lost, and an isocyanate is formed. Treatment of the isocyanate with aqueous base produces the amine. There are several ways to make the acyl azide. Two are shown below.

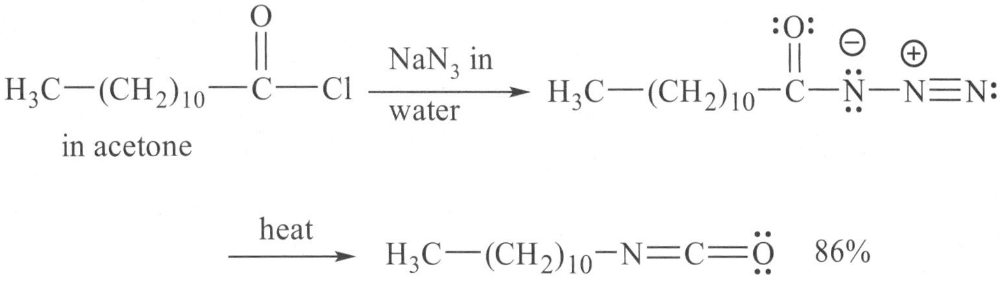

Sodium azide is added to an acid chloride to form the acyl azide. Once formed, the azide is heated in an inert solvent to form the isocyanate.

Acyl hydrazide

Acyl hydrazides can be made from esters. When treated with nitrous acid, the hydrazides are ozidized to the azides. Heat converts the azides into isocyanates, which, when heated with water and acid, makes the hydrochloride salt of the amine. This particular diamine has the common name putresine, which tells you something about its odor!

One important point about the Hofmann and Curtius rearrangements is that they are stereospecific. If the carbon bearing the carboxylic acid derivative is a stereogenic center, the amine is formed with retention of configuration. See the examples below.

Retention of configuration is expected because the bonds around the stereocenters never change their locations relative to each other during the rearrangement.

REVIEW EXERCISES FOR CHAPTER 11

1. Give IUPAC names for the following structures.

a.	b.

a.

b.

c.

d.

2. How could you prepare 1-hexanamine from each of the following compounds?

a.	b.	c.	d.

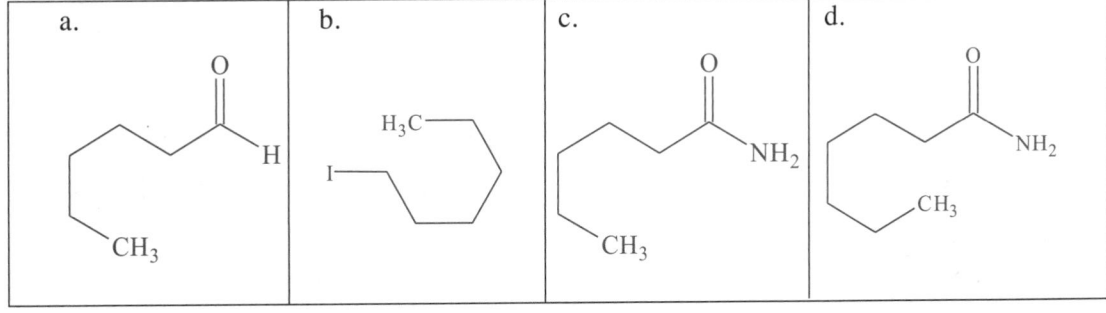

3. How could you prepare *N*-propylaniline from each of the following compounds?

a.

b.

c.

d.

4. Give a reasonable arrow-pushing mechanism for the following reaction.

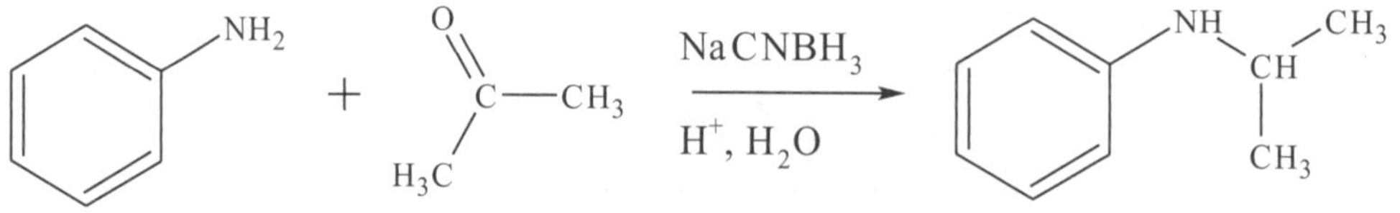

5. Give reasonable syntheses of the following products from indicated starting materials, plus any other reagents.

	Starting Material(s)	Product
a.	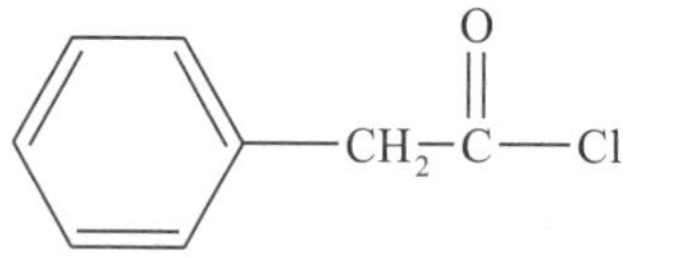	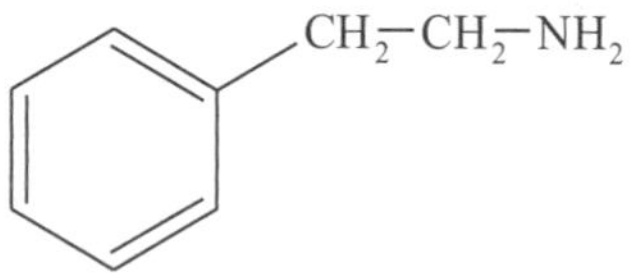
b.	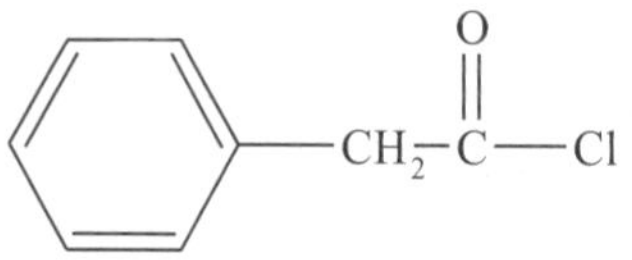	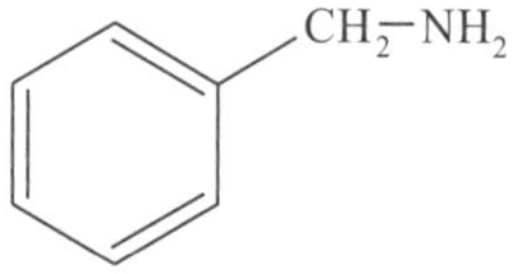
c.	Benzene 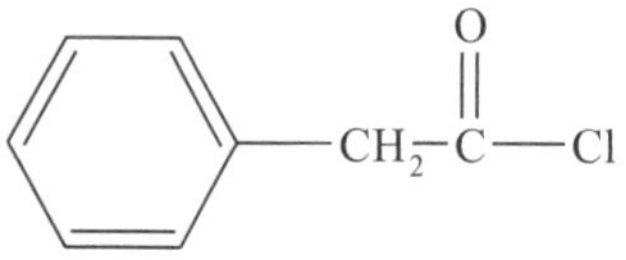	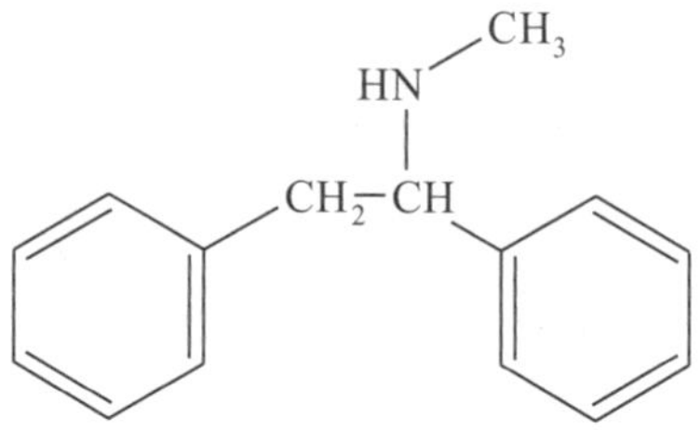
d.	$H_3C-CH_2-CH_2-CH_2-Br$	$H_3C-CH_2-CH_2-CH_2-CH_2-NH_2$
e.	$H_3C-CH_2-CH_2-CH_2-Br$	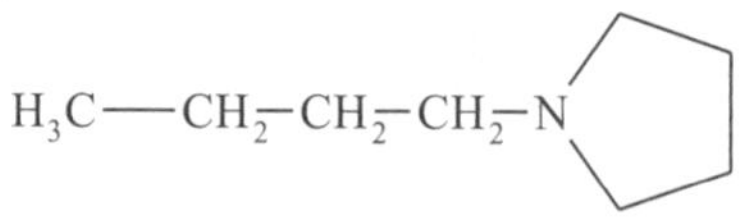
f.	Benzene	Fluorobenzene
g.	Aniline	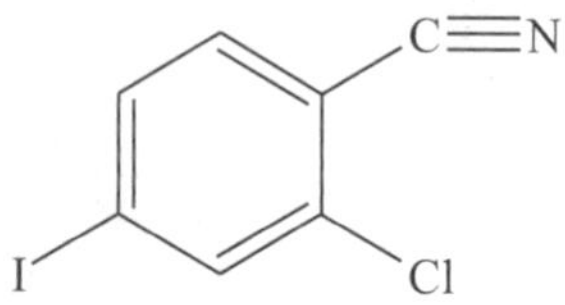

Aldehydes and Ketones

WHAT YOU WILL LEARN

In this chapter, you will learn:

- how to name aldehydes and ketones using IUPAC rules;
- how the strong and weak nucleophiles react with aldehydes and ketones, and the mechanisms by which they react;
- how to prepare aldehydes and ketones from a variety of functional groups;
- how the acidity of alpha-hydrogens allows aldehydes and ketones to react with halogens and other aldehyde molecules.

SECTIONS IN THIS CHAPTER

- Nomenclature

- Physical Properties

- General Reaction: Nucleophilic Addition to the C=O

- Reactions with Strong Nucleophiles

- Reactions with Weak Nucleophiles

- Oxidations of Aldehydes

- Reductions of Aldehydes and Ketones

- Preparations

- Acidity of the Alpha-Hydrogen

- Halogenation at the Alpha-Carbon

- Aldol Condensation

Aldehydes and ketones are the first of several functional groups that have a C=O (which is called a carbonyl). The general structure and some specific examples of aldehydes and ketones are shown below.

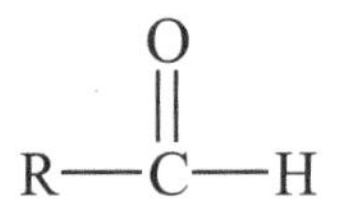	The generic structure of an aldehyde. R is some carbon-containing group, or a hydrogen.
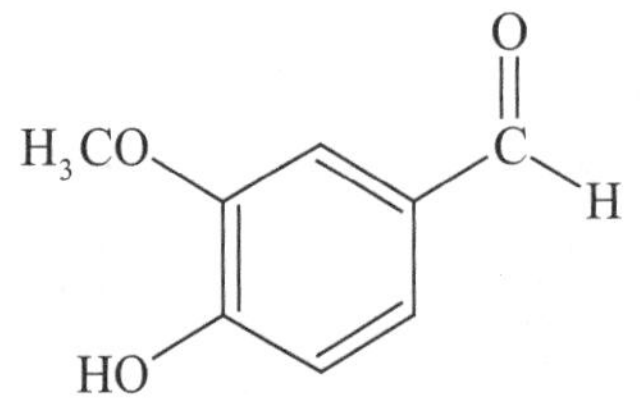	This is **vanillin**, the main component of vanilla flavor, which gives vanilla its characteristic odor.
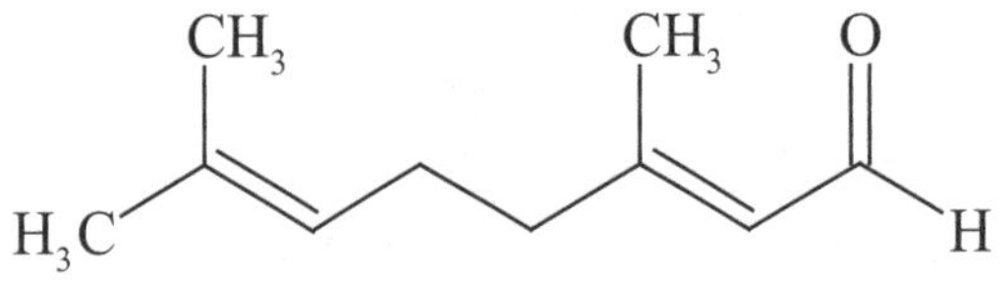	This is **geranial**, the major component of lemon grass oil. It has a strong lemon odor.
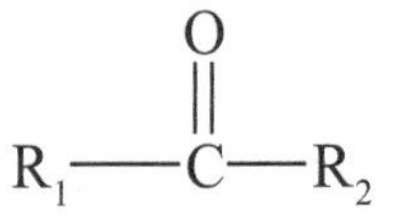	The generic structure of a ketone. R_1 and R_2 are carbon-containing groups.
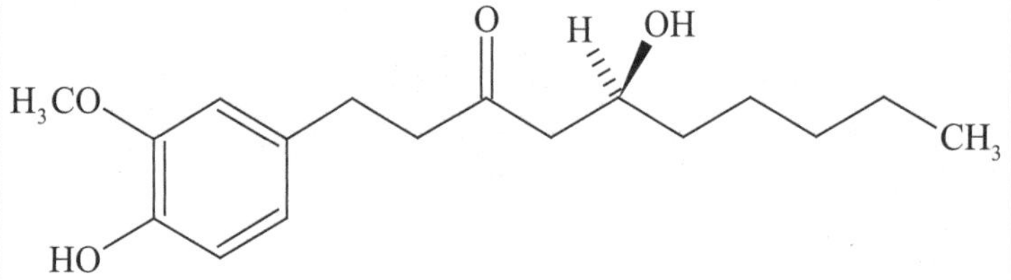	**Gingerol**, the major pungent compound in ginger oil.
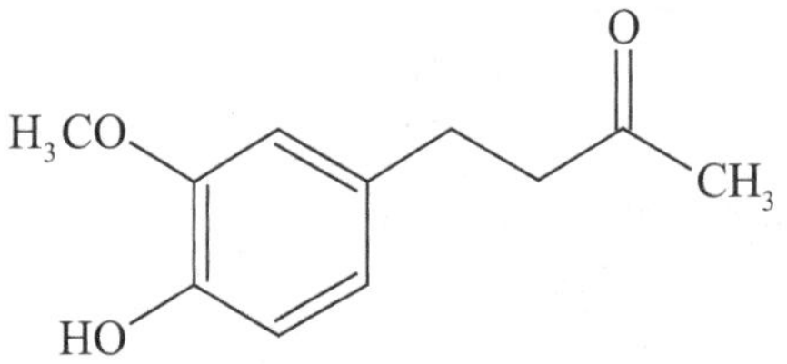	**Zingerone**, another ketone isolated from ginger.

A. Nomenclature

Aldehydes and ketones are higher in priority than any of the previous functional groups we have encountered. Aldehydes are just higher in priority than ketones. The IUPAC ending for an aldehyde is -*al*, and for a ketone, it is -*one*. See the examples below.

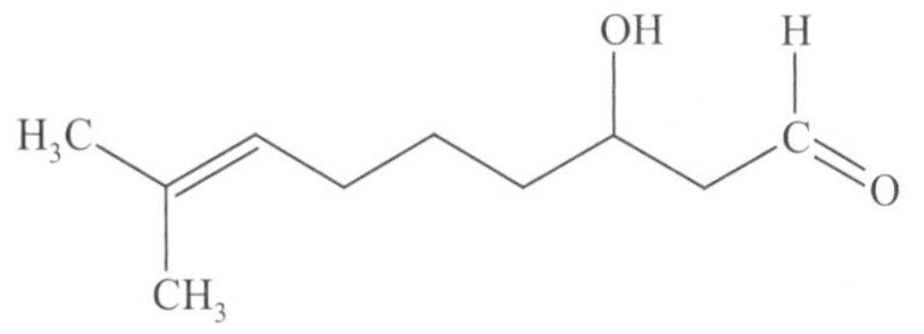	The longest chain is nine carbons. There is a C=C, so it is a non-7-enal (the aldehyde controls the numbering). There is a hydroxy group on carbon 3, and a methyl group on carbon 8, so it is **3-hydroxy-8-methylnon-7-enal**.
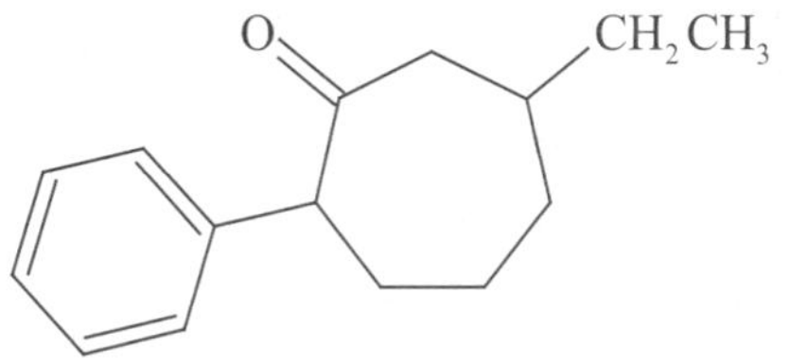	The ketone is in the seven-membered ring, so it is a cycloheptanone. We number counterclockwise because the phenyl group is the closest to the ketone. The name is **6-ethyl-2-phenylcycloheptanone**.
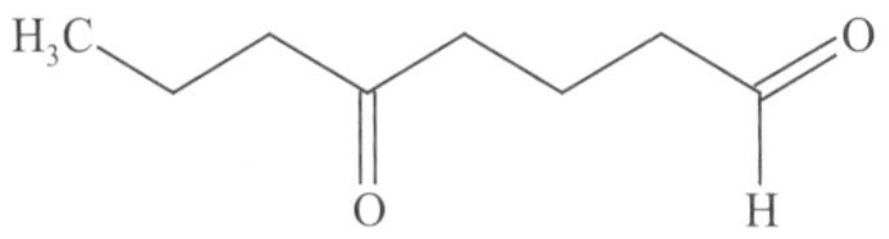	This has both an aldehyde and a ketone. The aldehyde takes precedence, so this is an octanal. The substituent is the double bonded O, which IUPAC calls *oxo*. The name is **5-oxooctanal**.

Some aldehydes and ketones have common names that are widely used, so you should recognize them. These are shown below.

Compound	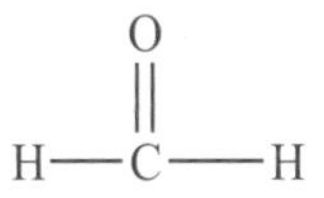	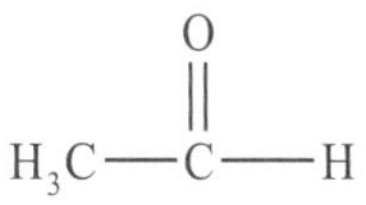	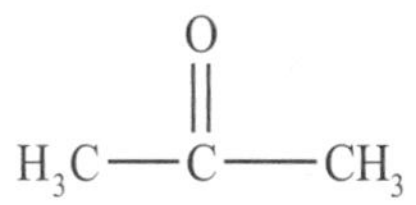
IUPAC name	Methanal	Ethanal	2-Propanone
Common name	Formaldehyde	Acetaldehyde	Acetone

B. Physical Properties

Because aldehydes and ketones have an oxygen, they can accept hydrogen bonds from water. Therefore, they are soluble in water, approximately as soluble as alcohols. Because aldehydes and ketones do not have any significantly partially positive hydrogens, they cannot be hydrogen-bond donors, and have lower boiling points than alcohols. However, because the C=O is polar, there are dipole–dipole attractions between two aldehydes or two ketones, so their boiling points are higher than alkanes. See the tables on the next page.

Name	Water-Solubility (g/100 mL of water)	Boiling Point (°C)
Methanal	55	–20
Ethanal	Miscible	21
Propanal	14	49
Butanal	7.1	75
Pentanal	1.4	103
Hexanal	0.56	131
Propan-2-one	Miscible	56
Butan-2-one	27.5	80
Pentan-2-one	4.0	102
Pentan-3-one	3.4	101
Hexan-2-one	1.4	128
Hexan-3-one	1.5	124

Comparison of Water-Solubilities and Boiling Points of Selected Compounds			
Name	Structure	Water-Solubility (g/100 mL of water)	Boiling Point (°C)
Butan-1-ol		7.4	118
Butanal		7.1	75
Pentane		0.0036	36
Butan-2-ol		8.3	100
Butan-2-one		27.5	80
2-Methylbutane		0.0048	28

C. General Reaction: Nucleophilic Addition to the C=O

When a nucleophile (Nu:) reacts with an aldehyde or ketone, the general reaction is addition of the nucleophile to the C=O. The nucleophile bonds to the carbon, and a hydrogen (usually) bonds to the oxygen. The general reaction is shown below.

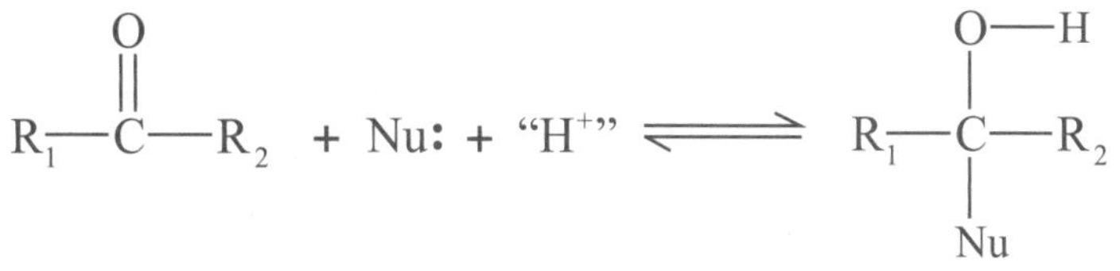

$$
\underset{\substack{\text{O} \\ \| \\ R_1-C-R_2}}{} + \text{Nu:} + \text{``}H^+\text{''} \rightleftharpoons \underset{\substack{\text{O-H} \\ | \\ R_1-C-R_2 \\ | \\ Nu}}{}
$$

The mechanism is slightly different depending on how strong the nucleophile is. Strong nucleophiles attack the partially positive carbonyl carbon, and push a pair of electrons up onto the oxygen. Protonation of the negative oxygen is the second step. This is shown below.

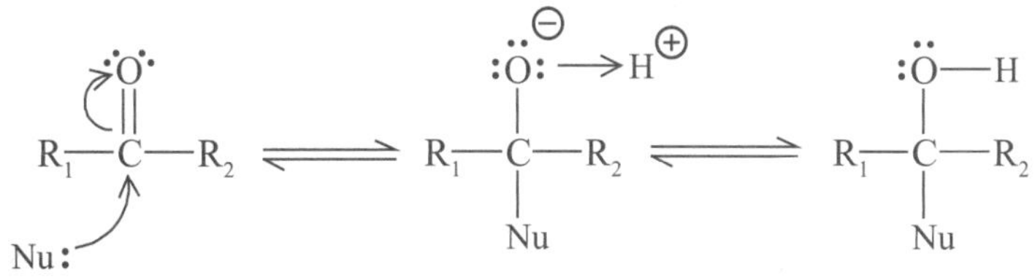

If the nucleophile is weaker, then the oxygen of the C=O is protonated first. Protonation makes the carbonyl carbon much more positive, and it is then attacked by the nucleophile. This is shown below.

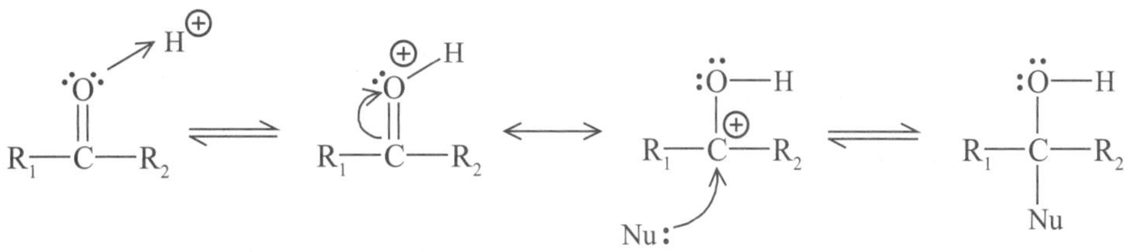

D. Reactions with Strong Nucleophiles

D.1. REACTIONS WITH ORGANOMETALLIC REAGENTS

Acetylide ions, Grignard reagents, and alkyl lithium reagents add to the carbonyl group of an aldehyde or ketone to form an alcohol. These reactions were discussed in Chapter 8.

Some examples are shown below.

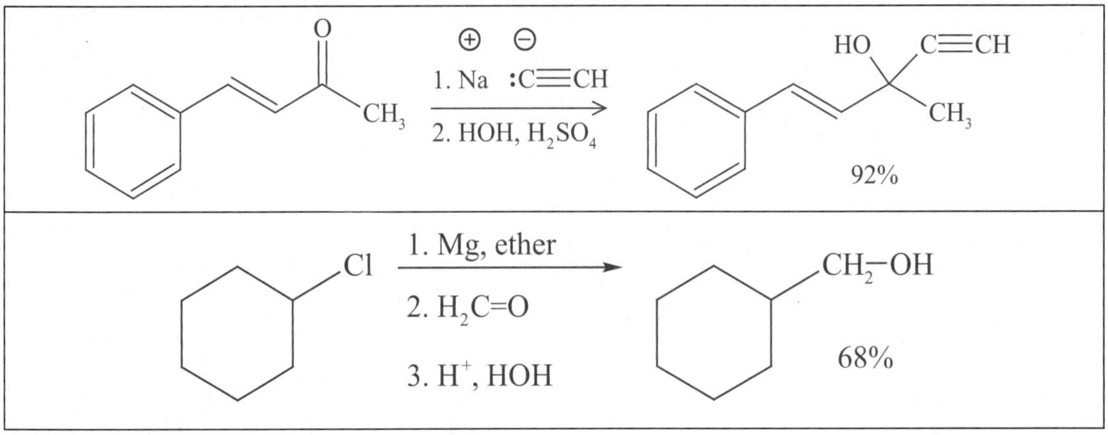

D.2. REACTION WITH CYANIDE

General Reaction:

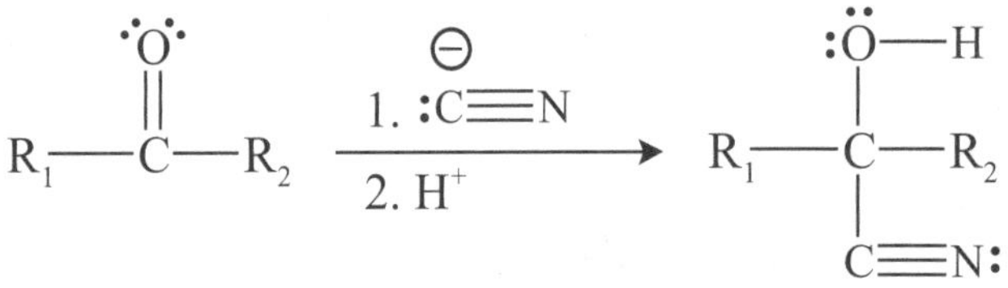

Specific Examples:

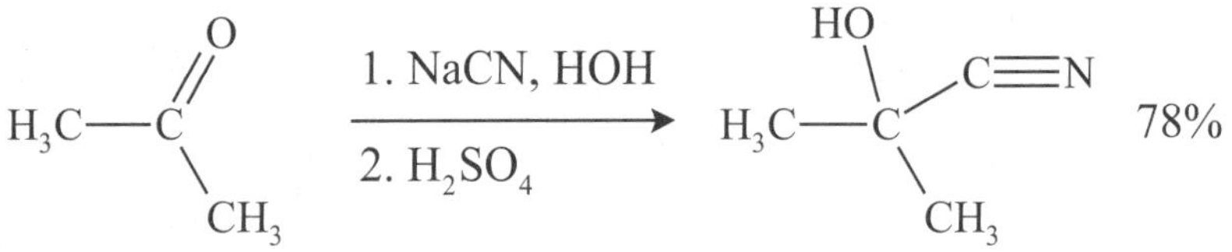

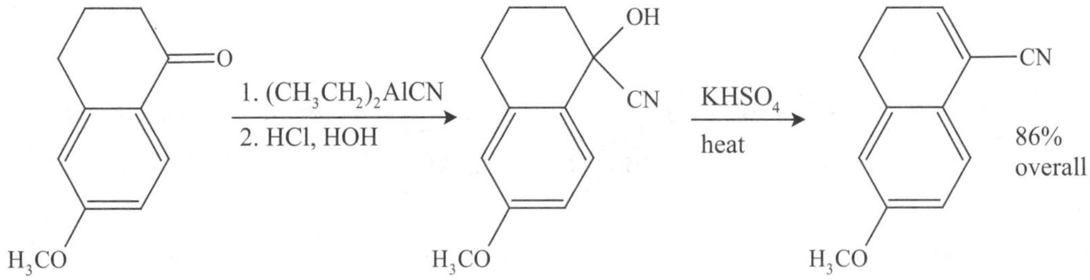

Cyanide is a strong nucleophile that adds to the C=O group of aldehydes and ketones. Protonation of the resulting alkoxide gives a hydroxynitrile, which is sometimes called a cyanohydrin. The second example uses diethylaluminum cyanide, which often gives better yields with aromatic ketones. Dehydration of the resulting hydroxynitrile gives the alkene.

D.3. REACTIONS WITH PRIMARY AMINES AND AMMONIA

General Reaction:

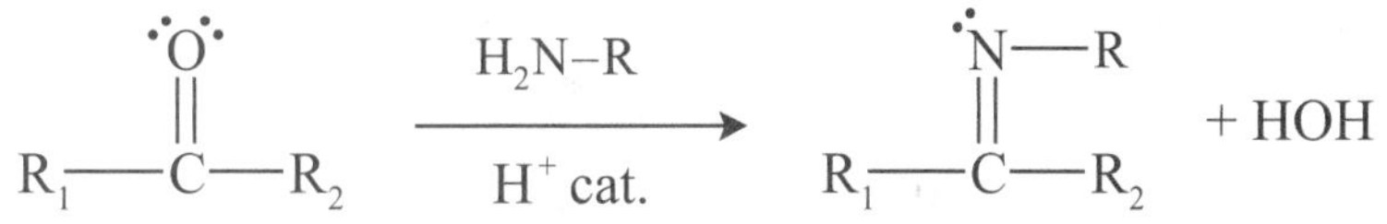

Specific Example:

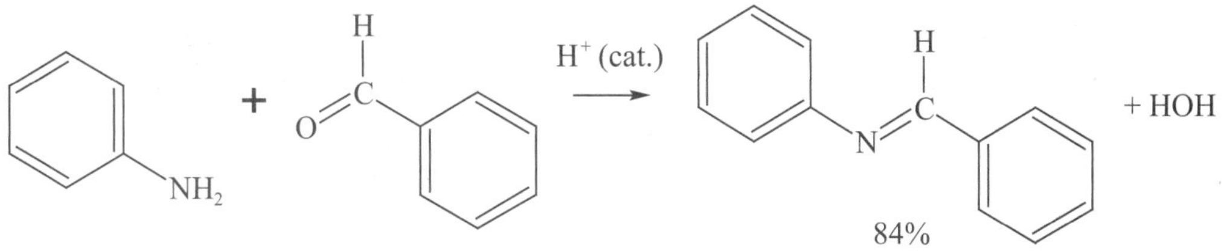

This reaction and mechanism were discussed in Chapter 11. The product C=N compounds are called **imines** or **Schiff bases**. Although the reaction does occur in the absence of acid, it is generally much slower without it.

D.4. THE WITTIG REACTION

General Reaction:

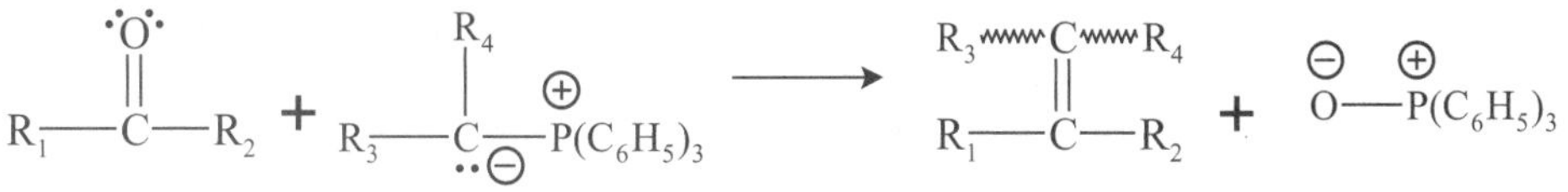

An ylid

Specific Examples:

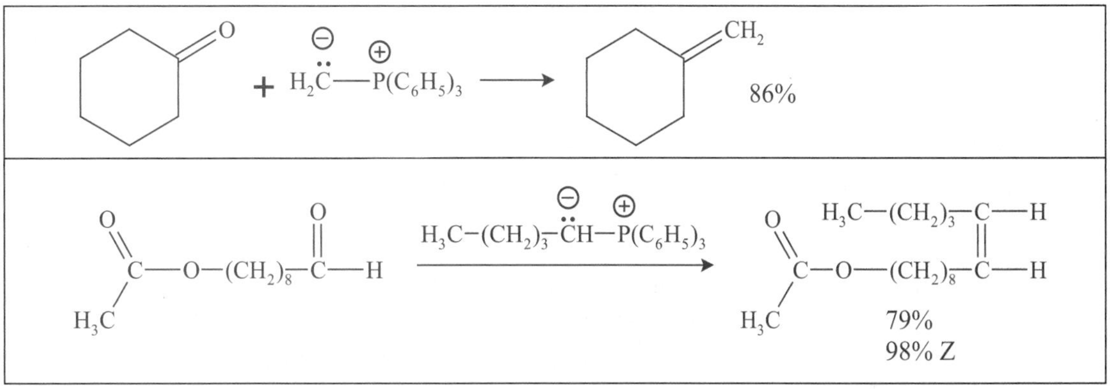

The Wittig reaction was discovered by German chemist Georg Wittig and his co-workers, who published their first paper on the reaction in 1953. Wittig was awarded the Nobel Prize in chemistry in 1979, along with H. C. Brown.

The mechanism of the reaction is shown below. Phosphorus is a third-row element, and can form stable compounds with five single bonds. "C_6H_5" is a phenyl group.

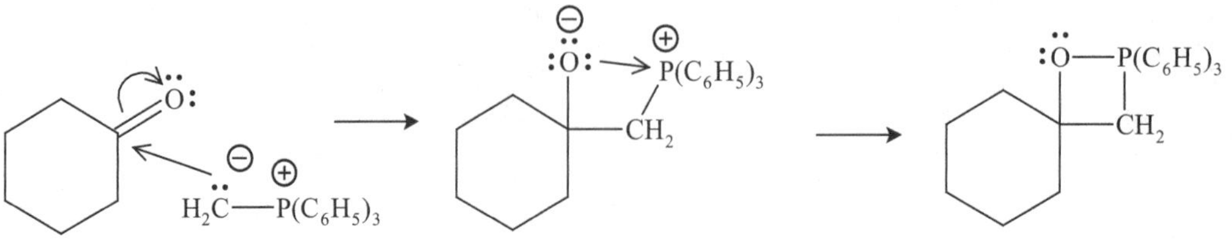

A "betaine"   An "oxaphosphetane"

The carbanion attacks the partially positive carbon of the ketone, pushing a pair of electrons up onto the oxygen, to form the betaine. The betaine subsequently forms an oxaphosphetane. Some oxaphosphetanes have been observed in low-temperature NMR studies.

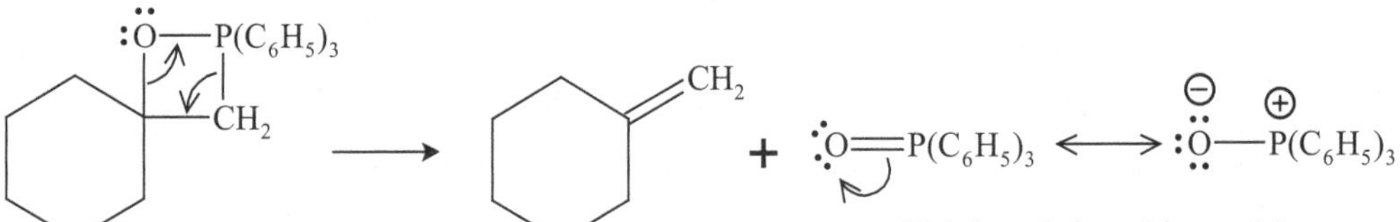

Triphenylphosphine oxide

The oxaphosphetane decomposes to form the alkene and triphenylphosphine oxide. The driving force for this is the formation of a very strong P–O bond, as well as relief of ring strain.

The ylids are formed by a two-step process. An alkyl halide is reacted with triphenylphosphine to form a phosphonium salt. The phosphonium salt is reacted with a very strong base, such as $NaNH_2$ or butyl lithium, to form the ylid. This is shown below.

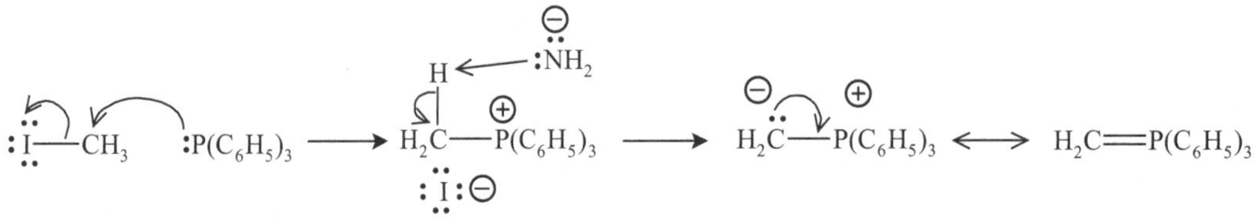

Phosphonium salt   Ylid

Triphenylphosphine does an S_N2 attack on the backside of the carbon to form the phosphonium salt. The strong base pulls off a hydrogen from the carbon to form the ylid. The ylid is stabilized by resonance.

In the second of the specific examples, the major alkene product was the (Z)-isomer. This is often true, especially with ylids derived from simple alkyl halides. With ylids derived from alkyl halides where the anion can be stabilized further by resonance, or if additional base is used, the (E)-isomer is often the major product. See the examples below.

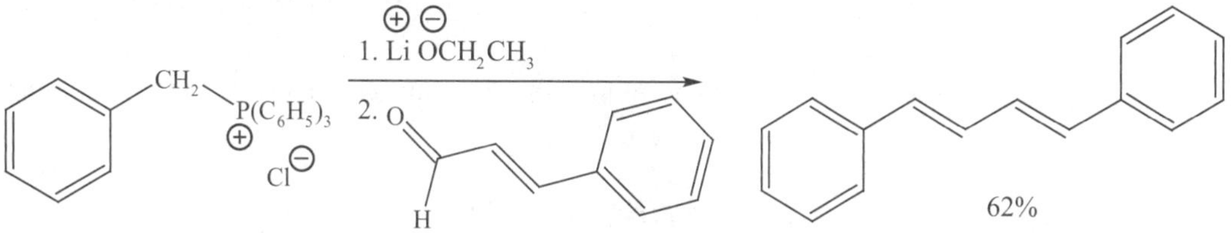

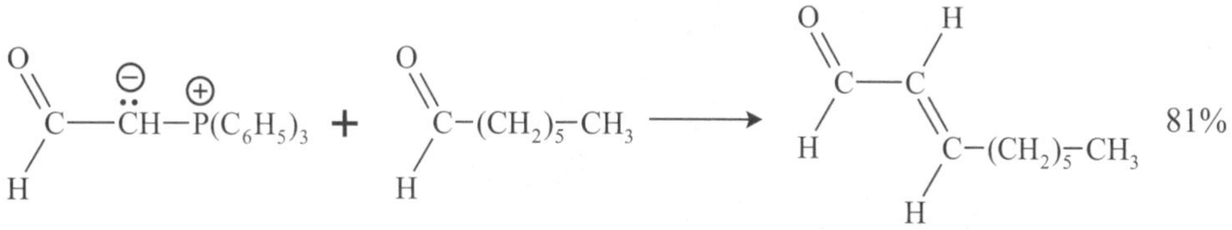

E. Reactions with Weak Nucleophiles

E.1. REACTION WITH WATER

General Reaction:

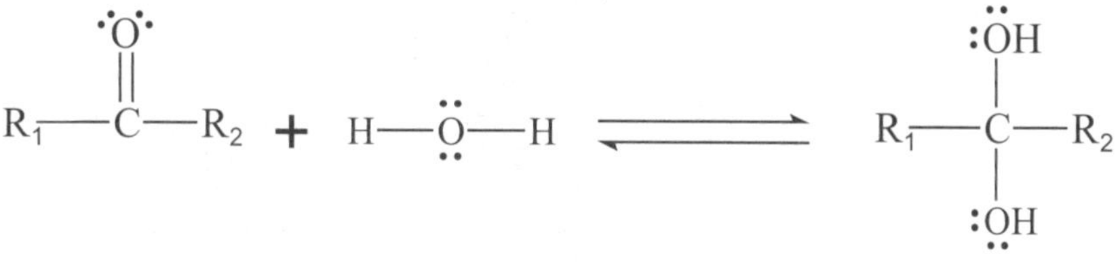

Specific Examples:

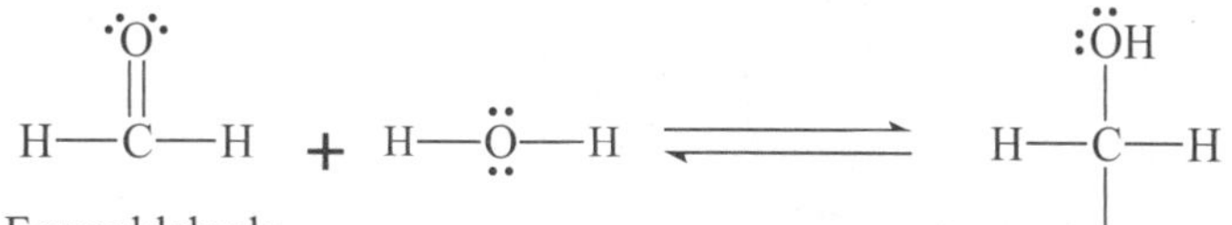

Formaldehyde

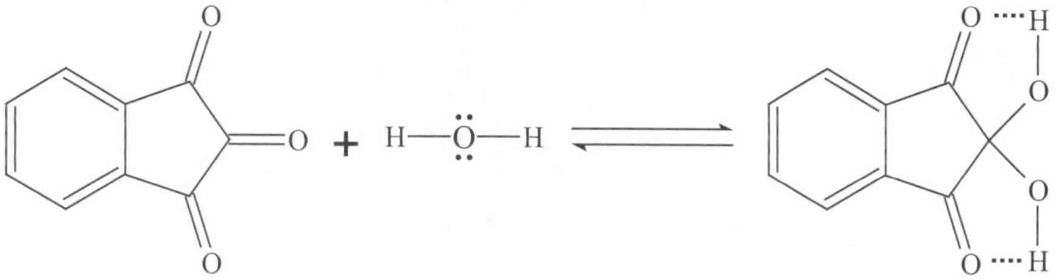

Ninhydrin, a dye

Aldehydes and ketones can add water to form a diol, which is called a hydrate. This is an equilibrium reaction, and for most aldehydes and ketones, the equilibrium lies mostly to the left: that is, there is very little hydrate at equilibrium. Certain aldehydes and ketones, such as the ones shown above, exist primarily as hydrates. In the case of ninhydrin, the formation of the hydrate is favored by formation of two hydrogen bonds.

This reaction is of little preparative importance, but it is important in understanding the oxidation of aldehydes, which is discussed in Section F.

E.2. REACTION WITH ALCOHOLS: FORMATION OF HEMIACETALS AND ACETALS

General Reaction:

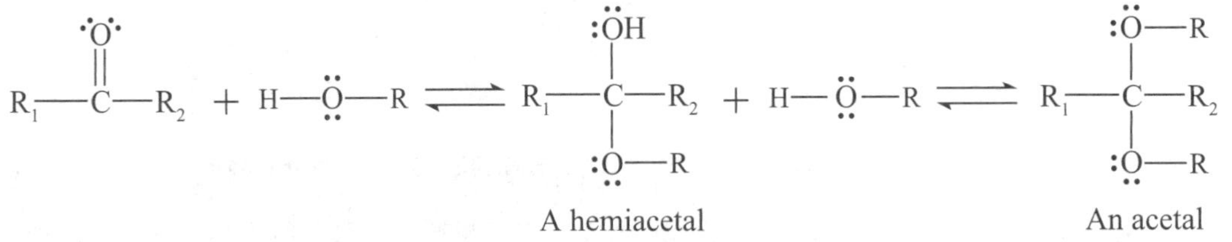

A hemiacetal   An acetal

Specific Examples:

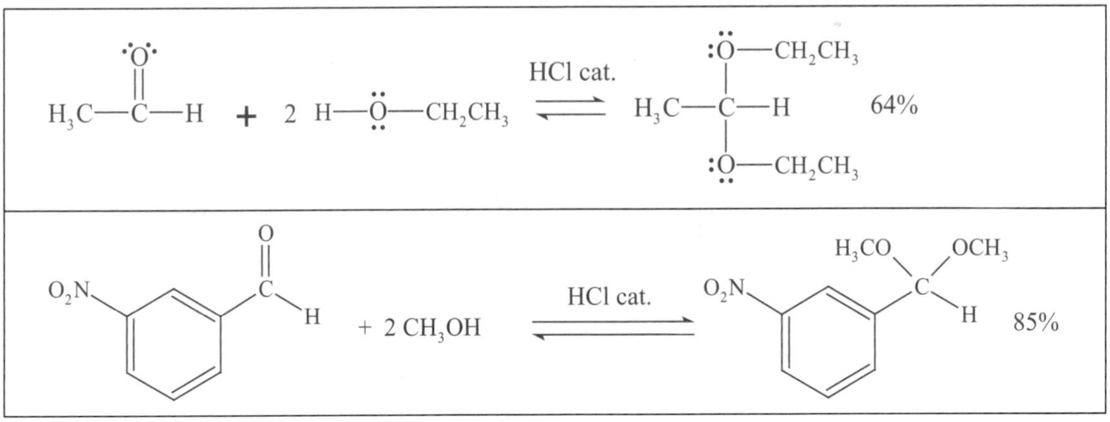

The reaction of an aldehyde or a ketone with an alcohol, usually in the presence of an acid catalyst, leads to the formation of a hemiacetal. Most hemiacetals, like hydrates, are relatively unstable and revert back to the C=O compound and the alcohol. If there is an excess of the alcohol, however, an acetal will be formed, especially if care is taken to remove the water formed as a by-product. The mechanism is shown below.

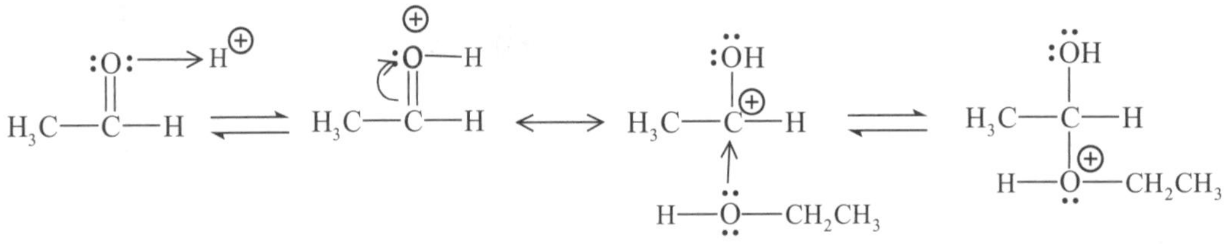

The acid catalyst protonates the C=O oxygen, which puts a positive charge on oxygen. A resonance form can be drawn to show that the carbon shares the positive charge. An alcohol molecule attacks the positive carbon.

the hemiacetal

The resulting product loses a proton to form the hemiacetal. Protonation of the OH group of the hemiacetal, followed by loss of water, forms the cation shown.

The cation is attacked by another alcohol molecule, which forms the protonated acetal. Loss of H⁺ gives us the acetal.

Acetal formation is reversible, which means that the acetal can be converted back to the aldehyde or ketone by treating it with water and an excess of acid. Why would you want to do this? Because acetals are essentially ethers, they don't react like aldehydes. Therefore, acetals can be used to disguise or "protect" aldehydes and ketones from nucleophiles, such as Grignard reagents. See the example below. This example also shows that a diol can be used to form an acetal, instead of two molecules of an alcohol.

$$H_3C-\overset{O}{\overset{\|}{C}}-CH_2-\overset{O}{\overset{\|}{C}}-O-CH_2CH_3 \; + \; \underset{OH \quad OH}{\overset{H_2C-CH_2}{|\qquad|}} \;\; \xrightarrow{\text{p-TsOH}} \;\; H_3C-\overset{\overset{H_2C-CH_2}{\overset{|\qquad|}{O\quad O}}}{\underset{\diagdown\diagup}{C}}-CH_2-\overset{O}{\overset{\|}{C}}-O-CH_2CH_3$$

64%

The ketone is reacted with 1,2-ethanediol ("ethylene glycol") to form an acetal.

$$H_3C-\overset{\overset{H_2C-CH_2}{\overset{|\qquad|}{O\quad O}}}{\underset{\diagdown\diagup}{C}}-CH_2-\overset{O}{\overset{\|}{C}}-O-CH_2CH_3 \;\; \xrightarrow[\text{2. HOH}]{\text{1. 2 }C_6H_5MgBr} \;\; H_3C-\overset{\overset{H_2C-CH_2}{\overset{|\qquad|}{O\quad O}}}{\underset{\diagdown\diagup}{C}}-CH_2-\overset{OH}{\underset{C_6H_5}{\overset{|}{C}}}-C_6H_5$$

84%

The ester is reacted with two equivalents of phenylmagnesium bromide (C_6H_5MgBr) to form a tertiary alcohol. If the ketone had not been protected as a acetal, the phenylmagnesium bromide would have reacted with it as well.

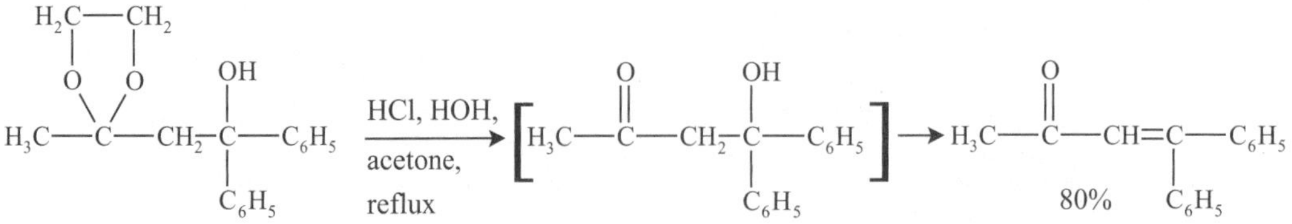

Heating the acetal with water and a trace of acid reverses the acetal-forming reaction and regenerates the ketone. Under these conditions, the tertiary alcohol eliminates and forms an alkene.

E.3. REACTIONS WITH THIOLS

General Reaction:

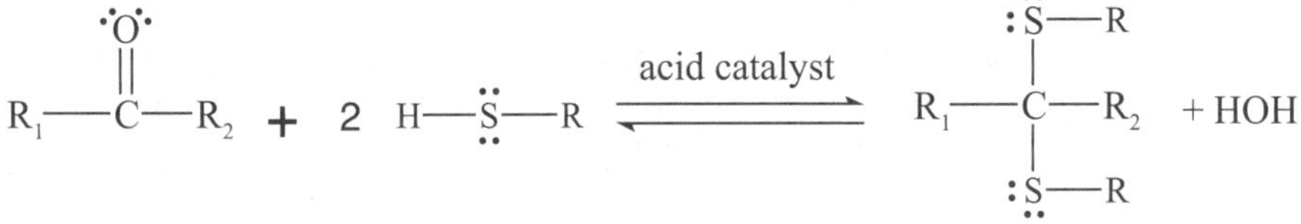

A dithioacetal

Specific Example:

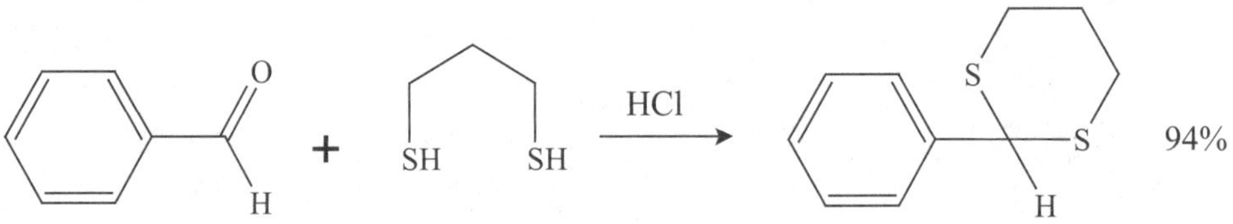

Thiols, especially dithiols, react with aldehydes and ketones to make dithioacetals. The reaction is similar to the reaction with alcohols in Section E.2. Dithioacetals are more stable than regular acetals, but can be removed with acid to regenerate the C=O, usually in the presence of a metal salt, such as $HgSO_4$ or $CuCl_2$. Dithioacetals can also be used in further synthetic reactions, as we will see later.

F. Oxidations of Aldehydes

General Reaction:

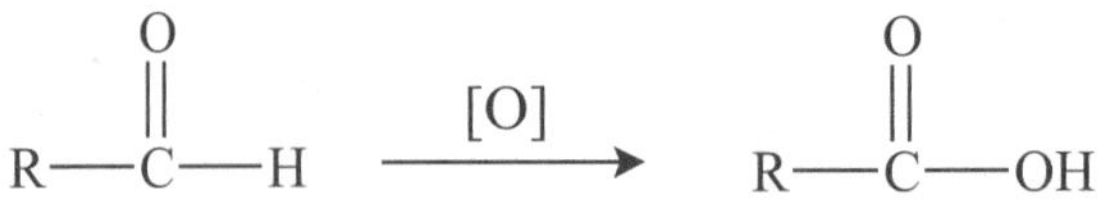

Specific Examples:

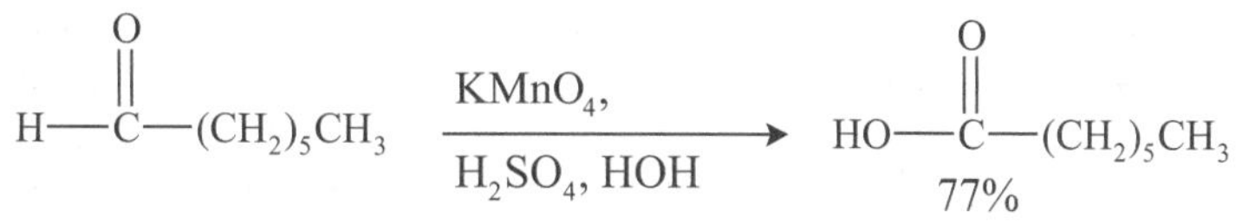

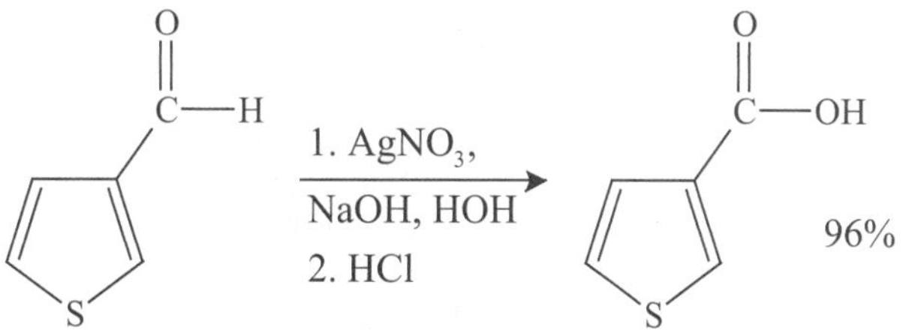

Aldehydes are easily oxidized to carboxylic acids. The same reagents that we used to oxidize primary alcohols to carboxylic acids (potassium permanganate, potassium dichromate, chromium trioxide in sulfuric acid) can be used. Aldehydes can be selectively oxidized to carboxylic acids by very weak oxidizing agents, such as silver nitrate in sodium hydroxide. Metallic silver is produced as the by–product. This reaction is the basis for the Tollen's test: Formation of a "silver mirror" shows the presence of an aldehyde.

G. Reductions of Aldehydes and Ketones

G.1. REDUCTIONS TO ALCOHOLS

General Reaction:

[H] stands for reduction, or a generic reducing agent.

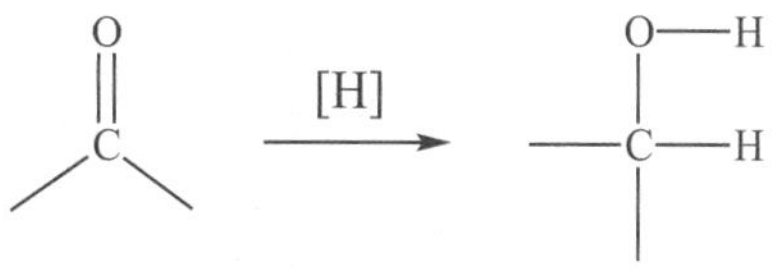

Specific Examples:

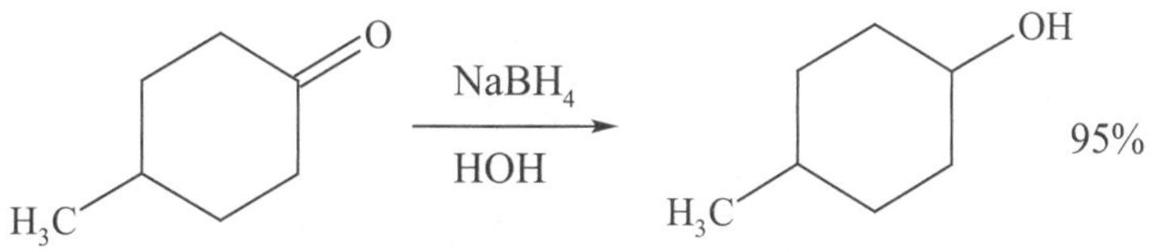

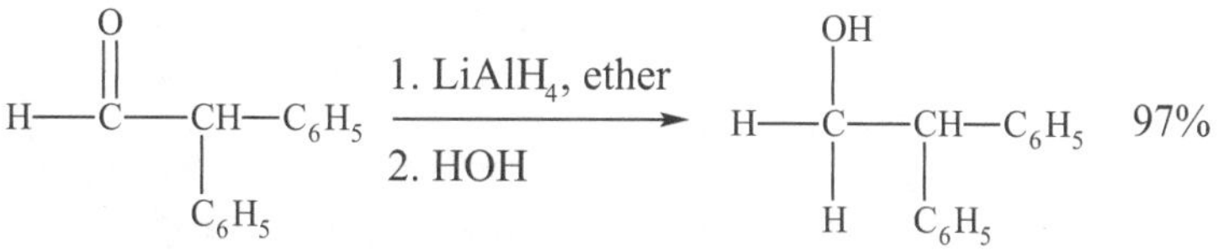

Aldehydes and ketones can be reduced to alcohols in high yields by hydride-reducing agents. This reaction was discussed in Chapter 8.

G.2. REDUCTION TO A CH₂

General Reaction:

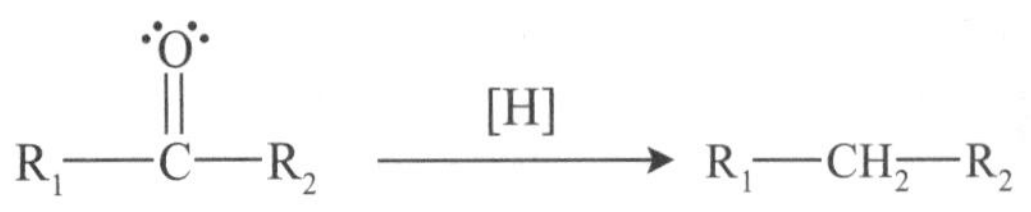

Specific Examples:

These examples are three classical and relatively general methods of reducing the C=O of an aldehyde or a ketone to a CH₂. The Clemmensen reduction involves using amalgamated zinc in strongly acidic conditions and was discussed in Chapter 10.

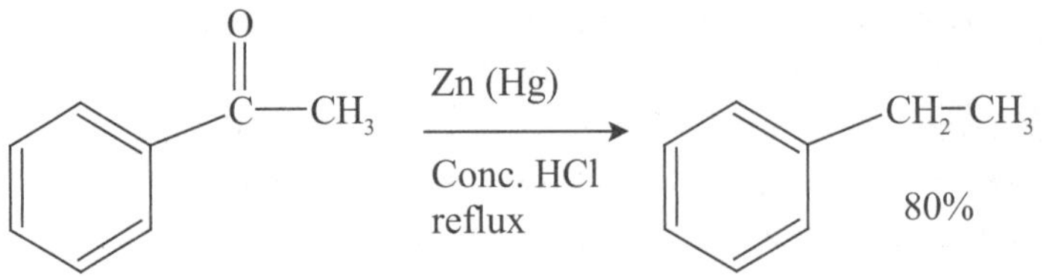

The reduction of dithioacetals with Raney Nickel (a specially prepared catalyst) in ethanol is a general procedure that involves essentially neutral conditions, both for the production of the dithioacetal and for the reduction. The following example illustrates that C=C's survive the desulfurization step. The one disadvantage, from a practical point of view, is that the thiols stink!

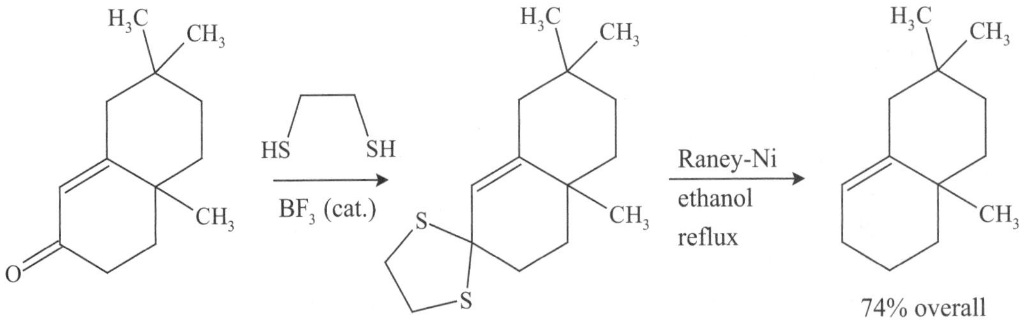

The Wolff–Kishner procedure involves strongly basic conditions, but it is somewhat more general in use than the Clemmensen procedure because high boiling alcohols are used as solvents, and more organic compounds are soluble in these solvents.

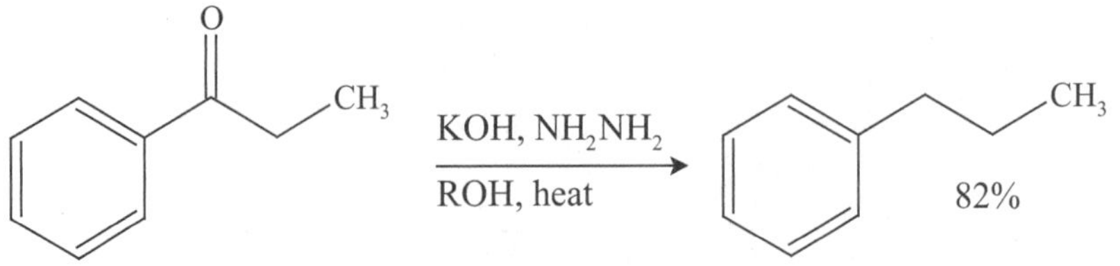

The mechanism of the reaction is shown below. C_6H_5 is a phenyl group.

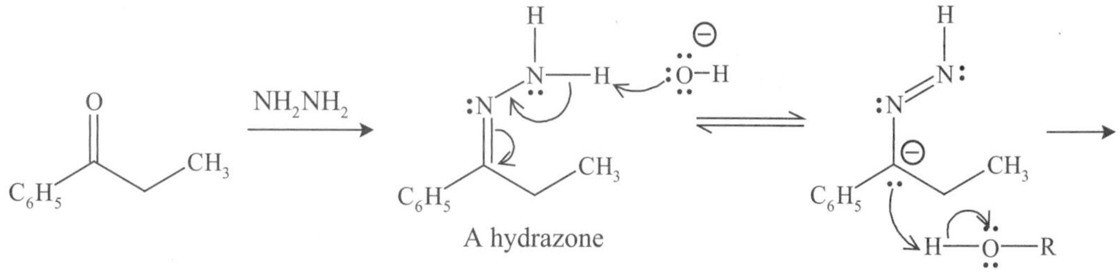

A hydrazone

The first step is the formation of the hydrazone, which was discussed in Chapter 11. Hydroxide pulls a hydrogen off the NH_2, which, after some electron pushing, puts a negative charge on the indicated carbon. The carbanion pulls a hydrogen off the alcohol solvent to form the next intermediate.

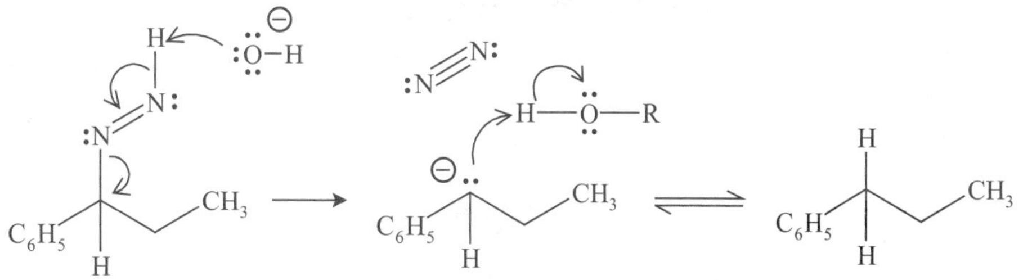

The last two steps are now repeated. Hydroxide pulls off the other N–H, and some electrons shift to form a nitrogen molecule and another carbanion. The carbanion pulls off another alcohol hydrogen to form the product.

A convenient variation of the Wolff–Kishner reduction that is useful in organic laboratories, is to convert the aldehyde or ketone into the semicarbazone first and then to add it to a hot solution of KOH in a high boiling alcohol. Hydroxide cleaves the semicarbazone to the hydrazone and then does the Wolff–Kishner reduction.

H. Preparations

A number of reactions are useful in preparing aldehydes and ketones. Many of them have been discussed in earlier chapters.

H.1. OXIDATION OF PRIMARY AND SECONDARY ALCOHOLS

The oxidation of primary and secondary alcohols was discussed in Chapter 8. This is probably the most general method of synthesis of these compounds. Secondary alcohols are oxidized to ketones by a variety of oxidizing agents. To stop the oxidation of a primary alcohol to an aldehyde, a chromium oxide in pyridine reagent is used. Some examples are shown below.

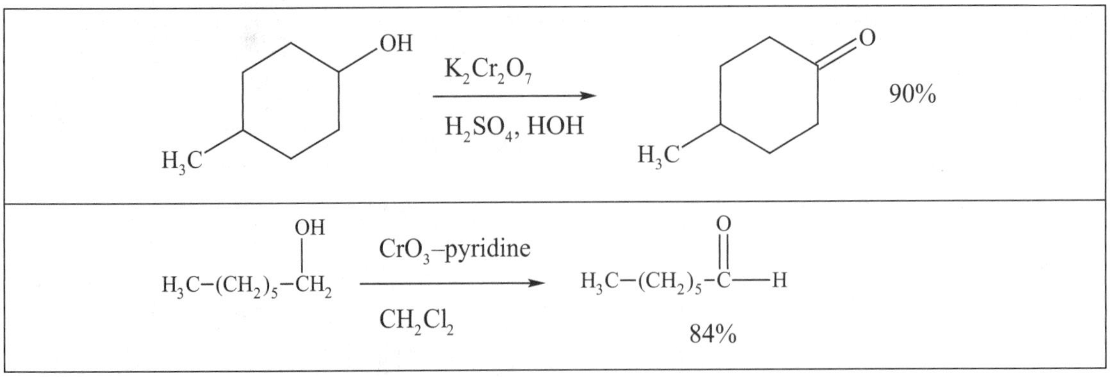

H.2. FRIEDEL–CRAFTS ACYLATION AND THE GATTERMANN REACTION

The Friedel–Crafts acylation and the Gattermann reaction were discussed in Chapter 10 and are common ways to introduce a ketone or an aldehyde onto an aromatic ring that does not have a strongly deactivating group bonded to it.

H.2.a. FRIEDEL–CRAFTS ACYLATION

General Reaction:

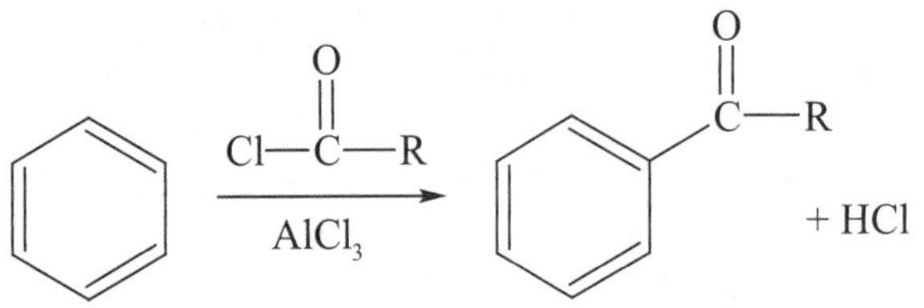

Specific Examples:

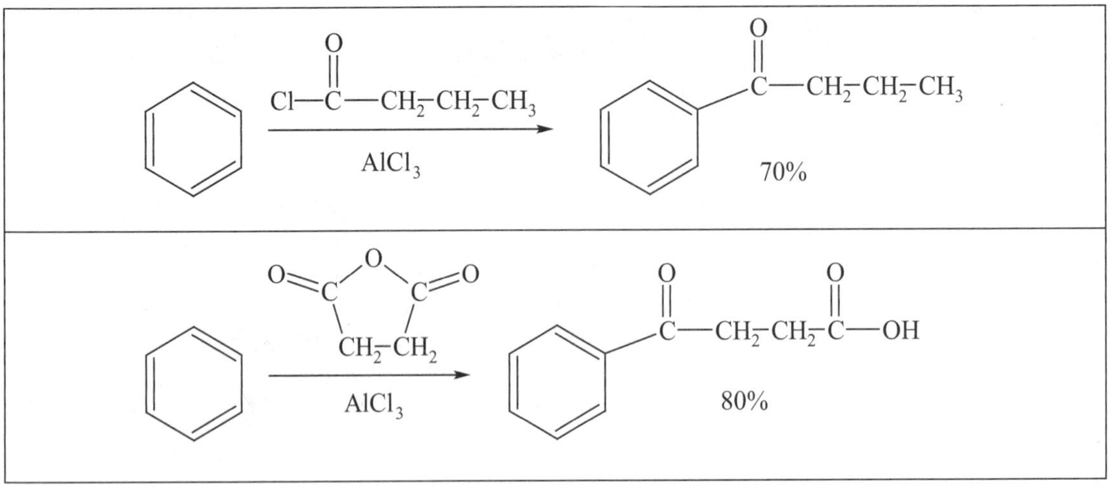

H.2.b. THE GATTERMANN REACTION

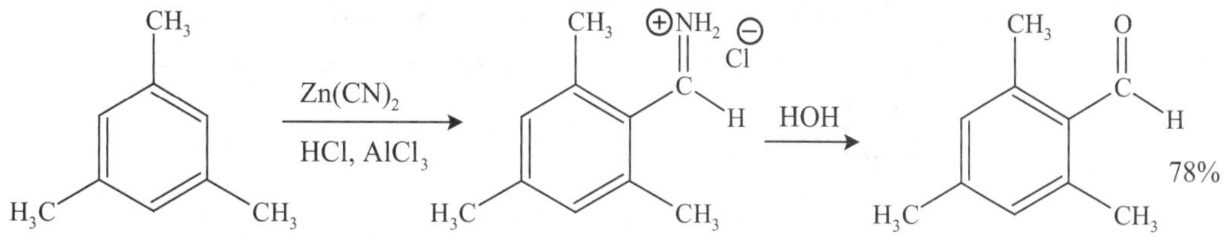

Remember, aldehydes cannot be prepared directly because HCOCl does not exist as a stable material. That is why the Gattermann reaction was used here.

H.3. KETONES BY HYDRATION OF ALKYNES

Preparing ketones by hydrating alkynes was discussed in Chapter 5. With the help of an acid catalyst in a Markovnikov fashion, water adds to form an enol, which tautomerizes to a ketone.

General Reaction:

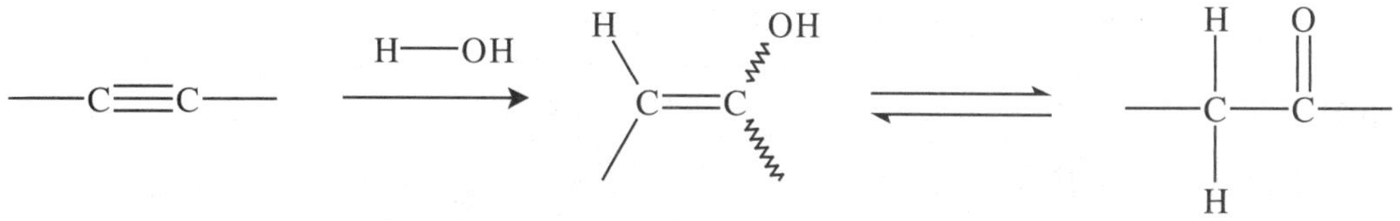

Specific Example:

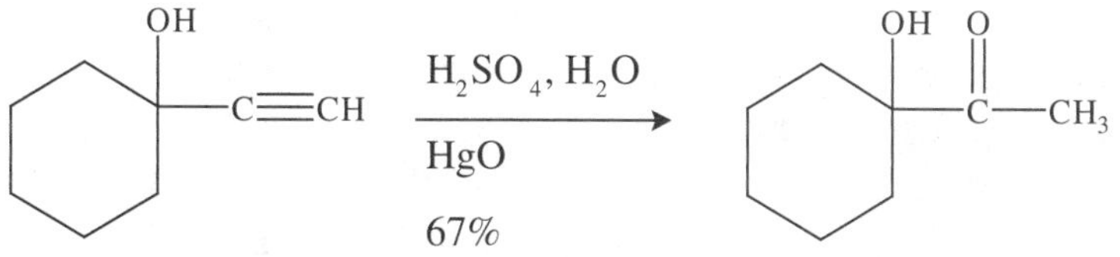

H.4. ALDEHYDES BY HYDROBORATION OF ALKYNES

Alkynes can be hydroborated by disubstituted boranes (R_2BH) to give vinyl boranes. This reaction is an anti-Markovnikov addition because boranes are electrophilic reagents. The vinylboranes can be oxidized to aldehydes. This reaction was discussed in Chapter 5. An example follows. This particular borane is called catecholborane.

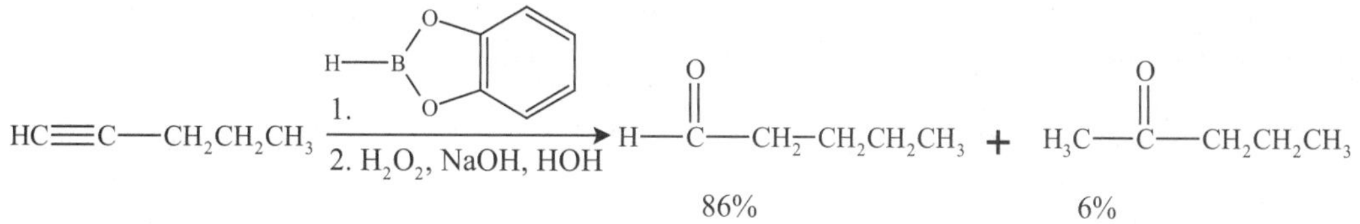

I. Acidity of the Alpha-Hydrogen

The carbon next to the C=O of an aldehyde or ketone is referred to as the alpha-carbon. A hydrogen attached to an alpha-carbon is called an alpha-hydrogen. Carbons and hydrogens farther from the C=O are indicated by the appropriate letter in the Greek alphabet. See the following example.

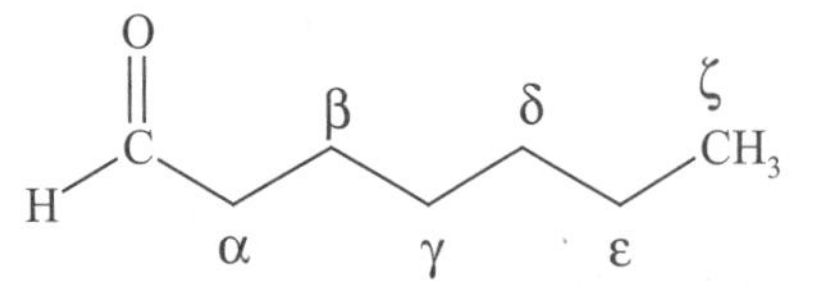

The first six letters of the Greek alphabet are alpha (α), beta (β), gamma (γ), delta (δ), epsilon (ε), and zeta (ζ). Sometimes, if a group is on the last carbon of the chain, it is referred to as omega (ω) because omega is the last letter of the Greek alphabet.

The alpha-hydrogens of aldehydes and ketones are slightly acidic, having pK_a's of about 19–20. This makes them weaker acids than alcohols or water. In the presence of hydroxide, an equilibrium is set up. The anion of an aldehyde or ketone is called an enolate ion and is resonance-stabilized. This is shown below for ethanal (acetaldehyde).

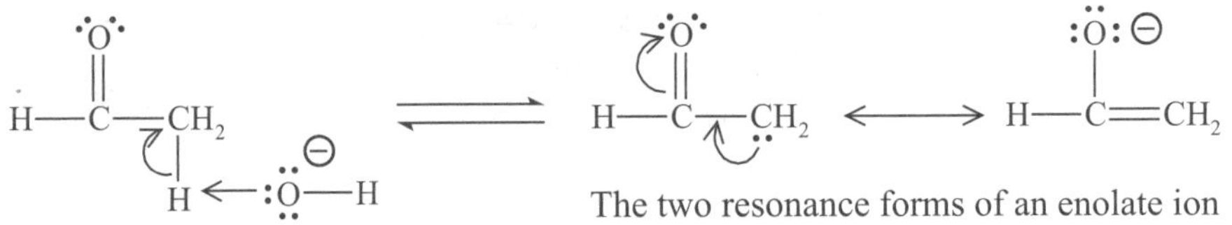

The two resonance forms of an enolate ion

An enolate is nucleophile and could react either at the oxygen or at the carbon. Because carbanions are more basic than oxygen anions, enolates generally react as carbon nucleophiles.

If the enolate ion were to pick up a hydrogen from water on the oxygen, we would form the enol. This is shown below.

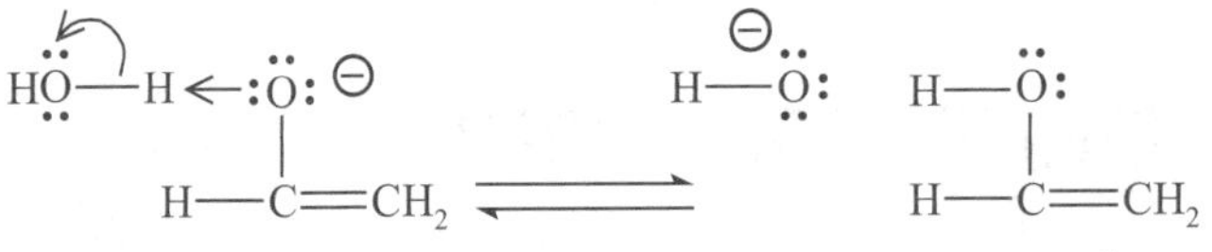

an enol

This tells us that aldehydes and ketones are in equilibrium with their enol forms. For most aldehydes and ketones, very little enol is formed at equilibrium: For ethanal, about 0.001% of the enol is formed. However, if the enol form can be stabilized by conjugation or other factors, it can be a significant portion of the equilibrium mixture. See the examples on the next page.

Compound		Percent Enol (Solvent)
		6.5% (H_2O) 23.1% (CCl_4)
In water, water can hydrogen-bond to the oxygens of the C=O's, making it less likely to enolize. In CCl_4, which cannot hydrogen-bond, enolization promotes hydrogen-bonding and stabilization by conjugation.		
		About 100%
Enolization is favored because the enol, phenol, is aromatic.		

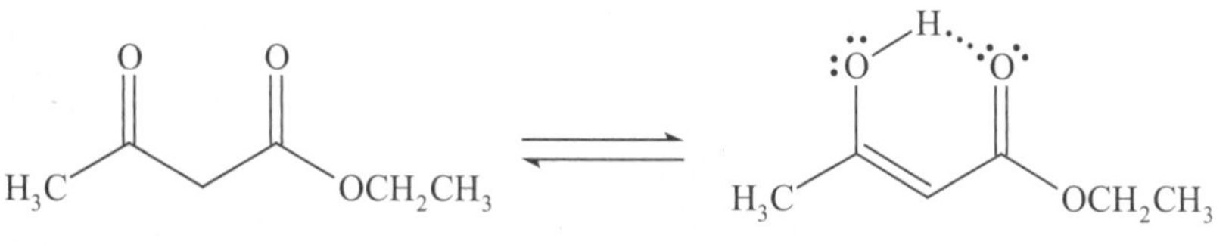

J. Halogenation at the Alpha-Carbon

J.1. ACID-CATALYZED HALOGENATION

If an aldehyde or a ketone is treated with Cl_2, Br_2, or I_2 in the presence of an acid, one alpha-hydrogen can be substituted with a halogen. See the examples below.

General Reaction:

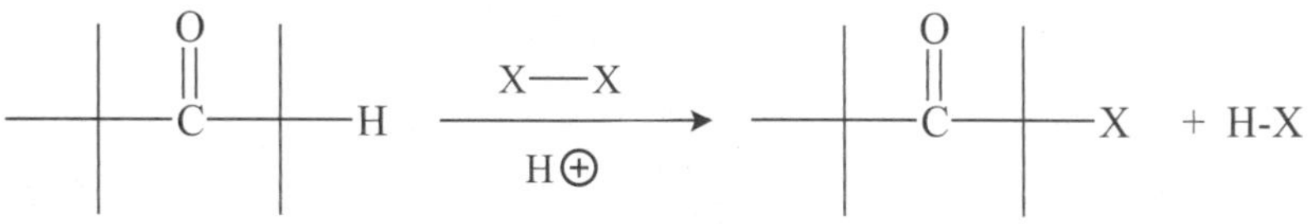

Specific Examples:

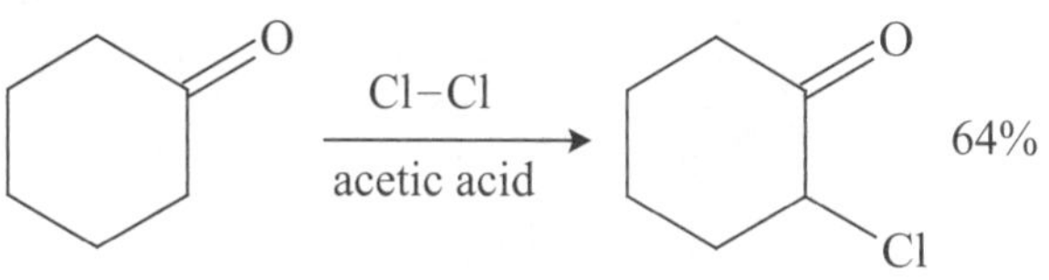

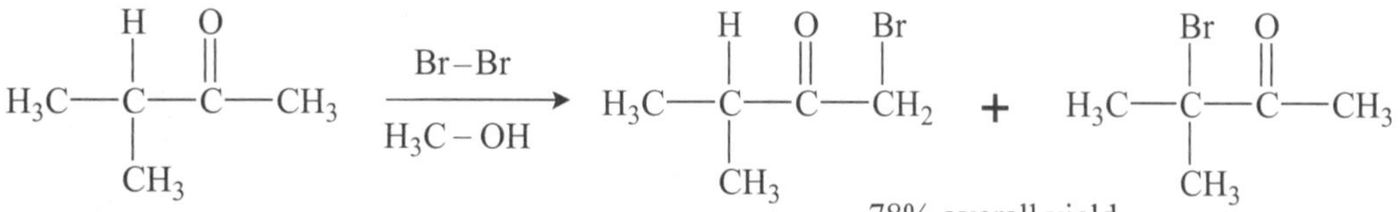

78% overall yield

95% this isomer   5% this isomer

As the second example shows, it is possible to halogenate the less substituted carbon of a ketone selectively. Because H–X is generated as a by–product, the reaction generates its own catalyst. The mechanism is shown below.

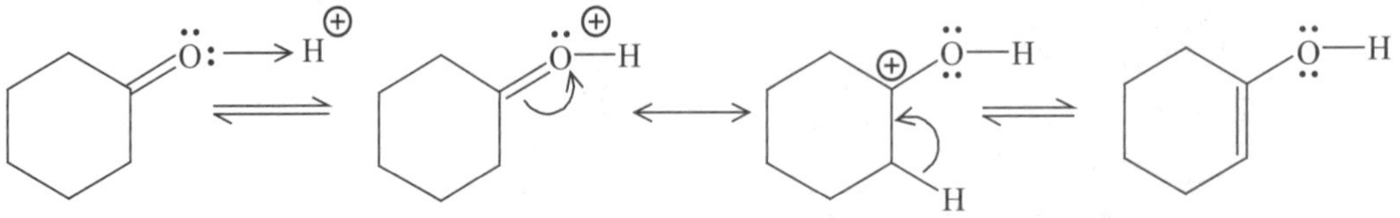

The acid catalyst protonates the carbonyl oxygen, which generates a resonance-stabilized cation. Loss of H^+ from an alpha-carbon produces the enol.

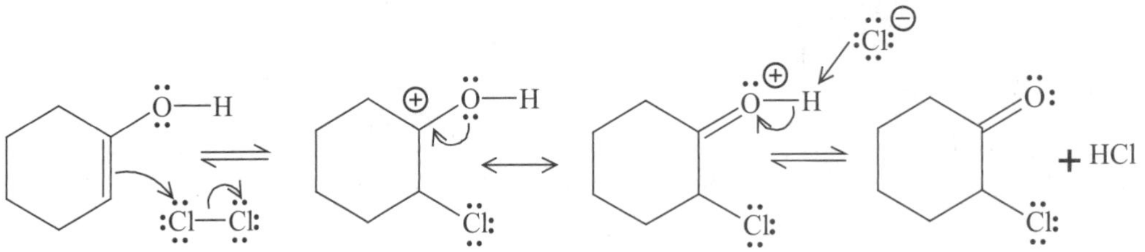

The electrons from the C=C attack Cl–Cl. The cation formed is stabilized by resonance. Loss of H^+ gives us the product chloroketone.

J.2. BASE-PROMOTED HALOGENATION: THE HALOFORM REACTION

In the presence of a base, it is usually not possible to halogenate an alpha-position once. This is because the product alpha-haloketone is a stronger acid than the starting ketone; the product reacts with the remaining halogen faster than the starting material. However, this reaction can be used as a way to prepare carboxylic acids from methyl ketones. See the following examples.

General Reaction:

$$R{-}\overset{\overset{\large O}{\|}}{C}{-}CH_3 \xrightarrow[2.\ H^+]{1.\ 3\ X_2,\ 4\ \overset{\ominus}{OH}} R{-}\overset{\overset{\large O}{\|}}{C}{-}OH + 3\ HOH + 3\ \overset{\ominus}{X} + \underset{\text{(Haloform)}}{HCX_3}$$

Specific Examples:

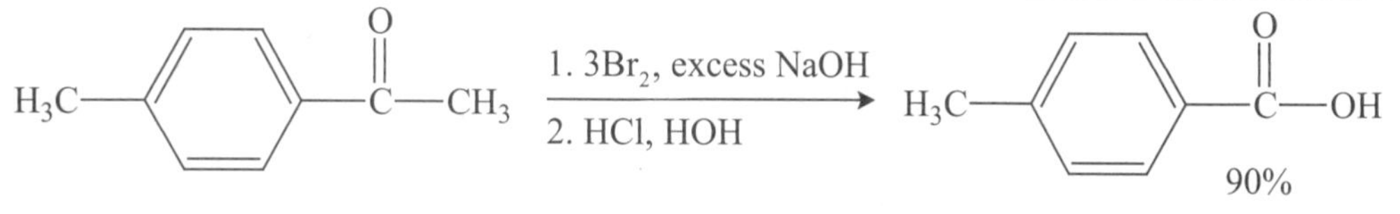

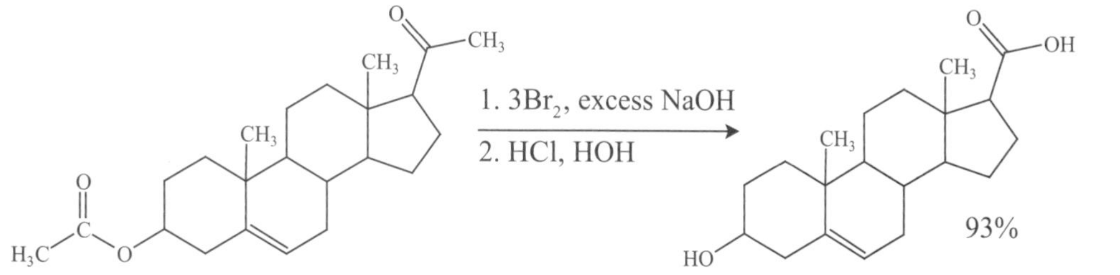

In this case, the methyl ketone is converted into a carboxylic acid. The ester is also hydrolyzed by the base to an alcohol and acetic acid. This reaction is covered in Chapter 13.

If iodine is used as the halogen, the by–product is CHI_3 (iodoform), which is a light yellow solid. If an unknown compound is treated with excess I_2 and sodium hydroxide, the formation of a yellow precipitate, iodoform, indicates the presence of a methyl ketone or ethanal. Unfortunately, secondary alcohols can be oxidized under these conditions to ketones, so any alcohol of the general structure $CH_3CH(OH)$–R also gives a positive test. However, a simple treatment of the unknown with Jones' Reagent ($CrO_3/HOH/H_2SO_4$) will easily distinguish a secondary alcohol from a ketone. (How?) An unknown can be classified fairly easily as a methyl ketone using these two tests.

K. Aldol Condensation

General Reaction:

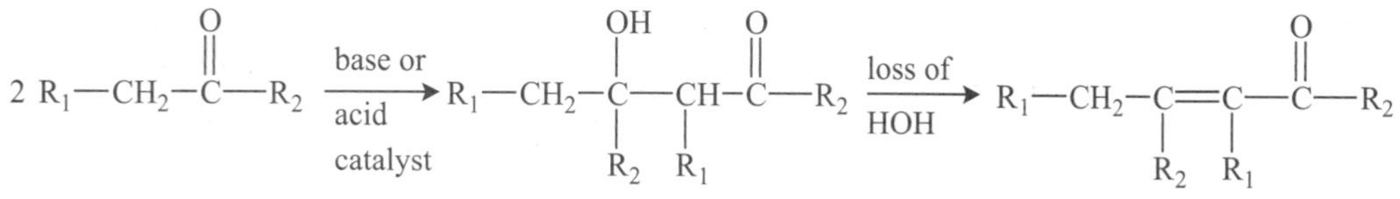

An "aldol"

Specific Examples:

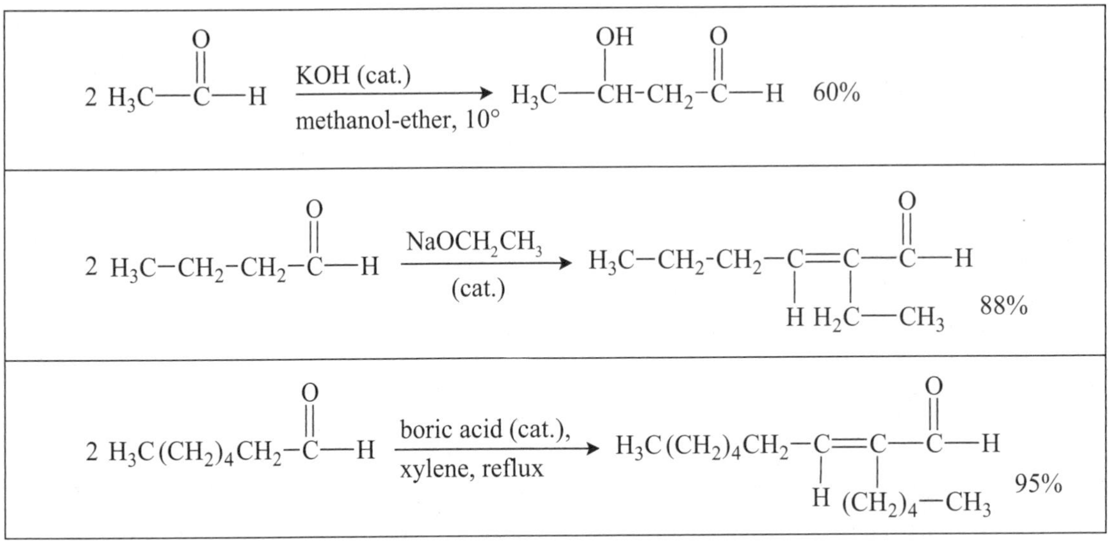

If two molecules of an aldehyde with an alpha-hydrogen are reacted together with a catalytic amount of a strong base, one of two products can form, depending upon the reaction conditions. The initial product is the **aldol** (an aldehyde-alcohol). If the aldehyde has an additional alpha-hydrogen, upon standing or heating, water is lost to form the conjugated aldehyde. Most ketones, other than acetone, do not readily undergo this reaction. The mechanism of the base-catalyzed reaction is shown below.

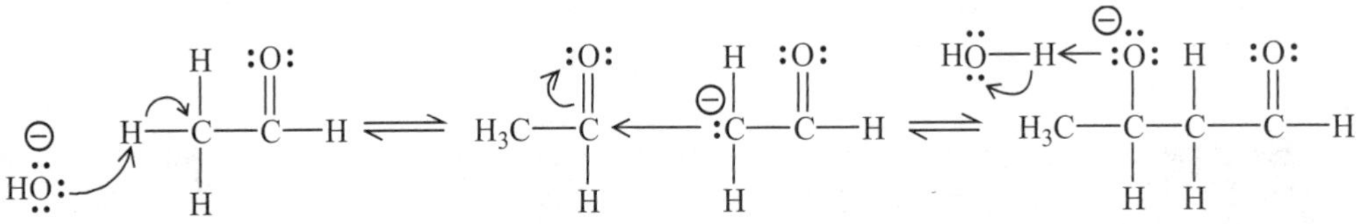

The base pulls an alpha-hydrogen off one molecule of the aldehyde. The aldehyde anion attacks the carbonyl carbon of the other aldehyde molecule to form the new carbon–carbon bond. The resulting oxygen anion pulls a hydrogen off water to make the aldol product.

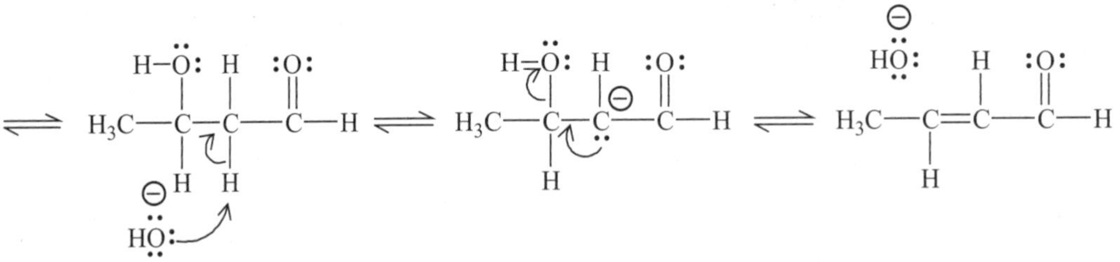

The base pulls off another alpha-hydrogen to make another anion. The anion forms a C=C by pushing out HO⁻ on the adjacent carbon. Notice that each time we use a hydroxide ion, we reform one later, so the reaction is truly catalytic in base.

What do you think happens if you use two different aldehydes, such as ethanal and propanal?

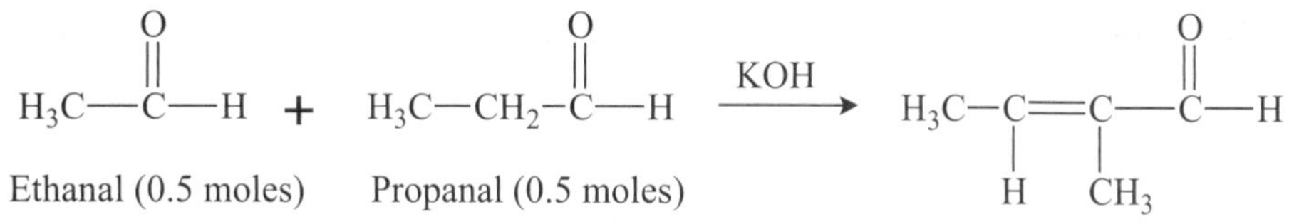

$$H_3C-\overset{\overset{\displaystyle O}{\|}}{C}-H \quad + \quad H_3C-CH_2-\overset{\overset{\displaystyle O}{\|}}{C}-H \quad \xrightarrow{\text{KOH}} \quad H_3C-\underset{\underset{\displaystyle H}{|}}{C}=\underset{\underset{\displaystyle CH_3}{|}}{C}-\overset{\overset{\displaystyle O}{\|}}{C}-H$$

Ethanal (0.5 moles)   Propanal (0.5 moles)

This reaction has been reported in the chemical literature, but the isolated yield is only 25–30%. What other products do you think could have formed?

Because both products have alpha-hydrogens, KOH could have pulled an alpha-hydrogen off either aldehyde. The resulting carbanion could react with the first aldehyde molecule it came in contact with, either an ethanal or a propanal. Therefore, a mixture of four different aldol products could form, which could subsequently dehydrate to four different conjugated products. This is shown below.

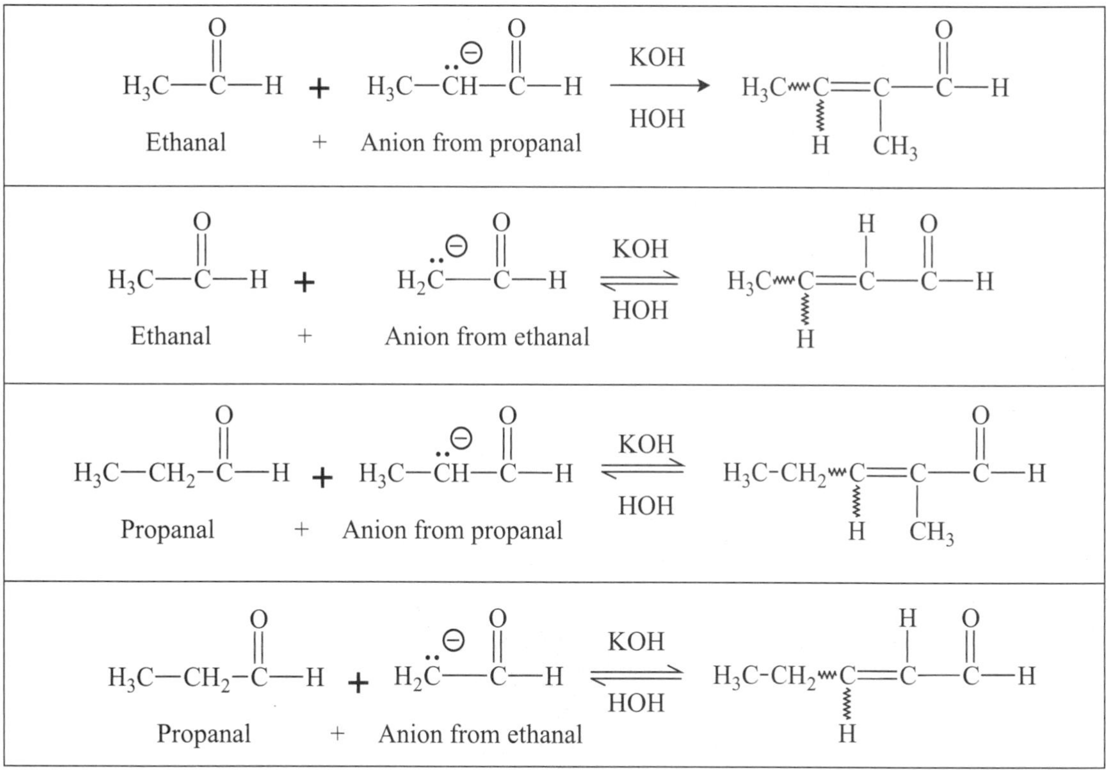

We started with equal amounts of each of the aldehydes, so we should isolate approximately equal amounts of each product. The boiling points are different enough that the products can be separated with a very good fractionating column, but this is not a separation I would like to do! So generally, you don't try to react together two aldehydes that both have alpha-hydrogens.

One variation of this reaction uses a ketone with alpha-hydrogens reacting with an aldehyde that does *not* have alpha-hydrogens. This reaction, called the Claisen–Schmidt reaction, often gives good yields of the conjugated ketone products. Some examples are shown below.

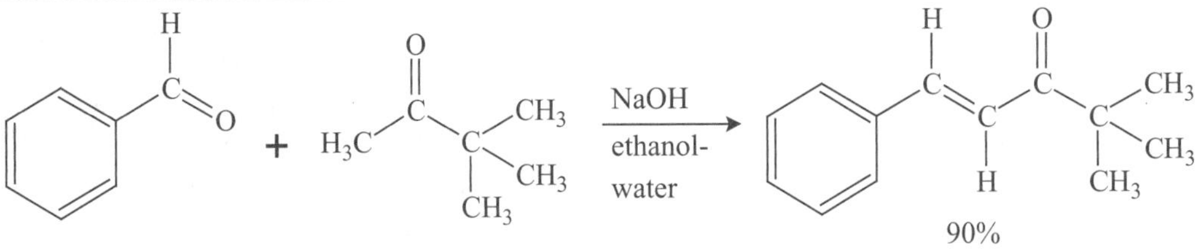

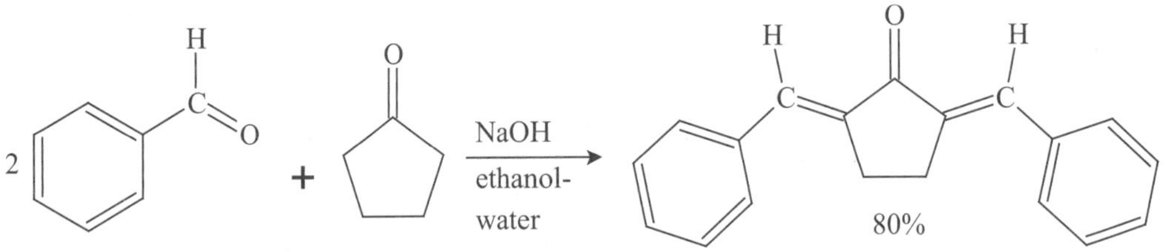

In this case, cyclopentanone has alpha-hydrogens on both sides of the C=O, so a benzaldehyde can react on each side.

We will look at more examples of reactions of enolate ions in Chapter 14.

REVIEW EXERCISES FOR CHAPTER 12

1. Give IUPAC names for the following structures.

a.	b.	c.

2. Show how these products can be made by the reaction of an aldehyde or ketone with an appropriate reagent and reaction conditions.

a.	b.	c.
(Two different ways, with *different types of reactions*)		
d.	e.	f.

3.   Show how you would make the following compounds starting with the indicated materials.

Starting Materials	Product
a. Benzaldehyde, plus any alcohols with one to four carbons	
b. Benzaldehyde, plus any alcohols with one to four carbons.	
c. All carbons in the product must come from cyclopentene	
d.	

Carboxylic Acids and Derivatives

WHAT YOU WILL LEARN

In this chapter, you will learn:

- how to recognize carboxylic acids and derivatives;
- how to name carboxylic acids and derivatives using IUPAC rules;
- how carboxylic acids and derivatives can be interconverted;
- how to draw the variations on the nucleophilic acyl substitution mechanism;
- how carboxylic acids and derivatives can be reduced.

SECTIONS IN THIS CHAPTER

- Structures

- Nomenclature of Carboxylic Acids

- Physical Properties of Carboxylic Acids

- Acidity of Carboxylic Acids

- The General Reaction of Carboxylic Acid Derivatives: Nucleophilic Acyl Substitution, Addition–Elimination, or "Up, Down, and Out"

- Conversions of Carboxylic Acids into Carboxylic Acid Derivatives

- Naming of Carboxylic Acid Derivatives

- Interconversions of Carboxylic Acids and Derivatives

- Reductions of Carboxylic Acids and Derivatives

A. Structures

Carboxylic acids are commonly found in nature, as well as in pharmaceuticals and other synthetic materials. Sometimes carboxylic acids are written as RCOOH, or RCO_2H, to save space. A number of other related materials are derived from carboxylic acids, by replacing the OH with a halogen, an OR group, or an NRR' group. These compounds are called carboxylic acid derivatives. Some examples of these molecules are shown below. Although nitriles are not actually carboxylic acid derivatives, they are included here because of their similarities to amides.

Functional Group	Generic Structure	Real Example
Carboxylic acid	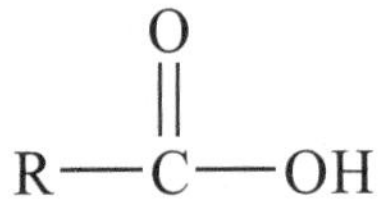	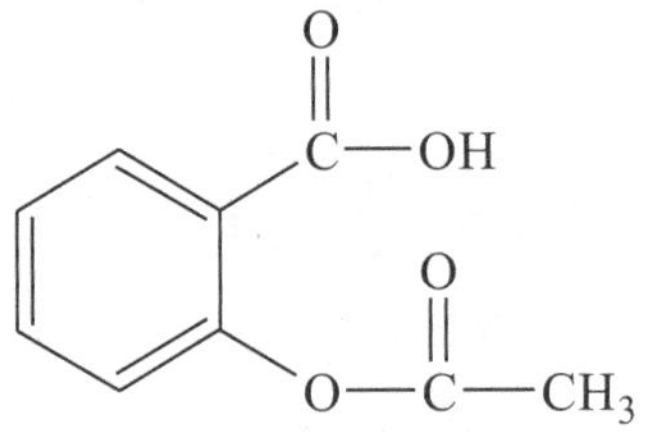 Aspirin: also has an ester
Acid chloride	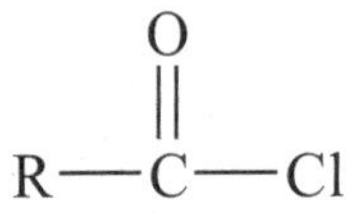	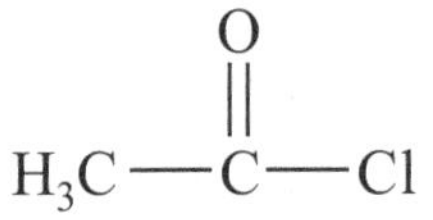 Ethanoyl chloride or "acetyl chloride"
Anhydride	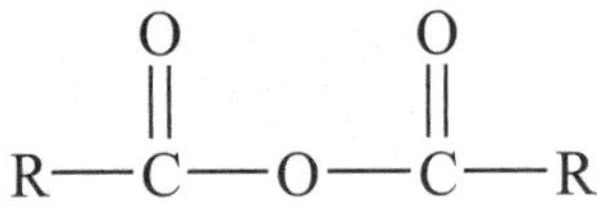	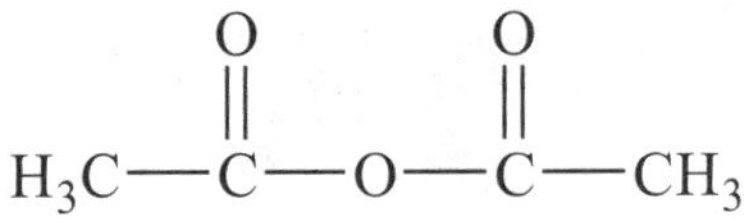 Ethanoic anhydride, or "acetic anhydride"
Ester	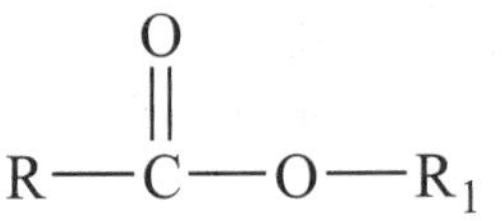	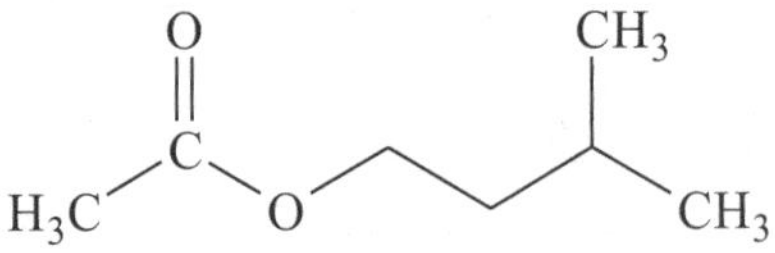 "Isoamyl acetate": a component in banana oil
Amide	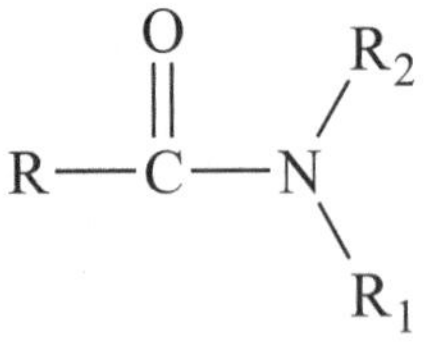	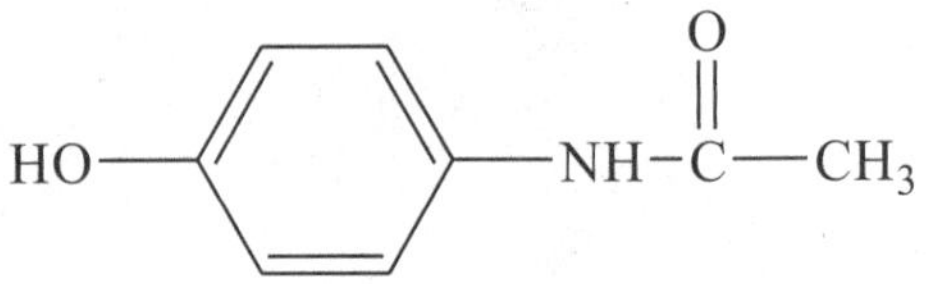 Acetaminophen: pain-relieving ingredient in Tylenol©
Nitrile	R—C≡N	H_3C—C≡N "Acetonitrile," a common solvent

B. Nomenclature of Carboxylic Acids

Carboxylic acids are named just like aldehydes, except that the functional group ending is *-oic acid* (acid is a separate word). Carboxylic acids are higher in naming priority than any other functional group we have named until now. See the examples below.

(structure of octanoic acid)	Since there are eight carbons in the chain, this is **octanoic acid**.
(structure with O—CH₃, O₂N)	A benzene ring with an acid group on it is called benzoic acid. There is a methoxy group and a nitro group also attached. We number counterclockwise from the carbon where the acid group is attached, so the name is **2-methoxy-5-nitrobenzoic acid**.
(structure with HO, OH, H, NH₂)	This has two carboxylic acid groups, so it is a pentanedioic acid. There is an amino group on the second carbon from the right. This carbon is also a stereocenter, so we have to specify whether it is R or S. The correct name is **(*S*)-2-aminopentanedioic acid**. This is also called glutamic acid, which is one of the 20 common amino acids found in proteins.

Carboxylic acids also have common names, which are often used in the chemical literature, as well as on bottles of chemicals. These are given below.

Carboxylic Acid	IUPAC Name	Common Name
H–COOH	Methanoic acid	Formic acid
CH_3–COOH	Ethanoic acid	Acetic acid
CH_3CH_2COOH	Propanoic acid	Propionic acid
$CH_3(CH_2)_2COOH$	Butanoic acid	Butyric acid
$CH_3(CH_2)_3COOH$	Pentanoic acid	Valeric acid
$CH_3(CH_2)_4COOH$	Hexanoic acid	Caproic acid
$CH_3(CH_2)_5COOH$	Heptanoic acid	Enanthic acid
$CH_3(CH_2)_6COOH$	Octanoic acid	Caprylic acid
$CH_3(CH_2)_8COOH$	Decanoic acid	Capric acid
$CH_3(CH_2)_{10}COOH$	Dodecanoic acid	Lauric acid
$CH_3(CH_2)_{12}COOH$	Tetradecanoic acid	Myristic acid
$CH_3(CH_2)_{14}COOH$	Hexadecanoic acid	Palmitic acid
$CH_3(CH_2)_{16}COOH$	Octadecanoic acid	Stearic acid

The six-, eight-, and ten-carbon acids give goats their characteristic odor, and all of the common names are derived from the Latin word, *caper*, for goat.

The dicarboxylic acids also have well-known common names, as shown below.

Dicarboxylic Acid	IUPAC Name	Common Name
HOOC–COOH	Ethanedioic acid	Oxalic acid
HOOC–CH_2–COOH	Propanedioic acid	Malonic acid
HOOC–$(CH_2)_2$–COOH	Butanedioic acid	Succinic acid
HOOC–$(CH_2)_3$–COOH	Pentanedioic acid	Glutaric acid
HOOC–$(CH_2)_4$–COOH	Hexanedioic acid	Adipic acid
HOOC–$(CH_2)_5$–COOH	Heptanedioic acid	Pimelic acid

A mnemonic device to remember the first letters of the common names of these acids is, "<u>O</u>h <u>M</u>y, <u>S</u>uch <u>G</u>ood <u>A</u>pple <u>P</u>ie!" I am not aware of a similar mnemonic for monocarboxylic acids.

C. Physical Properties of Carboxylic Acids

Compared to the other functional groups we have looked at, carboxylic acids have higher boiling points and greater water-solubilities than other compounds of similar molecular weight. See the table below.

Carboxylic Acid	Boiling Point (°C)	Water-Solubility (g/100 mL of water)
H–COOH	101	Miscible
CH_3–COOH	117	Miscible
CH_3CH_2COOH	141	Miscible
$CH_3(CH_2)_2COOH$	164	Miscible
$CH_3(CH_2)_3COOH$	187	5.0
$CH_3(CH_2)_4COOH$	205	1.1
$CH_3(CH_2)_5COOH$	223	0.25
$CH_3(CH_2)_6COOH$	239	0.07

The following table compares these properties for several compounds of similar shape and molecular weight.

Compound	Boiling Point (°C)	Water-Solubility (g/100 mL of water)
$H_3C-CH_2-CH_2-\overset{\overset{\displaystyle CH_3}{\displaystyle\vert}}{CH}-CH_3$	60	0.0015
$H_3C-CH_2-CH_2-\overset{\overset{\displaystyle O}{\displaystyle\vert\vert}}{C}-CH_3$	102	5.5
$H_3C-CH_2-CH_2-\overset{\overset{\displaystyle CH_3}{\displaystyle\vert}}{CH}-OH$	119	16.6
$H_3C-CH_2-CH_2-\overset{\overset{\displaystyle O}{\displaystyle\vert\vert}}{C}-OH$	164	Miscible

The higher water-solubility can be explained by the fact that carboxylic acids have more possible places than any other functional group where it can hydrogen-bond with water. The higher boiling point can be explained by the fact that each pair of carboxylic acid molecules is hydrogen-bonded together twice: This is called a hydrogen-bonded dimer. See below.

	Water can potentially hydrogen-bond five different ways to one molecule of a carboxylic acid. These are shown in the drawing to the left.
	The carbonyl oxygen of each molecule is hydrogen-bonded to the O–H of the other molecule.

D. Acidity of Carboxylic Acids

Carboxylic acids are the most acidic of the common functional groups. There are at least two reasons for this. First, the C=O oxygen is pulling electrons away from the O–H. This weakens the O–H bond, which would make the molecule more likely to give up a hydrogen. Second, the anion formed by loss of H+ is stabilized by resonance. This helps shift the equilibrium more to the right. These are illustrated below.

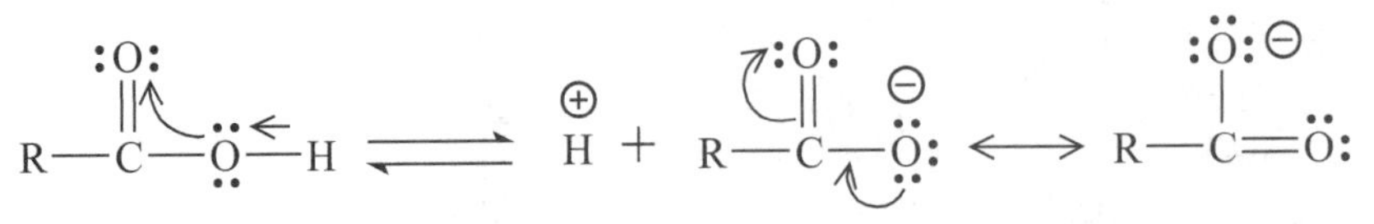

| Inductive effect of the carbonyl oxygen weakening the O–H bond | Resonance stabilization of the carboxylate ion |

The pK_a values of several carboxylic acids are shown in the following table.

Structure	pK_a
CH_3–COOH	4.76
CH_3–CH_2–COOH	4.87
CH_3–CH_2–CH_2–COOH	4.82

Structure	pK_a
CH_3O–CH_2–COOH	3.54
Cl–CH_2–COOH	2.86
Cl_3C–COOH	0.64

Acetic acid, with a pK_a of 4.76, is our standard for comparison. Adding an alkyl group raises the pK_a slightly, due to the inductive effect of an alkyl group, which donates electrons toward the COOH group. This makes the hydrogen of the O–H less likely to ionize. An electronegative atom pulls electrons toward itself by the inductive effect, making the molecule more acidic, as shown by methoxyacetic acid and chloroacetic acid. More than one electronegative atom increases the effect, as shown by trichloroacetic acid. The effect of the location of an electronegative atom is shown on the next page.

Structure	pK$_a$	Comments
CH$_3$—CH$_2$—CH$_2$—COOH	4.82	As the chlorine gets further and further away from the O–H of the carboxylic acid, the inductive effect becomes weaker, and the effect on the pK$_a$ of the carboxylic acid is lessened.
CH$_3$—CH$_2$—CH—COOH 　　　　　│ 　　　　　Cl	2.86	
CH$_3$—CH —CH$_2$—COOH 　　　│ 　　　Cl	4.05	
H$_2$C—CH$_2$—CH$_2$—COOH 　│ 　Cl	4.52	

There are similar effects for substituted benzoic acids. See the following table.

Structure	pK$_a$	Comments
	4.19	This is benzoic acid, our standard of comparison.
	3.48	The nitro group is strongly electron-withdrawing, so it pulls electron density away from the carboxylic acid group. This makes *meta*-nitrobenzoic acid a stronger acid than benzoic acid.

Structure	pK_a	Comments
	3.41	The nitro group withdraws electrons from the benzene ring by induction, which, in turn, makes the carboxylic acid group more acidic. The para isomer is slightly more acidic than the meta isomer. This is probably the result of a resonance effect, which makes the carbon on the benzene ring attached to the carboxylic acid somewhat positive. There is no analogous resonance form possible with *meta*-nitrobenzoic acid.
	4.09	Methoxy is an electron-withdrawing group, due to the the electronegativity of the oxygen. Therefore, *meta*-methoxybenzoic acid is more acidic than benzoic acid. Methoxy is less electron-withdrawing than nitro, so the effect is not as great.
	4.47	*Para*-methoxybenzoic acid is a weaker acid than benzoic acid. This is due to the fact that the methoxy group is able to donate electrons toward the carboxylic acid group directly by resonance. This makes the carboxylic acid group more negative and less likely to lose a proton.

E. The General Reaction of Carboxylic Acid Derivatives: Nucleophilic Acyl Substitution, Addition-Elimination, or "Up, Down, and Out"

General Reaction:

$$
\underset{\substack{R—C—Z}}{\overset{:O:}{\|}} \; + \; :Nu \longrightarrow \underset{\substack{R—C—Nu}}{\overset{:O:}{\|}} \; + \; :Z
$$

Z is the leaving group, and Nu is the nucleophile. At face value, this looks like an S_N2 substitution reaction; however, the mechanism is very different. The mechanism of this reaction will be used, in one form or another, to convert carboxylic acids into several of the derivatives and to interconvert the carboxylic acid derivatives. I call this mechanism "Up, Down, and Out," as I will illustrate below.

$$
\underset{\substack{R—C—Z \\ \uparrow \\ :Nu}}{\overset{:O:}{\|}} \; \xrightarrow{\text{"up"}} \; \underset{\substack{R—C—Z \\ | \\ Nu}}{\overset{:\ddot{O}: \ominus}{|}}
$$

The nucleophile attacks the partially positive carbon and pushes an electron pair "up" onto the carbonyl oxygen. This forms an intermediate with four groups attached to what was the carbonyl carbon. This intermediate is called the **tetrahedral intermediate**, because of its shape. This is the same type of intermediate we formed when a strong nucleophile attacked the C=O of an aldehyde or a ketone.

$$
\underset{\substack{R—C—Z \\ | \\ Nu}}{\overset{:\ddot{O}: \ominus}{|}} \; \xrightarrow{\text{"down and out"}} \; \underset{\substack{R—C—Nu}}{\overset{:O:}{\|}} \; + \; :Z
$$

An electron pair from the oxygen comes "down" to reform the C=O. As this happens, the bond to the leaving group Z is broken, and the electron pair is pushed "out" onto Z. This completes the substitution reaction.

This mechanism generally doesn't occur with carboxylic acids directly. The reason for this is that the nucleophile can often act as a base and pull off the acidic hydrogen. We will see some variations of this mechanism as we prepare certain carboxylic acid derivatives.

F. Conversions of Carboxylic Acids into Carboxylic Acid Derivatives

The main reactions of carboxylic acids are converting them into other carboxylic acid derivatives. These are described below.

F.1. PREPARATION OF ACID CHLORIDES

Carboxylic acids are converted into acid chlorides by heating the acid with thionyl chloride ($SOCl_2$), phosphorus trichloride (PCl_3), or phosphorus pentachloride (PCl_5). An example is shown below.

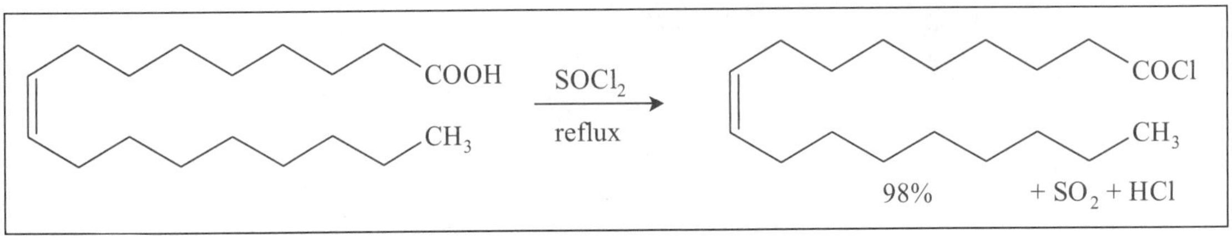

The reaction with thionyl chloride is convenient in a lab setting because the by–products are gases. The mechanism with thionyl chloride is shown below.

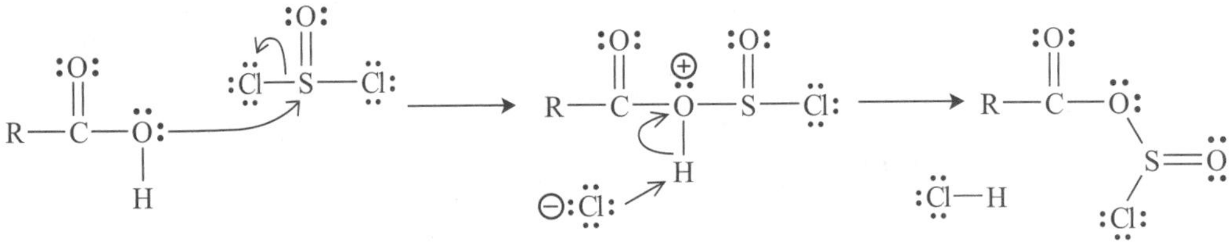

The acid attacks thionyl chloride and pushes out a chloride. The chloride pulls off the acid hydrogen to form HCl and a mixed anhydride of the acid with thionyl chloride.

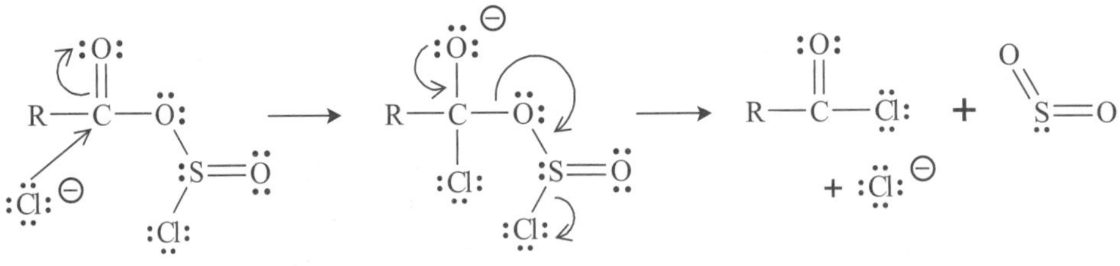

A chloride attacks the C=O and pushes an electron pair up onto the carbonyl oxygen. The C=O is reformed, with production of SO_2 and chloride as shown.

The mechanisms with PCl_5 and PCl_3 are similar.

F.2. PREPARATION OF ANHYDRIDES

Although carboxylic acids can be converted into anhydrides in a number of ways, very few of these reactions are actually done in academic labs. One of these is the reaction of a salt of a carboxylic acid with an acid chloride. An example is shown below.

General Reaction:

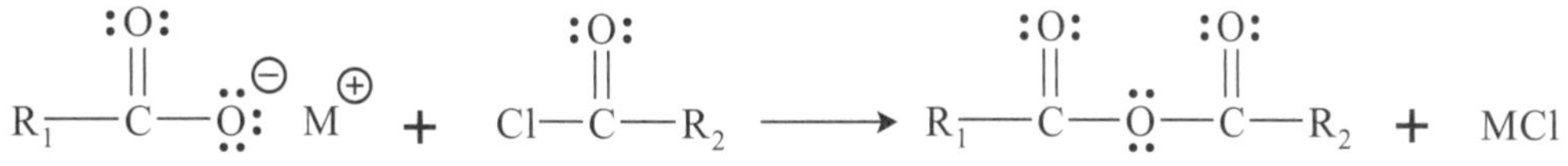

$$R_1-\overset{\overset{\displaystyle :O:}{\|}}{C}-\overset{..}{\underset{..}{O}}:^{\ominus} M^{\oplus} + Cl-\overset{\overset{\displaystyle :O:}{\|}}{C}-R_2 \longrightarrow R_1-\overset{\overset{\displaystyle :O:}{\|}}{C}-\overset{..}{\underset{..}{O}}-\overset{\overset{\displaystyle :O:}{\|}}{C}-R_2 + MCl$$

(M is a metal)

Specific Example:

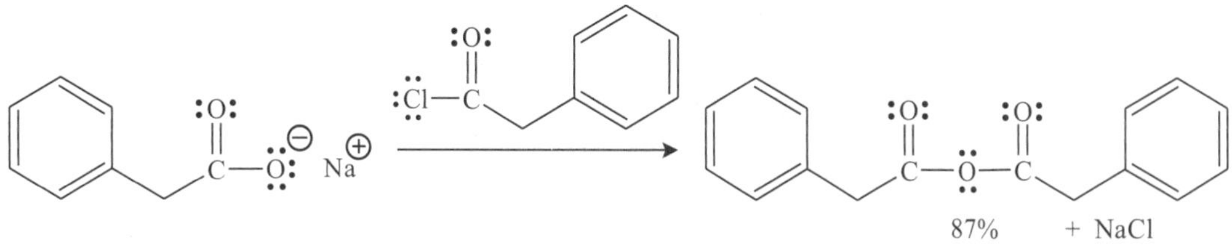

87%   + NaCl

The mechanism of the reaction is the standard nucleophilic acyl substitution mechanism.

F.3. PREPARATION OF ESTERS

General Reaction:

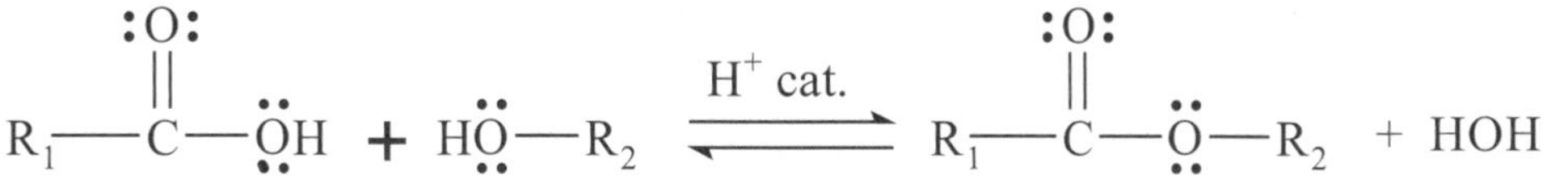

$$R_1-\overset{\overset{\displaystyle :O:}{\|}}{C}-\overset{..}{O}H + H\overset{..}{\underset{..}{O}}-R_2 \underset{}{\overset{H^+ \text{ cat.}}{\rightleftharpoons}} R_1-\overset{\overset{\displaystyle :O:}{\|}}{C}-\overset{..}{\underset{..}{O}}-R_2 + HOH$$

Specific Example:

excess   60-70%

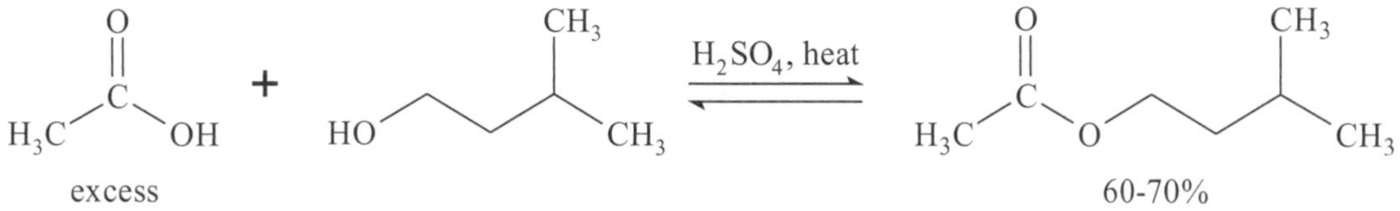

In this reaction, often referred to as the Fischer esterification (in honor of Nobel Prize-winning sugar chemist Emil Fischer), a carboxylic acid and an alcohol are heated in the presence of an acid catalyst to form an ester and water. This reaction is an equilibrium reaction, with an equilibrium constant of about 4. Therefore, to drive the reaction toward products, an excess of one of the reactants is used. Alternatively, water can be removed in some cases by distillation, which will also shift the reaction toward the ester. The mechanism of the reaction is the acid catalyst version of the standard mechanism and is shown below.

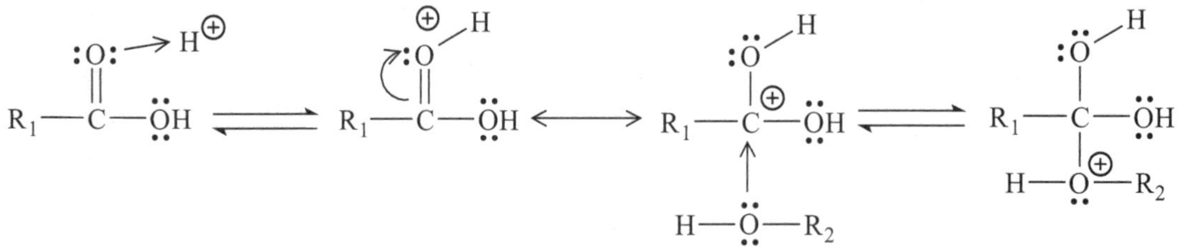

The acid catalyst protonates the carbonyl oxygen, which forms a resonance-stabilized cation. The carbonyl carbon has a significant amount of positive charge. The alcohol oxygen attacks the carbonyl carbon. The net result of the last two electron movements is the "up" step of the mechanism.

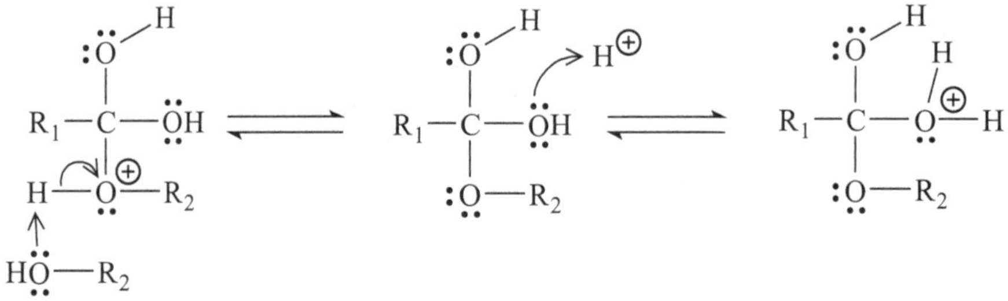

A proton is lost from the original alcohol to form the tetrahedral intermediate. One of the OH's is then protonated, to form a good leaving group.

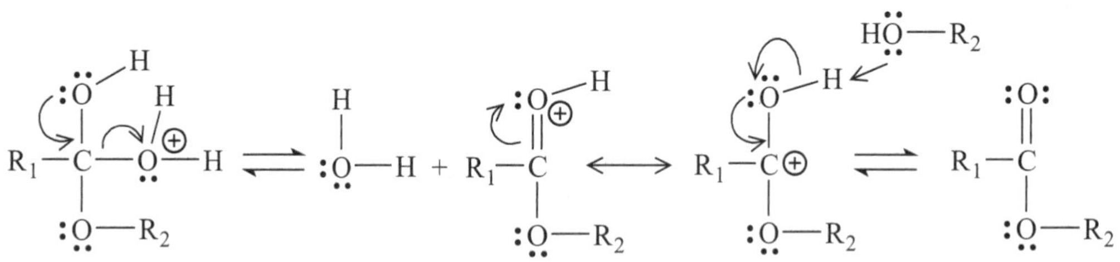

Water is lost with the "down and out" step, followed by a loss of a proton to give us the ester. We have reformed a proton, so this reaction is catalytic in acid.

This mechanism starts off very similarly to that for the formation of an acetal from a ketone. Instead of adding a second alcohol molecule, the C=O is reformed to make the ester. Also, this reaction is reversible. If we add water and acid to an ester, we can make a carboxylic acid and an alcohol. This will be discussed later.

F.4. PREPARATION OF AMIDES

Amides are not generally prepared directly from carboxylic acids: Usually, amides are prepared from acid chlorides, anhydrides, or esters. However, if a carboxylic acid is heated with an amine at a high temperature, an amide can be formed. Some examples follow.

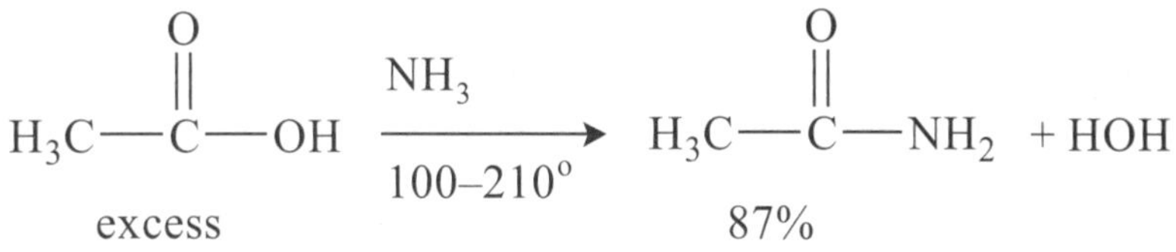

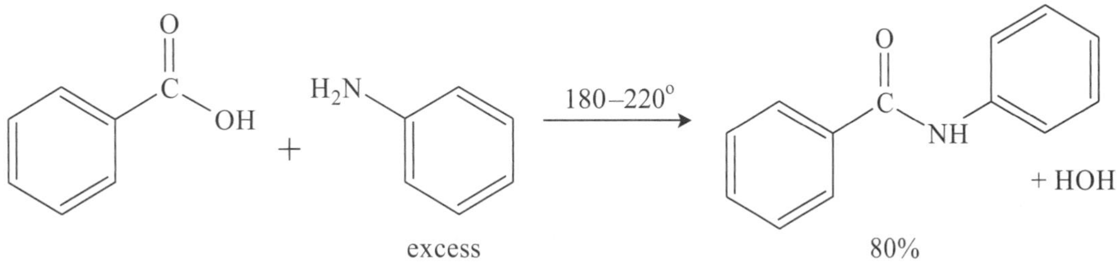

The mechanism of this reaction is not well understood. Presumably, a salt is formed first, but how water is subsequently lost to form the amide is unclear.

G. Naming of Carboxylic Acid Derivatives

G.1. ACID CHLORIDES

Acid chlorides are primarily used as intermediates in the preparation of other carboxylic acid derivatives. They are prepared from carboxylic acids, as was described earlier. The IUPAC ending for an acid chloride is *-oyl chloride*. Some examples are shown below.

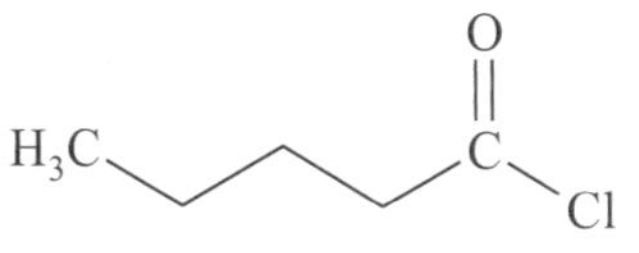	This is a five-carbon acid chloride, so the name is **pentanoyl chloride**.
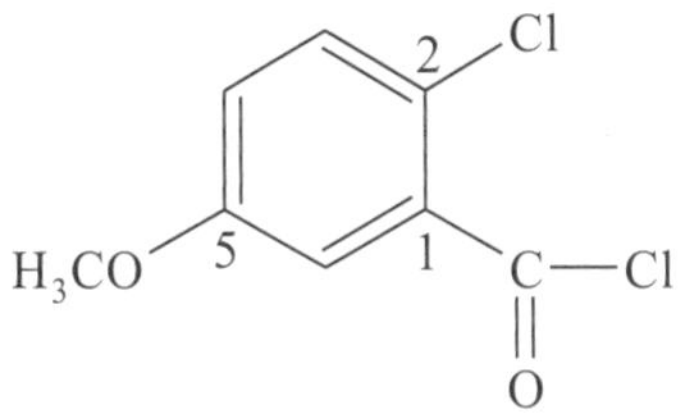	This is a derivative of benzoic acid, so it is a benzoyl chloride. It is numbered as shown to give the lower set of numbers, so it is **2-chloro-5-methoxybenzoyl chloride**.

Of all of the carboxylic acid derivatives, these are the most reactive. They react exothermically with nucleophiles of all types. The reactions of acid chlorides and all the other carboxylic acid derivatives will be discussed in Section I.

G.2. ANHYDRIDES: NAMING

In much the same way as acid chlorides, anhydrides are primarily used as intermediates to prepare other carboxylic acid derivatives. A few anhydrides are found in nature. Symmetrical anhydrides are named by dropping the *acid* of the carboxylic acid name, and replacing it with *anhydride*. Here are some examples of anhydride naming.

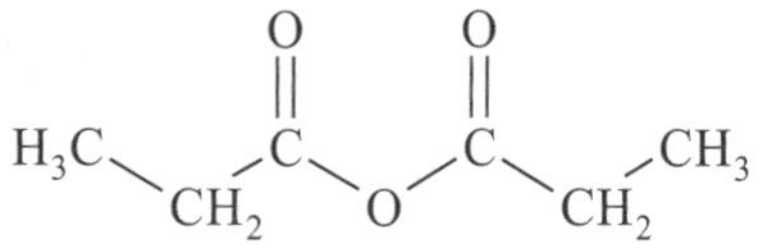	This is the anhydride prepared (in theory) from two molecules of the propanoic acid, so it is called **propanoic anhydride**.
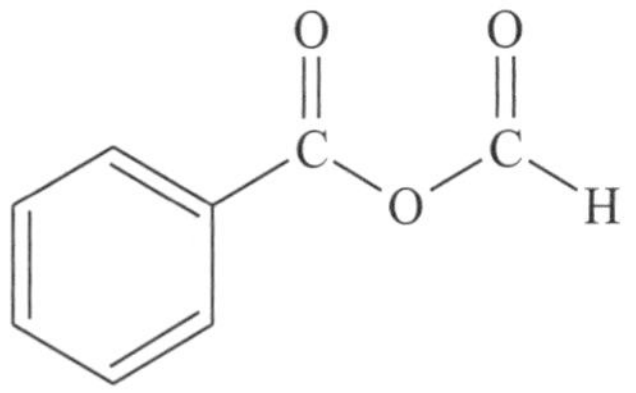	This is a mixed anhydride, prepared (in theory) from two different carboxylic acids. The two acids were benzoic and methanoic acids, so the anhydride name is **benzoic methanoic anhydride**.
	This is the anhydride derived from butanedioic acid, so it is **butanedioic anhydride**, which is also known as succinic anhydride.

Anhydrides are less reactive than acid chlorides, but they are more reactive than other acid derivatives.

G.3. ESTERS: NAMING

Esters commonly found in nature are odor and flavor components. As discussed earlier, esters are commonly prepared from carboxylic acids and alcohols. Esters names have two words: the first word comes from the alcohol portion, and the second word is derived from the acid portion. The first word is the carbon group named as a substituent. The second word has the ester ending, *-oate*. Some examples are shown below.

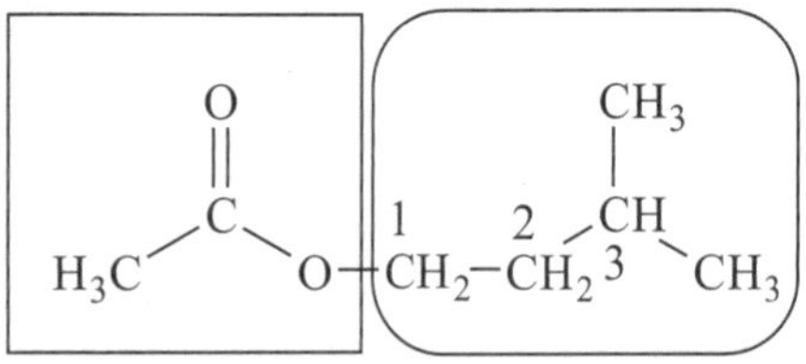 Acid portion Alcohol portion	The alcohol portion is named as 3-methylbutyl. The acid portion is ethanoic acid, which is changed to ethanoate. Therefore, the IUPAC name is **3-methylbutyl ethanoate**. This is the major odor-producing chemical in bananas.
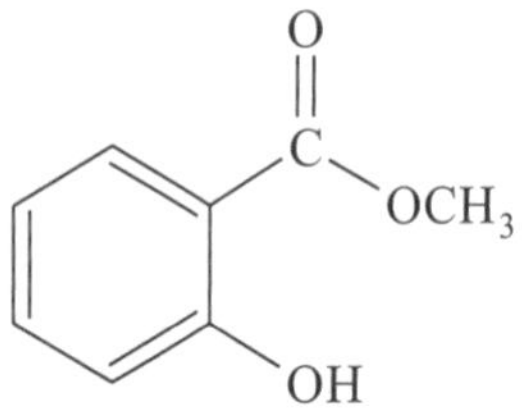	The alcohol portion is simply a methyl group. The acid portion would be 2-hydroxybenzoic acid. The ester name is **methyl 2-hydroxybenzoate**. This is the major odor-producing chemical in oil of wintergreen.

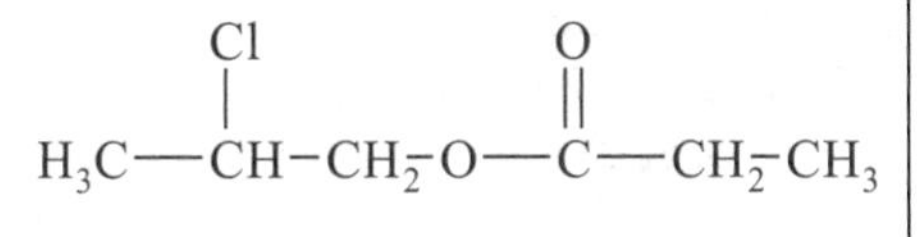	This ester is **2-chloropropyl propanoate**.
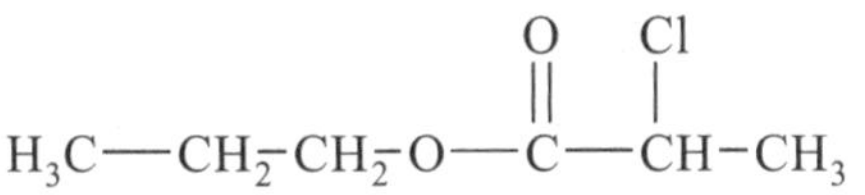	This ester is **propyl 2-chloropropanoate**. It is important to put the substituent (2-chloro in this case) in the proper part of the name.

G.4. AMIDES: NAMING

Amides are the least reactive of the carboxylic acid derivatives. Amides are found in peptides and other biological molecules. The IUPAC ending for an amide name is *-amide*. Some examples are shown below.

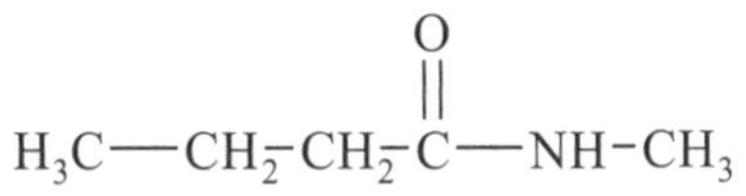	This is derived from a four-carbon carboxylic acid, so it is a butanamide. There is a methyl group on the nitrogen, so the name of this compound is ***N*-methylbutanamide**.
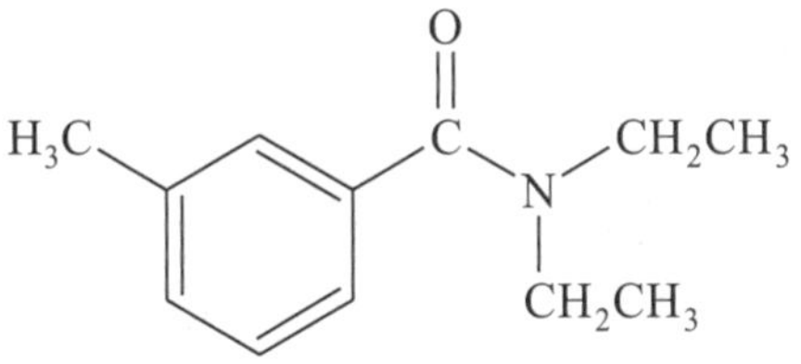	This is a benzoic acid derivative, so it is a benzamide. There is a methyl group on the third carbon of the benzene ring and two ethyl groups on the nitrogen. The IUPAC name is ***N,N*-diethyl-3-methylbenzamide**.
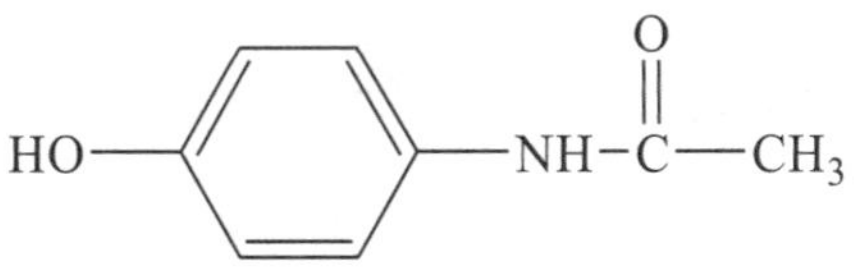	The acid portion is ethanamide. There is an OH group on the 4 position of the benzene ring, so this is ***N*-(4-hydroxylphenyl)ethanamide**.

G.5. NITRILES: NAMING

Nitriles are mostly found in industrial products; they are rarely found in nature. Some of them are useful solvents. Nitriles are named by putting the ending *-nitrile* after the hydrocarbon name. As a substituent, a nitrile is called a cyano group. The nitrile is just below all the carboxylic acid derivatives in priority. See the following examples.

CH₃ on a nine-carbon chain with a nitrile, H₃C chain ending in C≡N	This is a nine-carbon chain with a nitrile on the end. There is a methyl group on the sixth carbon, so this is **6-methylnonanenitrile**.
N≡C—C—C—O—CH₂—CH₃ with CH₃ groups and O	This is an ester, so it has the highest priority. The alcohol part is ethyl. The acid part is a substituted three-carbon chain. The name therefore is **ethyl 2-cyano-2-methylpropanoate**.

The following table summarizes the naming priorities of all the functional groups we have covered. There are additional functional groups that are higher in priority than carboxylic acids.

Priority Group	Functional Group	Ending As Highest Priority	Substituent Name
1	Carboxylic acid	-oic acid	Carboxy
2	Anhydride	anhydride	—
3	Ester	-oate	Alkoxycarbonyl
4	Acid chloride	-oyl chloride	Chlorocarbonyl
5	Amide	-amide	Amido
6	Nitrile	-nitrile	Cyano
7	Aldehyde	-al	Oxo (formyl only for the entire aldehyde group)
8	Ketone	-one	Oxo
9	Alcohol	-ol	Hydroxy
10	Thiol	-thiol	Mercapto
11	Amine	-amine	Amino
12	Alkene	-ene	-en-*
13	Alkyne	-yne	-yn-*
14	Benzene	benzene	Phenyl
15	Alkane	-ane	Alkyl
16	Ether		Alkoxy**
	Halogens		Halo**
	Nitro		Nitro**

* As substituents *within* the longest chain of carbons, the –*an*- of an alkane name is replaced with –*en*- or –*yn*-, as appropriate.
** Ethers, halogens, and nitro groups are *always* substituents.

H. Interconversions of Carboxylic Acids and Derivatives

As discussed earlier, the general reaction of carboxylic acid derivatives is nucleophilic acyl substitution, as illustrated below.

$$R-\overset{\overset{\ddot{O}:}{\|}}{C}-\ddot{Z}: \;\; + \;\; :Nu \;\; \longrightarrow \;\; R-\overset{\overset{\ddot{O}:}{\|}}{C}-Nu \;\; + \;\; :\ddot{Z}:^{\ominus}$$

For each of the carboxylic acid derivatives, Z is a different atom or group of atoms. See below.

Derivative	Acid Chloride	Anhydride	Ester	Amide
Z	Cl^-	$RCOO^-$	RO^-	R_2N^-
Base strength	Weakest ⟵		⟶	Strongest
Leaving group ability	Best ⟵		⟶	Worst

Since chloride is the weakest base and the best leaving group, acid chlorides are the most reactive toward nucleophiles. Anhydrides are less reactive, followed by esters, with amides being the least reactive. As a result, more and more vigorous reaction conditions are required to cause the less reactive carboxylic acid derivatives to react. We will start with the most reactive, acid chlorides, and work down to the least reactive, amides. Even though acid chlorides can be used to prepare all the other derivatives, amides cannot be used to prepare any of the others directly. However, all the acid derivatives can be used to prepare carboxylic acids. The reactions of the carboxylic acid derivatives, and the conditions needed, are shown next.

H.1. REACTIONS OF ACID CHLORIDES

Acid chlorides generally react with nucleophiles at room temperature or lower. The by–product of these reactions is usually HCl, so bases are often used to neutralize the HCl formed. See the following examples.

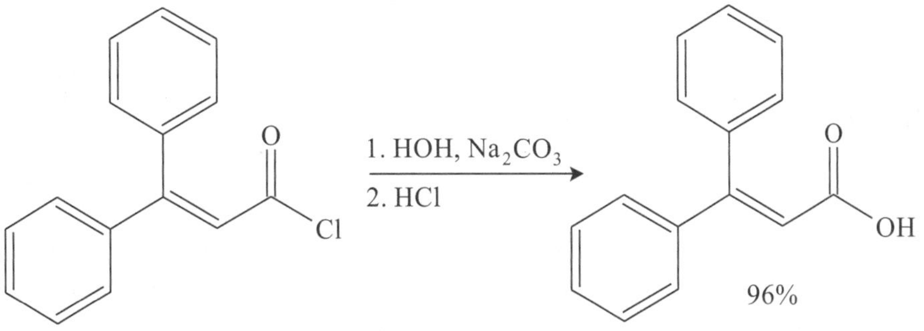

96%

This reaction is not often done because acid chlorides are usually prepared from the carboxylic acids.

This exothermic reaction was cooled as the acid chloride was added to the ethanol.

In this case, the acid chloride is insoluble in water, so it reacts with the amine preferentially. The NaOH neutralizes the HCl formed.

H.2. REACTIONS OF ANHYDRIDES

Anhydrides are less reactive than acid chlorides, so an acid catalyst is often used with weak nucleophiles to speed up the reactions.

This reaction is not usually done because anhydrides are normally prepared from carboxylic acids.

I often do this reaction as a demonstration in lecture, with a digital thermometer in a test tube. I add the alcohol and the anhydride to the test tube, and the temperature doesn't increase, which shows that no reaction is apparently occurring. I then add a drop of sulfuric acid, and the temperature rapidly rises to 70–80°C! After a few minutes, I pour the reaction mixture into water: The ester layer separates, with its characteristic banana odor, showing that a reaction occurred.

I also do this demonstration during the same lecture. In this case, the temperature rises to about 80°C when the aniline is added, without using a catalyst. This demonstrates (hopefully) that amines are stronger nucleophiles than alcohols. This is also a common lab experiment in most organic labs because the product, acetanilide, can be used to teach recrystallization techniques.

H.3. REACTIONS OF ESTERS

Esters are less reactive than anhydrides, so more vigorous conditions are needed to convert them into acids and amides. Esters are often hydrolyzed to carboxylic acids with a base, which forms the salts of the acids, and are then acidified to form the carboxylic acid. See below.

Refluxing the ester in dilute sodium hydroxide forms the sodium salt of the acid. Acidification of the solution of the salt produces the acid.

The starting triester is called trimyristin and can be isolated from nutmeg. Because it is insoluble in water, it is hydrolyzed in a mixture of ethanol and water, in which it is soluble.

This reaction is called transesterification. In this case, excess ester is heated with an alcohol in the presence of an acid catalyst. The lower boiling alcohol and excess ester are removed by distillation. In other cases, excess alcohol is used.

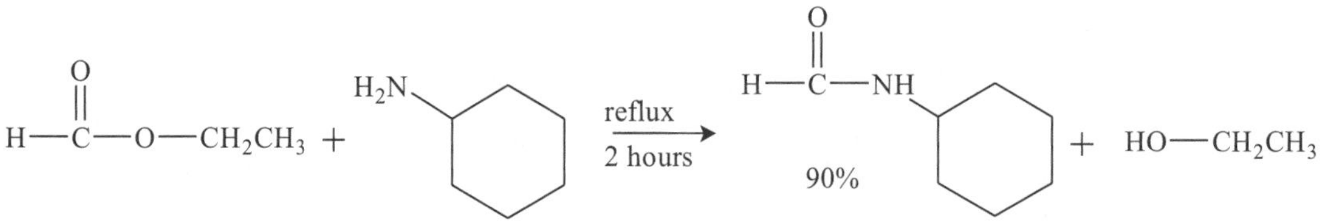

This is one of the easier and better ways to prepare amides of methanoic acid (commonly called formic acid) because methanoyl chloride and methanoic anhydride are unknown compounds. Because esters are less reactive than anhydrides, the amine must be heated with the ester for a reaction to occur.

H.4. REACTIONS OF AMIDES AND NITRILES

Amides are generally heated for several hours in strong acid or base solution to hydrolyze them to carboxylic acids. Nitriles react similarly. Some examples are shown below.

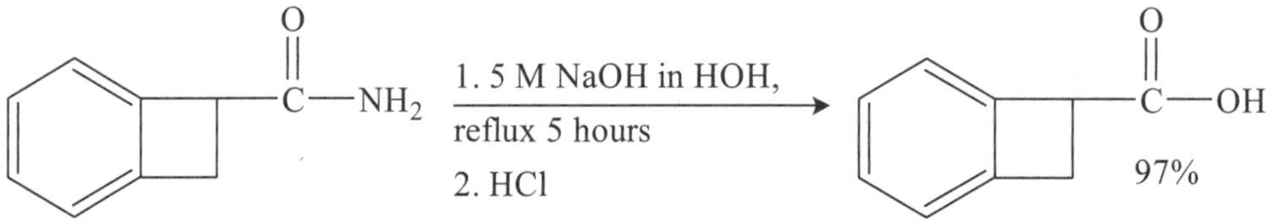

Hydrolysis under basic conditions initially produces the salt of the carboxylic acid. Acidification gives the carboxylic acid.

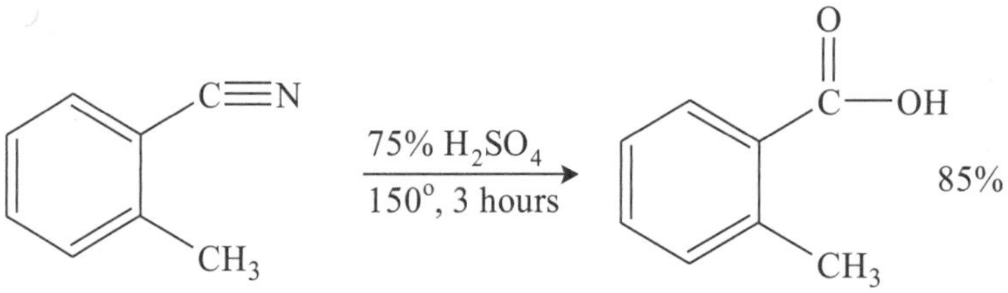

Strongly acidic conditions cause complete hydrolysis of the nitrile to the carboxylic acid.

One other reaction of nitriles is quite useful. If a nitrile is treated with an alkene or a tertiary alcohol in the presence of a strong acid at room temperature, a substituted amide is the product. This reaction is called the Ritter reaction: Some examples are shown here.

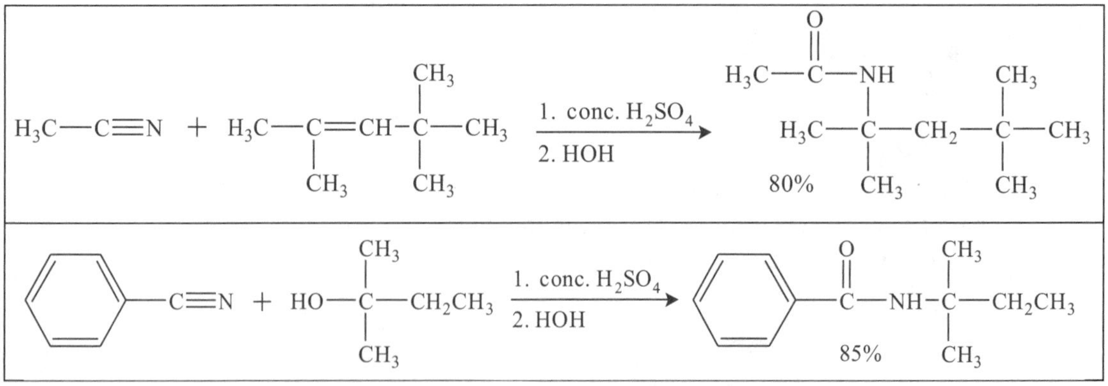

It has been established that the mechanism does not involve hydrolysis of the nitrile to the amide. Instead, a cation is formed, which is then attacked by the nitrile nitrogen. Water then adds to the resulting intermediate. The mechanism follows. Ph stands for phenyl.

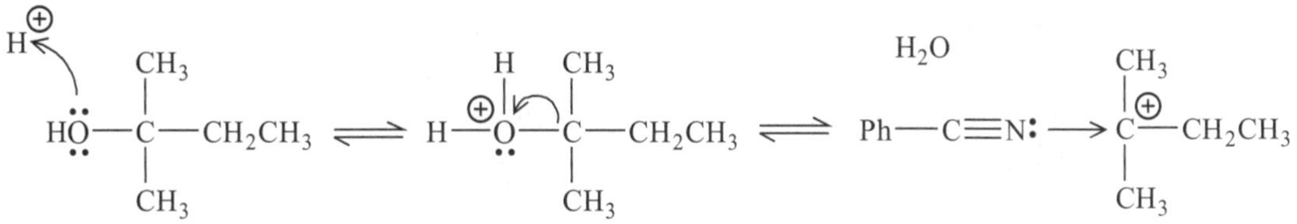

The alcohol is protonated, and water is lost, to form a cation. The cation is attacked by the nitrogen of the nitrile.

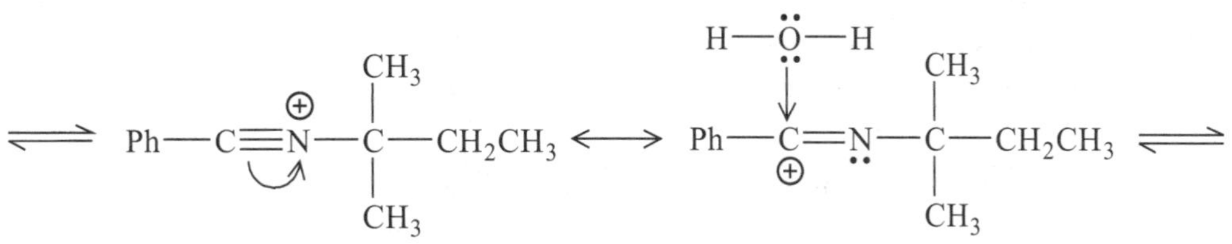

The positive charge on the nitrogen is shared by the carbon. This carbon is attacked by a water.

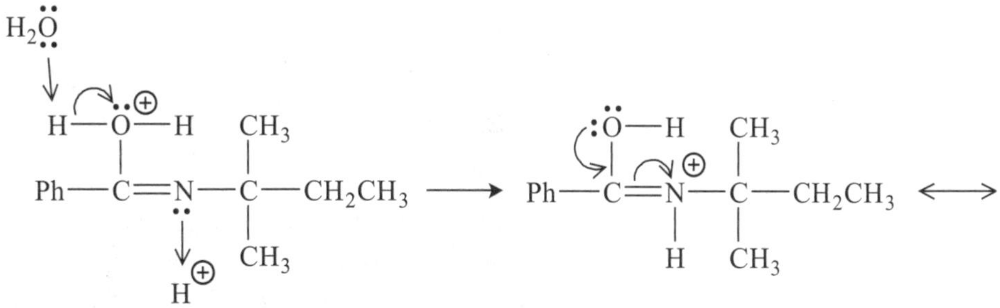

A hydrogen is lost from the water molecule to form an OH, and the nitrogen picks up another hydrogen from the acid. The positive charge on nitrogen is again shared by the carbon and, more importantly, the oxygen, as shown below.

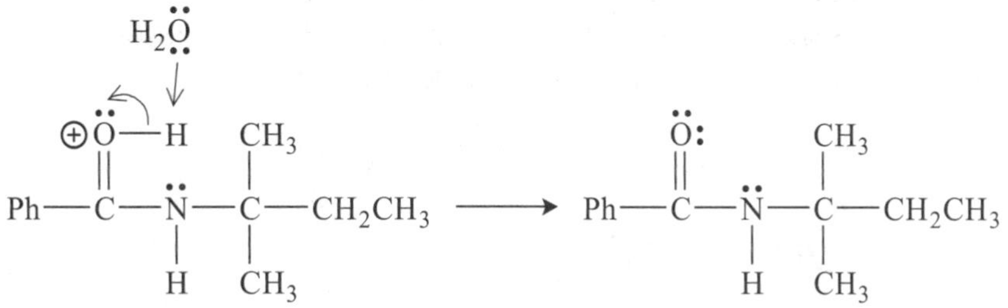

This resonance form is the protonated amide product. Loss of H+ gives us the amide.

I. Reductions of Carboxylic Acids and Derivatives

All the carboxylic acid derivatives can be reduced by lithium aluminum hydride (LiAlH$_4$). Carboxylic acids, acid chlorides, anhydrides, and esters produce primary alcohols as products, while amides and nitriles form amines. Carboxylic acids and amides can also be reduced by borane (BH$_3$). Nitriles may also be reduced by hydrogen and a catalyst. Some examples are shown here.

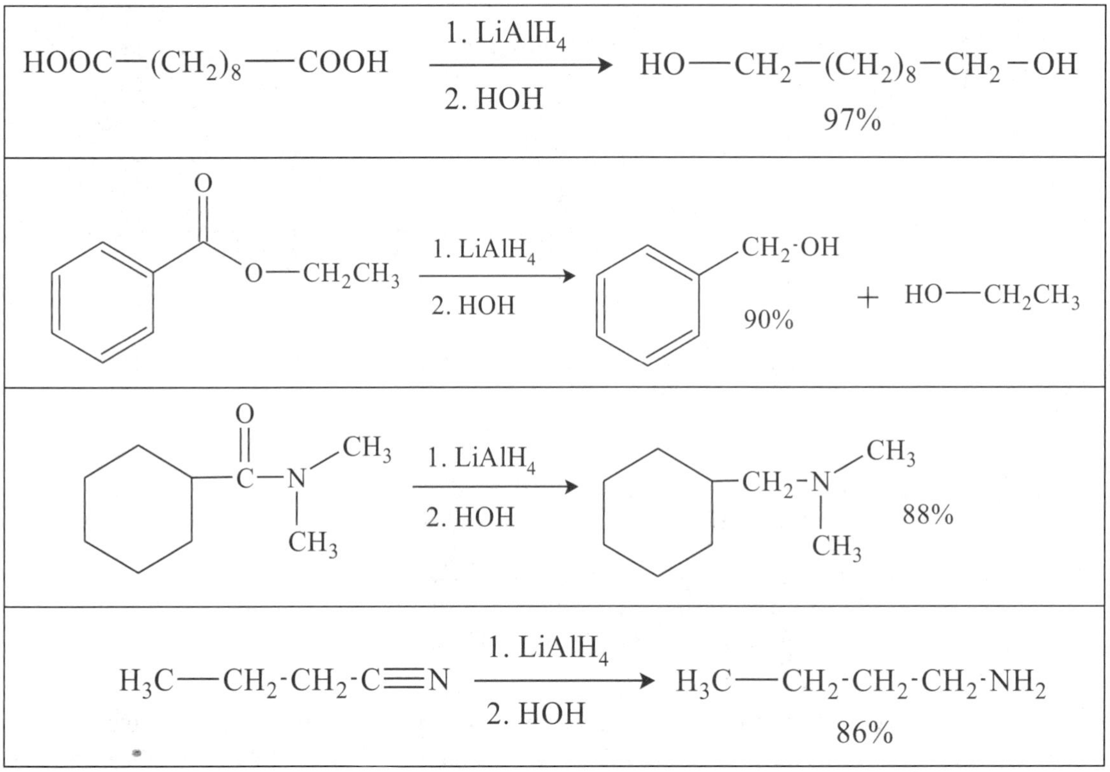

$$\text{HOOC}-(\text{CH}_2)_8-\text{COOH} \xrightarrow[\text{2. HOH}]{\text{1. LiAlH}_4} \text{HO}-\text{CH}_2-(\text{CH}_2)_8-\text{CH}_2-\text{OH}$$

97%

1. LiAlH$_4$
2. HOH

CH$_2$·OH

90% + HO—CH$_2$CH$_3$

1. LiAlH$_4$
2. HOH

88%

$$\text{H}_3\text{C}-\text{CH}_2\cdot\text{CH}_2\cdot\text{C}\equiv\text{N} \xrightarrow[\text{2. HOH}]{\text{1. LiAlH}_4} \text{H}_3\text{C}-\text{CH}_2\cdot\text{CH}_2\cdot\text{CH}_2\cdot\text{NH}_2$$

86%

The mechanisms of reductions of acid chlorides, anhydrides, and esters are similar, and one example is shown below. It is the basic nucleophilic acyl substitution mechanism, followed by nucleophilic addition to the resulting aldehyde.

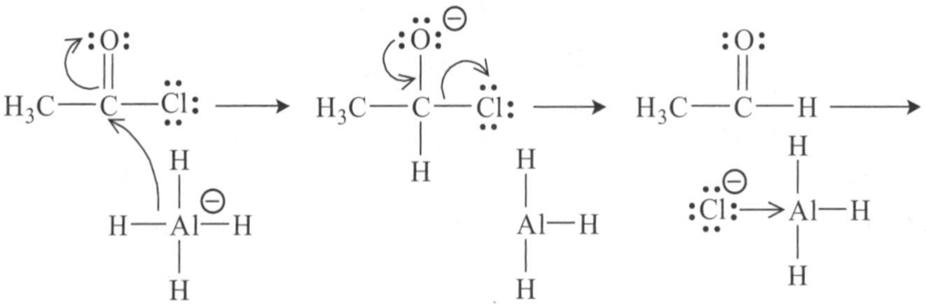

A hydride attacks the carbonyl carbon and pushes an electron pair up onto the oxygen. The electron pair comes down to reform the C=O and pushes the chloride ion out to form an aldehyde. The chloride complexes with the AlCl$_3$ to form a new tetracoordinated aluminum hydride, AlClH$_3$.

The AlClH$_3$ attacks the aldehyde to form an alkoxide. When water is added, the alkoxide pulls a hydrogen off of water to form the alcohol.

REVIEW EXERCISES FOR CHAPTER 13

1. Give IUPAC names for each of the following structures.

   a.

   b.

   c.

   d.

2. Show how 3-ethylpentanoic acid *can be converted into* each of the following compounds. More than one step may be required.
   a. 3-Ethylpentanoyl chloride
   b. Ethyl 3-ethylpentanoate
   c. 3-Ethyl-*N*-phenylpentanamide
   d. *N*,3-Diethylpentanamine
   e. 3-Ethyl-1-(4-methoxyphenyl)-1-pentanone
   f. 3,5-Diethyl-3-pentanol

3. Show how 3-ethylpentanoic acid *could be prepared from* each of the following starting materials. More than one step may be required.
   a. 3-Ethylpentanoyl chloride
   b. Ethyl 3-ethylpentanoate
   c. 3-Ethyl-*N*-phenylpentanamide
   d. 3-(Bromomethyl)pentane
   e. 3-Ethylpentanenitrile

4. Give a reasonable arrow-pushing mechanism for the following reaction.

5. Show how you could prepare each of the following molecules, using benzene and benzoic acid as your only sources of benzene rings.

   a.

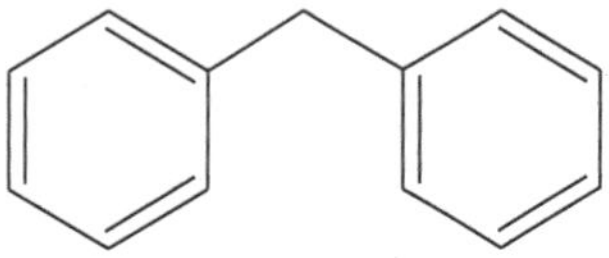

   b.

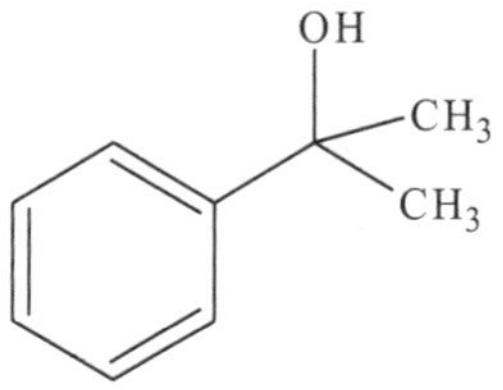

   c.

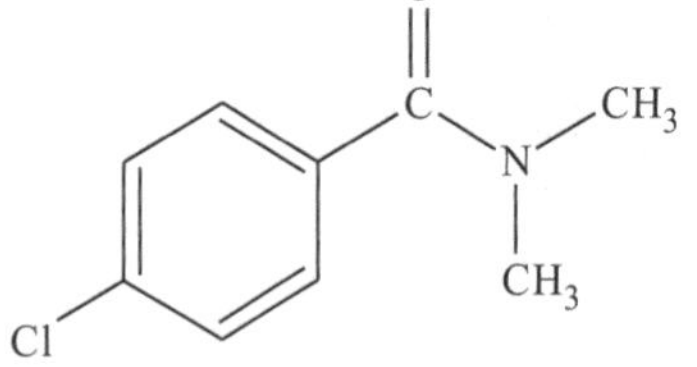

6. Rearrange the following compounds in order of decreasing acidity (most acidic on the left to least acidic on the right). Give approximate pK_a's of all the compounds.

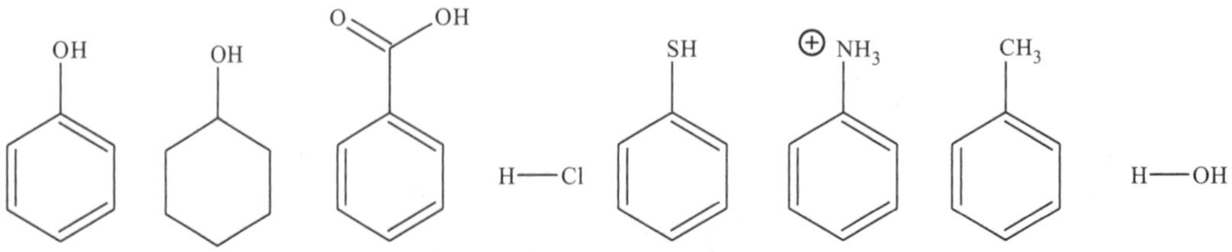

More Chemistry of Enols and Enolates

WHAT YOU WILL LEARN

In this chapter, you will learn:

- how to identify potentially acidic alpha-hydrogens;
- how to prepare substituted acetones and acetic acids via alkylation reactions;
- how to prepare beta-ketoesters and related compounds via condensation reactions;
- how to use conjugate addition reactions;
- how the Robinson Annelation combines an aldol reaction with a conjugate addition to make a six-membered ring.

SECTIONS IN THIS CHAPTER

- Acidities of Various Types of C–H Bonds

- Alkylation of Malonic Esters and Acetoacetic Esters

- Knoevenagel Condensation and Related Reactions

- The Claisen Condensation and Related Reactions

- Conjugate Addition Reactions (The Michael Reactions)

- Robinson Annelation

- Alkylations and Acylations of Ketones

In Chapter 12, one of the reactions of aldehydes and ketones we discussed was the aldol condensation. It is one of several important carbon–carbon bond-forming reactions that involve enolate ions. We also looked at the Wittig reaction, which also involves a resonance-stabilized carbanion attacking a C=O. In this chapter, we will look at some more of these reactions and show how they can be used in synthesis of reasonably complex molecules.

A. Acidities of Various Types of C–H Bonds

Hydrogens on carbons next to C=O's and C≡N's are more acidic than similar hydrogens not adjacent to these groups. This is due to resonance stabilization of the resulting anions. The pK_a's of some representative compounds are shown below.

Conjugate Acid	pK_a	Conjugate Base
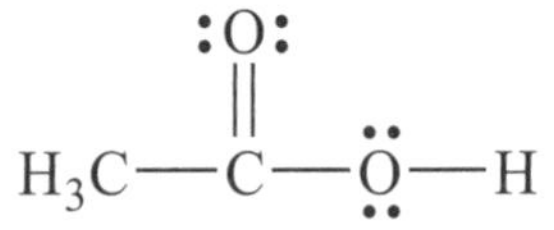	4.76	
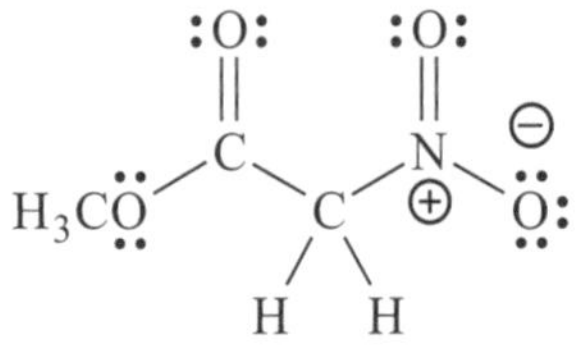 ("Methyl nitroacetate")	5.8	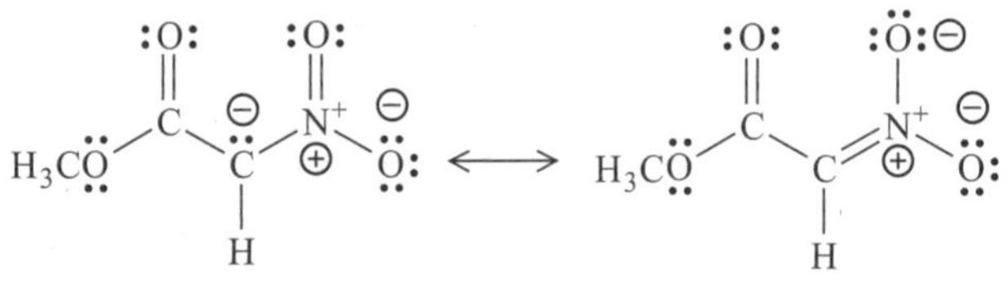 (plus other resonance forms)
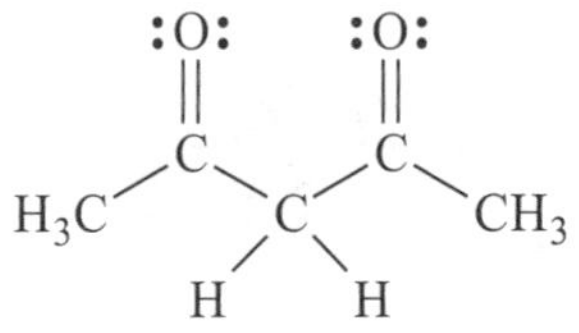	9.0	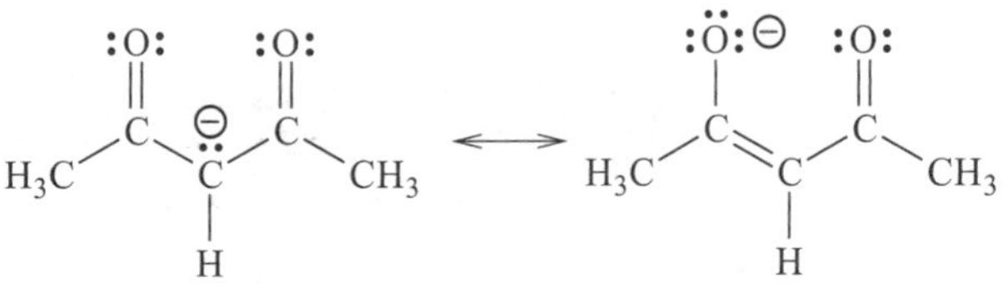 (plus other resonance forms)
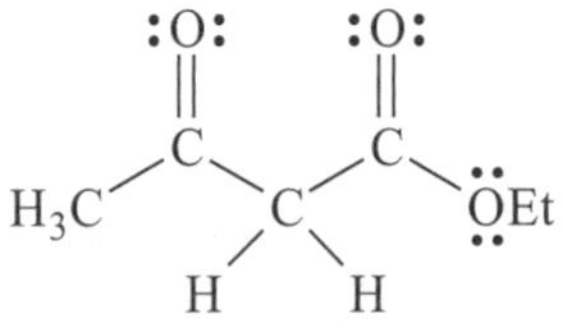 ("Ethyl acetoacetate")	11	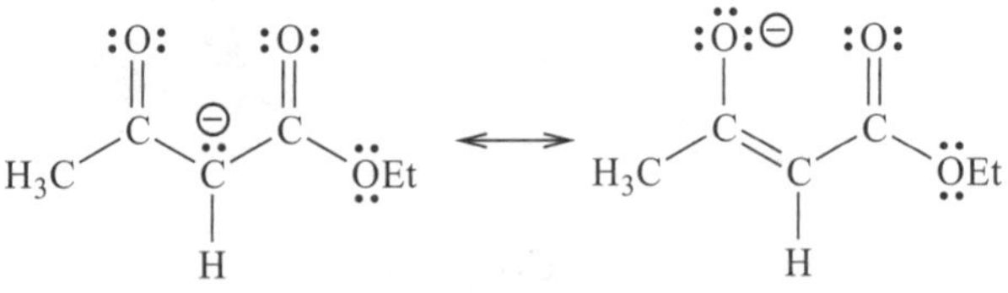 (plus other resonance forms)
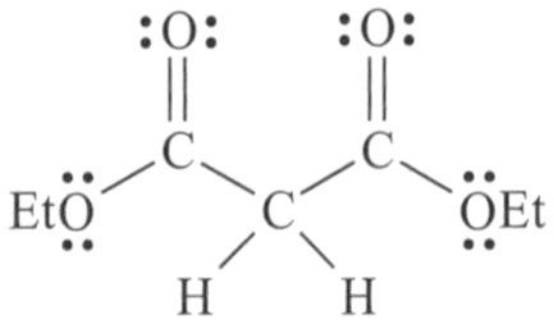 ("Diethyl malonate")	13	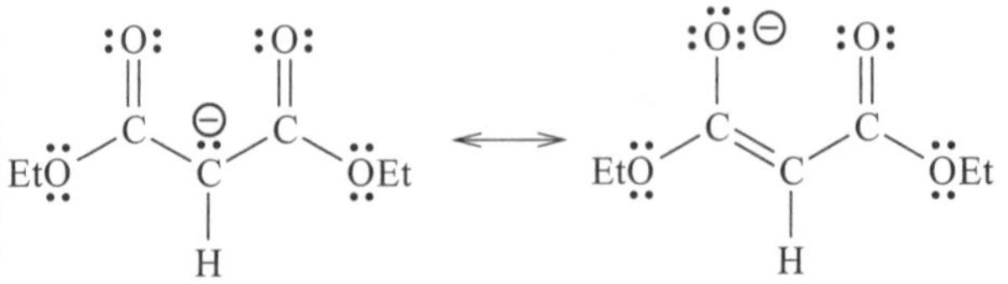 (plus other resonance forms)

Conjugate Acid	pK_a	Conjugate Base
H—Ö—H	15.7	⊖ :Ö—H
	16	⟷ 4 more resonance forms
 ("Acetone)	20	

Many of the conjugate bases in this table are stabilized by resonance (which is part of the reason why the conjugate acids are rather acidic). In many of these resonance structures, the negative charges are shared by a carbon and one or more oxygens. To promote the formation of carbon–carbon bonds, reaction conditions are chosen to maximize reaction at the carbanion center, rather than at the oxygen. Considerations of these reaction conditions are beyond the scope of most beginning organic courses, although we will look at this to a small degree in Section G. Suffice it to say that the choice of base, solvent, temperature, and order of addition of the reagents are all important. As we go through each section in this chapter, I will give some typical reaction conditions and discuss why these conditions are used.

Compounds such as diethyl malonate and ethyl acetoacetate are referred to as "active methylene" compounds because the hydrogens in the CH_2 between the carbonyl groups are acidic and can potentially be replaced by other groups.

B. Alkylation of Malonic Esters and Acetoacetic Esters

Esters of malonic acid (propanedioic acid) and acetoacetic acid (3-oxobutanoic acid) can be deprotonated with strong bases, and the resulting anions can be alkylated with primary alkyl halides. This is usually done in connection with a sequence of reactions, which results in the formation of a substituted acetic acid or a substituted acetone, respectively. Some examples are shown below.

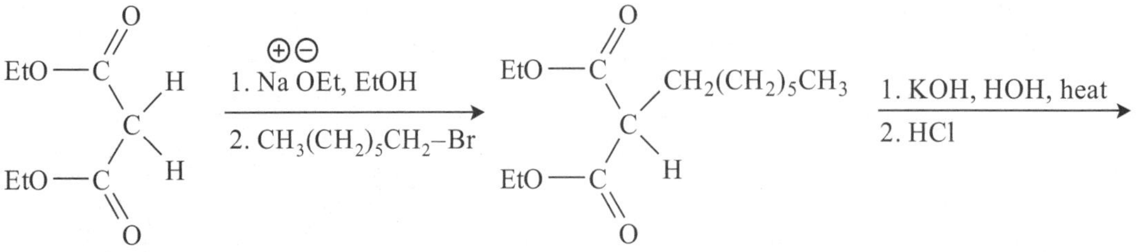

Diethyl malonate, or
"malonic ester"

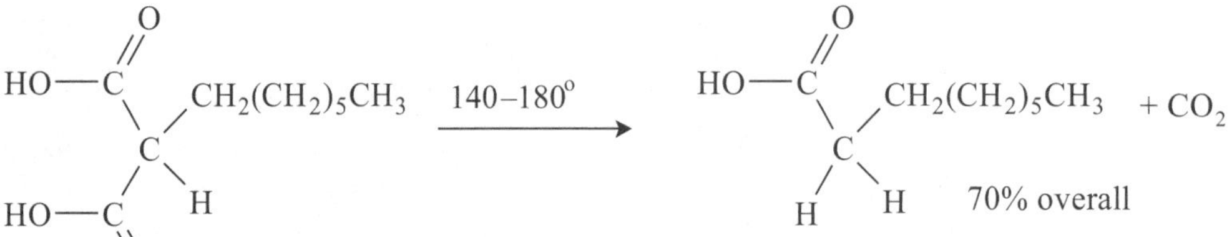

Nonanoic acid, which can be looked
at as a substituted acetic acid

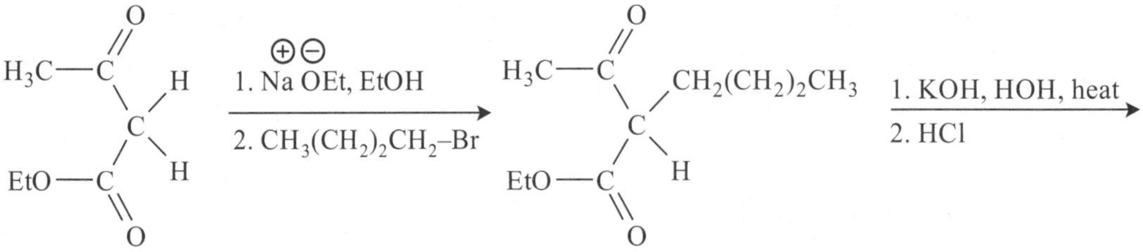

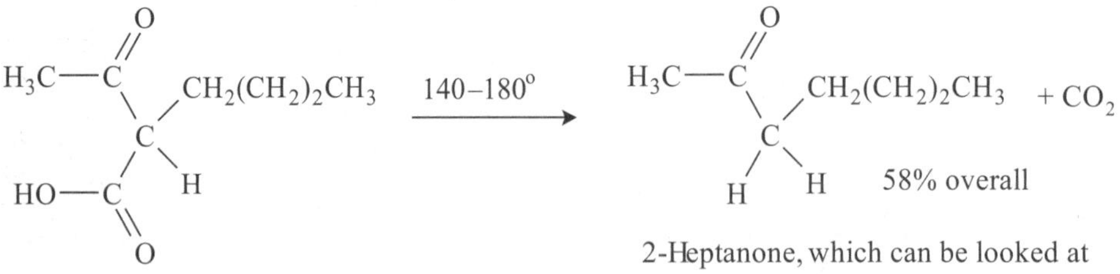

2-Heptanone, which can be looked at
as a substituted acetone

The mechanisms of these reactions are the same. The first is discussed below.

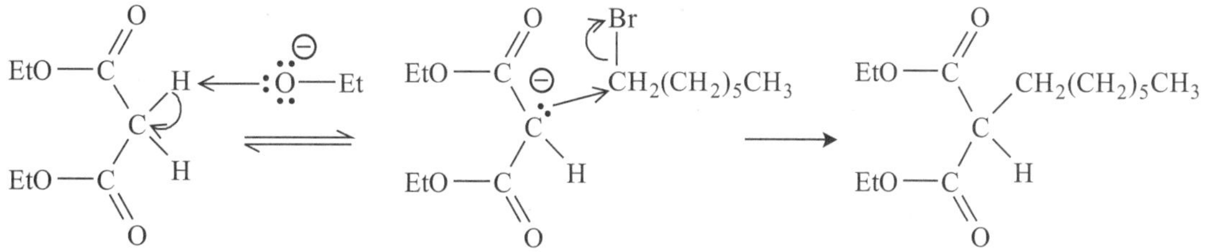

Ethoxide pulls off one of the acidic hydrogens of diethyl malonate to form a resonance-stabilized anion. The anion attacks the alkyl halide in an S_N2 manner to form the alkylated product.

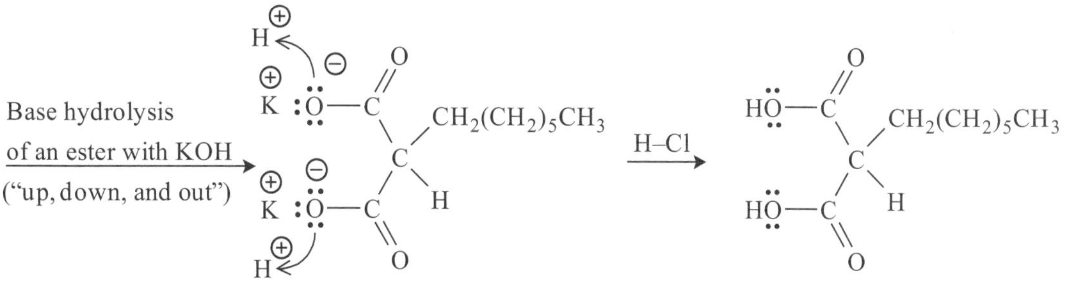

The esters are hydrolyzed in a base using the standard nucleophilic acyl substitution mechanism ("up, down, and out") discussed in Chapter 13. The resulting carboxylate anions are protonated by adding a strong acid, such as HCl.

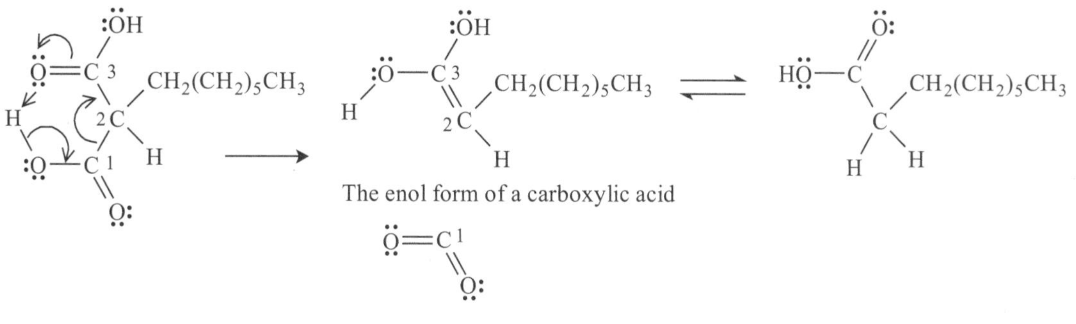

The enol form of a carboxylic acid

If a carboxylic acid is two carbons away from another carbonyl, the carboxylic acid can be removed by heating the compound to about 140°C. The reaction is thought to involve the concerted movement of electrons shown above, which produces CO_2 and the enol form of the carboxylic acid. The enol form readily tautomerizes to the acid form, which is much more stable.

In the preceding examples, sodium ethoxide is used as the base. Ethoxide is a strong enough base to remove completely one of the acidic hydrogens. Another commonly used base is NaH (sodium hydride), which is often used in *N*, *N*-dimethylformamide (DMF) as a solvent. DMF is a polar solvent that cannot donate hydrogen bonds. It can dissolve most organic compounds, and many inorganic ones. Other similar solvents are dimethylsulfoxide (DMSO) and hexamethylphosphoric triamide (HMPA). The structures of these solvents are shown on the next page.

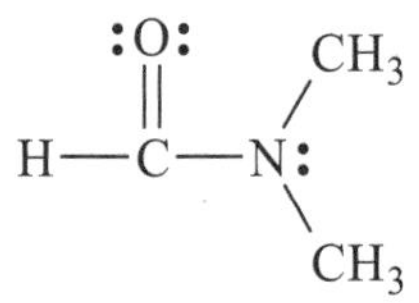 DMF	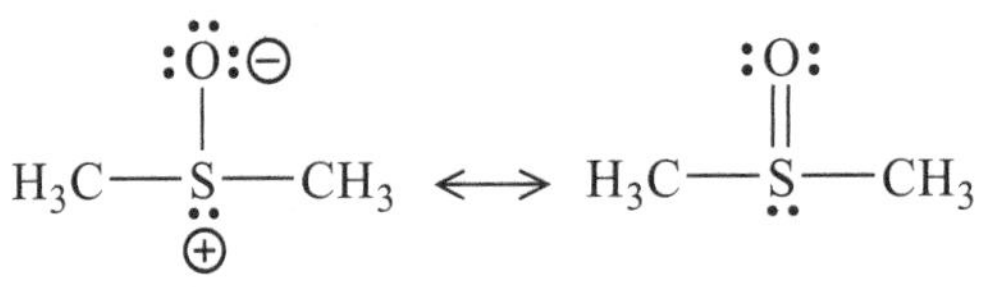 DMSO, shown as two resonance structures	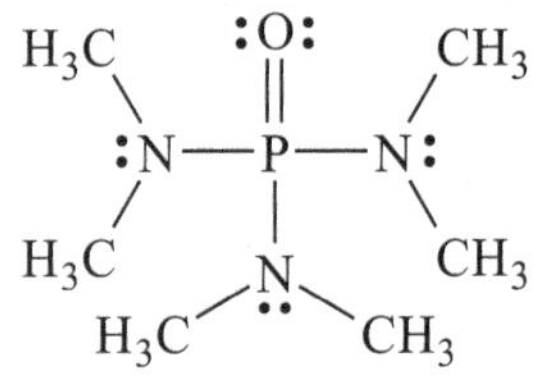 HMPA

Because there are two acidic hydrogens in diethyl malonate and ethyl acetoacetate, they can be sequentially removed and replaced with alkyl groups. An example is shown below.

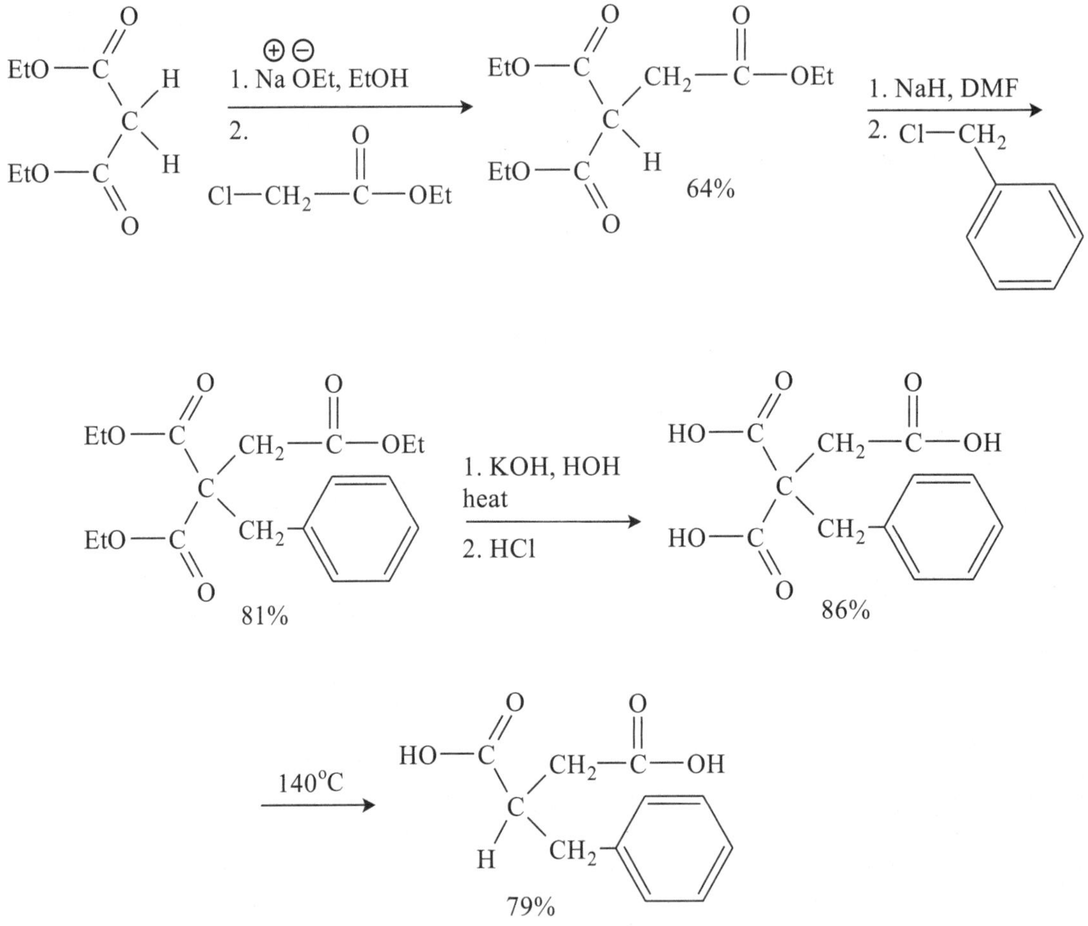

In this example, diethyl malonate was first alkylated with a chloroester, to make the triester. The triester was then alkylated with benzyl chloride. Hydrolysis of the esters formed the triacid, which was decarboxylated to form the product diacid. Notice that only one of the carboxylic acids from diethyl malonate was lost because they were the only ones two carbons away from another carbonyl. This reaction sequence was the first four steps of my Ph.D. thesis project.

You may have noticed that the yield of the first alkylation step is not too great: It is often in the 50–75% range. The reason for this is that dialkylation of diethyl malonate or ethyl acetoacetate is a competing reaction. The following reaction shows what happened in the first step of my Ph.D. thesis work. The mechanism for how it happened follows.

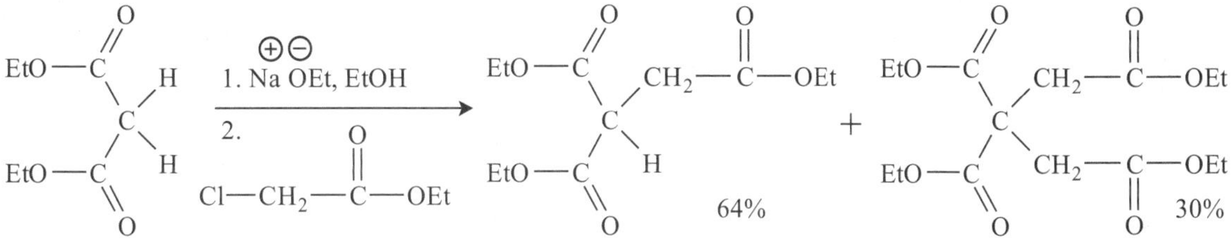

The yields are based on ClCH$_2$COOEt. There was also some unreacted diethyl malonate.

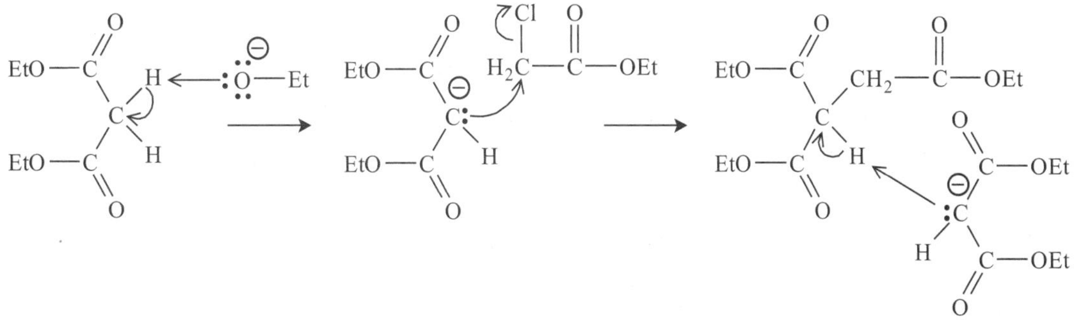

Just as before, ethoxide pulls off one of the hydrogens of diethyl malonate to form the carbanion, which subsequently attacks the alkyl halide to form the triester. However, not all the carbanion reacts instantaneously with the alkyl halide. Some of the carbanion functions as a base and pulls the remaining acidic hydrogen off the triester.

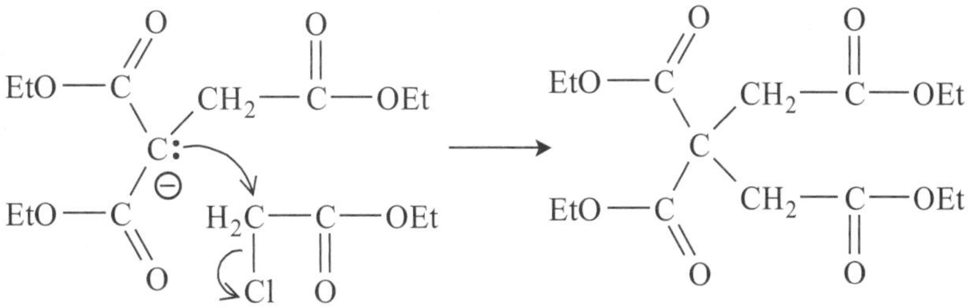

The triester carbanion reacts with more of the alkyl halide to form the dialkylated product, which is a tetraester. Unfortunately for me, the tetraester was useless as far as my project was concerned.

Dialkylation is an unfortunate side reaction in these reactions and is difficult to minimize. Therefore, other reaction sequences are sometimes used to get around this difficulty.

C. The Knoevenagel Condensation

The Knoevenagel condensation is the reaction of an aldehyde or a ketone with an active methylene compound. The product is conjugated, due to loss of water after addition of the active methylene compound to the carbonyl. Weaker bases are often used to avoid side reactions of the aldehydes and ketones, such as the aldol condensation. Some examples are shown below.

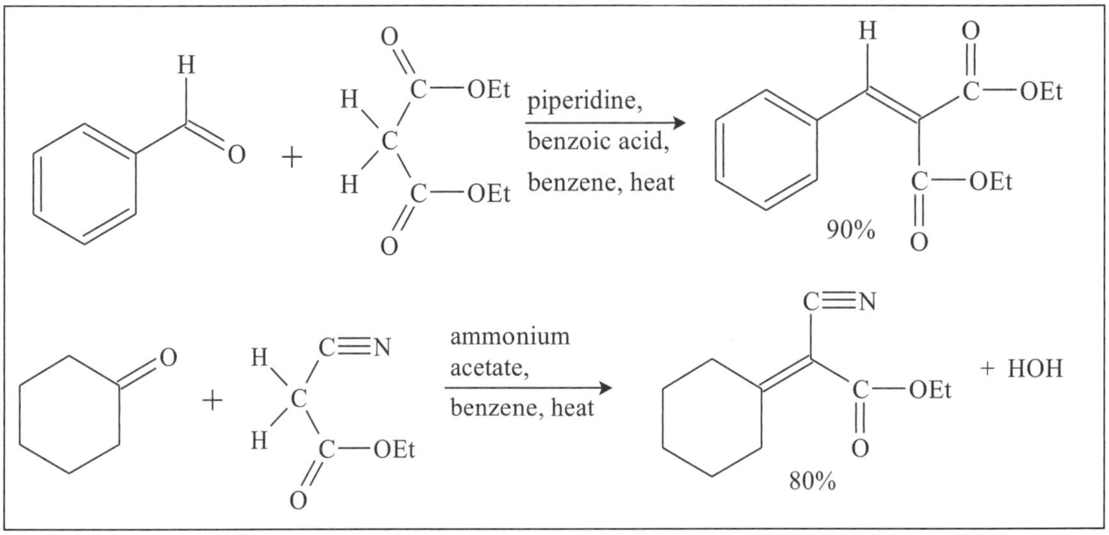

Benzene and toluene are often used as solvents in these reactions because they form **azeotropes** (constant boiling mixtures) with water. The azeotrope can be distilled out of the reaction, and the extent of the reaction can be monitored by measuring the volume of water formed (assuming you started with dry reactants). The mechanism is similar to what we've seen before, and is shown below.

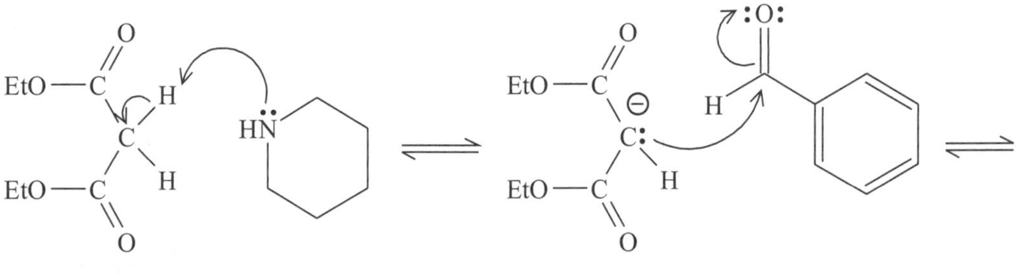

Piperidine pulls off one acidic hydrogen, and the resulting carbanion attacks the carbonyl carbon of benzaldehyde.

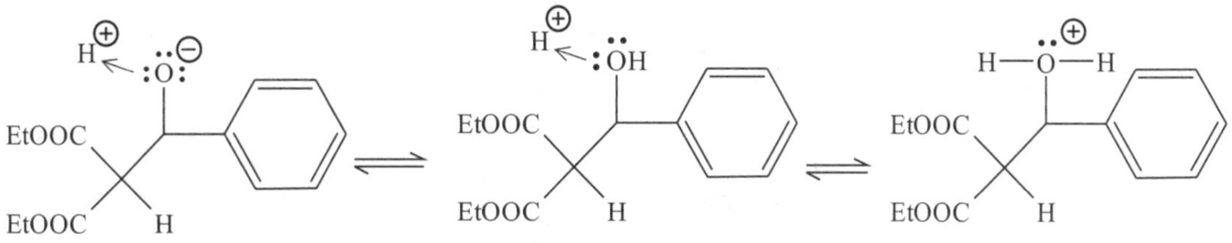

The resulting anion is protonated by acid to form the alcohol, and then it is protonated again.

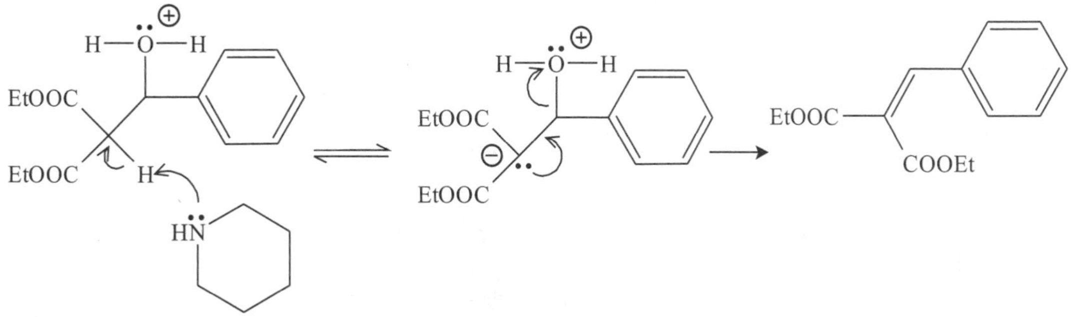

Piperidine pulls off the other acidic hydrogen, followed by formation of the C=C with loss of water, to form the product.

The combination of a Knoevenagel condensation, followed by catalytic reduction of the C=C, can be used effectively to monoalkylate diethyl malonate and to avoid the problem of dialkylation, as shown below.

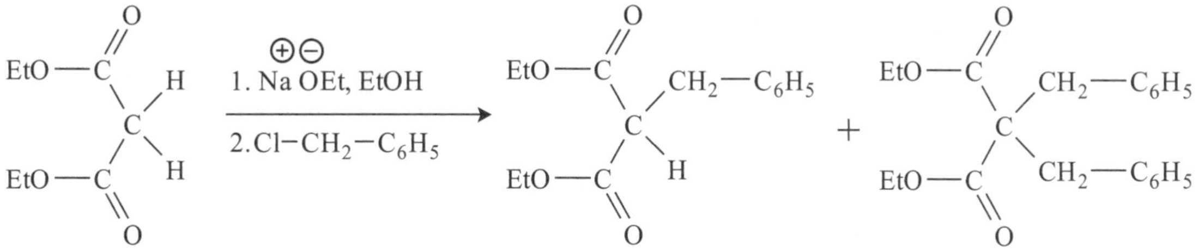

Alkylation of diethyl malonate gives a mixture of mono- and dialkylated products.

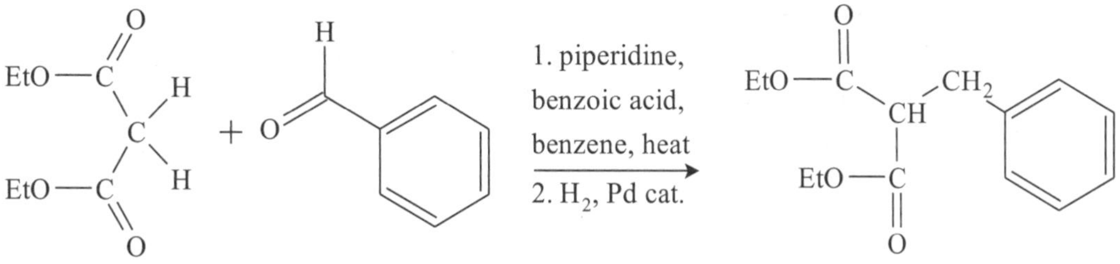

This sequence gives only the monoalkylated product.

D. The Claisen Condensation and Related Reactions

The Claisen condensation is the reaction of two esters with a strong base, to form a beta-ketoester. Some examples are shown below.

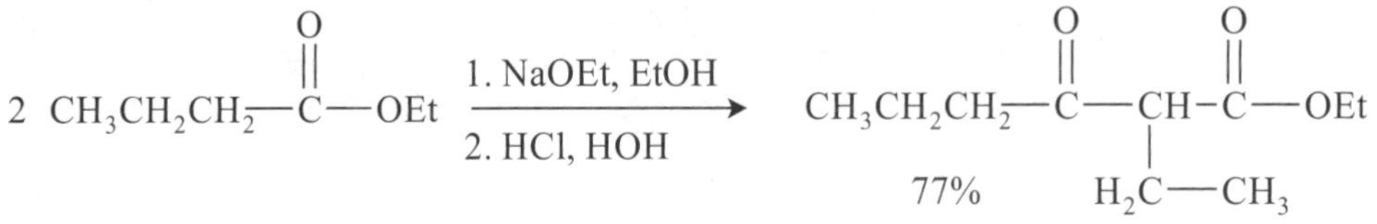

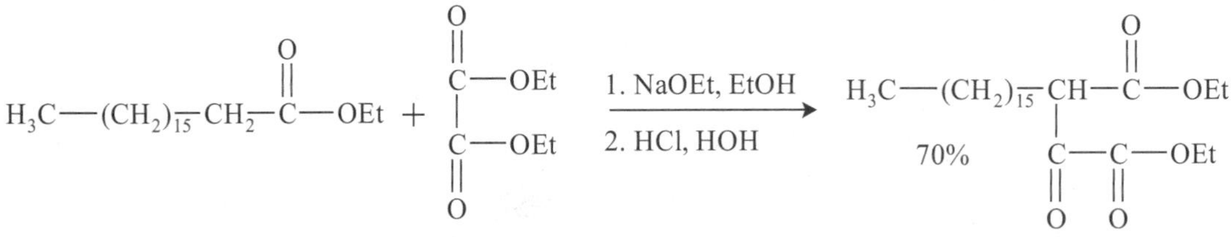

$$2 \ CH_3CH_2CH_2\!-\!\overset{\overset{\displaystyle O}{\|}}{C}\!-\!OEt \quad \xrightarrow[\text{2. HCl, HOH}]{\text{1. NaOEt, EtOH}} \quad CH_3CH_2CH_2\!-\!\overset{\overset{\displaystyle O}{\|}}{C}\!-\!\underset{\underset{\displaystyle H_2C\!-\!CH_3}{|}}{CH}\!-\!\overset{\overset{\displaystyle O}{\|}}{C}\!-\!OEt$$

$$77\%$$

$$H_3C\!-\!(CH_2)\overline{_{15}}\,CH_2\!-\!\overset{\overset{\displaystyle O}{\|}}{C}\!-\!OEt \ + \ \begin{matrix} \overset{\overset{\displaystyle O}{\|}}{C}\!-\!OEt \\ | \\ \underset{\underset{\displaystyle O}{\|}}{C}\!-\!OEt \end{matrix} \quad \xrightarrow[\text{2. HCl, HOH}]{\text{1. NaOEt, EtOH}} \quad \begin{matrix} H_3C\!-\!(CH_2)\overline{_{15}}\,CH\!-\!\overset{\overset{\displaystyle O}{\|}}{C}\!-\!OEt \\ | \\ \underset{\underset{\displaystyle O}{\|}}{C}\!-\!\overset{\overset{\displaystyle O}{\|}}{C}\!-\!OEt \end{matrix}$$

$$70\%$$

The second ester does not have any alpha-hydrogens, so it is attacked by the carbanion derived from the first ester.

The mechanism of the first reaction follows.

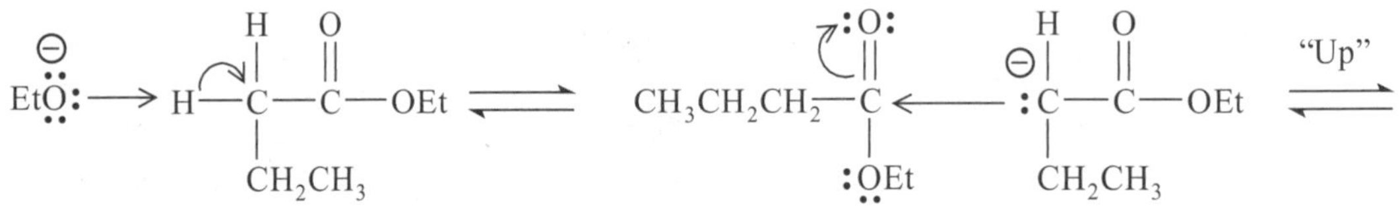

$$Et\overset{..}{\underset{..}{O}}{:}\!\longrightarrow\! H\!-\!\overset{\overset{\displaystyle H}{|}}{\underset{\underset{\displaystyle CH_2CH_3}{|}}{C}}\!-\!\overset{\overset{\displaystyle O}{\|}}{C}\!-\!OEt \ \rightleftharpoons \ CH_3CH_2CH_2\!-\!\overset{\overset{\displaystyle :O:}{\|}}{\underset{\underset{\displaystyle :OEt}{|}}{C}}\!\longleftarrow\! {:}\overset{\overset{\displaystyle H}{|}}{\underset{\underset{\displaystyle CH_2CH_3}{|}}{C}}\!-\!\overset{\overset{\displaystyle O}{\|}}{C}\!-\!OEt \quad \text{"Up"} \ \rightleftharpoons$$

The base pulls an alpha hydrogen off the ester to form a carbanion, which attacks the C=O of another ester and pushes an electron pair "up" onto the oxygen.

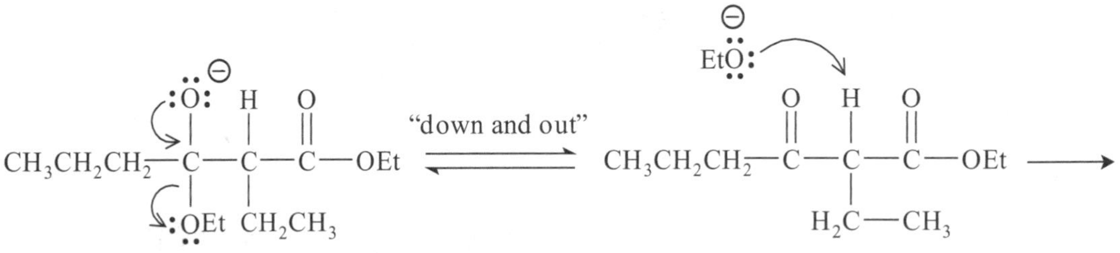

$$CH_3CH_2CH_2\!-\!\overset{\overset{\displaystyle :\overset{..}{O}:}{|}}{\underset{\underset{\displaystyle :OEt}{|}}{C}}\!-\!\overset{\overset{\displaystyle H}{|}}{\underset{\underset{\displaystyle CH_2CH_3}{|}}{C}}\!-\!\overset{\overset{\displaystyle O}{\|}}{C}\!-\!OEt \quad \xrightarrow{\text{"down and out"}} \quad CH_3CH_2CH_2\!-\!\overset{\overset{\displaystyle O}{\|}}{C}\!-\!\overset{\overset{\displaystyle H}{|}}{\underset{\underset{\displaystyle H_2C\!-\!CH_3}{|}}{C}}\!-\!\overset{\overset{\displaystyle O}{\|}}{C}\!-\!OEt \longrightarrow$$

The nucleophilic acyl substitution is completed ("down and out"), which forms the beta-ketoester. The hydrogen between the two C=O's of the beta-ketoester is more acidic than the original ester, so it reacts with the ethoxide present to form the anion shown. Therefore, the Claisen condensation requires a full equivalent of a base.

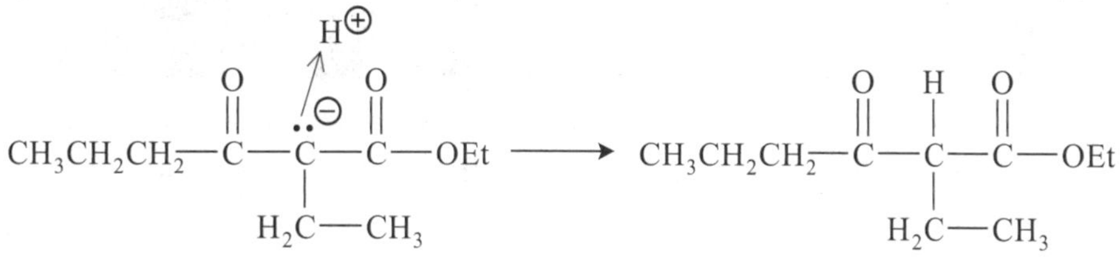

This anion is highly resonance-stabilized and, thus, less reactive than the original ester anion. The reaction stops here until acid is added to form the product.

The **Dieckmann condensation** is just a cyclic version of the Claisen condensation. If a molecule has two ester groups appropriately arranged so that they can form a five-, six-, or seven-membered ring, it will usually cyclize to form a beta-ketoester. Sometimes larger rings can be formed, if the conditions are controlled. Some examples are shown below.

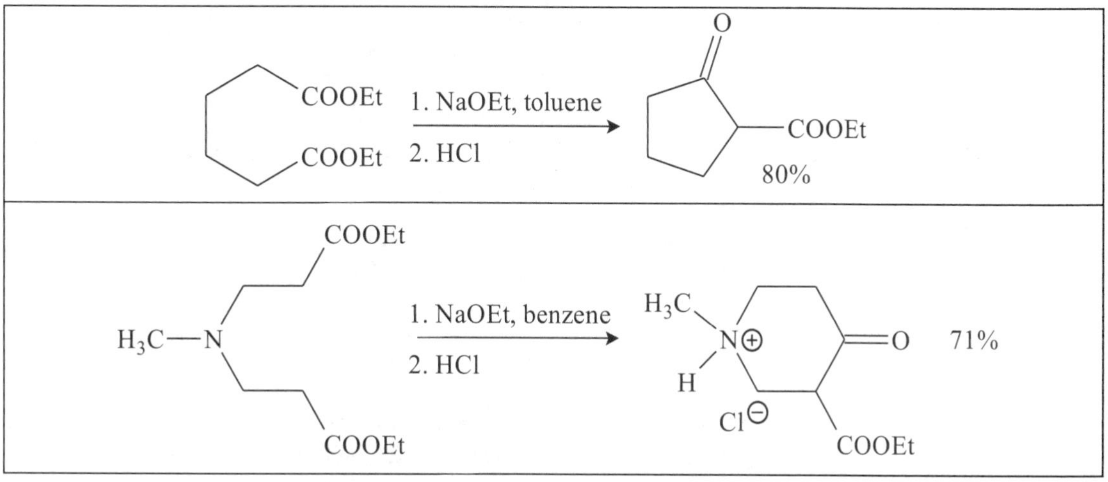

Just like ethyl acetoacetate, these beta-ketoesters can be converted into ketones by hydrolysis of the esters and decarboxylation. The beta-ketoesters can also be alkylated first, followed by ester hydrolysis and decarboxylation, to make alkylated ketones. See the example below.

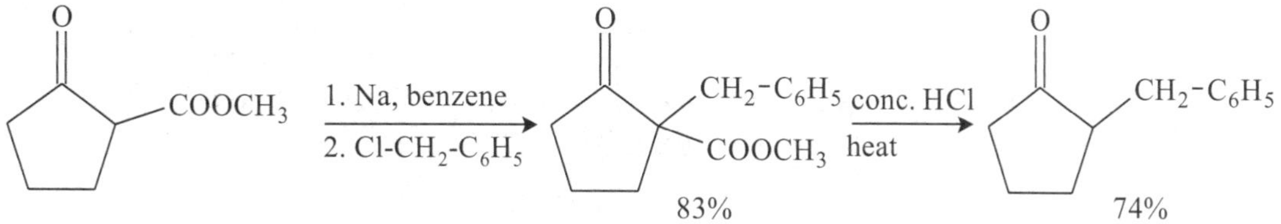

E. Conjugate Addition Reactions (The Michael Reaction)

Conjugated aldehydes, ketones, esters, and nitriles can react with nucleophiles at the beta-carbon of the C=C to yield addition products. Some examples follow.

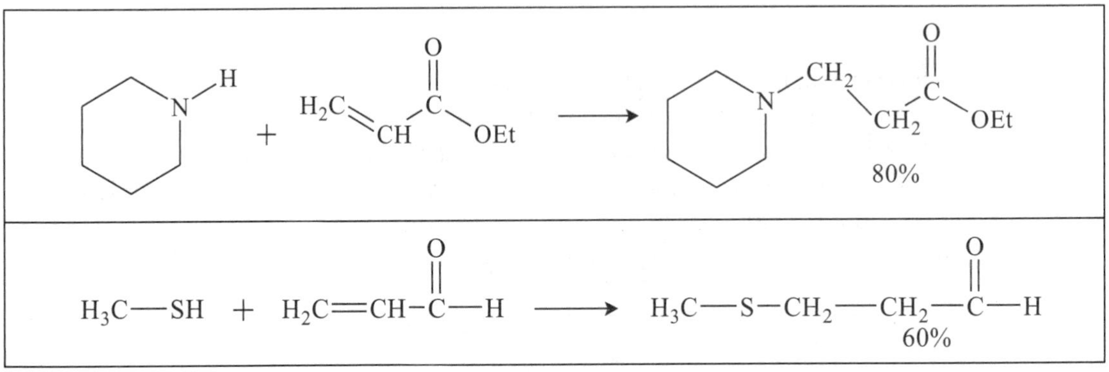

This reaction looks a little unusual because nucleophiles don't normally add to C=C's. However, the conjugation of the C=C influences the polarity of the carbons. The beta-carbon becomes significantly positive in character. See below.

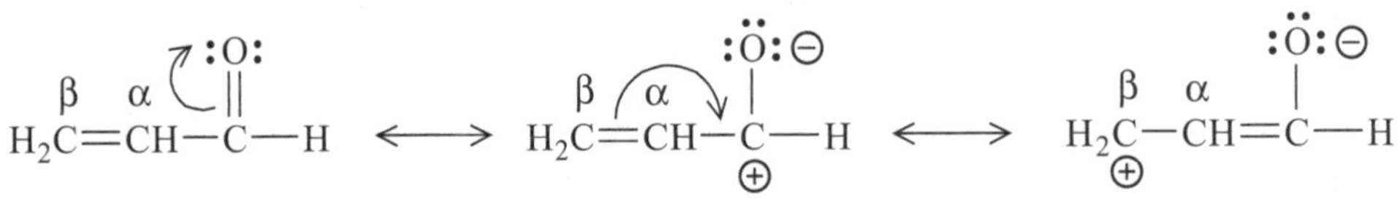

Therefore, nucleophiles, especially more weakly basic ones, such as amines and thiols, preferentially react at the beta-carbon. The mechanism is shown below.

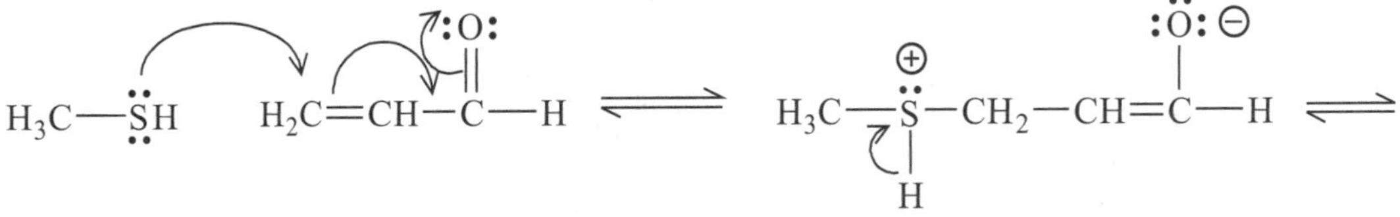

The sulfur attacks the beta-carbon and pushes the electron pair toward the carbonyl carbon. The pair of electrons in the pi bond of the C=O is pushed on the oxygen The hydrogen is lost from the sulfur as H⁺.

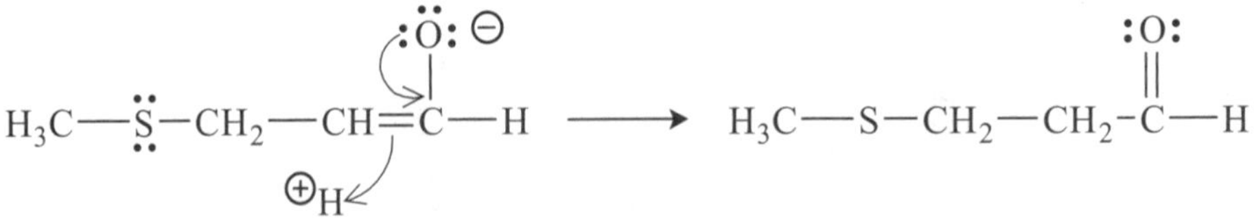

An electron pair on the oxygen moves down to reform the C=O. This pushes an electron pair to the alpha-carbon, which picks up a proton to form the final product.

Enolate ions also add to the beta-carbon of a conjugated carbonyl compound by a similar mechanism. Some examples of this are shown below. If more than one equivalent of conjugated carbonyl compound is used, then multiple adducts can be prepared.

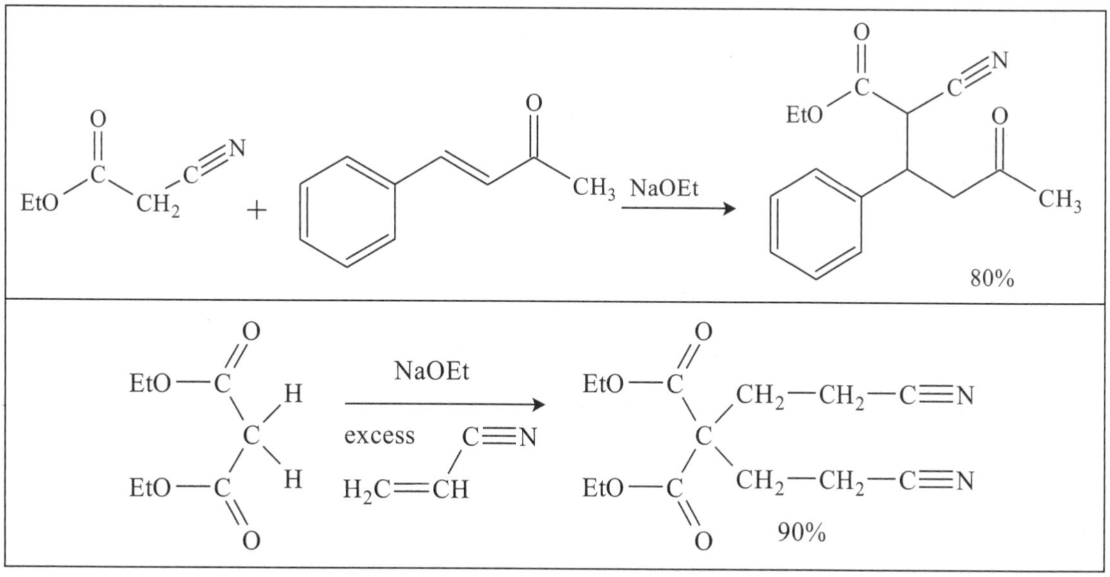

80%

90%

F. Robinson Annelation

The combination of a conjugate addition, followed by an aldol condensation to form a ring, is called a Robinson annelation (**annelation** is just a fancy word for ring-forming reaction). See the next example.

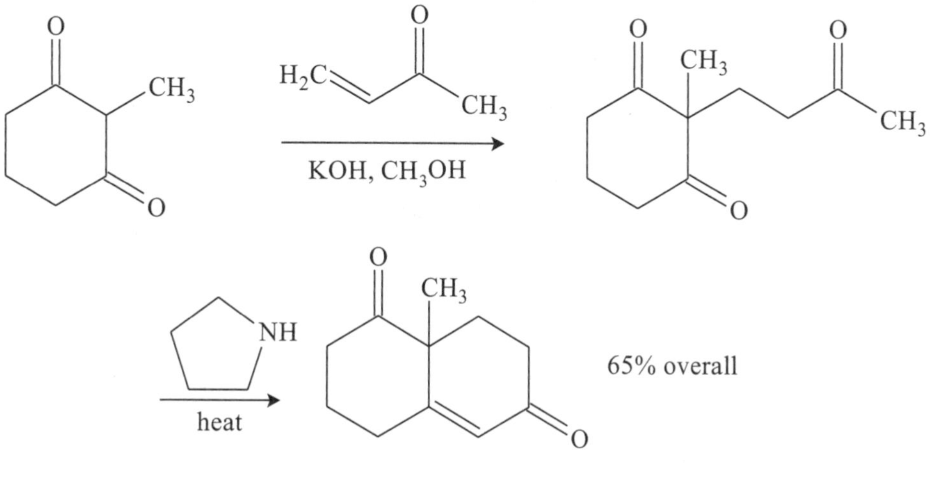

KOH, CH$_3$OH

65% overall

heat

G. Alkylations and Acylations of Ketones

G.1. DIRECT ALKYLATION OF KETONES

Ketones cannot be alkylated using bases such as sodium hydroxide or sodium ethoxide because these bases don't completely remove an alpha-hydrogen from a ketone and because these bases react with alkyl halides. To alkylate a ketone, a very strong base, such as sodium amide ($NaNH_2$) or lithium diisopropylamide (LDA, $Li N(CH(CH_3)_2)_2$), must be used, to remove an alpha-hydrogen completely. The ketone is added slowly to the solution of the base at low temperature. This allows the ketone to react completely with the base to form the enolate because no excess ketone is present. This procedure works best with ketones that are symmetric, so that a mixture of enolates is not formed. See the following examples.

Even though all the appropriate precautions were taken, some dialkylation occurred. Fortunately, the products can be separated by distillation readily.

Even though a bulky base like LDA preferentially removes the hydrogen from the less hindered carbon 6 of 2-methylcyclohexanone, some alkylation at the 2-position occurs. Some dialkylated product is also formed.

As you can see from these examples, direct alkylation of ketones is possible, but it does not always occur cleanly. A number of other procedures have been developed to monoalkylate ketones indirectly.

G.2. ALKYLATION AND ACYLATION OF ENAMINES

If a ketone is reacted with a secondary amine in the presence of an acid catalyst, an enamine is formed. An example of this reaction, and the mechanism, are shown below.

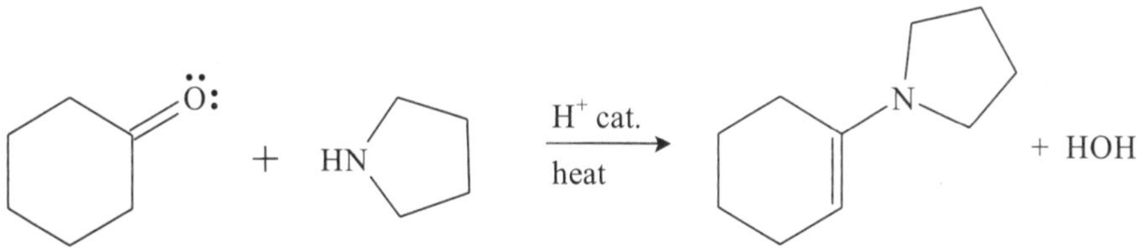

"Pyrrolidine"   The pyrrolidine enamine of cyclohexanone

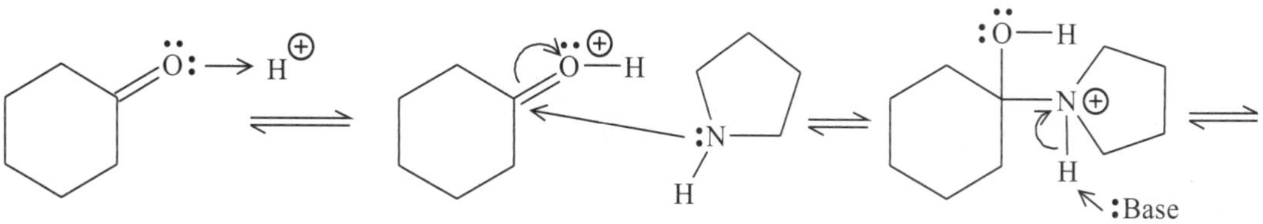

Acid protonates the ketone carbonyl. The amine attacks the carbonyl carbon. Loss of H⁺ from the nitrogen gives the amino-alcohol intermediate.

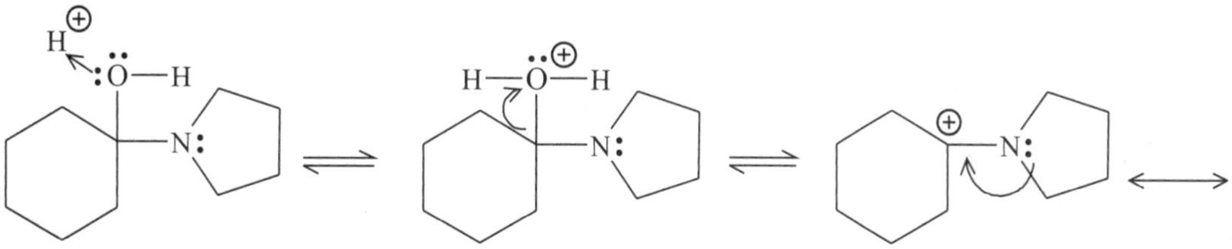

The OH is protonated, and water is lost to form a cation, which is resonance-stabilized. Loss of a proton forms the enamine, as shown here.

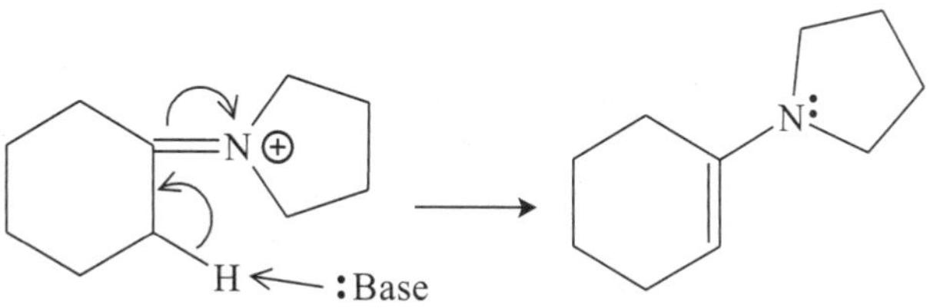

Enamines can be alkylated on carbon by reactive alkyl halides and acid chlorides. After hydrolysis, the products are alkylketones and 1,3-diketones, respectively. Some examples are shown below.

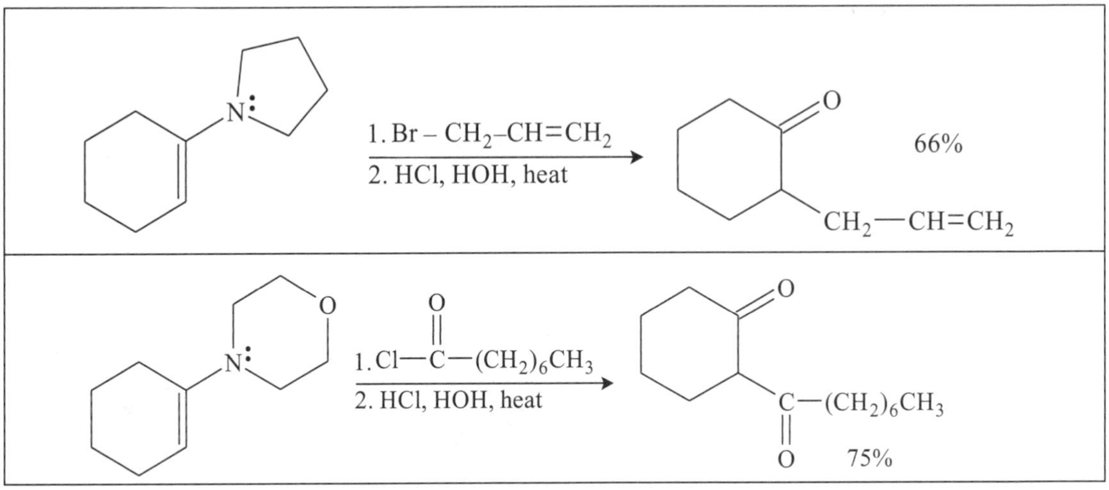

These reactions do not form dialkylated products. The mechanism of the first example is shown below.

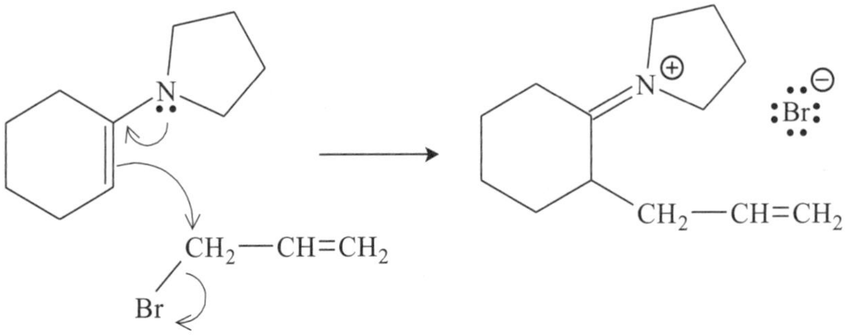

The enamine double bond attacks the alkyl halide, as does an S_N2 reaction. The nitrogen electron pair forms a double bond to help stabilize the structure. The resulting iminium salt is hydrolyzed in acid to from the ketone. This mechanism is essentially the reverse of the mechanism we used to form the enamine.

REVIEW EXERCISES FOR CHAPTER 14

1. Show how you could prepare each of the following products from the indicated starting material.

Starting Material	Product
a.	
b.	
c.	
d.	

Starting Material	Product
e. 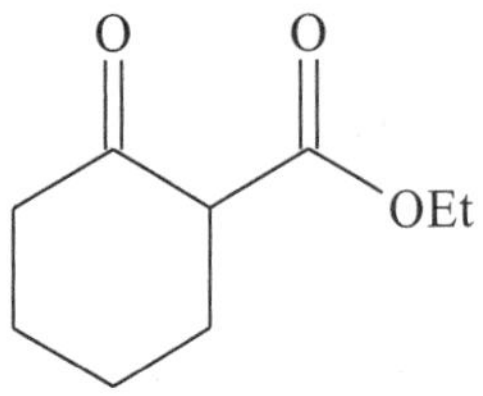 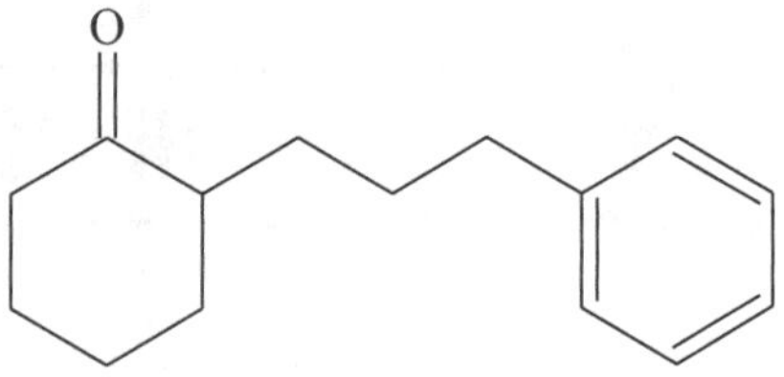	

2. If a beta-diketone, such as the one below, were heated with a concentrated solution of NaOH in water, the indicated ketoacid would be formed. Draw a reasonable arrow-pushing mechanism for this reaction, and explain why the carbon–carbon bond is cleaved.

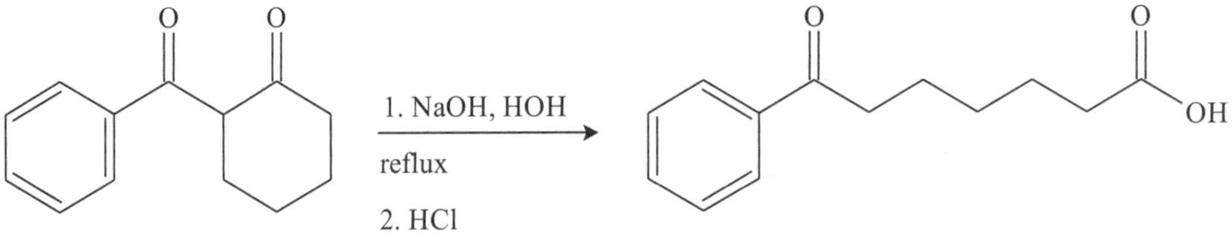

1. NaOH, HOH
reflux

2. HCl

3. The following sequence of reactions is from a synthesis of some of the components of pepper spray (H. Kaga, M. Miura, and K. Orito, *Synthesis* **1989**, 864–66). Supply the needed reagents or products. Some of these reactions are from previous chapters.

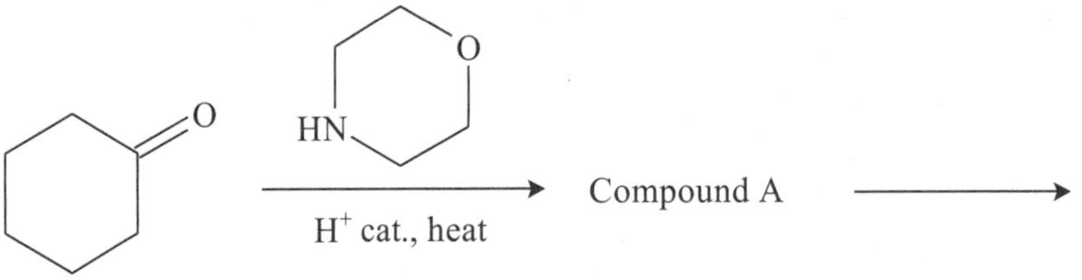

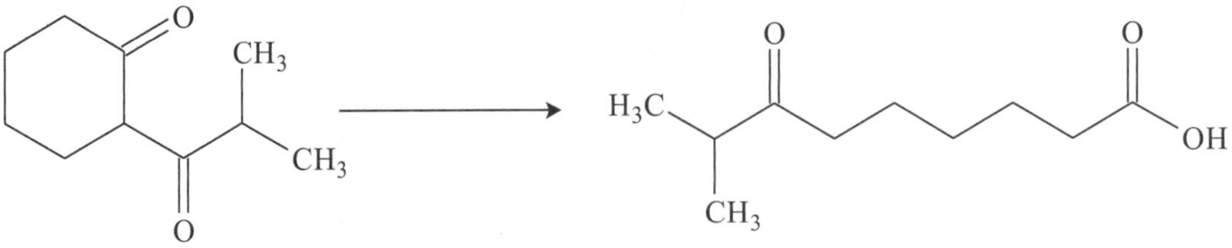

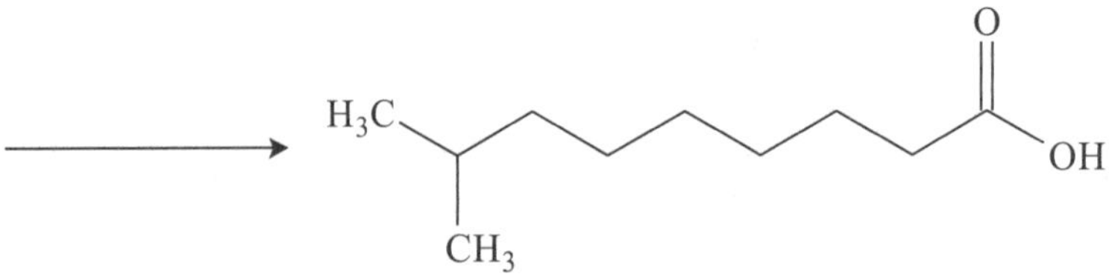

Compound A

H^+ cat., heat

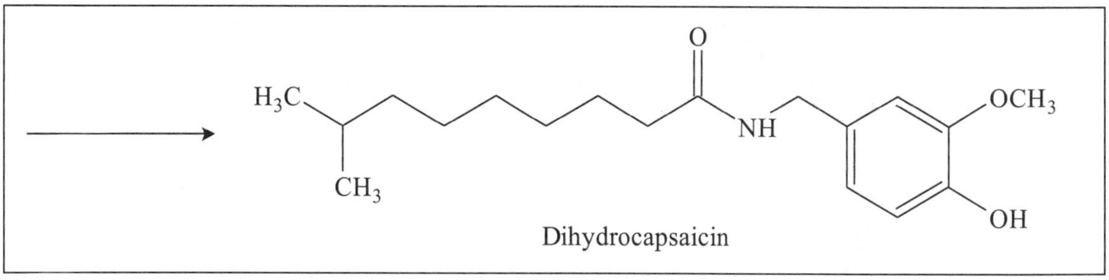

H₃C ... NH ... OCH₃ ... OH

Dihydrocapsaicin

4. Draw the structures of the products formed *after each step* of the following reaction sequence.

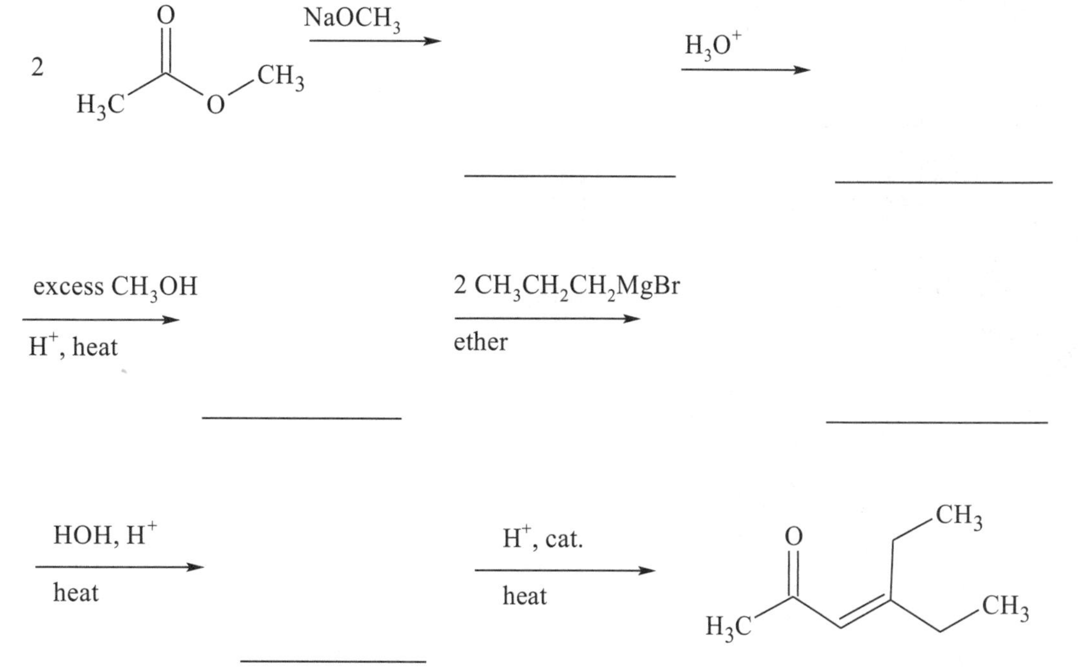

2   (structure with O, H₃C, O—CH₃)   →[NaOCH₃]   →[H₃O⁺]

_____   _____

→[excess CH₃OH / H⁺, heat]   →[2 CH₃CH₂CH₂MgBr / ether]

_____   _____

→[HOH, H⁺ / heat]   →[H⁺, cat. / heat]   (structure with O, H₃C, CH₃, CH₃)

_____

Spectroscopy

WHAT YOU WILL LEARN

In this chapter, you will learn:

- how to do simple interpretation of infrared, mass and nuclear magnetic resonance spectra;
- how to use this information to determine the structure of an unknown compound.

SECTIONS IN THIS CHAPTER

- Infrared Spectroscopy

- Nuclear Magnetic Resonance Spectroscopy

- Mass Spectrometry

- Putting It All Together to Solve Problems

This chapter will focus on interpretation of spectra, rather than theory about spectroscopy or how spectra are obtained. Most organic texts have good explanations of these other areas. To me, the best way to learn how to interpret spectra is to look at spectra, rather than a table with a list of numbers. I have included a number of spectra in each section to help you see what spectra look like, and what the various peaks in them mean. The spectra shown here were run by the author.

A. Infrared Spectroscopy

A.1. INTRODUCTION

Infrared (IR) spectroscopy primarily shows us what kinds of bonds exist in organic molecules. Specific types of bonds absorb characteristic wavelengths of IR energy. When they do, the bonds can stretch or bend, and this absorption of energy can be detected. Therefore, IR spectroscopy is used primarily to detect functional groups.

IR spectra can be obtained in a number of formats. The IR spectrum of ethyl acetate is shown below.

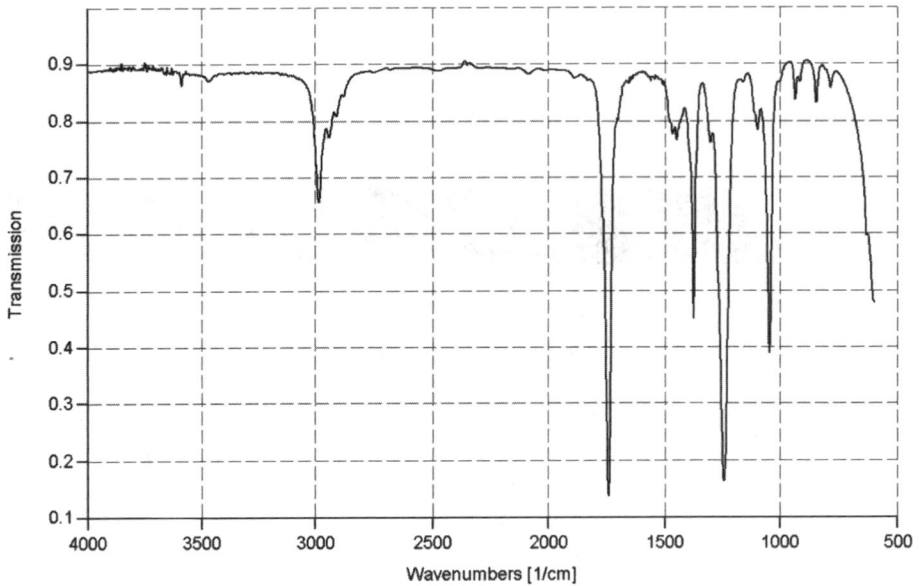

The horizontal axis is wavenumbers in units of cm^{-1}. Wavenumbers are calculated by the following formula:

$$\text{wavenumber (cm}^{-1}) = 10,000/\text{wavelength (μm)}$$

The vertical axis is transmission (T), or percentage transmittance ($\%T$). This is the percentage of the energy transmitted through the sample at a particular wavenumber. A low value of transmission or percentage transmittance corresponds to a high absorption of energy at that wavenumber. For example, in the preceding spectrum, the peak at 1240 cm^{-1} has a transmission of 0.12, or a $\%T$ of 12. This means very little energy at that wavenumber has been transmitted through the sample. And most of the energy at that wavenumber was absorbed.

A.2. INTERPRETATION OF IR SPECTRA

In general, I encourage students to look at spectra from right to left, and to look at the bigger peaks first, to get some idea of what functional groups are present. In a test situation, I suggest they spend no more than 30 seconds looking at an IR spectrum, and jotting down what the bigger peaks could be. Here is a general guide for interpretation of an IR spectrum. Units are in cm^{-1}:

Alcohol OH: 3300 (broad)

sp C–H: 3300 (narrow)

sp^2 C–H: 3200–3000

sp^3 C–H: 3000–2800

Aldehyde C–H: two peaks at 2800 and 2700 — often weak

C≡C: 2300–2100 — usually weak, unless one C is bonded to an H

C≡N: 2300–2100 — usually strong

C=O: 1800–1600 — usually strong

C=C: 1650–1600

Benzene C=C: 1600 and 1500

C–O: 1300–1050

Let's look at some sample spectra and interpret them.

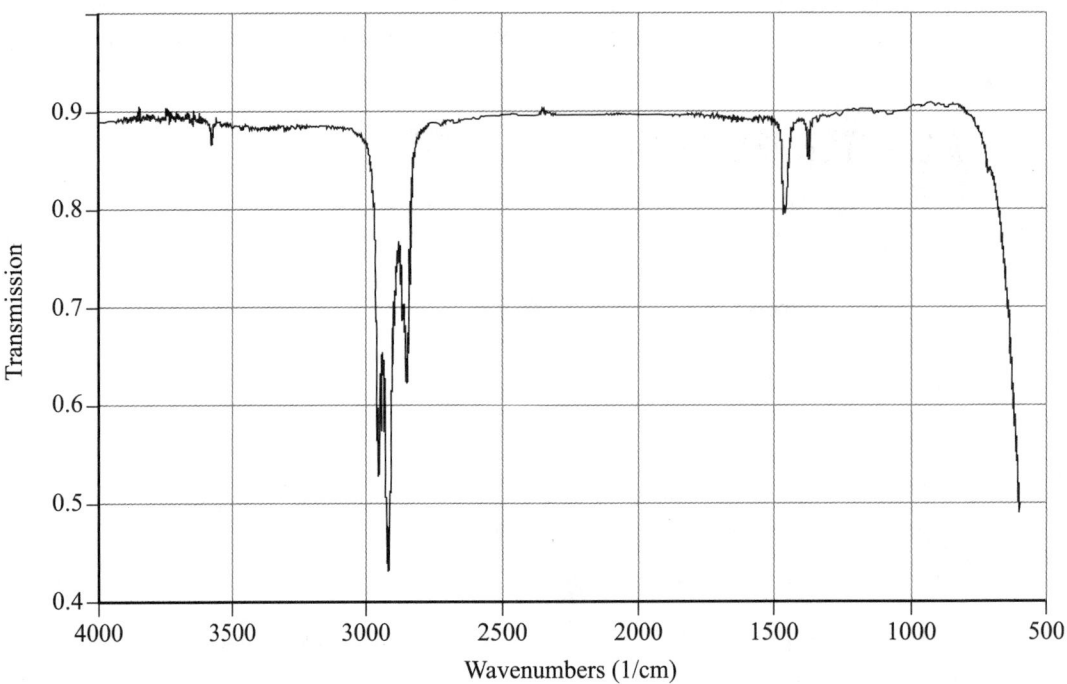

Peaks (cm^{-1})	Meaning
3000–2800	sp^3 C–H stretching vibrations
1500–1350	CH$_2$ and CH$_3$ bending vibrations

Because many organic compounds contain alkane portions, I've started you off with an alkane. You will see many of these peaks in other spectra.

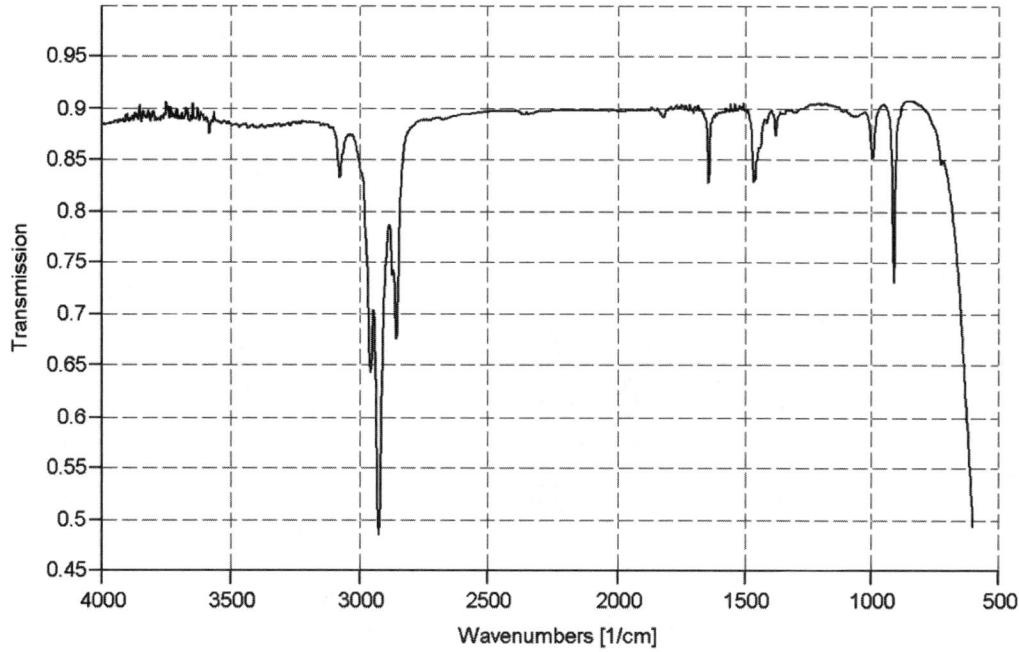

Peaks (cm⁻¹)	Meaning
3050	sp^2 C–H stretch
3000–2800	sp^3 C–H stretching vibrations
1650	C=C stretch
1450	CH$_2$ bending vibrations
1000, 910	Out-of-plane sp^2 C–H bending vibrations for a terminal alkene

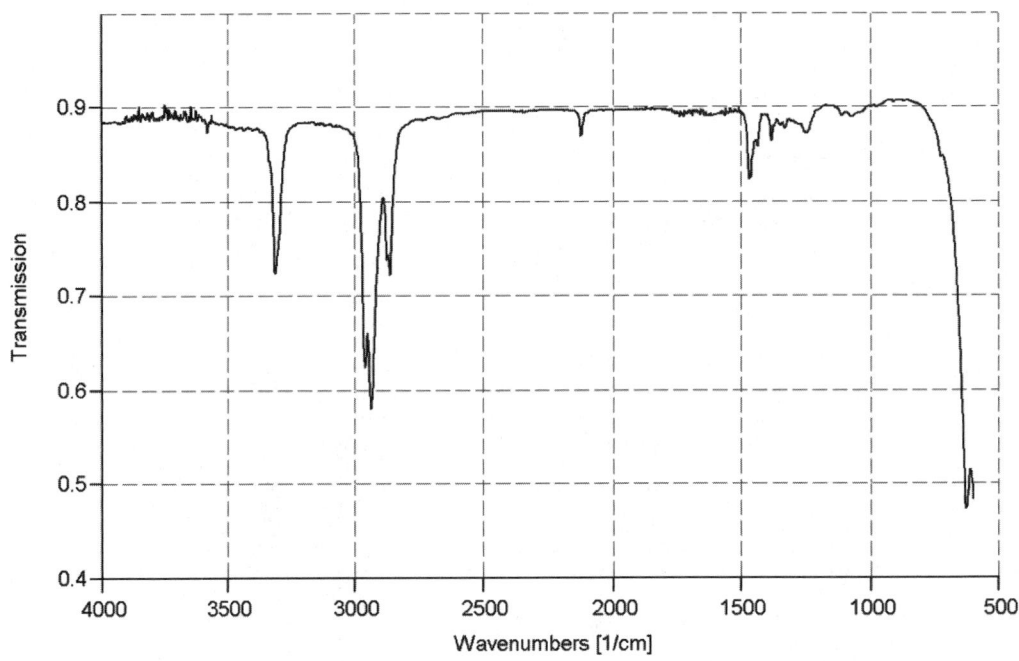

Peaks (cm⁻¹)	Meaning
3300	*sp* C–H stretch. This peak is usually narrow.
3000–2800	*sp*³ C–H stretching vibrations.
2120	C≡C stretch. This peak is often weak.

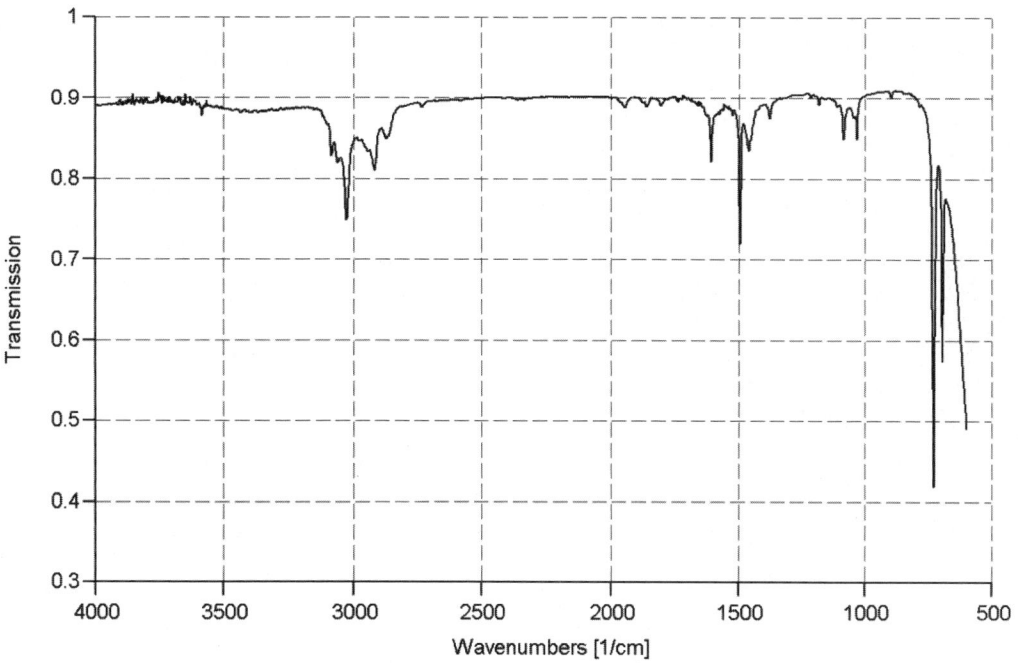

Peaks (cm^{-1})	Meaning
3200–3000	sp^2 C–H stretches
3000–2800	sp^3 C–H stretching vibrations
1600, 1500	Aromatic C=C stretches
740, 690	Out-of-plane C–H bending vibrations for a monosubstituted benzene

2-Propanol	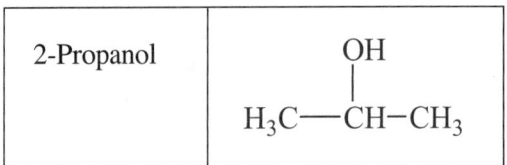

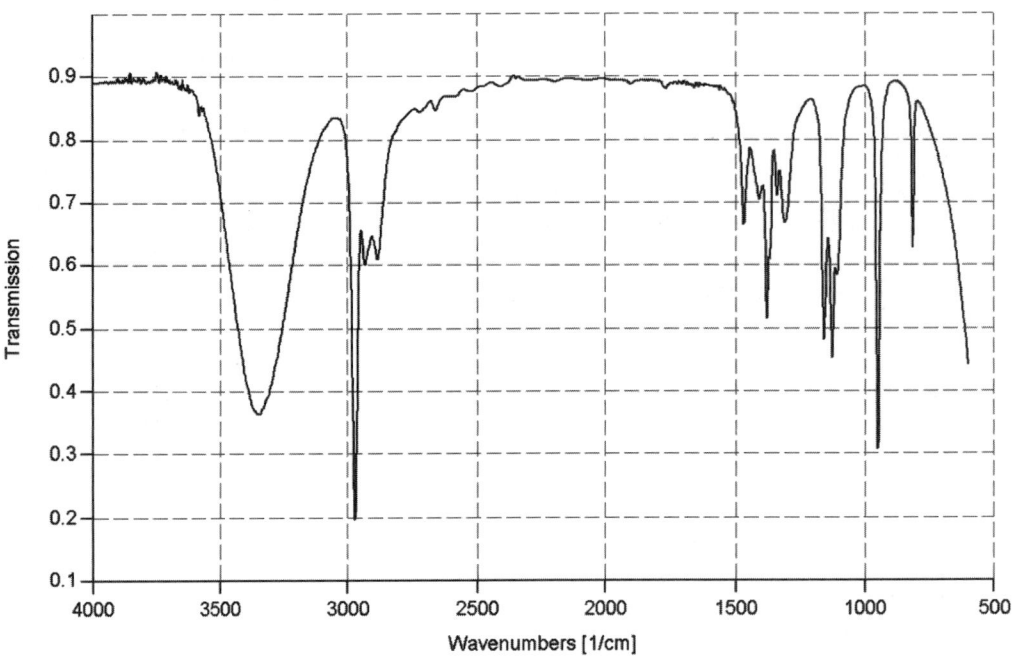

Peaks (cm⁻¹)	Meaning
3300	The O–H stretch. This peak is large because stretching the O–H bond increases the bond moment. It is broad because there is hydrogen-bonding between O–H groups, which weakens the strength of the O–H bond somewhat.
3000–2800	sp^3 C–H stretching vibrations
1500–800	Various C–H bending vibrations and C–C stretching vibrations. The C–O stretch is about 1100.

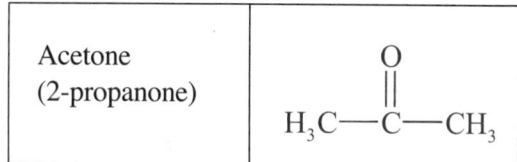

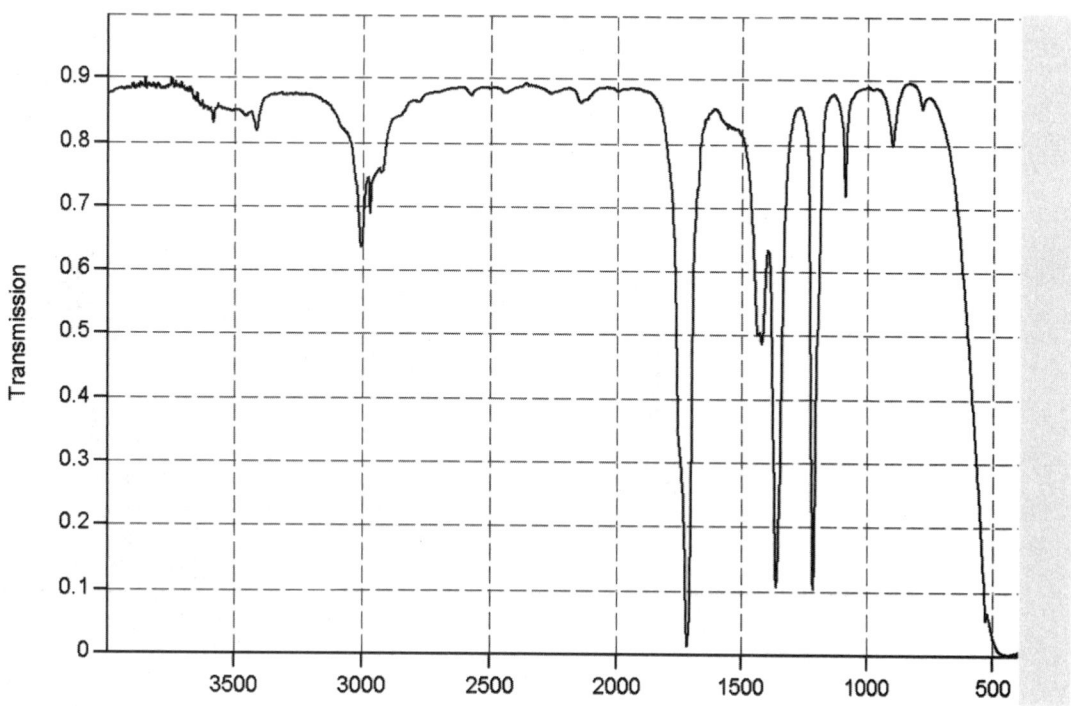

Peaks (cm⁻¹)	Meaning
3500	There is no O–H in acetone. However, acetone does dissolve water, and unless samples are carefully dried, some water O–H stretches will show up.
3000–2800	sp^3 C–H stretching vibrations.
1700	The C=O stretch. It is the largest peak in the spectrum.
1500–800	Various C–H bending vibrations and C–C stretching vibrations.

The absorbance around 500 cm⁻¹ is not a peak in acetone; it is instead an absorbance of an impurity in KBr. IR spectra are commonly obtained as thin films between two KBr disks.

Ethyl acetate	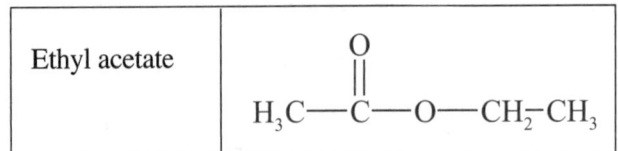

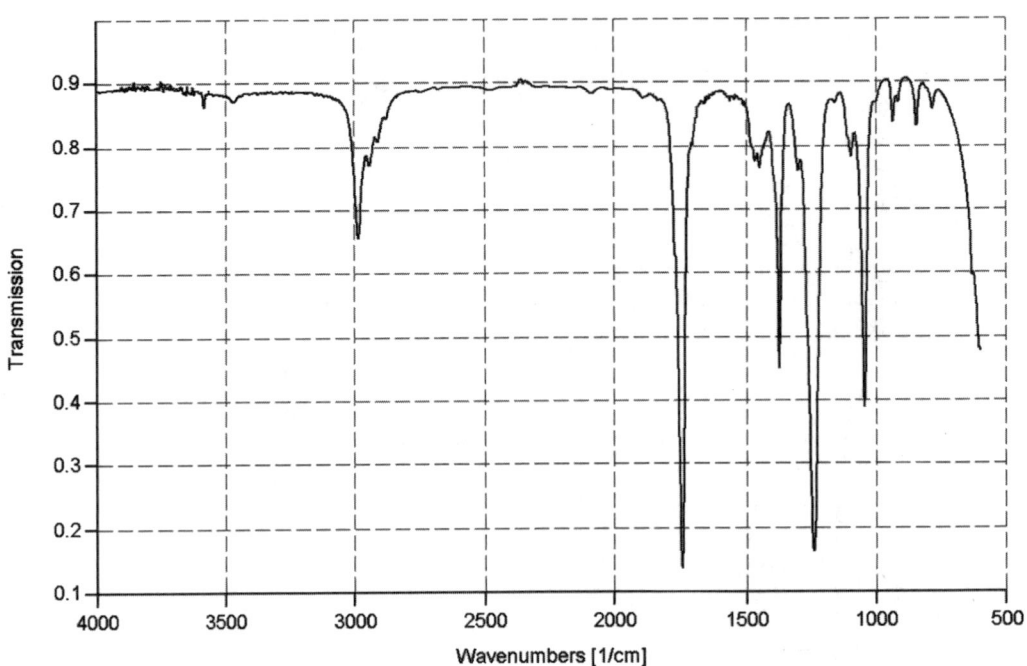

Peaks (cm⁻¹)	Meaning
3000–2800	sp^3 C–H stretching vibrations.
1740	The C=O stretch. This is usually one of the strongest peaks in the spectrum.
1240	The O=<u>C</u>–<u>O</u>–C stretch. This is very characteristic of acetate esters.
1050	The O=C–<u>O</u>–<u>C</u> stretch. This will vary in position, depending on how substituted the carbon is.

Propanoic acid	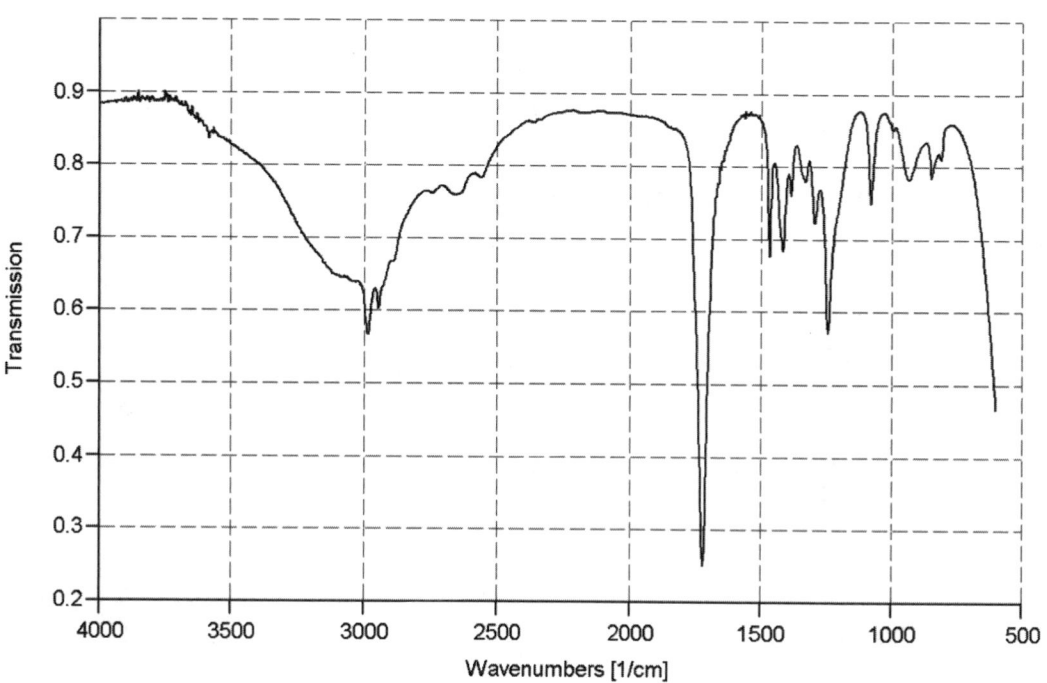

Peaks (cm⁻¹)	Meaning
3500–2500	This big, broad peak is the hydrogen-bonded O–H peak of the acid. Acids form hydrogen-bonded dimers, as shown above.
3000–2900	The little spikes on the bottom of the broad O–H peak are the sp^3 C–H stretches.
1740	The C=O stretch.
1240	The C–O stretch.

1-Butanamine	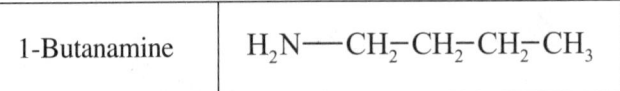

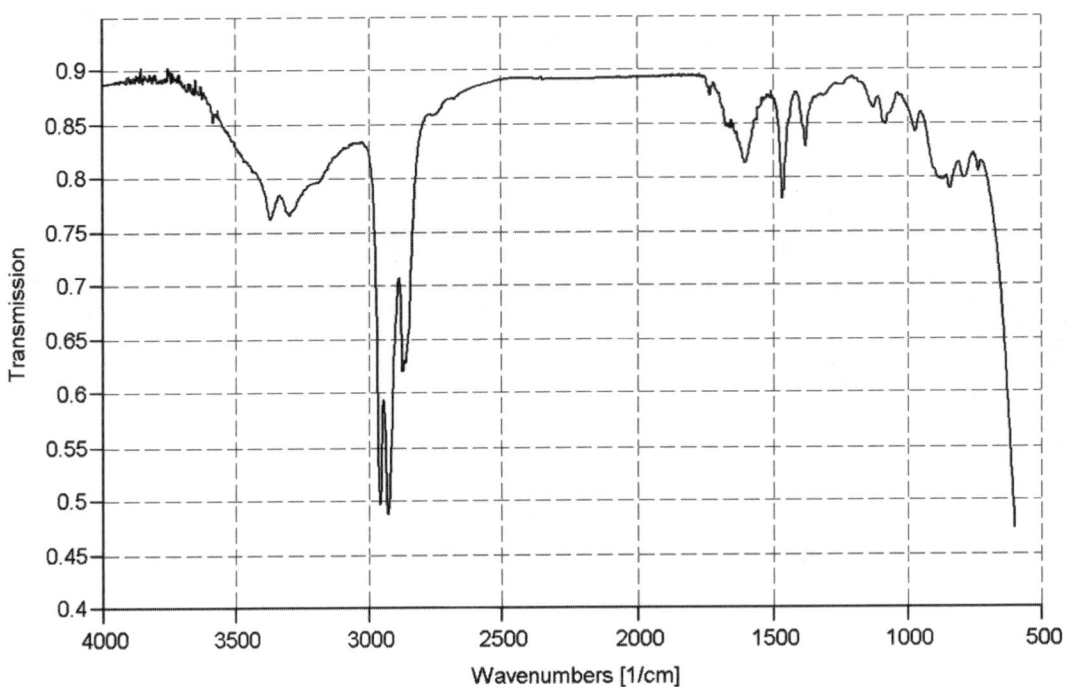

Peaks (cm⁻¹)	Meaning
3350	The stretches for the NH_2. An NH_2 shows up as two peaks, most of the time. A secondary amine or amide only has one peak in this area. These are weaker than O–H stretches because an N–H is a less polar bond than an O–H.
3000–2800	sp^3 C–H stretching vibrations.
1600	The N–H bending vibrations. These are usually broad and weak, so they can be distinguished from a C=O (strong) and a C=C (weak, but narrow), which also can show up in this region.

Obviously, I can't teach you everything about IR in nine spectra, but you should be getting the idea of what an IR spectrum looks like. If you need to determine the structure of an unknown from its spectra on an exam, you should spend between 20 and 30 seconds looking at the IR spectrum, and pick out the key functional groups. Then go on to the NMR spectra because they can give you more information. Let's look at NMR now.

B. Nuclear Magnetic Resonance Spectroscopy

Since its humble beginnings in the 1940s as a curiosity, NMR has developed into organic chemists' most important tool for determining the structure of organic molecules, in my opinion.

NMR can be applied to a wide variety of atoms, including ^{1}H, ^{13}C, ^{19}F, ^{31}P, and ^{14}N. In this book, I will focus on ^{1}H, which is called proton NMR or PMR, and ^{13}C, which is called carbon NMR or CMR.

HOW IT WORKS (GREATLY SIMPLIFIED!)

Protons (and certain other nuclei) act as if they are spinning charges. A spinning charge generates a magnetic field vector. These vectors are normally randomly oriented. If a sample is placed in a strong magnetic field, the vectors will either align themselves in the same direction as the field, or in the opposite direction. The two orientations are slightly different in energy, with a slight excess being aligned in the same direction as the applied magnetic field. If we put energy into the system, we can excite some of the spins from the lower energy state to the upper and can measure the amount of energy required to do this ("flip its spin"). This energy is near the radio frequency range (MHz). How is this useful to us? The environment a nucleus is in (what atoms it is bonded to or is close to), affects how much energy is required to flip its spin. I will leave further discussion of theory to your organic text and focus on the interpretation of spectra.

NMR can tell us four basic pieces of information.

B.1. WE CAN USE NMR TO TELL US HOW MANY DIFFERENT TYPES OF HYDROGENS OR CARBONS THERE ARE IN A COMPOUND

We can look at a structure and see how many different types of carbons and hydrogens there should be, on the basis of symmetry in the molecule. The proton NMR spectrum should have a signal for each type of hydrogen, and the carbon NMR spectrum should have a signal for each type of carbon. Some examples are given below. Lines for the planes of symmetry are drawn through the structures.

Compound	Ethane	Propane	Cyclohexane	2-Propanone
Structure	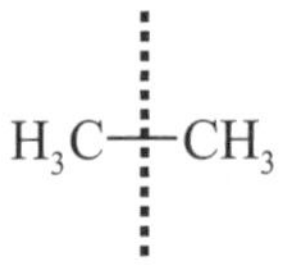 $H_3C\!-\!CH_3$	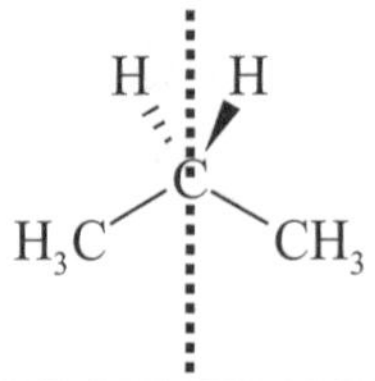 H H H_3C CH_3	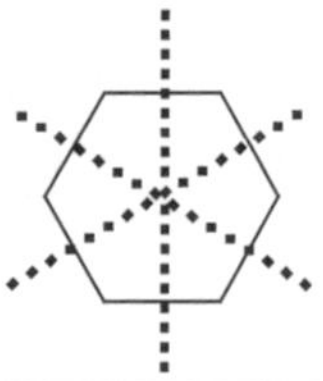 (hexagon structure)	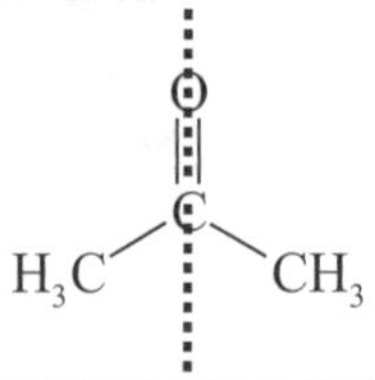 H_3C CH_3
No. of types of C	Just one: both methyls are the same.	Two: the CH_2, and the two methyls, which are the same.	One: there are three planes of symmetry, so all the C's are the same.	Two: the methyls and the carbonyl C.
No. of types of H	Just one: both methyls are the same.	Two: the CH_2 and the methyls, which are the same.	One: there are three planes of symmetry, so all the CH_2's are the same.	Just one: both methyls are the same.

What do NMR spectra look like? The proton and carbon NMR spectra of 2-propanone are shown below.

The horizontal scale is parts per million (ppm), sometimes called delta (δ). Notice the peak in the proton spectrum at 2.08 ppm. This is the peak for the hydrogens. All proton spectra are referenced to the position of the H peak in $(CH_3)_4Si$ (tetramethylsilane, or TMS), which is arbitrarily assigned the value of 0 ppm in proton spectra. Often, TMS is added to a sample, and you will see a TMS peak in a spectrum. The ppm value is calculated as follows:

ppm = [10^6 × (number of Hertz required to flip spin of a type of H)
   −(number of Hertz required to flip spin of TMS's H's)]/
   nominal number Hz for magnetic field strength used for type of atom

The ppm scale is used so that the scale is not dependent upon the size of the magnet used. The first NMRs had small magnets, so that only 30 or 60 MHz of energy for hydrogen was required. The newer NMR spectrometer at my former institution has a bigger magnet and requires 300 MHz. The largest magnets now are 1200 MHz. The larger the magnetic field, the better able to separate very similar types of H's.

Proton NMR spectrum of 2-propanone

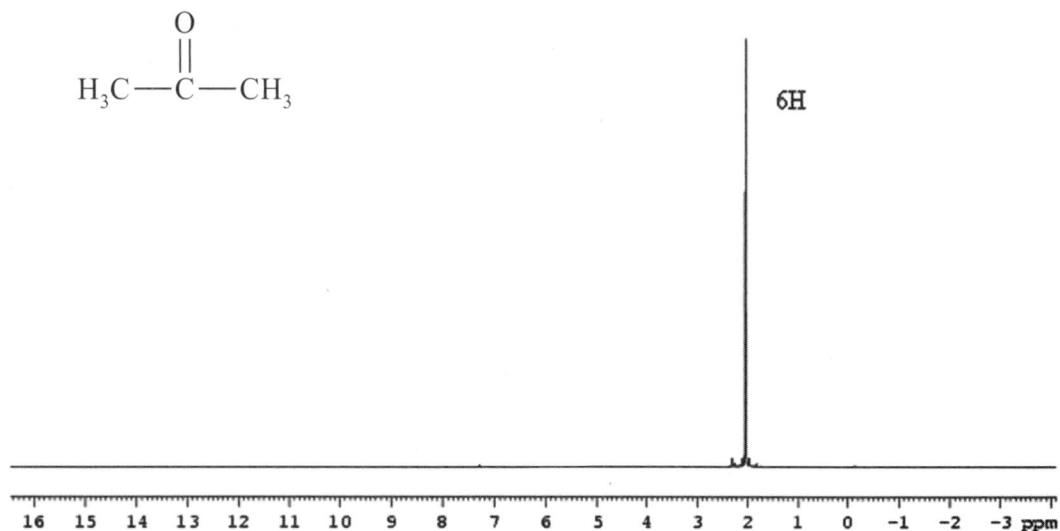

Carbon NMR spectrum of 2-propanone

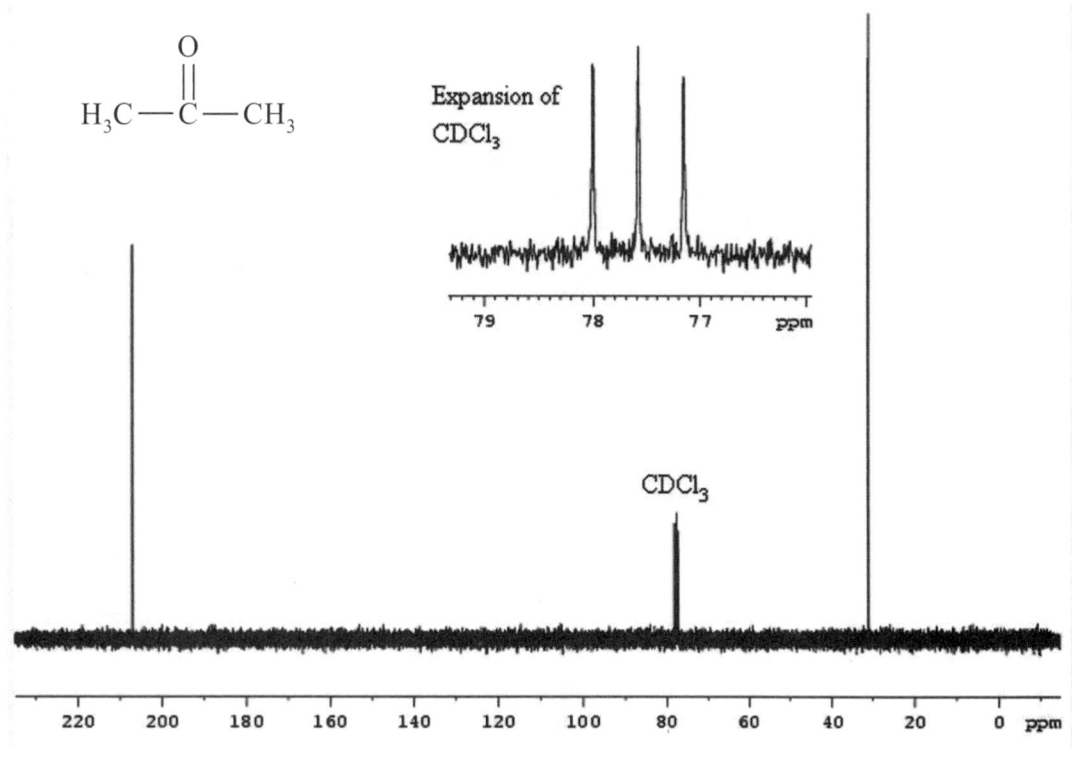

   In the carbon spectrum, we see a peak about 206 ppm, and a peak at 30 ppm, for the two types of C's. The peaks at about 77 ppm are from the deuterated chloroform ($CDCl_3$) solvent. Deuterium is ^2H, an isotope of hydrogen. Deuterium-containing solvents are often used because D doesn't show up in proton spectra, and the newer instruments use the D frequency of the solvent as a reference.

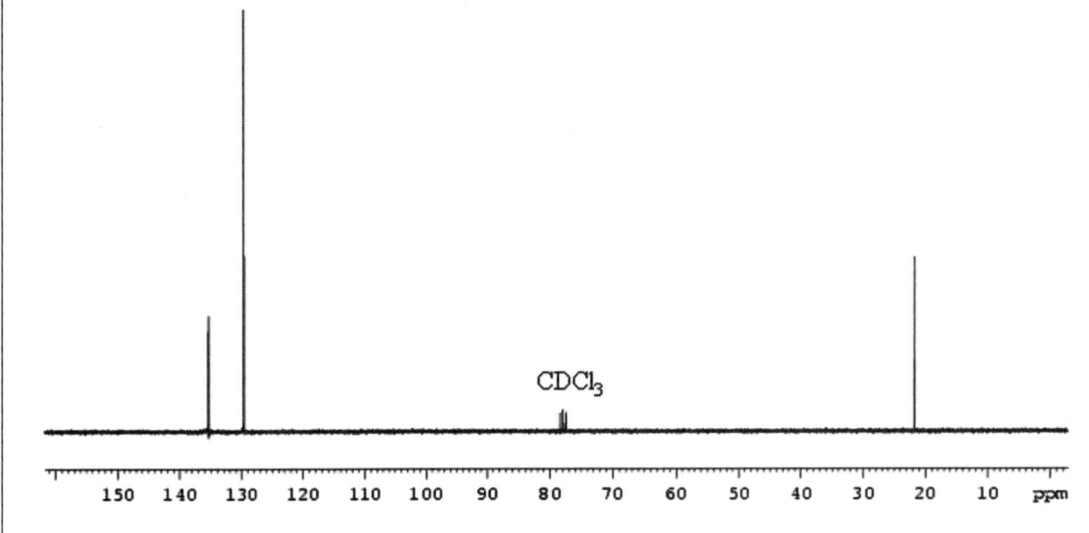

How many different types of C's and H's are there in 1,4-dimethylbenzene (*p*-xylene)? There are two planes of symmetry: one vertical and one horizontal. The vertical plane tells us the two CH₃'s are the same, as well as the two carbons without H's. All the CH carbons are equivalent as well, so there are three types of carbons. All the H's on the benzene ring are equivalent, and the two methyls are equivalent, so there are two types of hydrogens. The spectra are shown below.

There are three carbon peaks, for the three types of C's.

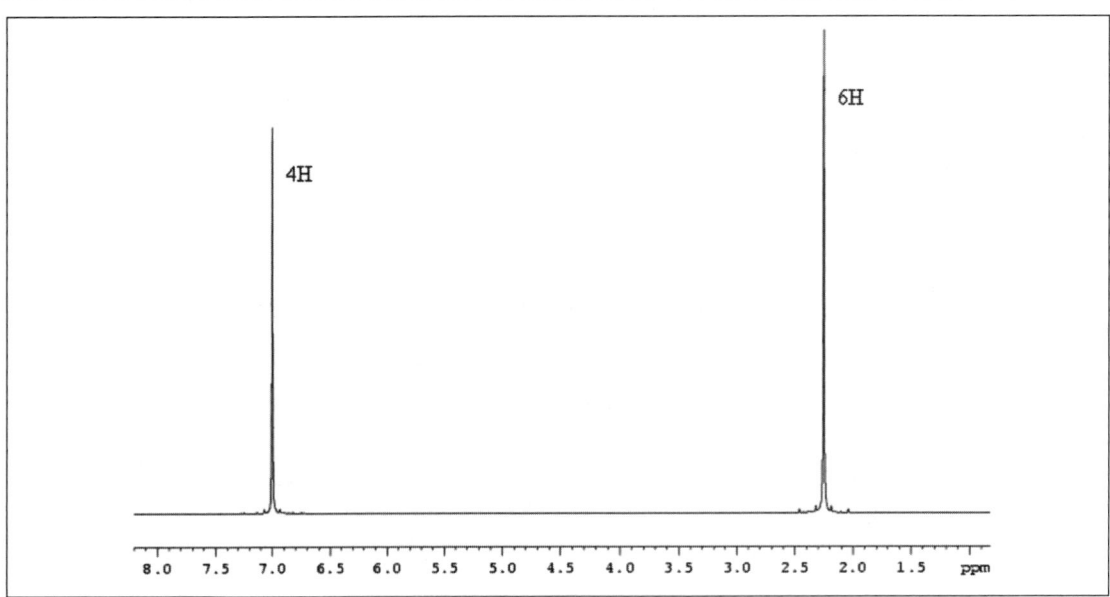

There are two peaks for the two types of H's.

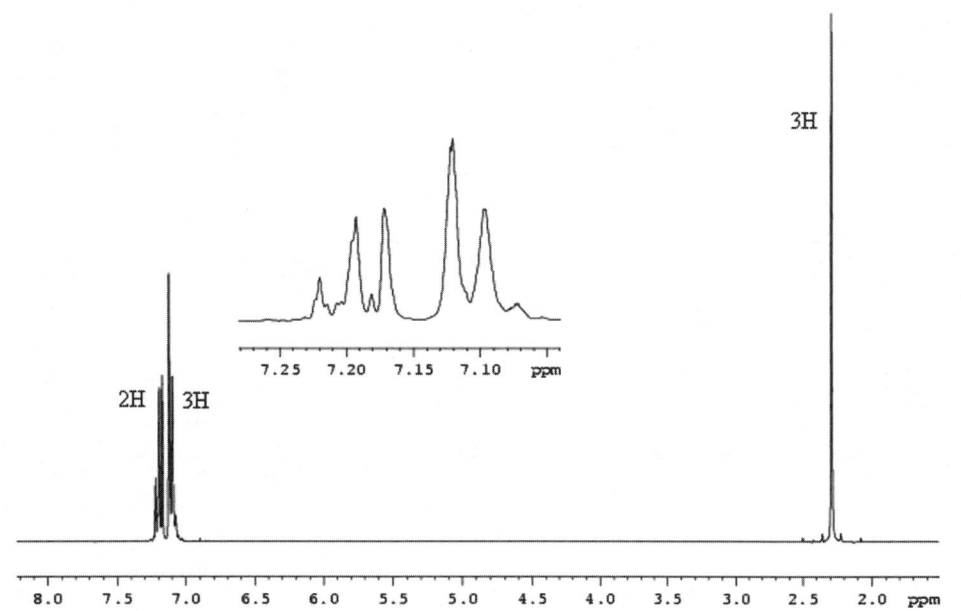

We don't always see all the peaks we expect. Consider toluene. We expect four types of H's, and five types of C's. However, the benzene H's aren't different enough (at 300 MHz) to distinguish each type. We do see five types of C's.

There are three types of benzene hydrogens, but they are not all separated out: One type (the meta hydrogens, Hm) is the peaks at about 7.2 ppm, and the other two types are the set of peaks at 7.1 ppm. A larger magnetic field instrument should separate the peaks for the ortho and para hydrogens. The methyl is the peak at 2.3 ppm.

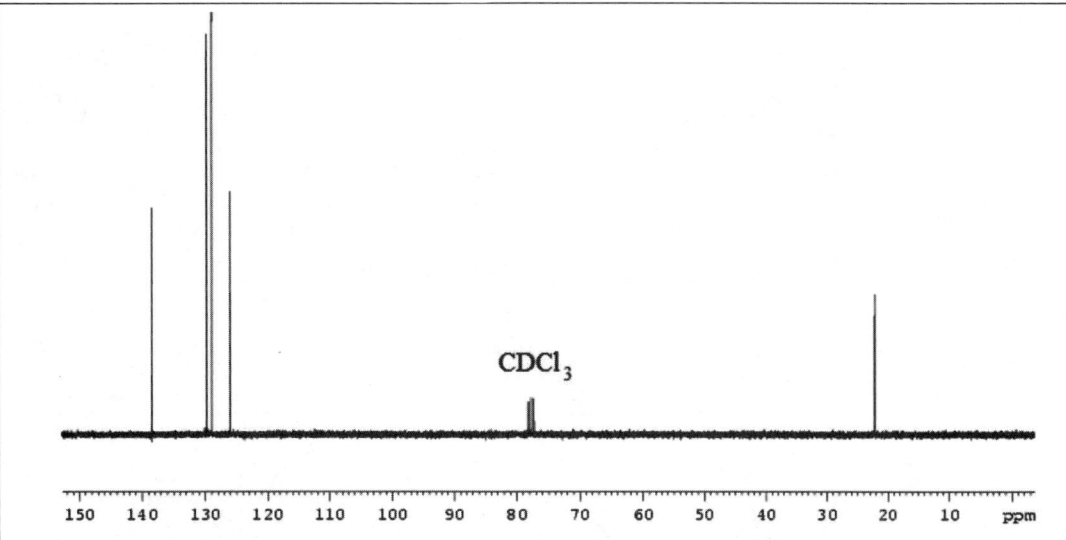

We see the four benzene carbons between 140 and 120 ppm and the methyl carbon at 22 ppm.

Octane has four types of H's and C's. But look at the spectra below. The carbon spectrum shows the expected four peaks, but there are only two sets of peaks in the proton spectrum. All the CH_2's are very similar, so they show up as only one peak. So, NMR isn't perfect, but it is helpful.

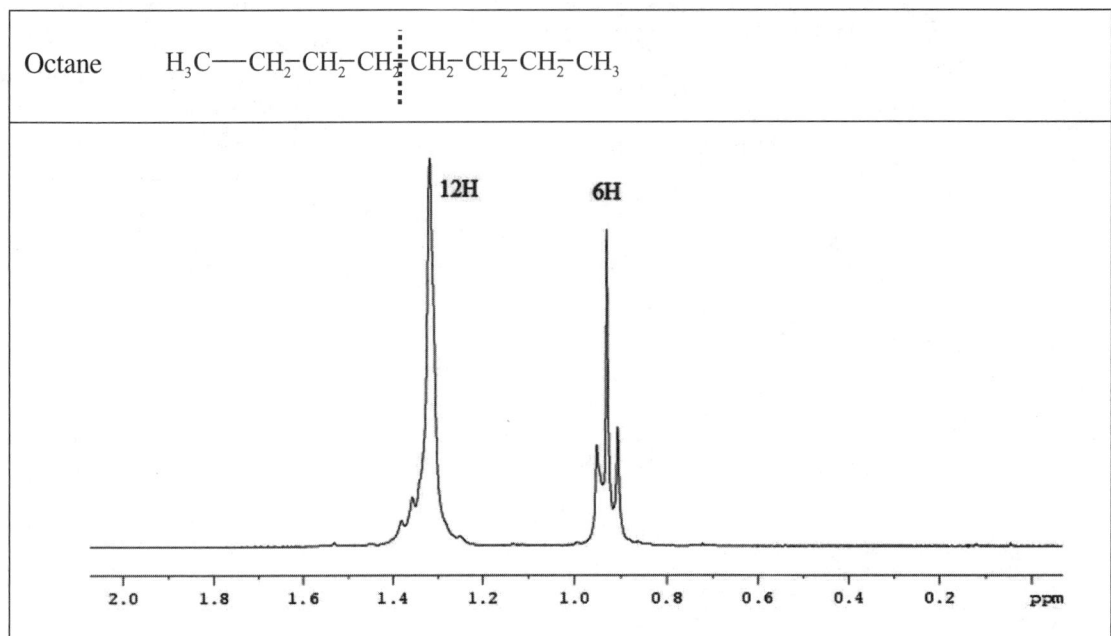

The set of peaks at 0.9 ppm is for the two methyls. The six CH_2's are all very similar and show up as the peak at 1.3 ppm.

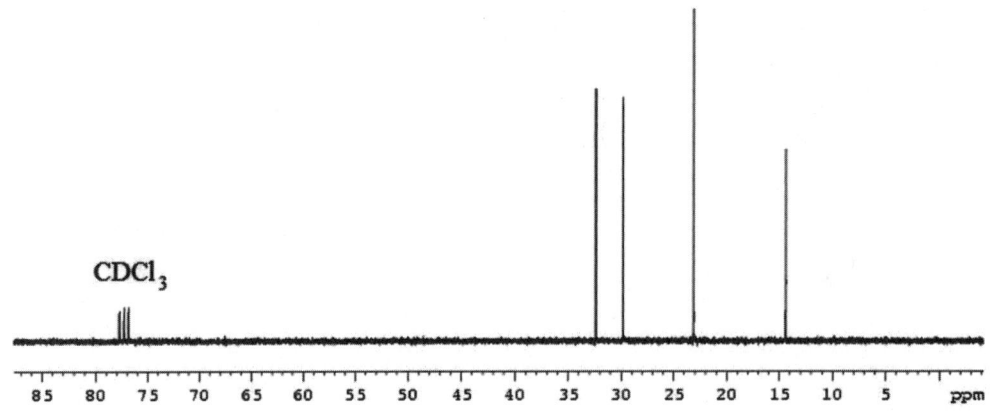

The four types of carbons show up as separate peaks.

B.2. THE POSITION OF THE PEAK (ITS PPM VALUE, OR "CHEMICAL SHIFT") TELLS US SOMETHING ABOUT WHAT TYPE OF ATOM IT IS AND WHAT IT'S BONDED TO

PROTON NMR

Position of Peak	Possible Interpretation
0–2 ppm	H's on sp^3 C's bonded to sp^3 C's (alkane type H's)
1.5–3.0 ppm	H's on sp^3 C's bonded to sp^2 C's
2.5–5 ppm	H's on sp^3 C's bonded to electronegative atoms
4.5–6.5 ppm	H's on simple alkenes
6.5–9 ppm	H's on benzene rings
9–10.5 ppm	H's on aldehydes
10–12 ppm	Carboxylic acid H's (if no water present)
Other OH's and NH's are variable.	

CARBON NMR

Position of Peak	Possible Interpretation
0–50 ppm	sp^3 C's bonded to sp^3 C's
20–80 ppm	sp^3 C's bonded to sp^2 C's and sp^3 C's bonded to electronegative atoms.
65–90 ppm	sp C's (alkynes)
100–160 ppm	Alkene and benzene sp^2 C's
160–210 ppm	C=O C's

The numbers above are general guidelines, and there are always exceptions. For example, $CHCl_3$ has a peak at 7.24 ppm in the proton spectrum. Why? Multiple halogens on the same atom influence the position more than just one halogen.

B.3. THE AREA UNDER A PEAK CAN TELL US THE RELATIVE NUMBER OF EACH TYPE OF H. CARBON IS USUALLY NOT VERY ACCURATE. JARGON TERM: "INTEGRATION"

There is only one type of hydrogen in 2-propanone, so we really don't know how many hydrogens there are. The "6H" at the top of the peak was typed in after the spectrum was run. In the carbon spectrum, the peak for the two CH_3's is bigger as the C=O peak. The relative sizes of the peaks depends upon the exact parameters used to obtain a carbon spectrum. It is possible to obtain accurate carbon spectra, but it takes much longer to run them.

In the proton spectrum of 1,4-dimethylbenzene, the peak on the right is actually 1.5 times as big as the one on the left. So all we really know is the ratio of the hydrogens is 1:1.5. Because we can't have 1.5 H's, we know the real ratio must be 2:3, 4:6, 6:9, . . ., but we don't know which it is without more information. In the carbon spectrum, the big peak is the four benzene carbons that are each bonded to a hydrogen.

B.4. PROTON NMR ALSO CAN TELL US HOW MANY H'S ARE ON THE ADJACENT C'S TO A TYPE OF H

Consider diethyl ether. How many types of hydrogens are there? Look at the spectrum below.

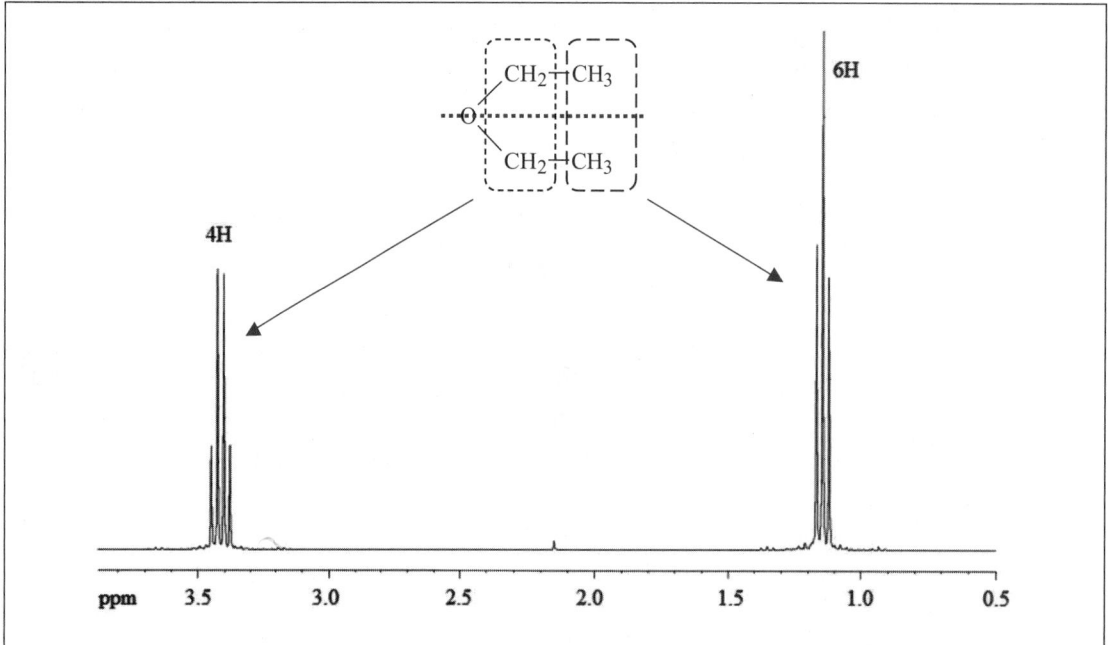

The signals for each type of H are being split into several peaks. Why? The hydrogens on the adjacent carbon are influencing the amount of magnetic field that the observed hydrogen experiences. Because the adjacent hydrogens can affect the magnetic field strength in more than one way, more than one peak is observed. There is a simple rule to predict how many peaks you should observe for a given type of hydrogen. The number of peaks is equal to 1 + the number of adjacent H's. This is sometimes referred to as the **N + 1 rule**. The CH$_2$ is adjacent to three H's, so the CH$_2$ is split into four peaks. The CH$_3$ is adjacent to two H's, so the CH$_3$ is split into three peaks.

The relative size of each of the split peaks is approximated by Pascals's triangle, which is shown below. Each row of the triangle shows the approximate heights of the peaks in the pattern.

No. of Peaks	Relative Intensities of the Peaks	Jargon Term of This Number of Peaks
1	1	Singlet
2	1 1	Doublet
3	1 2 1	Triplet
4	1 3 3 1	Quartet
5	1 4 6 4 1	Quintet or pentet
6	1 5 10 10 5 1	Sextet
7	1 6 15 20 15 6 1	Heptet or septet

For example, in a three-peak pattern, the middle peak is twice as high as the outside peaks. Sometimes, very small peaks on the outside of a splitting pattern are hard to see. A multiple-line pattern, which is hard to classify as one of the above, is just called a **multiplet**.

So why don't we see coupling in carbon NMR? The magnetically active isotope of C is ^{13}C, which is only about 1% natural abundance. The chance of two ^{13}C's being adjacent to each other is small, so we don't see them split each other.

Why don't hydrogens split carbons? Actually, they do! We can acquire carbon spectra that show splitting by H. However, doing so makes all the peaks smaller and often makes the spectra very complex looking. Therefore, carbon spectra are acquired in such a way that we don't see splitting of the peaks by the hydrogens.

Actually, the reason that the deuterated solvents show more than one peak in the C NMR spectrum is due to the splitting of the C by the deuterium atoms attached. The $N + 1$ rule is really the **2nI + 1 rule**, where n is the number of adjacent nuclei and I is the nuclei's nuclear spin. For H, $I = \frac{1}{2}$, so $2nI = n$. For deuterium, $I = 1$. Therefore, one deuterium on a C splits the C into three peaks, two deuteriums splits the C into five peaks, and three deuteriums splits the C into seven peaks.

This splitting is given the jargon term "coupling," and the amount of splitting is called the coupling constant (abbreviated as J). For noncyclic alkanes, the size of the coupling constant is about 7 Hz, for two hydrogens on adjacent carbons. Depending what the hybridization state of the atom is, the coupling constant can vary.

Note that you can't have just one coupled hydrogen; something else must be split. Other atoms, such as ^{19}F and ^{31}P, can split hydrogens on adjacent (and further away) atoms. If we looked at the ^{19}F- or ^{31}P-NMR spectra, we would see the F or P split by the hydrogen.

Look at the spectrum for 1,4-dimethylbenzene, shown above: Notice that there is no splitting here, even though there are adjacent hydrogens on the benzene ring. This is because splitting by equivalent H's is not observed.

In the proton spectrum of octane, shown previously, we see three peaks for the methyl groups' signal because the methyl is adjacent to a CH_2. The CH_2's are essentially a single peak because all the CH_2's are very similar, and the coupling to the methyl is obscured by the broadness of the peak.

The proton spectrum of toluene was also shown previously. The single peak is obviously the CH_3. The two sets of split peaks are the benzene ring H's. The meta hydrogens (Hm) are the set of three peaks at 7.3 ppm. The ortho and para hydrogens show up in about the same place, so the two patterns overlap.

Let's look at some NMR spectra for representative molecules. We will look at both the proton and carbon spectra because there are similarities between the two.

The spectra of butanoic acid are shown on the next page. How do you interpret the proton spectrum? the C spectrum?

Butanoic acid

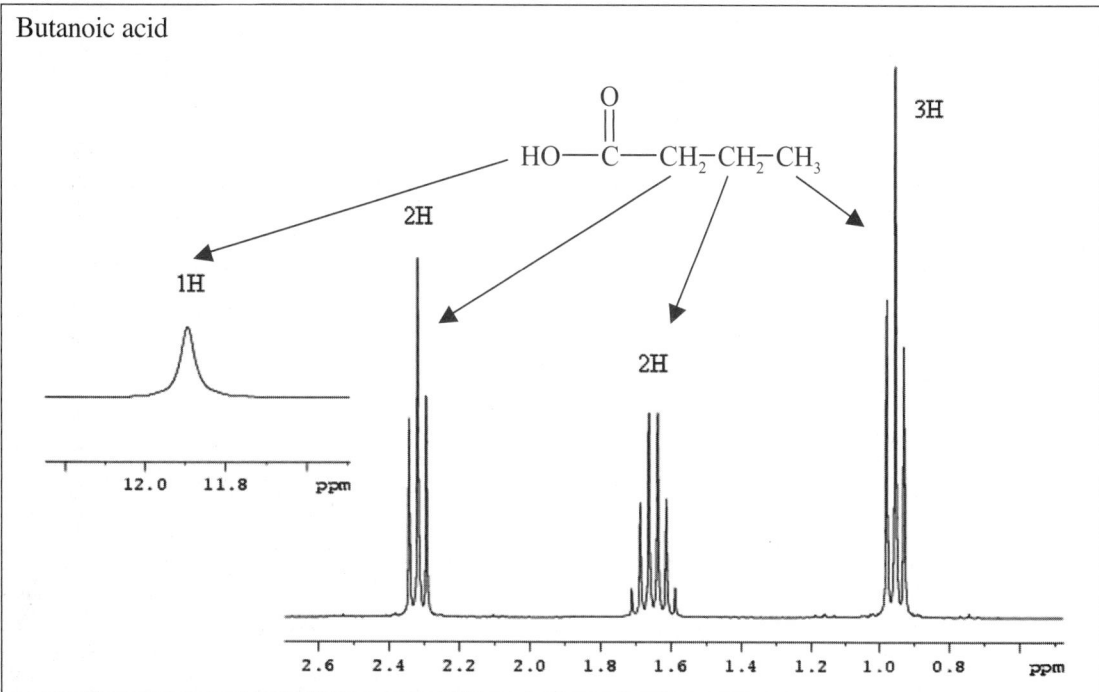

The OH is at 11.9 ppm, which is typical for a carboxylic acid. The CH_2 next to the C=O is at 2.3 ppm. It is three peaks because it is adjacent to two hydrogens. The other CH_2 is the set of six peaks at 1.65 ppm. It is six peaks because it is adjacent to a total of five nearly equivalent hydrogens. The methyl is the three-peak pattern at 0.95 ppm. It is three peaks because it is adjacent to a CH_2. In general, the closer a CH is to an electronegative atom, the further to the left in the spectrum it shows up.

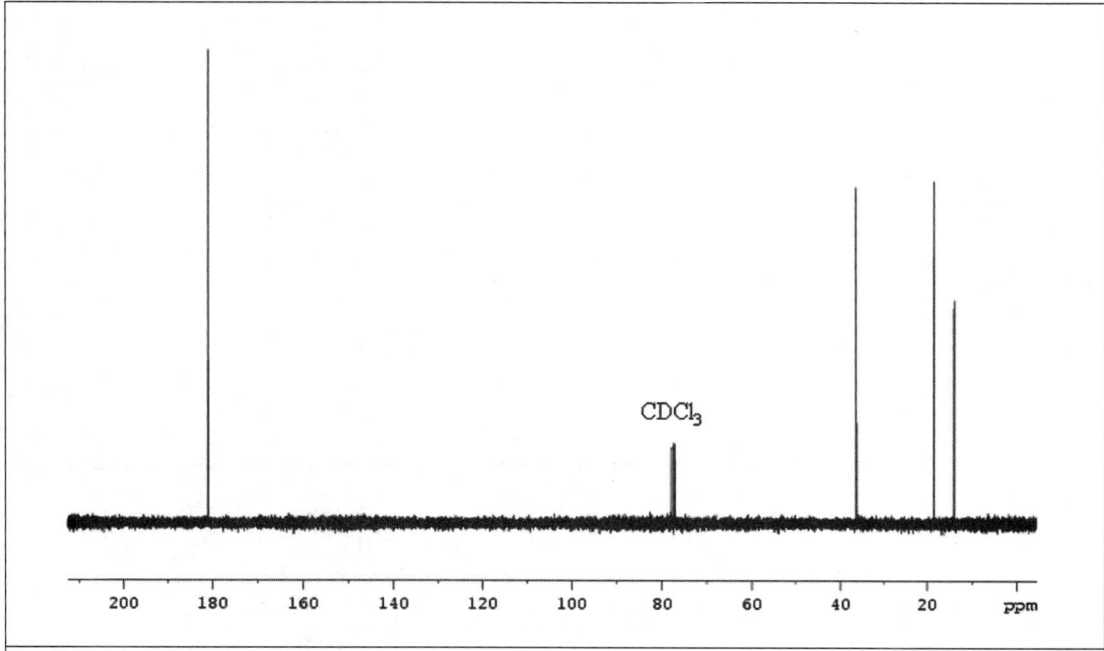

The C=O carbon is at 180 ppm. The three sp^3 carbons are between 10 and 40 ppm.

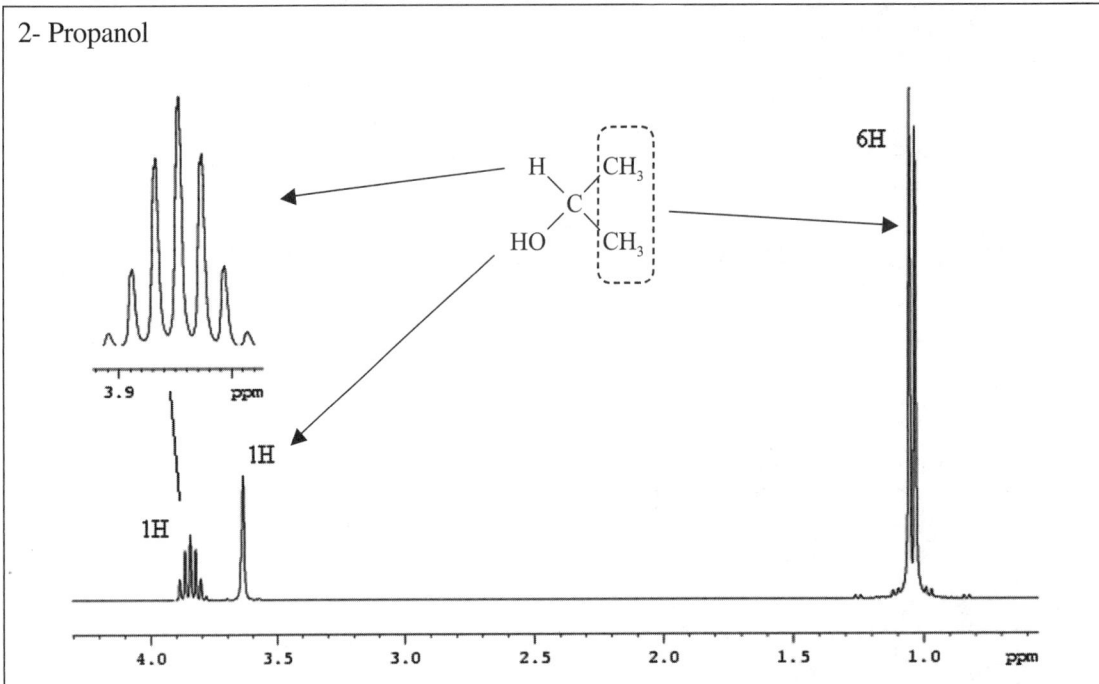

2- Propanol

The two methyls are equivalent. They are adjacent to one hydrogen, so they show up at two peaks. The C–H is adjacent to the six methyl hydrogens, so it shows up as seven peaks: The expansion was necessary to see the outer two peaks clearly. The O–H shows up as a broad peak. OH's and NH's rarely split C–H's.

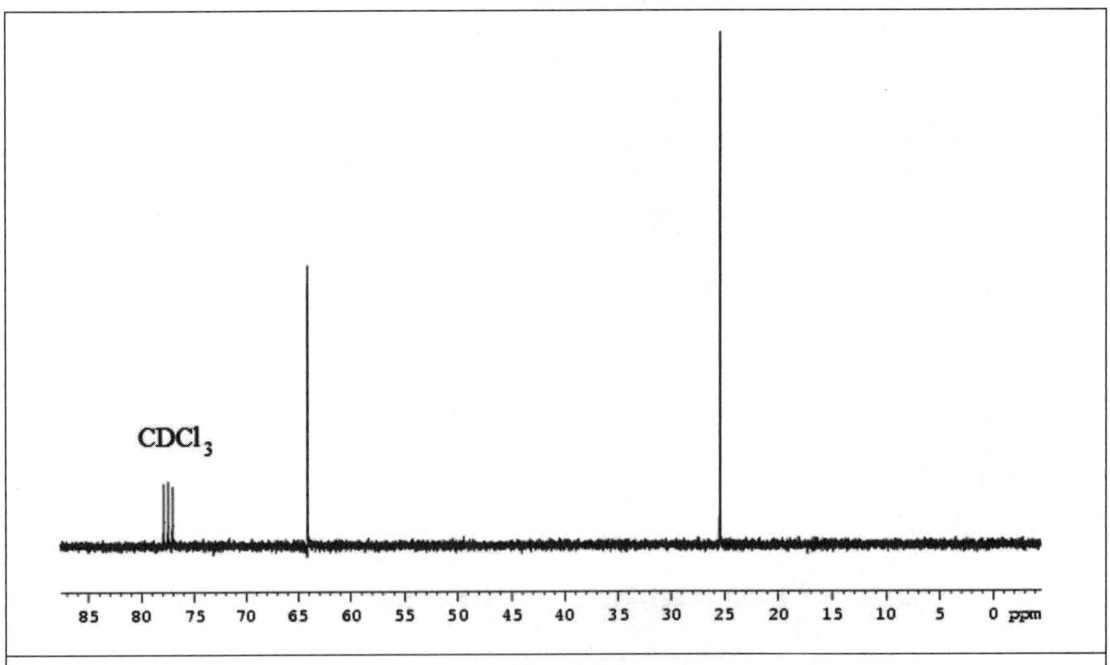

The methyl carbons are the tall peak at 25 ppm. The carbon bonded to the oxygen is at 64 ppm.

Hexanenitrile

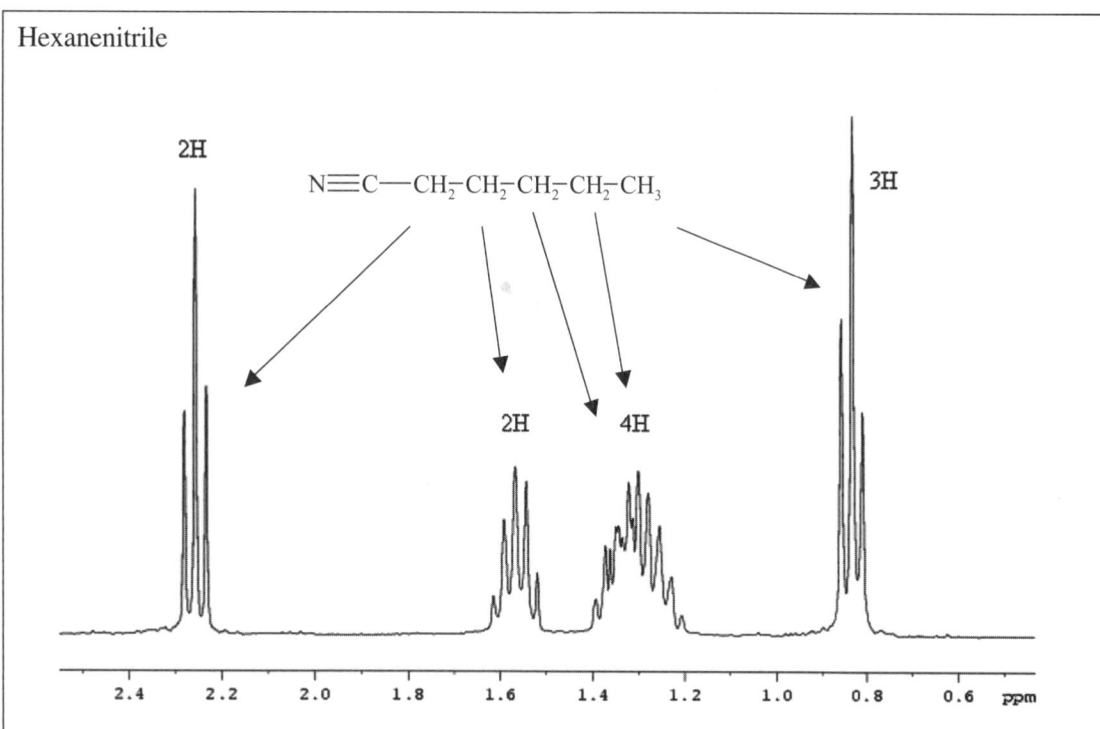

The CH_2 closest to the nitrile shows up at 2.25 ppm and is a triplet because it is adjacent to two hydrogens. The next CH_2 is adjacent to four hydrogens, so it is a pentet. The next two CH_2's are not totally resolved at 300 MHz, so they show up as a four-hydrogen multiplet about 1.3 ppm. The methyl is the triplet at 0.8 ppm.

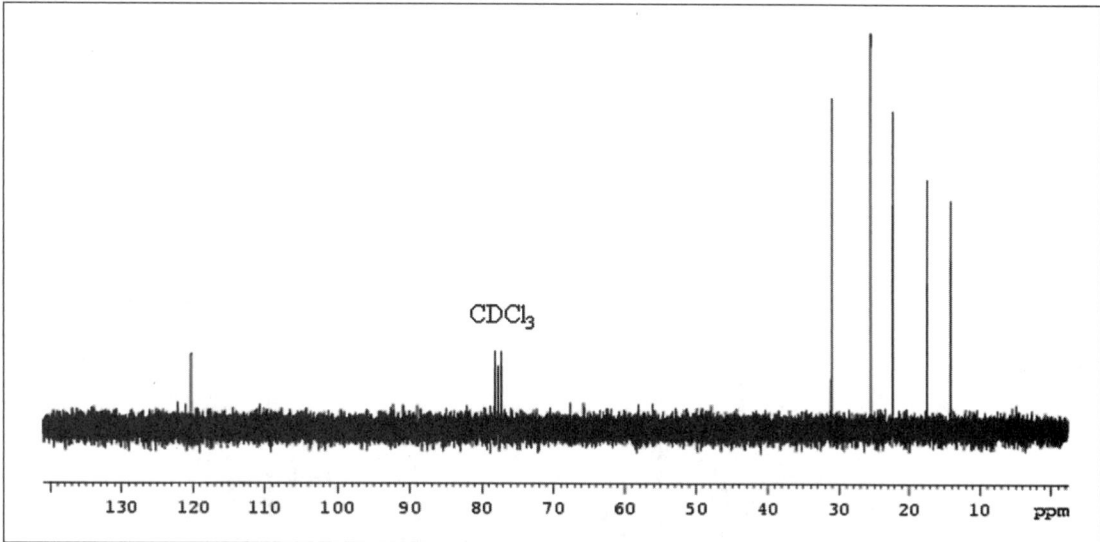

The $C\equiv N$ carbon shows up at 120 ppm. Unfortunately, this is where C=C carbons show up as well. However, the IR spectrum should clearly indicate that we have a $C\equiv N$. How? The five sp^3 carbons are the peaks at 10–35 ppm.

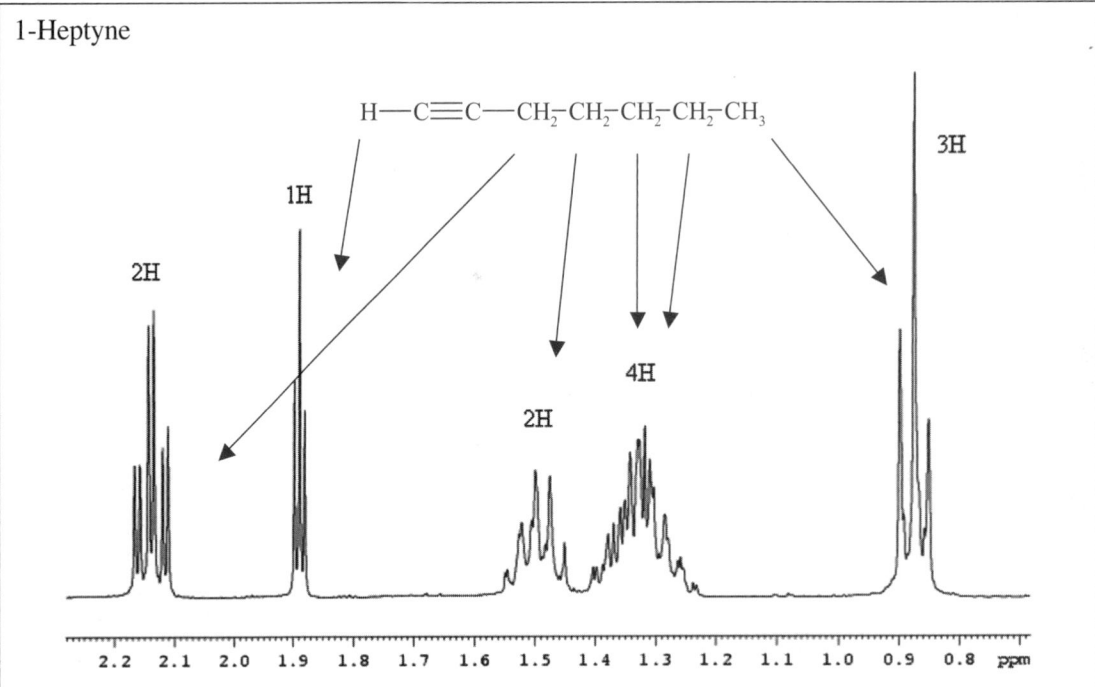

1-Heptyne

This is the most complex proton spectrum we've looked at. The triplet at 0.9 ppm is the methyl. The one H peak at 1.9 ppm is the alkyne H. It is a very closely split triplet (coupling constant = 2.65 Hz). Because there are no adjacent H's, it must be being split by the CH_2 two atoms away. This is referred to as long-range coupling, and coupling constants are smaller. The CH_2 at 2.15 ppm is being split by the alkyne H and the adjacent CH_2. Because the coupling constants are different, we don't see the traditional $N + 1$ pattern. Instead, the adjacent CH_2 splits it into a triplet ($J = 7$ Hz); then, each of those peaks is split by the alkyne H with $J = 2.65$ Hz. The next CH_2 is the distorted five-line pattern at 1.5 ppm. The other two CH_2's are the peaks at about 1.3 ppm.

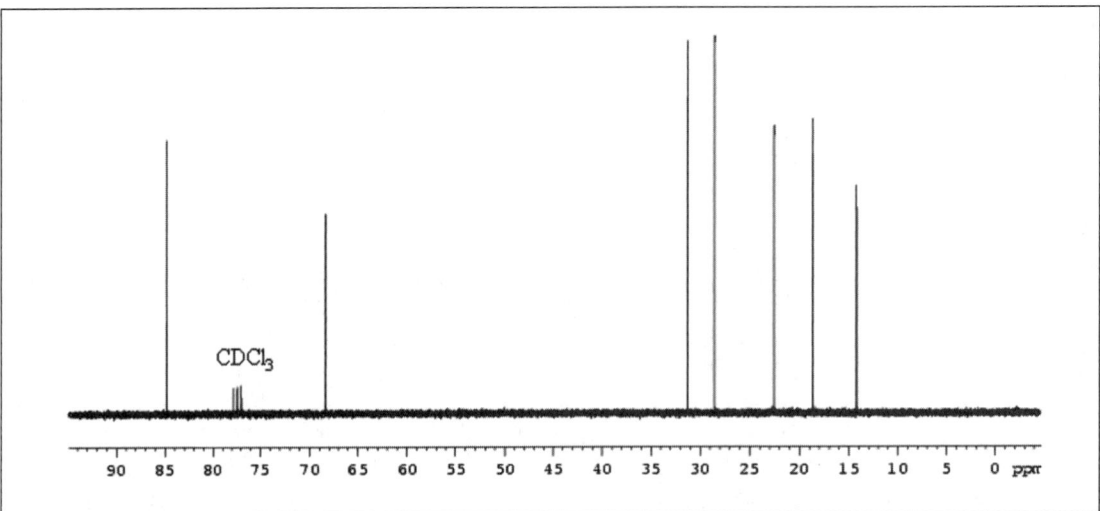

The carbon NMR spectrum shows seven peaks. The two alkyne C's are at 85 and 68 ppm. The five sp^3 carbons are the peaks between 10 and 35 ppm.

Cyclohexene

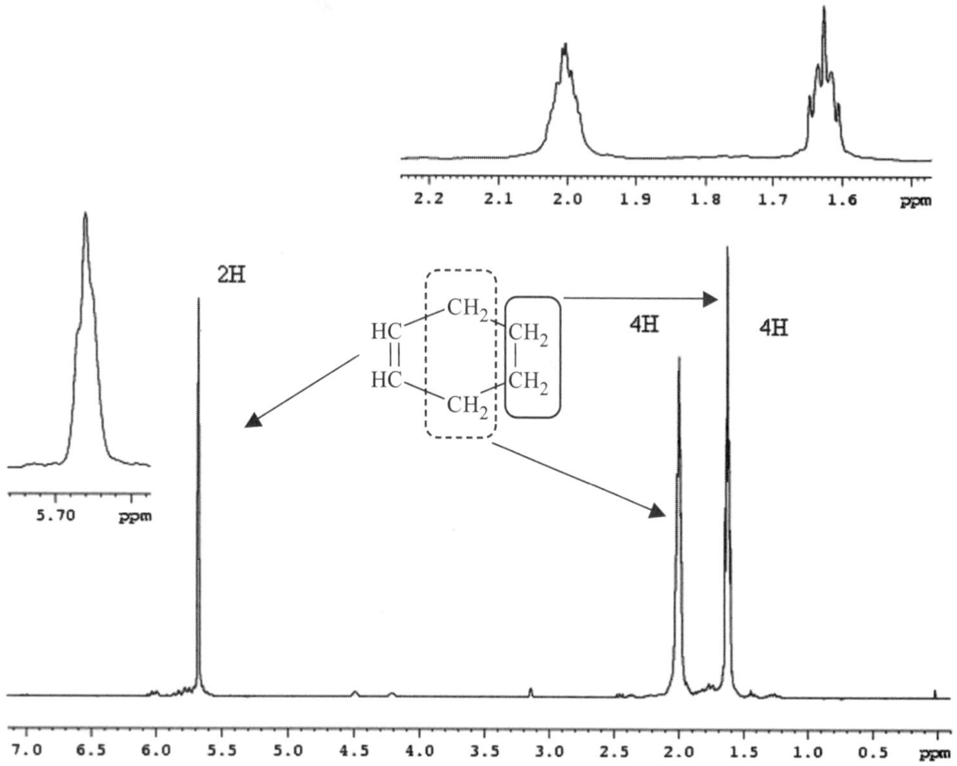

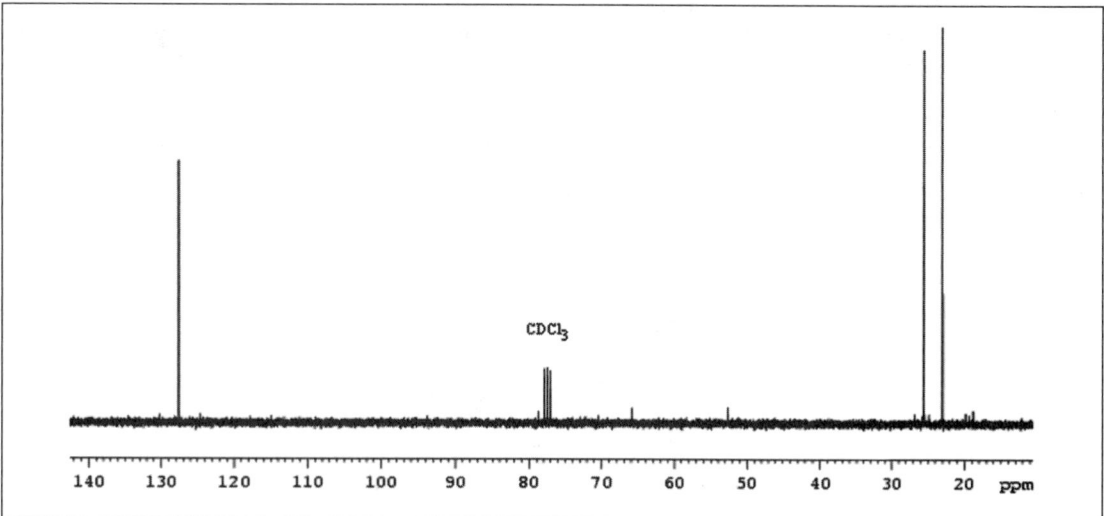

There are three types of hydrogens, due to symmetry. When you have a ring, sometimes the splitting patterns don't show up well. This is due to the carbon–carbon single bonds not being free to rotate, and the splitting can vary depending on the angles that the hydrogens are relative to one another. All of these patterns would be described as multiplets.

There are three types of C's, due to symmetry. The alkene carbons are at 127 ppm, and the sp^3 carbons are the two peaks between 20 and 30 ppm.

C. Mass Spectrometry

Again, I will refer you to your textbook for a complete discussion of the theory. In a common form of mass spectrometry (MS) called **electron impact**, the sample is inserted into a beam of electrons. The electron beam knocks off an electron from the molecule, usually from an unshared pair of electrons, or from a pi bond, if one is present. The molecule that has lost an electron (which is now a radical cation) usually will fall apart ("fragment" is the jargon term) into a cation and a radical. Generally, relatively stable cations are formed. This is illustrated here with 2-butanone.

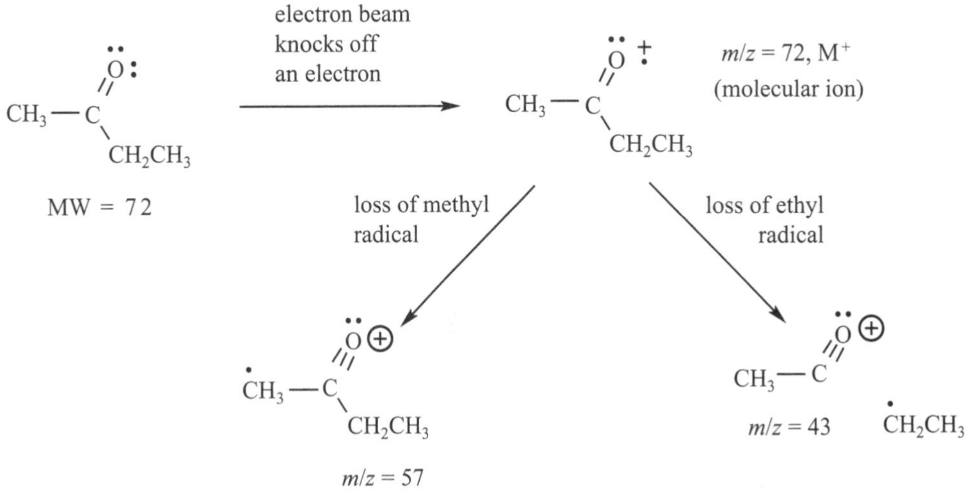

The term m/z means mass to charge ratio. Most ions have only one charge, so m/z equals the molecular weight of the ion. The ion produced after the first electron is removed is called the "molecular ion," and is sometimes abbreviated as M^+. The molecular ion fragments to form relatively stable cations, which are detected. The more stable cations are usually the most abundant fragments in the mass spectrum. The largest ion is called the **base peak**. The cation fragments can be detected. The mass spectrum of 2-butanone is shown below. The vertical scale is intensity: The more common fragments are the most intense. The horizontal scale is the mass/charge ratio. In general, you should only be concerned with the largest peaks in a mass spectrum. The mass spectra shown in this book are drawings, rather than actual spectra.

I usually only have students look for three things in a mass spectrum. You can get a lot more information from mass spectrometry, but for beginning students, these three things are enough.

1. The molecular weight. This comes from the molecular ion of the spectrum. Sometimes the molecular ion is not seen. Usually, this occurs if it can fall apart into a very stable cation. We will see some examples of this as we go along.

2-Butanone, MW=72

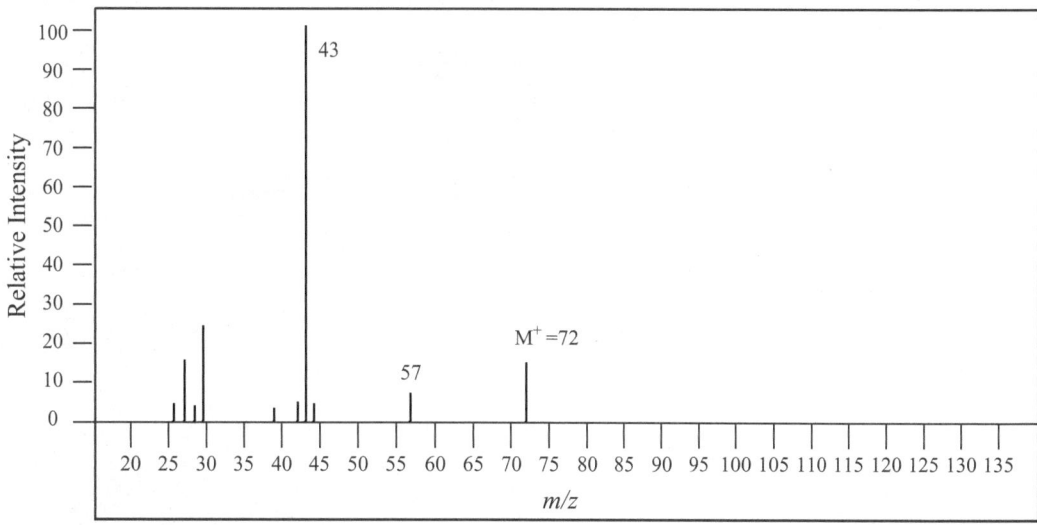

Other important data we can obtain from mass spectrometry:

2. A compound with an odd number of nitrogens will have an odd molecular weight, usually. This is always true if the compound contains only C, H, N, and O. If other atoms are present, then the presence of a nitrogen must be deduced by other methods.

3. Compounds containing a chlorine or a bromine show two peaks two mass units apart for fragments that contain a chlorine or bromine. This is because chlorine is composed of two major isotopes, ^{35}Cl and ^{37}Cl. About 75% of Cl is ^{35}Cl, and 25% is ^{37}Cl. Therefore, the size of the two peaks will have a ratio of about 3:1. The two isotopes of bromine occur in about equal amounts, so the two peaks will be about the same size.

2-Chloro-2-methylpropane, MW = 92 (and 94)

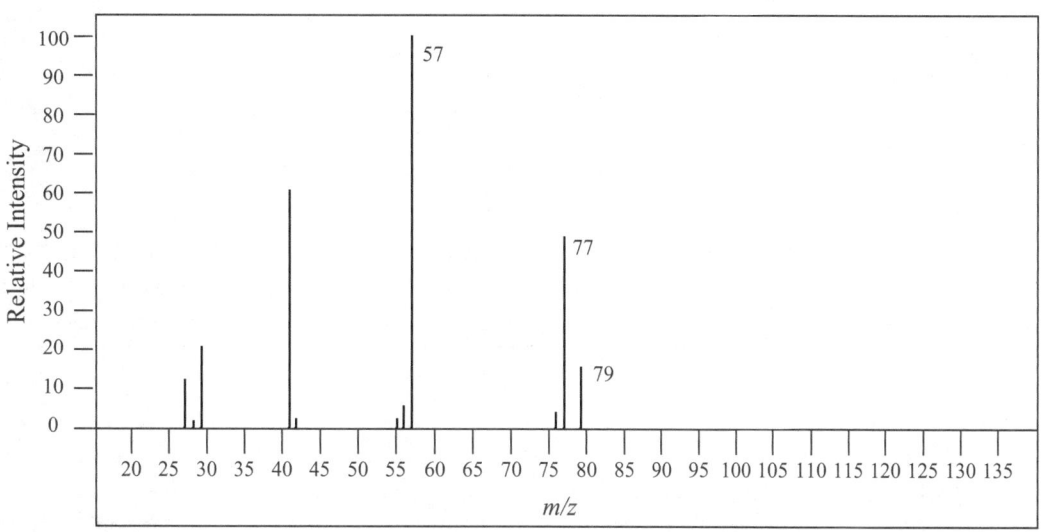

The molecular ion peaks are very tiny. However, the peaks at $m/z = 77$ and 79 are in a $3 : 1$ ratio, which indicates a chlorine. The peak at 57 is a *t*-butyl cation, which is very stable for a cation. See the following fragmentations.

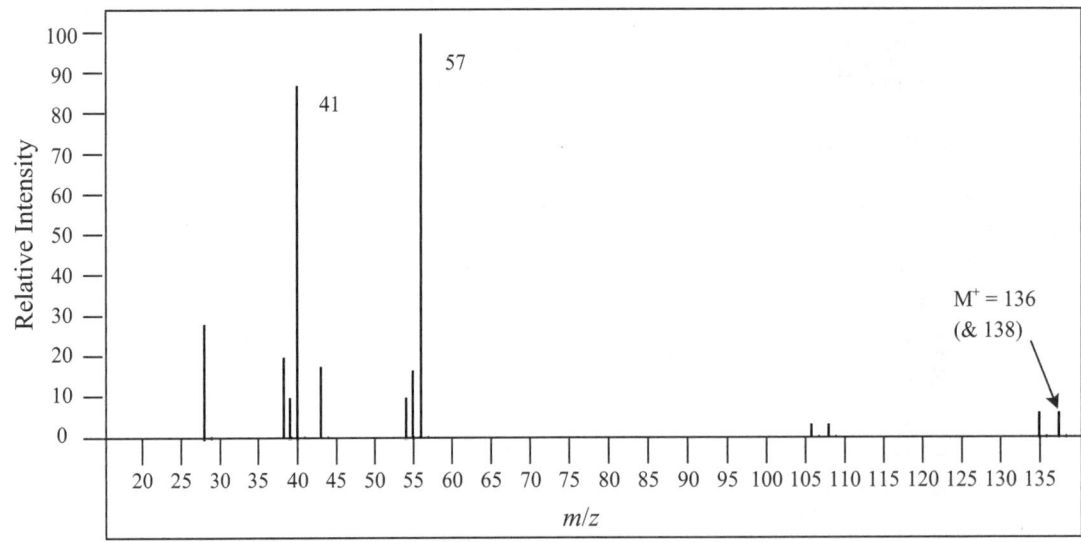

1-Bromobutane, MW = 136 (and 138)

In this case, we see the molecular ions (m/z = 136 and 138), as well as tiny bromine containing peaks at 107 and 109. The base peak is at 57, which corresponds to the loss of the bromine radical.

I have shown a number of mass spectra below. Try to figure out how the large fragment ions are formed.

Here is the spectrum of 2-propanol.

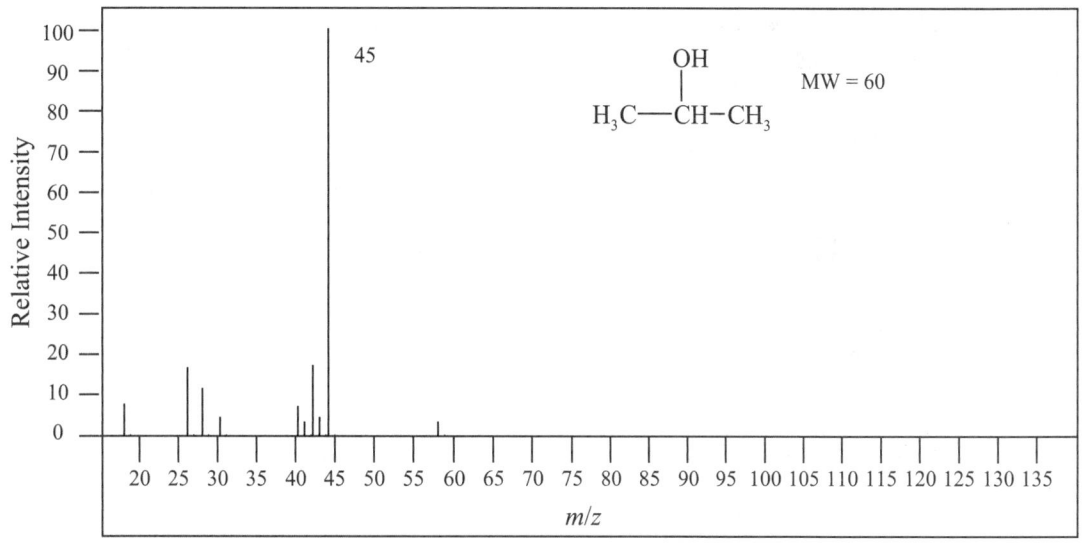

The molecular weight is 60, so that means 15 was lost to form the $m/z = 45$ ion. A loss of 15 mass units is usually a methyl group. The likely fragmentation is shown below. The $m/z = 45$ ion is stabilized by resonance.

Here is the spectrum of 1-butanamine.

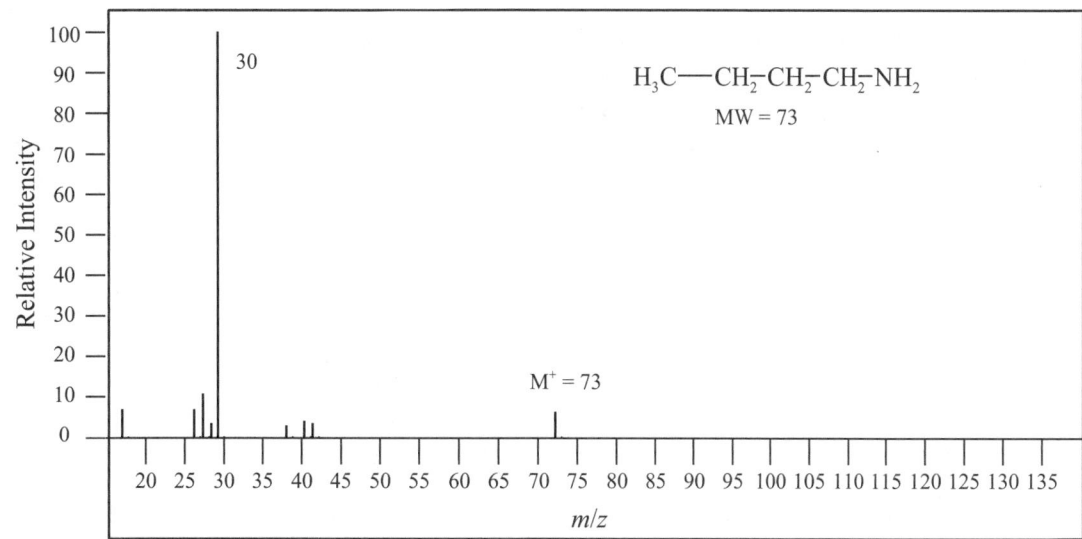

The molecular weight is odd, so we know it contains an odd number of nitrogens. Subtracting 43 from 73 equals 30. The 43 in molecular weight that is lost could be a propyl group. A likely fragmentation is the following. It is very similar to the alcohol fragmentation and also forms a resonance-stabilized cation.

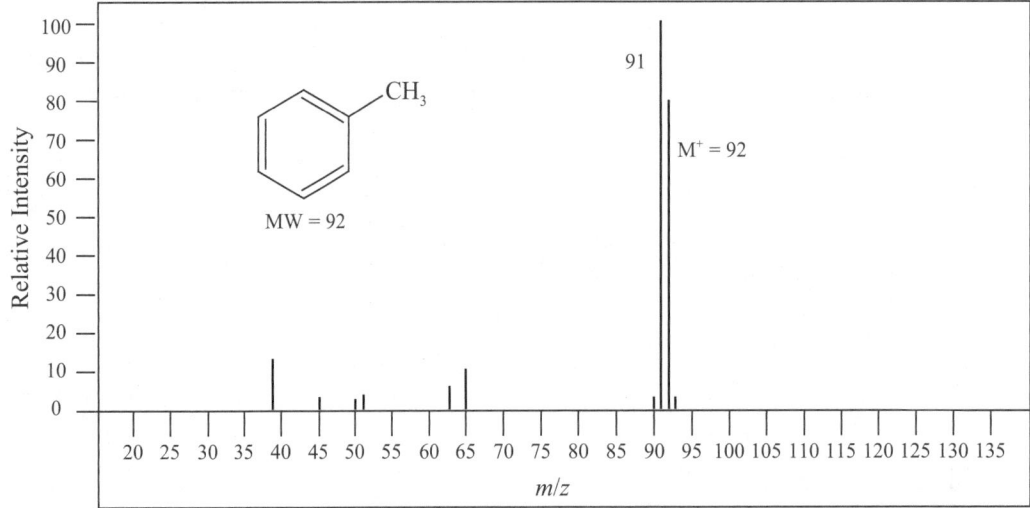

The mass spectrum of toluene is shown below.

Loss of an electron forms a relatively stable resonance-stabilized radical cation, with $m/z = 92$. Loss of a hydrogen atom forms the relatively stable $C_7H_7^+$ cation, which can exist either as a benzyl cation or a tropylium ion, both of which have several resonance structures. Any alkyl benzene will show similar peaks in its mass spectrum.

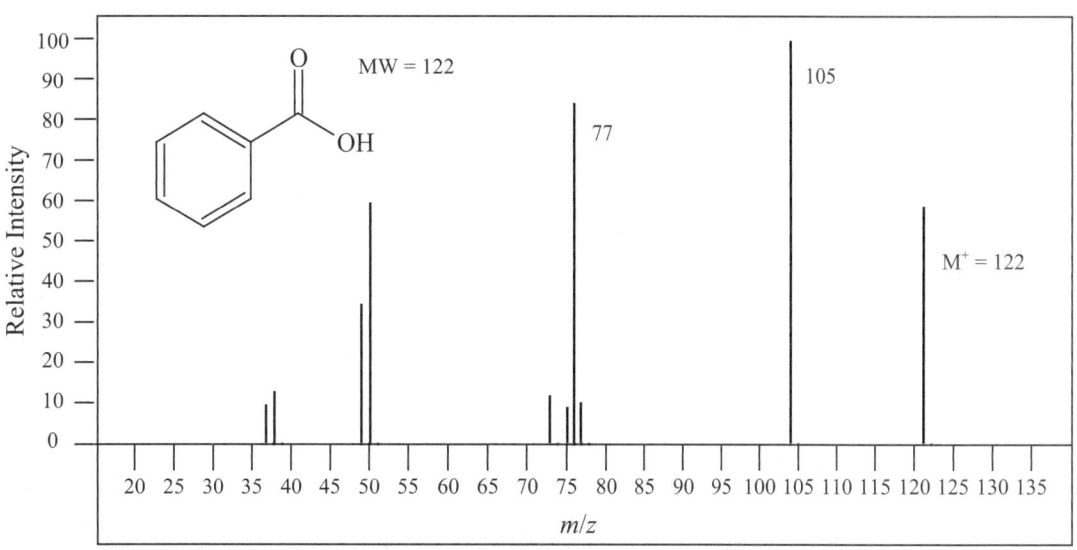

Benzoic acid

Carboxylic acids often show loss of 17, which is an OH, followed by loss of 28, which is CO (carbon monoxide). These fragmentations are shown below.

One common fragmentation of ketones was shown before, with 2-butanone. That is the cleavage of the carbon–carbon bond next to the carbonyl, which is called alpha-cleavage. With longer ketones, a fragmentation with rearrangement can occur, which is called the McLafferty rearrangement. These fragmentations are shown below in the spectrum of 2-heptanone.

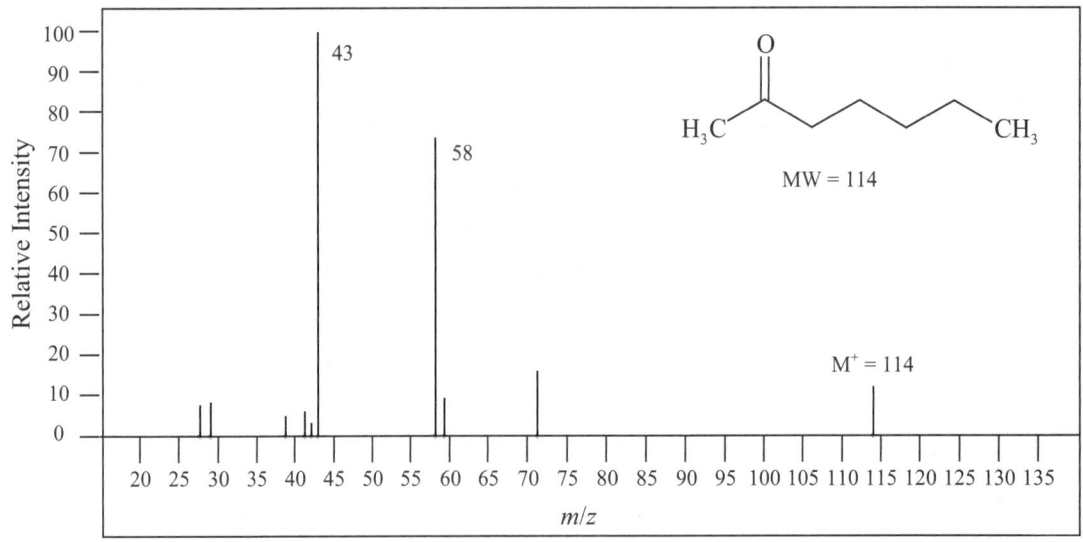

The McLafferty rearrangement can also occur with aldehydes, esters, amides, and nitriles with enough carbons to form the appropriate conformation.

Alkanes do not have pi bonds, so the initial loss of an electron is from a carbon–carbon single bond. In straight-chain alkanes, all the carbon-carbon single bonds are similar in energy, so you see fragments based on cleavage of virtually every single bond. The spectrum of heptane is shown below. In addition to the molecular ion at $m/z = 100$, the alkyl fragments at $m/z = 85, 71, 57, 43,$ and 29 are all present.

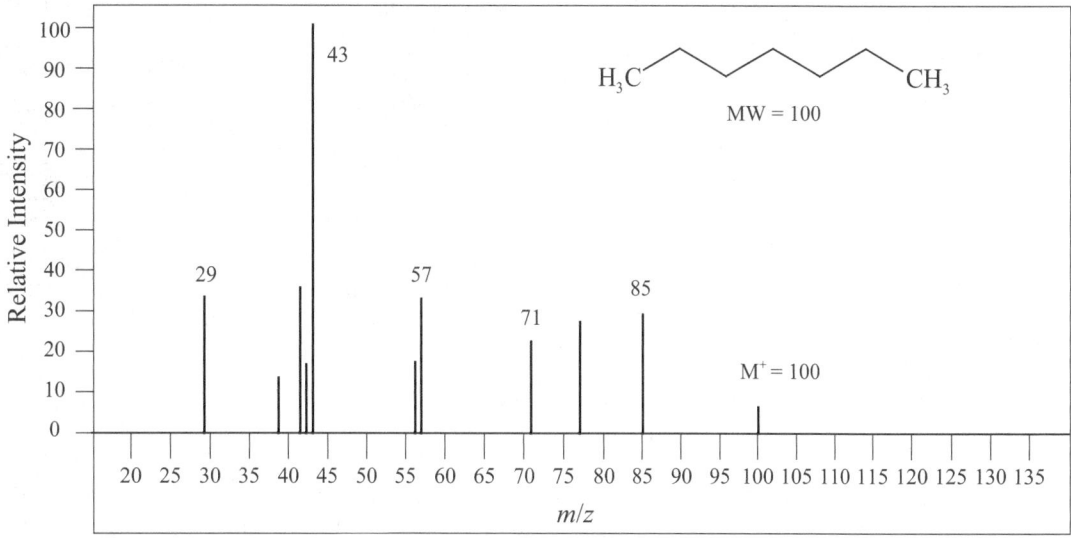

I am not expecting that you are now mass spectroscopy wizards, but you should be able to account for the identities of significant fragments in simple molecules.

D. Putting It All Together to Solve Problems

The goal of learning spectroscopy is to solve problems. A common real-life application of spectroscopy is running a reaction, isolating the product, and using spectroscopy to identify the product. Usually, you have a good idea what the product should be, so you know where to expect peaks in the various types of spectra. Sometimes, you don't see the peaks you are expecting, and you have to figure out what you did make. Since you know what you started with, you can usually figure it out. Another real-life application is determining the structure of an unknown compound. For example, scientists find plants that have some medicinal activity, such as anti-tumor properties. Chemists will systematically extract the plant with solvents and do various types of purifications to isolate the chemical responsible for the anti-tumor properties. Then they run spectra and try to determine the structure. In this case, the chemist often has no idea what to expect. Once a structure is proposed, other chemists may try to synthesize it, to see if that structure's spectra match those of the isolated compound, and more important, to see if that structure really has anti-tumor properties. If the synthesis is relatively short, and the compound really has anti-tumor properties, then it may be a lead for a new drug.

There are lots of approaches for determining the structure of an unknown. This is the one I teach my students, and it seems to work for most of them. If you find one that works for you better, use it.

I have given you a set of spectra for an unknown below.

1. Look at the mass spectrum. Look for evidence of bromine or chlorine from peak sizes. Check the molecular ion. If it is even, assume it probably has no nitrogens, or perhaps an even number of nitrogens.

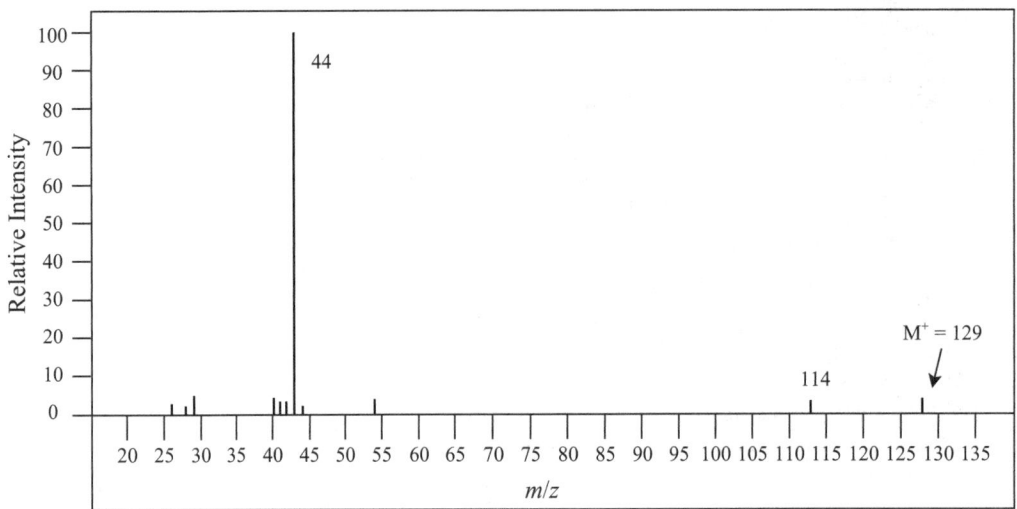

In this case, there are no obvious chlorine- or bromine-containing isotope peaks. The odd molecular ion points to a nitrogen.

2. Look at the IR spectrum for obvious functional group peaks.

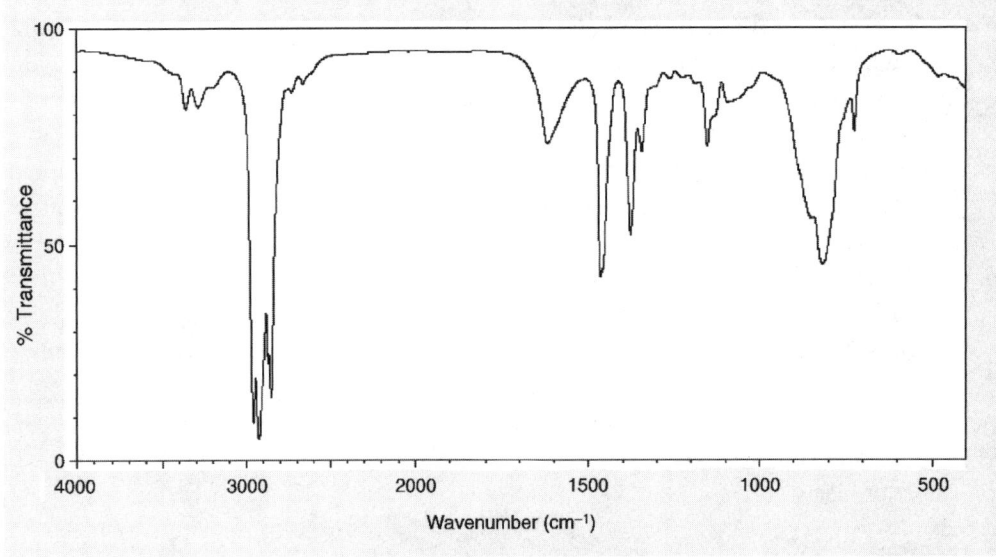

There is an NH_2 stretch at 3300 cm^{-1}, and an NH_2 bend at about 1600 cm^{-1}, along with sp^3 stretches between 3000 and 2800 cm^{-1}. This points to a primary aliphatic amine.

3. Then look at the carbon NMR spectrum to see the numbers of types of carbons.

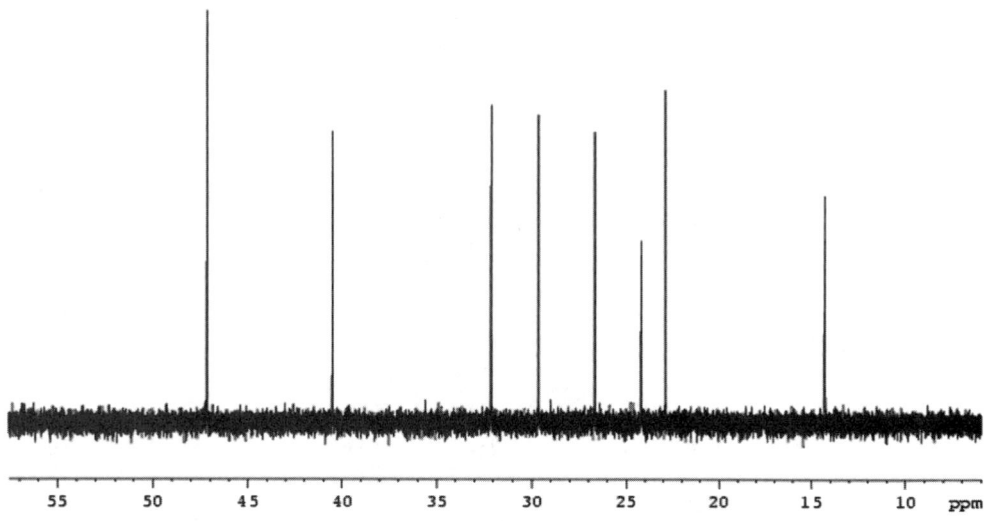

There are eight carbon peaks, so there are probably eight types of carbons. All these peaks are in the *sp*³ range.

4. Look at the proton NMR data, and pick out as much information as you can.

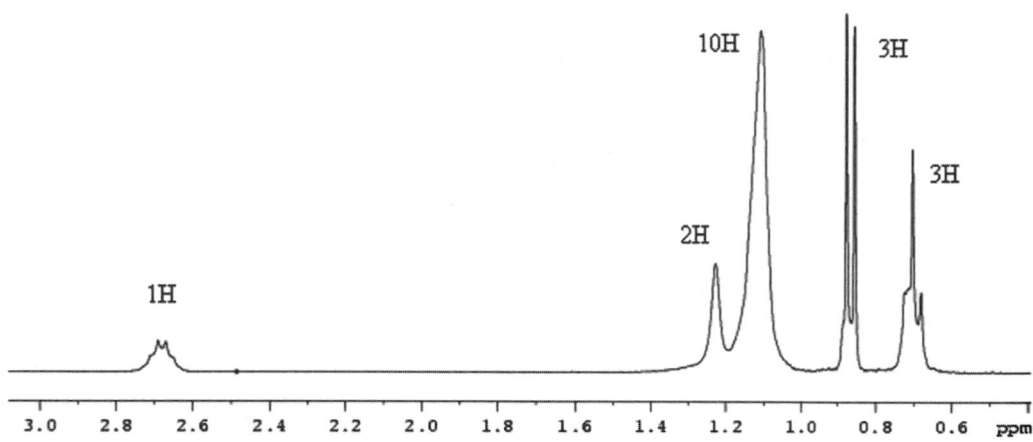

By counting the integration values, we know we have 19 hydrogens. If we add up atomic weights, 19 hydrogens + 8 carbons equals $19 + (8 \times 12) = 115$. The molecular weight was 129. $129 - 115 = 14$, which is probably one nitrogen. So the formula is most likely $C_8H_{19}N$. Now let's look at the proton NMR spectrum in more detail. There is a three-hydrogen peak at 0.7 ppm. It is a distorted triplet, which indicates two adjacent hydrogens. There is another three-hydrogen peak at 0.9 ppm that is a doublet, which means it has one adjacent hydrogen. These are probably two different methyl groups. The one-hydrogen multiplet at 2.7 ppm is probably the CH next to the NH_2. The ten-hydrogen peak at 1.1 ppm is probably five similar CH_2's. The two-hydrogen peak at 1.25 ppm is probably the NH_2. We now have a number of pieces: two methyls plus a $CH–NH_2$ plus five CH_2's. One possible structure is octan-2-amine. The big peak in the mass spectrum at $m/z = 44$ is probably formed as follows:

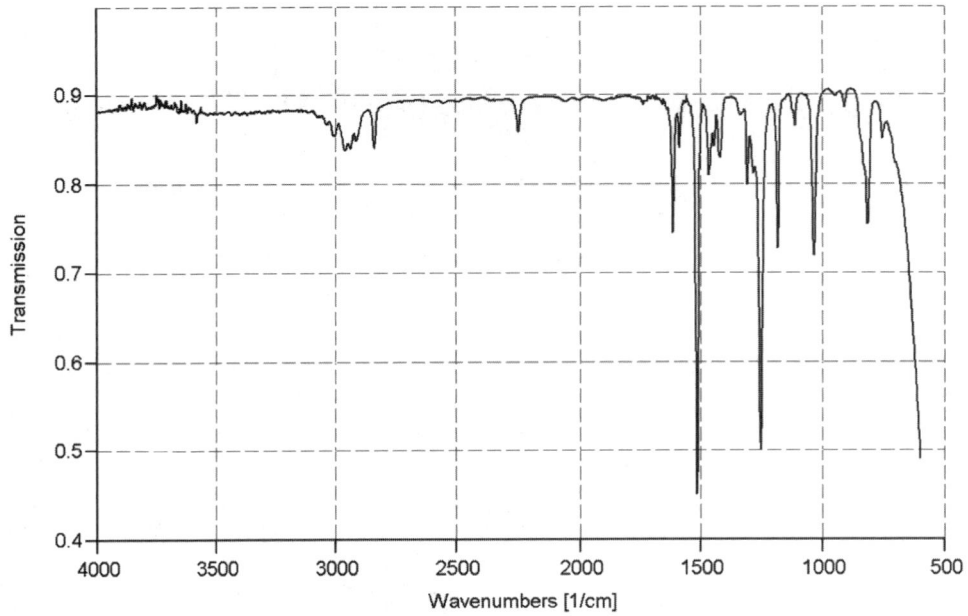

Let's try another one. Sometimes, a problem gives you the molecular formula and not the mass spectrum. A certain compound has the formula C_9H_9NO, and gives the spectra shown below. Let's look at the IR spectrum, shown below.

The small peak at about 2200 cm^{-1} could be a C≡N or a C≡C. The peaks at 1600 and 1500 cm^{-1} could be benzene C=C stretches. Now let's look at the carbon NMR spectrum.

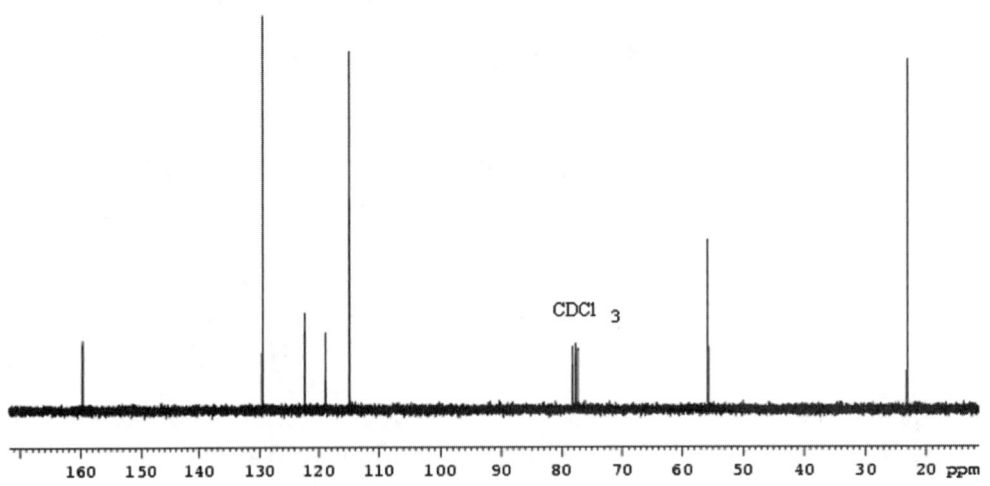

We see seven peaks. There are nine carbons, so there must be some symmetry. There are five peaks between 110 and 160 ppm. Because there are not six peaks, we can't have a 1,2- or 1,3-disubstituted benzene with different groups attached to the benzene ring. Why? If the benzene ring is monosubstituted or 1,4-disubstituted with different groups on the benzene ring, then we would expect four peaks. Because there are five peaks, we either have a C=C or a C≡N peak. There are two peaks for sp^3 carbons, at 55 and 24 ppm. Let's look at the proton NMR spectrum.

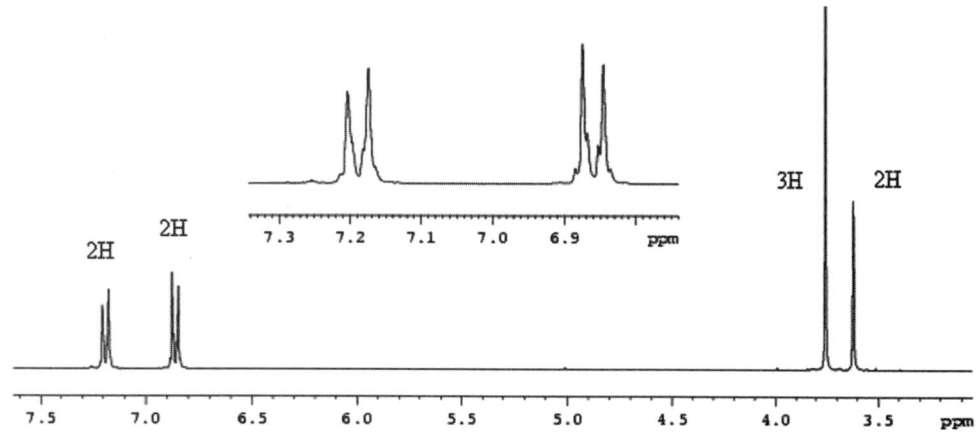

The two sets of peaks at 7.2 and 6.9 ppm are in the benzene region. They are distorted doublets that integrate for two hydrogens each, which is characteristic of a 1,4-disubstituted benzene, as long as the substituents are different. The only other peaks are two singlets. The one at 3.75 ppm integrates for three hydrogens and is probably a methyl group attached to an electronegative atom. The peak at 3.6 ppm is a CH_2. Because they are singlets, there are no hydrogens on the adjacent carbons. Let's look at the pieces we have.

Molecular formula	C_9H_9NO
1,4-disubstituted benzene	$-C_6H_4$
A CH_3 and a CH_2	$-C_2H_5$
Atoms remaining	C, N, O

If we assume there is a C≡N, then there is an oxygen to account for. It is probably an ether because we have no indication of an alcohol or a C=O. So, how can we put these pieces together? See the possibilities on the next page.

H₃C—O—〈benzene〉—CH₂–C≡N	This structure fits well. The CH₂ shows up at 3.6 ppm because it is next to both a benzene ring and the C≡N. The benzene hydrogens are all to the right of the normal benzene position (7.26) as expected because there are electron-donating groups (an O and a CH₂) attached to the benzene ring.
H₃C—O—CH₂—〈benzene〉—C≡N	This structure doesn't fit as well. Because the CH₂ is attached to the benzene ring and an oxygen, it should show up to the left of 3.5 ppm, probably closer to 5. The nitrile, being a strong electron-withdrawing group, would shift the positions of the hydrogens ortho to it to closer to 8 ppm in the proton NMR.

When you are solving spectral problems, try to pick out pieces. Once you have all of the pieces, then put them together in as many ways as make chemical sense and fit the spectra. In a beginning course, there are usually a limited number of ways to put pieces together. Remember, all atoms must fulfill the octet rule! If you propose a structure with a carbon making something other than four bonds, it is almost always going to be wrong!

REVIEW EXERCISES FOR CHAPTER 15

1. Some compounds give proton NMR spectra that consist of a single peak. This means there is a lot of symmetry in the structures of these compounds. Propose reasonable structures for the following formulas, that would produce only single peaks in the proton NMR spectrum. Remember, equivalent hydrogens do not split each other.
   a. C_8H_{18}, 0.9 ppm
   b. C_5H_{10}, 1.5 ppm
   c. $C_{12}H_{18}$, 2.2 ppm
   d. $C_4H_8O_2$, 3.7 ppm
   e. $C_4H_4N_2$, 8.6 ppm
   f. $C_4H_4N_2$, 2.9 ppm

2. The IR and NMR spectra for 2-methylpropyl propanoate (isobutyl propionate) are shown below.

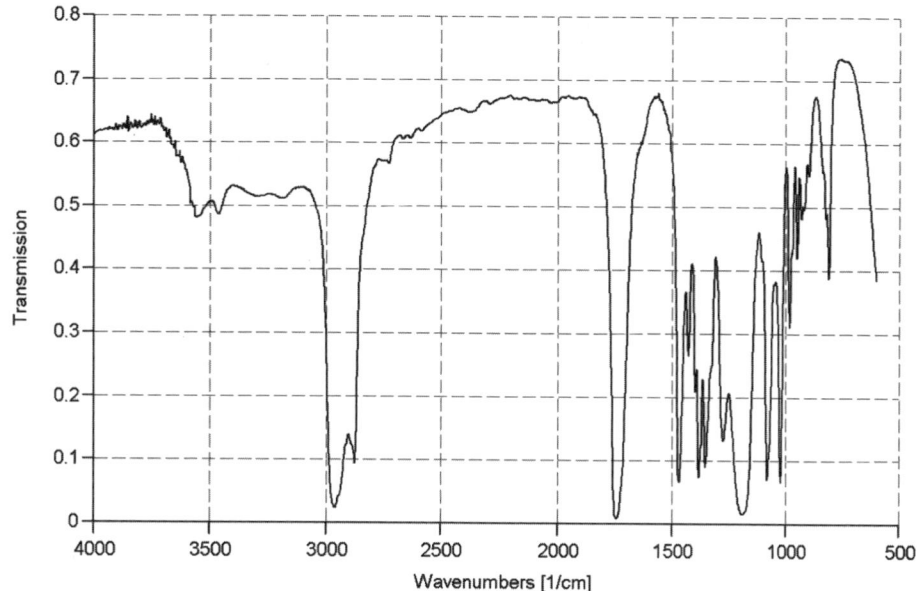

In the IR spectrum, what is producing:
a. The peak at 1750 cm^{-1}?
b. The peak at 1200 cm^{-1}?
c. The peak at 3000–2800 cm^{-1}?

How do you explain the coupling patterns in proton NMR spectrum? How do you explain the peak locations?

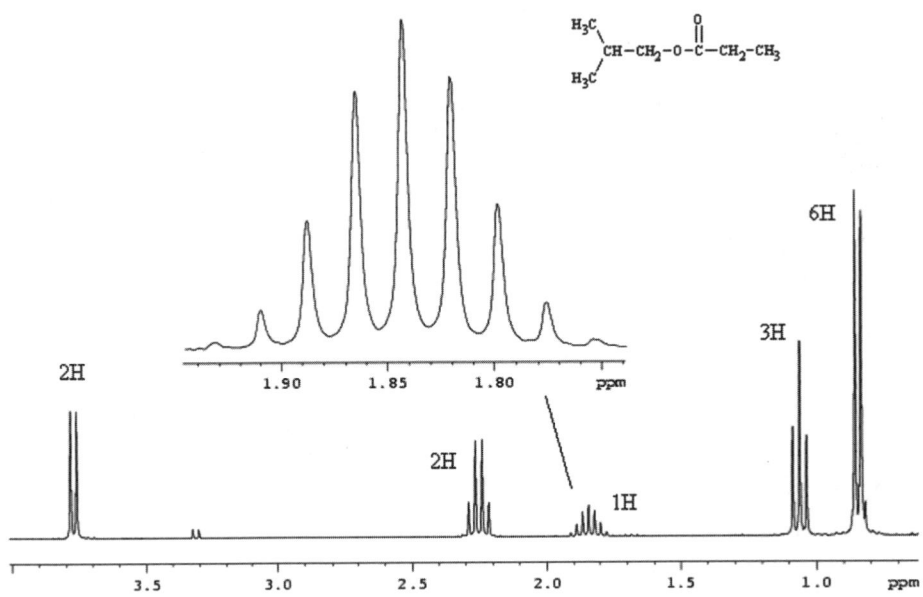

Why do we see only six peaks in the carbon spectrum? Which peak is the C=O carbon? Which peak is the C–O carbon?

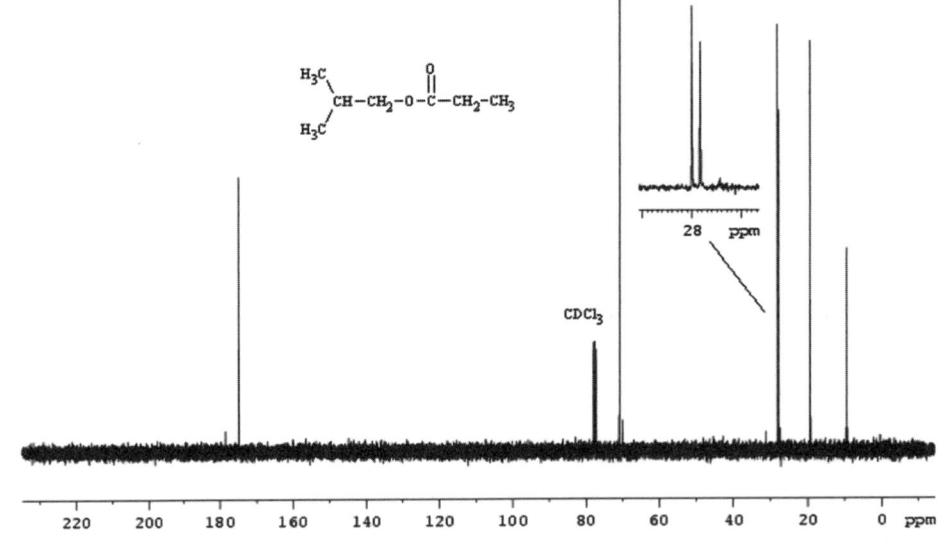

3. Propose a reasonable structure for the compound that produced the following set of spectra.

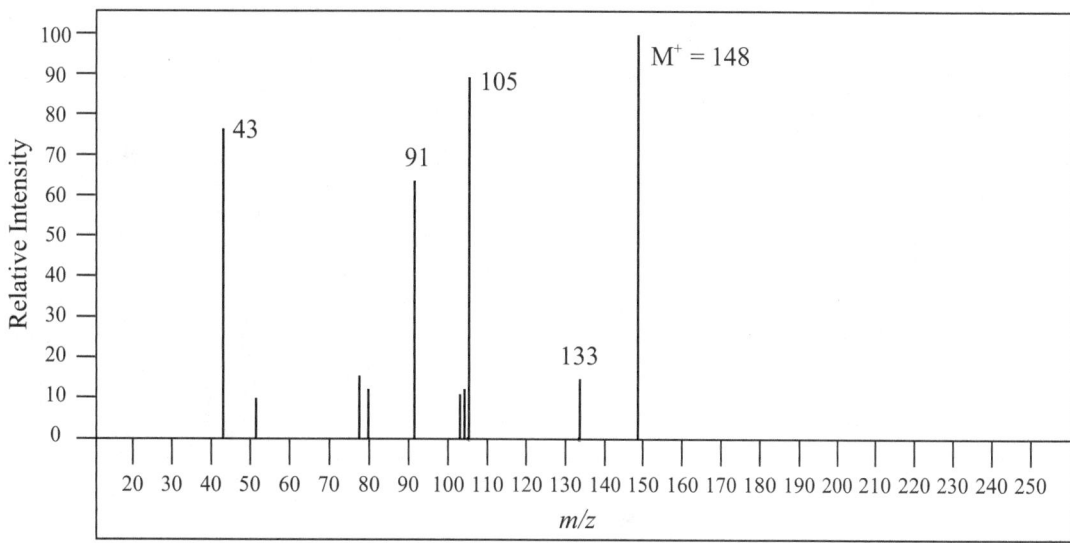

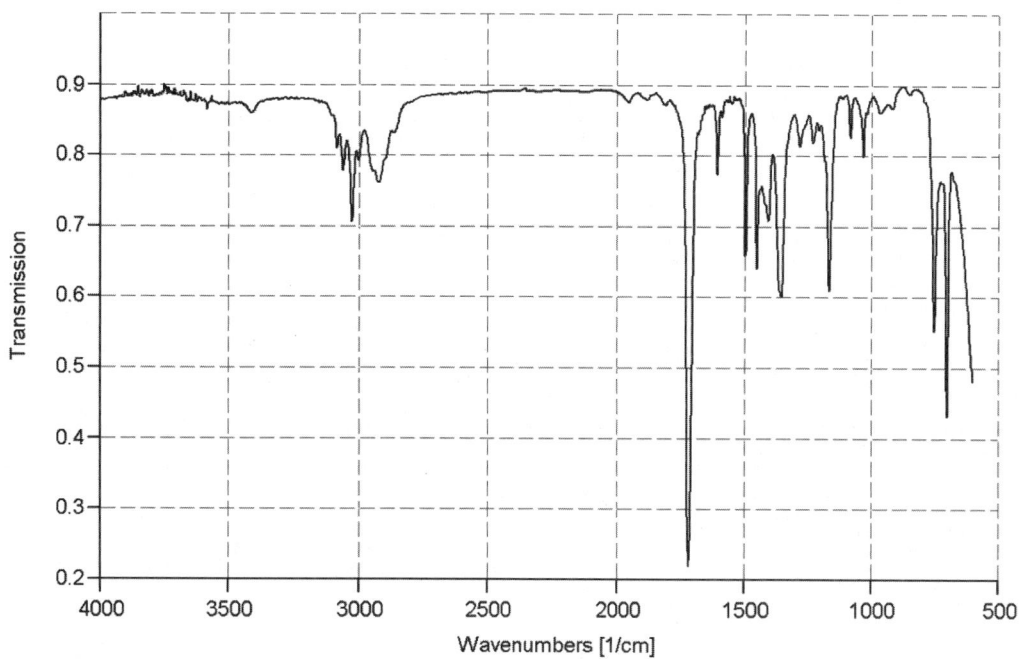

NMR spectra for Problem 4

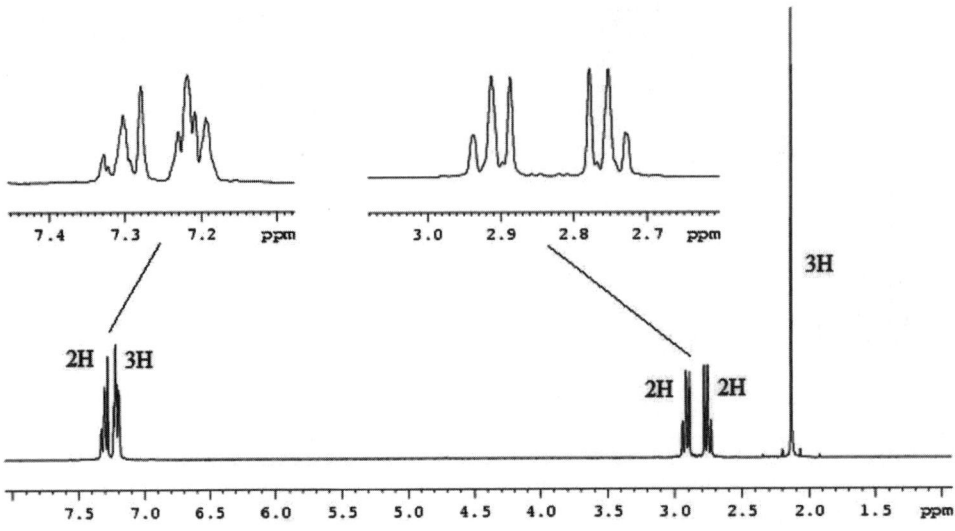

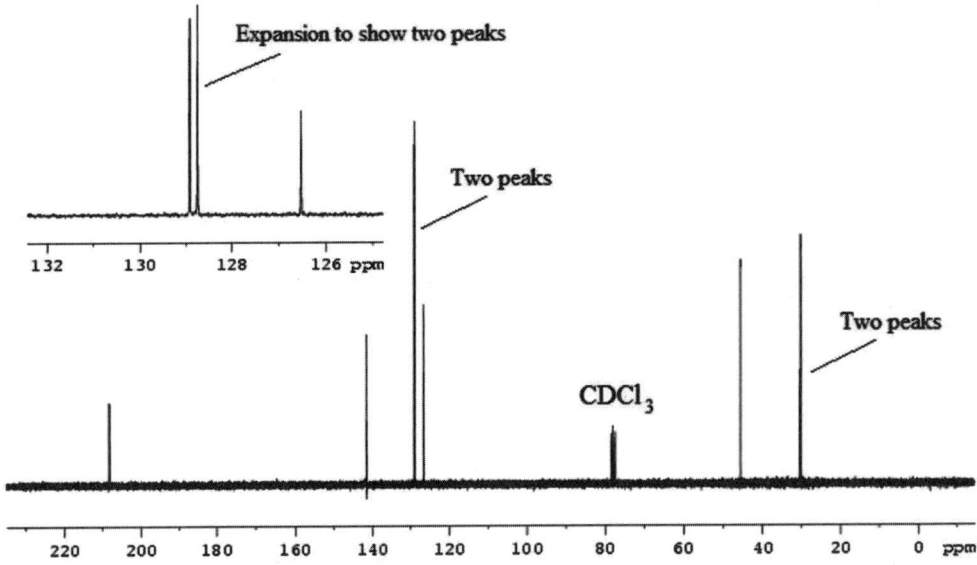

Answers to Review Exercises

Answers to Chapter 1 Review Exercises

1. There are many possible structures for each formula. Only two are shown for each.

   a. $C_4H_6O_2$

   b. $C_3H_3F_3N_2$

2. a. $\overset{\delta^+\ \delta^-}{C-Br}$

   b. $\overset{\delta^-\ \delta^+}{Cl-P}$

   c. $\overset{\delta^+\ \delta^-}{H-N}$

   d. $\overset{\delta^+\ \delta^-}{S-O}$

3. a.

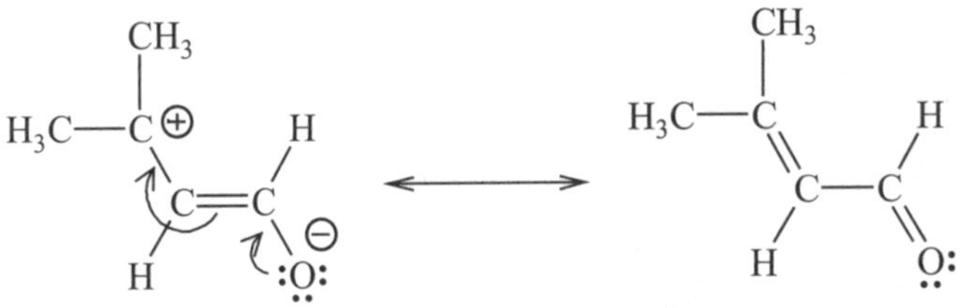

b.

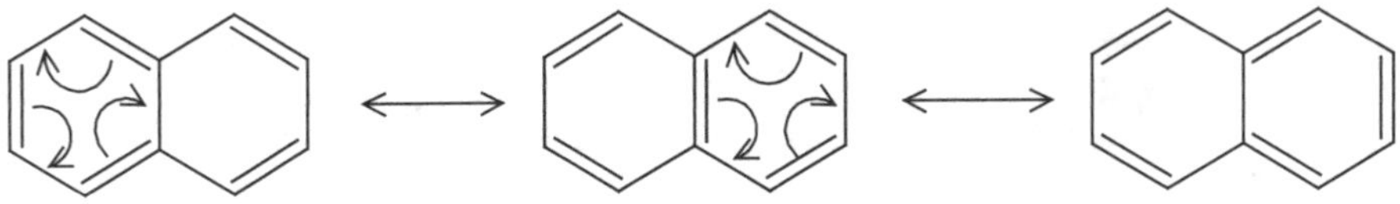

c.

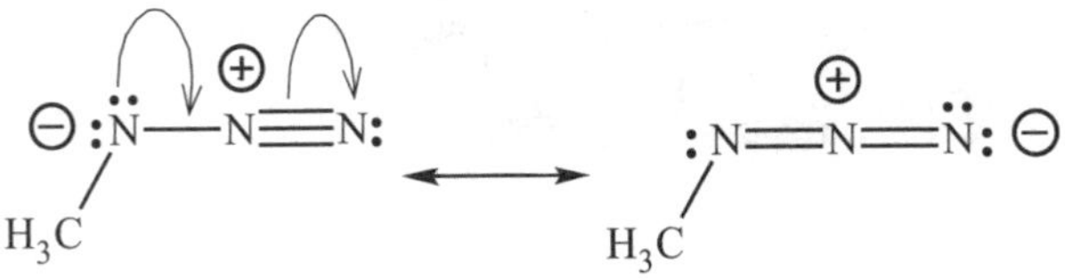

4. a.

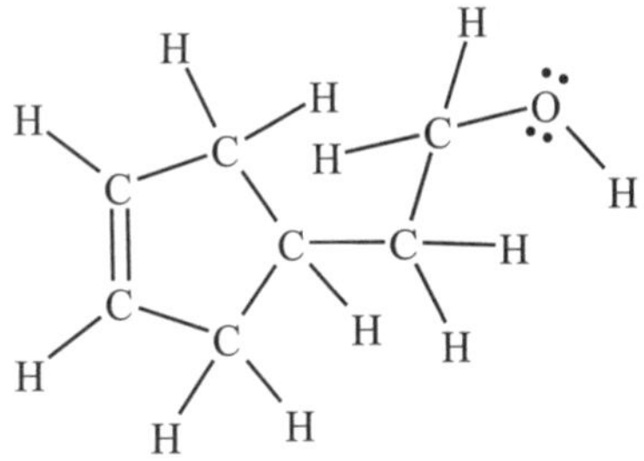

is the same as

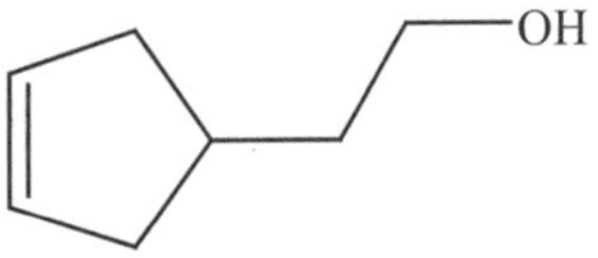

b.

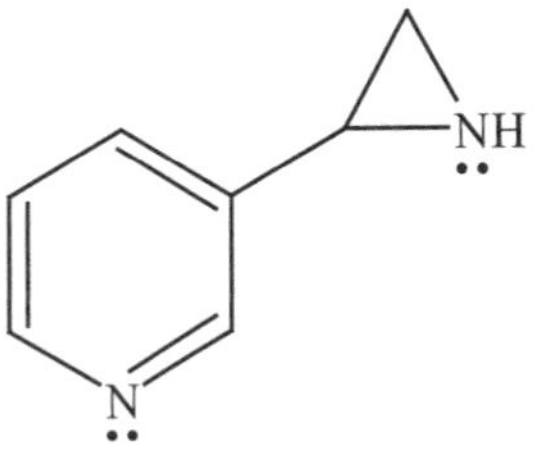

is the same as

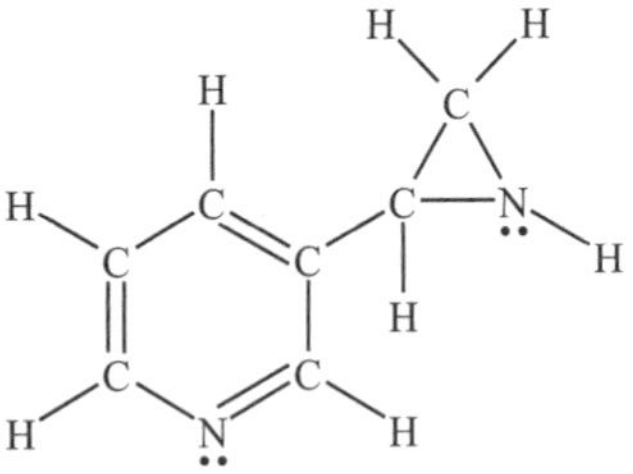

c.

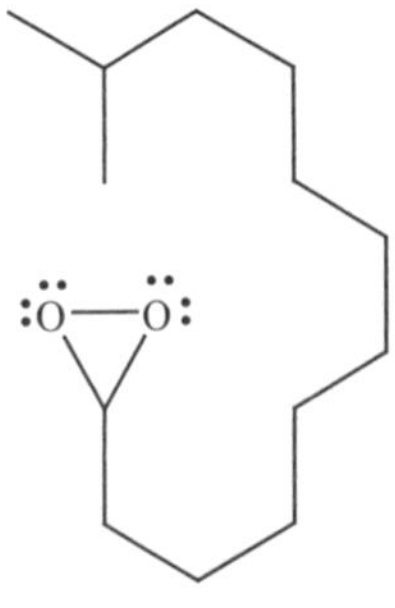

is the same as

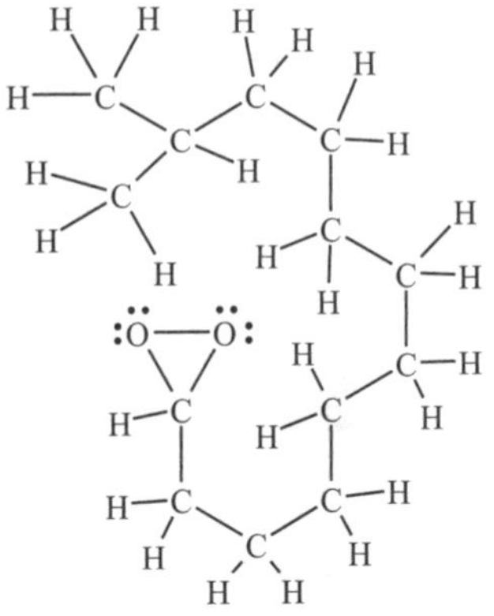

5. Water-solubility: Hexane has only C and H, so there are no significant partial charges in the molecule to attract hexane to water. Therefore, hexane is not very soluble in water. 1-Pentanol has a polar O–H group, with the H slightly positive and the O slightly negative. These partial charges attract water molecules by hydrogen-bonding-type interactions. These attractive forces make 1-pentanol slightly water-soluble.

Boiling point: Hexane has only C and H, so there are no significant partial charges in the molecule to attract hexane molecules to one another, other than the London attractive forces, which all molecules have. Therefore, hexane molecules require little energy to separate them from each other, and have a lower boiling point. 1-Pentanol has a polar O–H group, with the H slightly positive and the O slightly negative. These partial charges attract other 1-pentanol molecules by hydrogen-bonding interactions. These attractive forces hold 1-pentanol molecules together, and energy must be provided in the form of heat to separate the molecules from one another to get them into the vapor state. This additional heat required translates into a higher boiling point.

6. a. Codeine

b. Aspirin

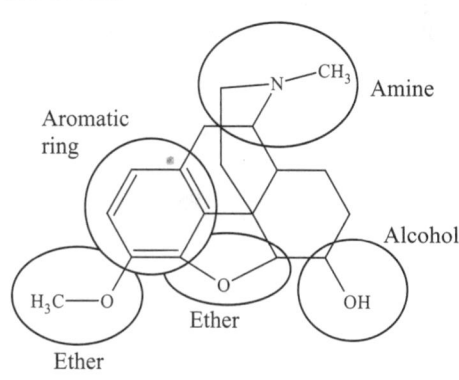

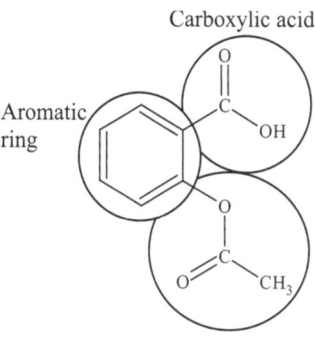

c. Amoxicillin

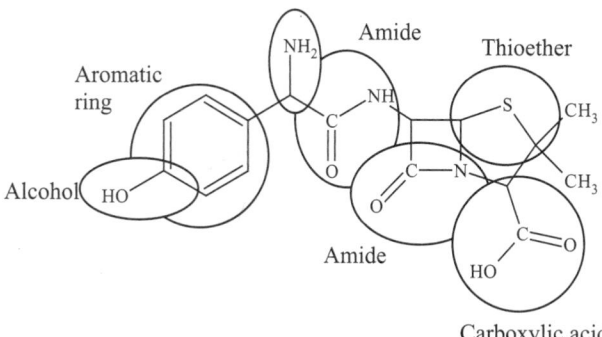

7. a.

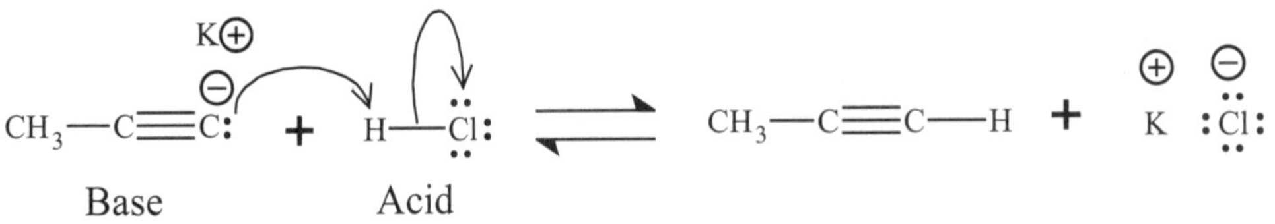

Base   Acid

b.

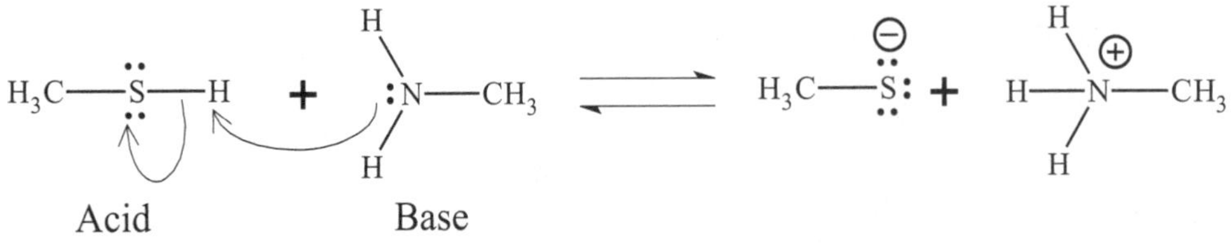

Acid   Base

Answers to Chapter 2 Review Exercises

1. a. 3,6,9,9-Tetramethyldodecane
   b. 1-Cyclopentyl-6-ethyl-2,9-diisopropylcyclohexadecane or
       1-Cyclopentyl-6-ethyl-2,9-bis(1-methylethyl)cyclohexadecane
   c. 1,3-di-*t*-Butylcyclopentane, or 1,3-bis(1,1-dimethylethyl)cyclopentane
   d. 3-Cyclohexyl-2,9-dimethyldodecane

2. a.

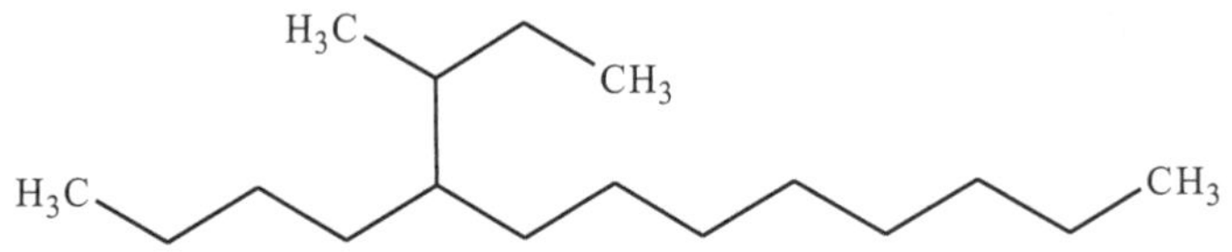

   b.

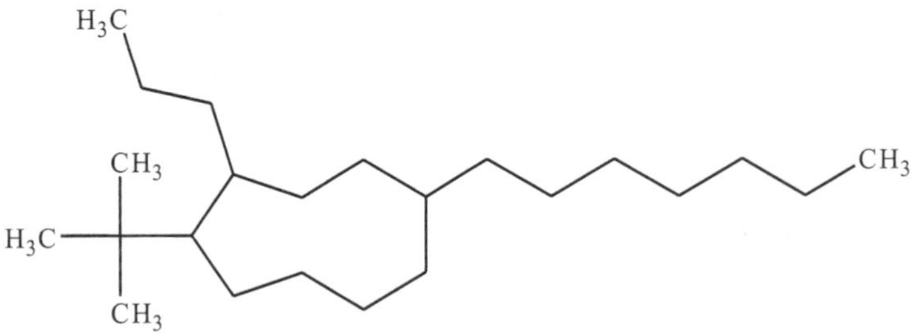

   c.

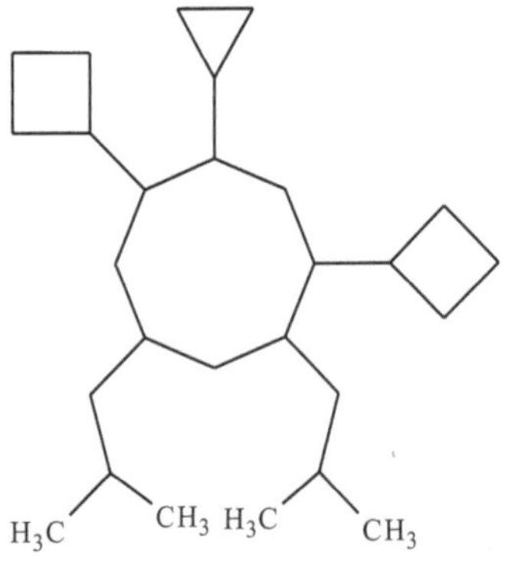

d.

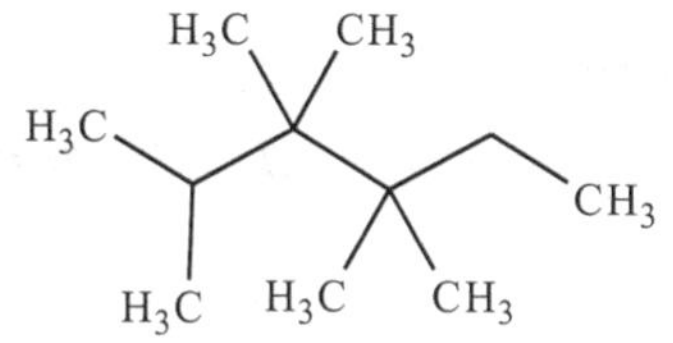

e.

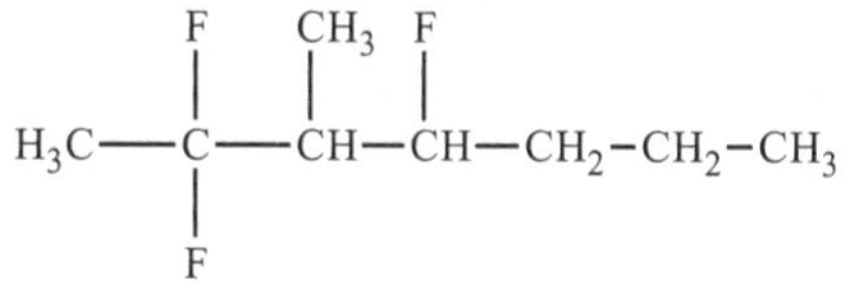

f. 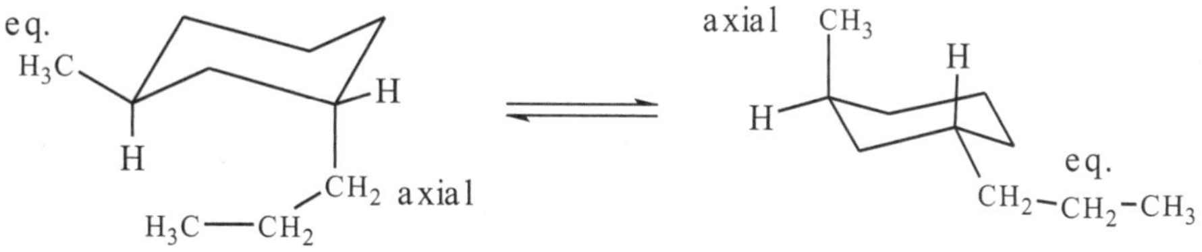   eq.

The right structure is more stable because the bigger propyl group is in the equatorial position.

3. The three Newman projections are shown below. I rotated the front carbon clockwise in 120° increments to generate the other two from the first one.

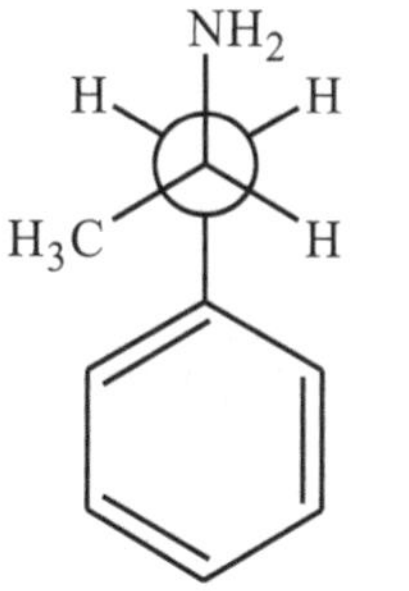 This form has the CH_3 gauche to the benzene ring and the NH_2 anti to the benzene.	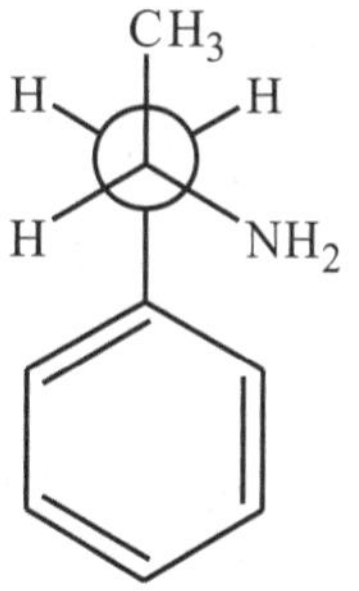 This form has the NH_2 gauche to the benzene ring and the CH_3 anti to the benzene.	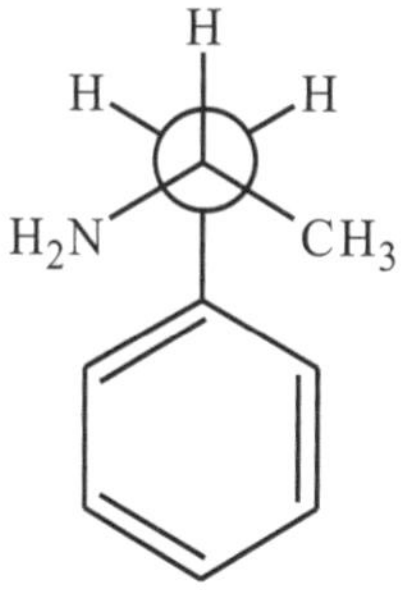 This form has both the NH_2 and the CH_3 gauche to the benzene ring, so it is less stable than the other two.

Because CH_3 and NH_2 are about the same size, the first two shapes are about equal in energy.

Answers to Chapter 3 Review Exercises

1. Each starred atom is bonded to four different groups of atoms.

2. a.

The groups are ranked as shown. A → B → C is clockwise, so the stereocenter is *R*.

The groups are ranked as shown. A → B → C is clockwise, so the stereocenter is *R*.

b.

The groups are ranked as shown. A → B → C is counterclockwise, so the stereocenter is *S*.

c.

The groups are ranked as shown. A → B → C is clockwise, so the stereocenter is *R*.

d.

This looks *R*, but the lowest priority group is coming at you. Therefore, it is not *R*, but *S*.

I flipped the structure over, so the lowest priority group is going away from you. Now it looks *S*.

3. a. Identical: there are no stereocenters.

b. Enantiomers: each stereocenter has the opposite configuration

c. Diastereomers: each stereocenter does <u>not</u> have the opposite configuration.

d. Structural isomers: the atoms are bonded together in a different order

Answers to Chapter 4 Review Exercises

1. a. Two hydrogens are added to the C=C.

b. Two bromine atoms are added in a *trans* fashion to the C=C.

c. H–OH is added in a Markovnikov fashion to the C=C.

d. H–OH is added in an anti-Markovnikov fashion to the C=C.

CH$_2$CH$_3$
''''H
,,\\OH
H

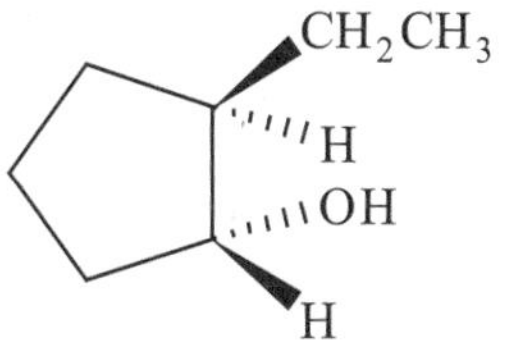

e. H–Cl is added in a Markovnikov fashion to the C=C.

CH$_2$CH$_3$
Cl
H
H

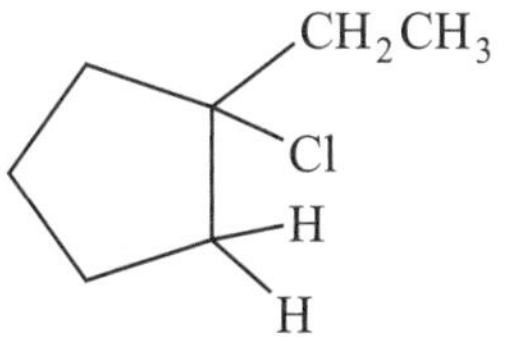

f. Two OH groups are added to the same side of the C=C.

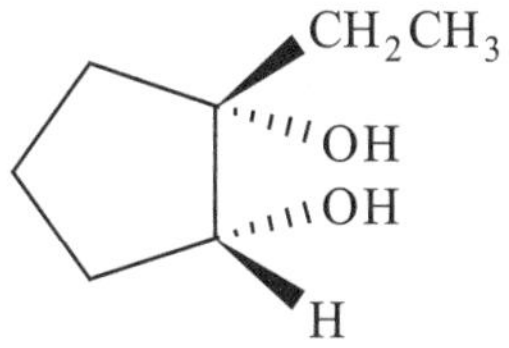

CH$_2$CH$_3$
''''OH
,,\\OH
H

g. The C=C is cleaved, with the top C of the C=C becoming a ketone, and the bottom C becoming a carboxylic acid.

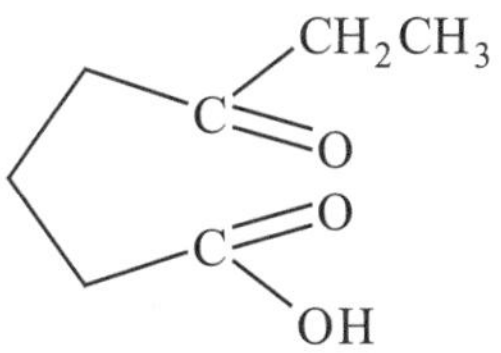

CH$_2$CH$_3$
C=O
C=O
OH

h. H–OH is added in a Markovnikov fashion to the C=C.

CH$_2$CH$_3$
OH
H
H

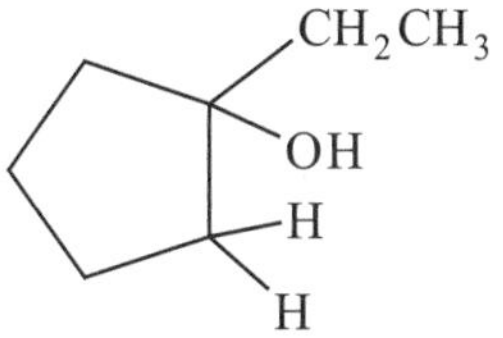

i. The C=C is cleaved, with the top C of the C=C becoming a ketone and the bottom C becoming an aldehyde.

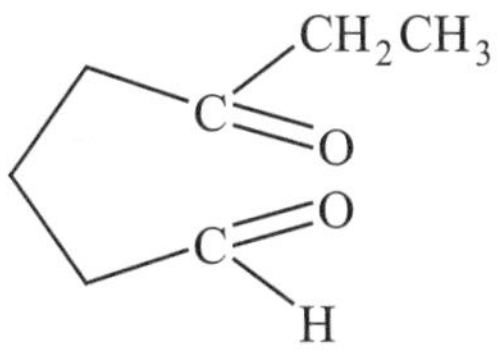

CH$_2$CH$_3$
C=O
C=O
H

j. This is the more direct way of forming an epoxide from a C=C.

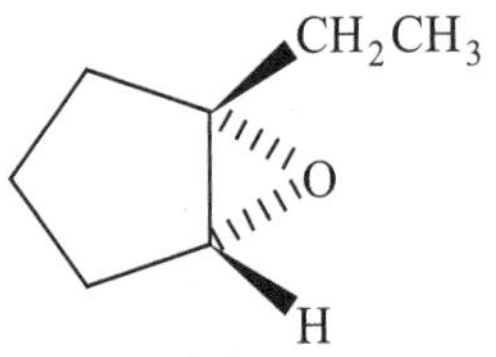

CH$_2$CH$_3$

O

H

k. H$^+$ protonates the epoxide, and water attacks from the opposite side to form the *trans* diol.

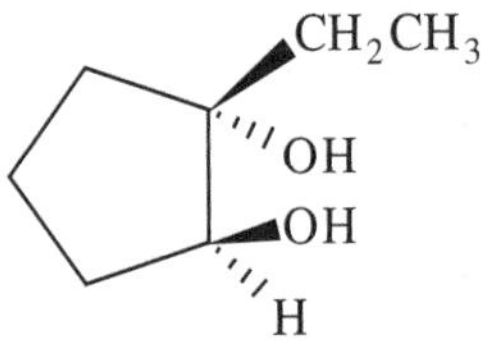

CH$_2$CH$_3$

OH

OH

H

l. This forms dibromocarbene, which adds to the C=C to form a cyclopropane.

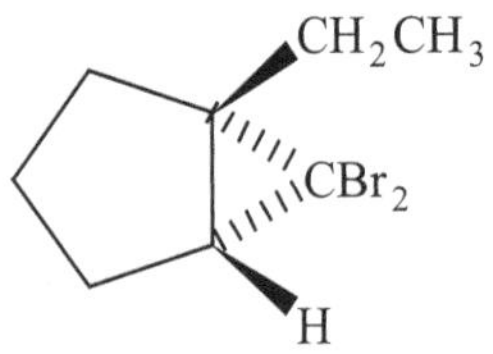

CH$_2$CH$_3$

CBr$_2$

H

2. a. Two hydrogens are added to the C=C.

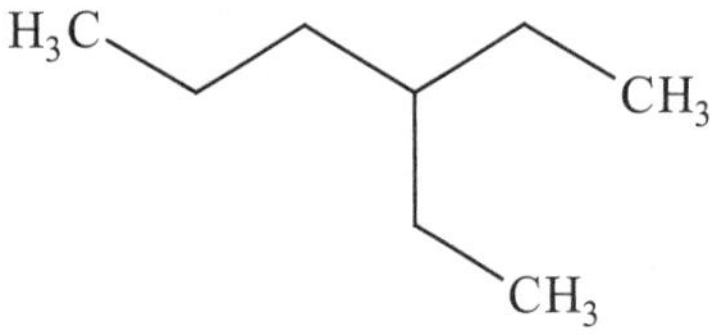

H$_3$C

CH$_3$

CH$_3$

b. Two bromine atoms are added to the C=C.

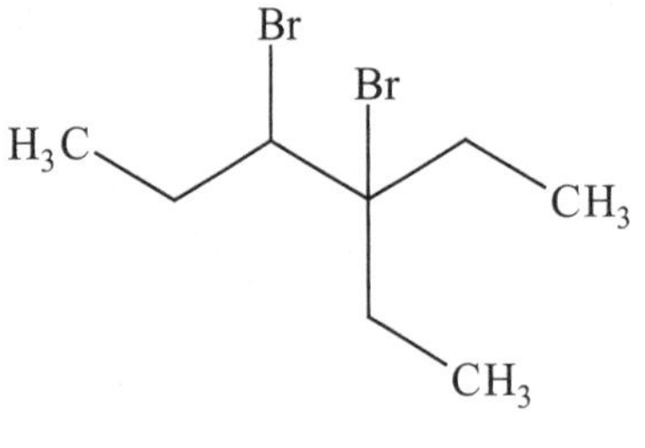

Br

Br

H$_3$C

CH$_3$

CH$_3$

c. H–OH is added in a Markovnikov fashion to the C=C.

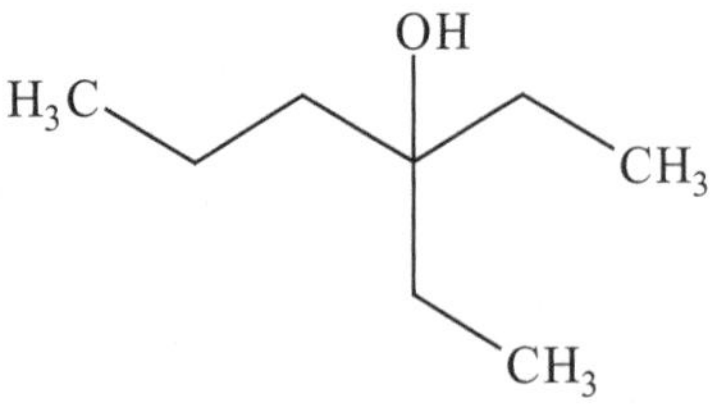

OH

H$_3$C

CH$_3$

CH$_3$

d. H–OH is added in an anti-Markovnikov fashion to the C=C.

e. H–Cl is added in a Markovnikov fashion to the C=C.

f. Two OH groups are added to the C=C.

g. The C=C is cleaved, with the one C of the C=C becoming a ketone and the other C becoming a carboxylic acid.

h. H–OH is added in a Markovnikov fashion to the C=C.

i. The C=C is cleaved, with the one C of the C=C becoming a ketone and the other C becoming an aldehyde.

j. This is the more direct way of forming an epoxide from a C=C.

k. H⁺ protonates the epoxide, and water attacks from the opposite side to form the diol.

l. This forms dibromocarbene, which adds to the C=C to form a cyclopropane.

3.

The methanol attacks from the back side of the cyclic bromonium ion because the Br is blocking the front side from attack.

4.

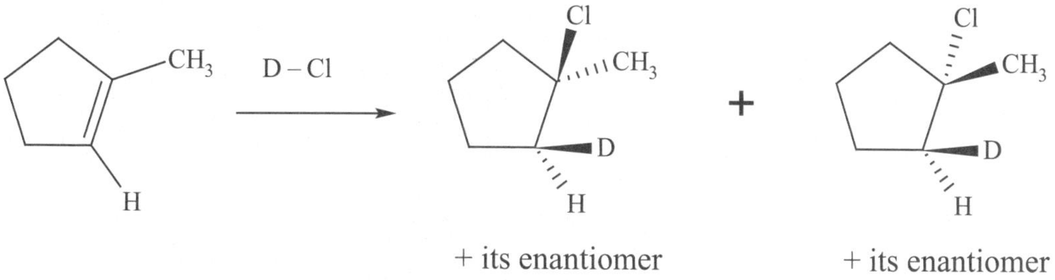

+ its enantiomer   + its enantiomer

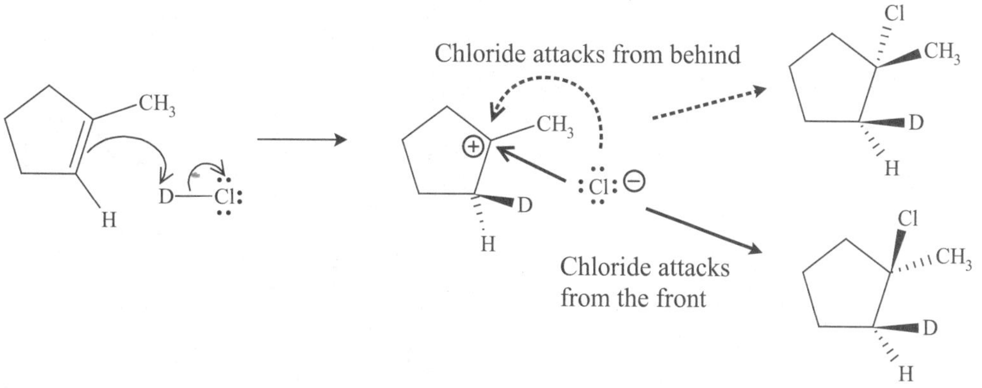

Chloride attacks from behind

Chloride attacks from the front

When the C=C attacks D–Cl, the D could end up coming at you (as shown) or going away from you. When chloride attacks the cation from the frontside or the backside, you can get the products. The enantiomers of these products are formed when chloride attacks the other possible carbocation intermediate.

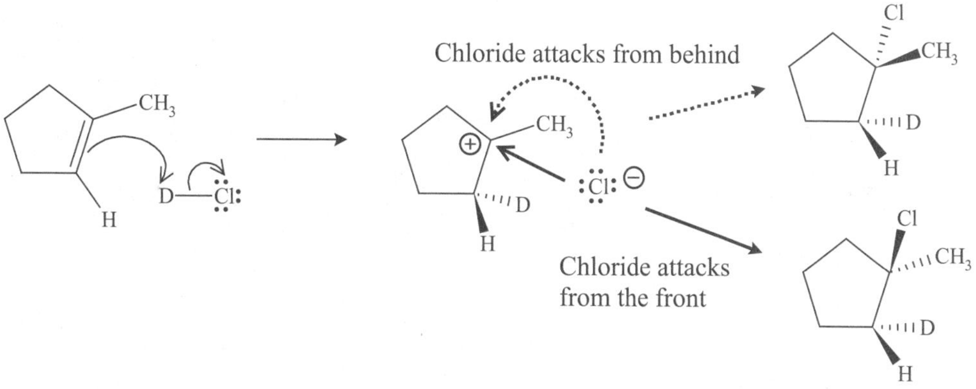

Chloride attacks from behind

Chloride attacks from the front

5. a.   (R)-1,5-dibromo-5-methylcyclohexene
   b.   (3E,7E)-4,7-diethylundeca-3,7-diene

6. a.

b.

c.

7. a.

Bromocyclohexane   Cyclohexane

We make an alkane by reducing a C=C with hydrogen and a catalyst. Alkenes are made from an alkyl halide by reacting the alkyl halide with a strong base.

b.

Cycloheptanol   Cyclohepta-1,3-diene

The trick to this one is seeing that you have to make the two C=C's from *trans*-1,2-dibromocycloheptane. If the Br's are trans, then E2 elimination reactions put the C=C's where we want them. We make *trans*-1,2-dibromocycloheptane by adding bromine to cycloheptene, which we make by dehydrating cycloheptanol.

Answers to Chapter 5 Review Exercises

1. a. 2-Cyclopentylhept-3-yne

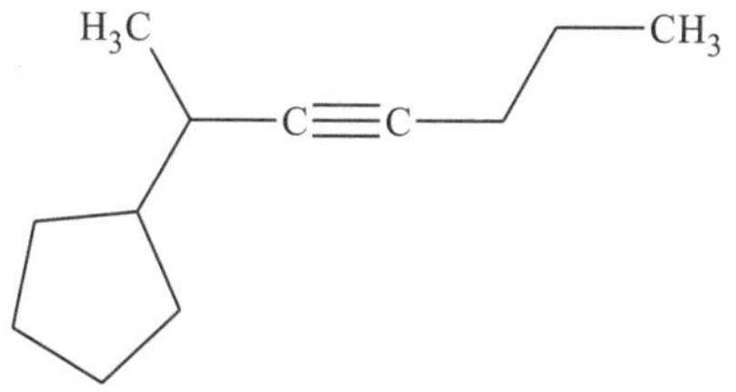

   b. 1-Ethynylcyclohexene

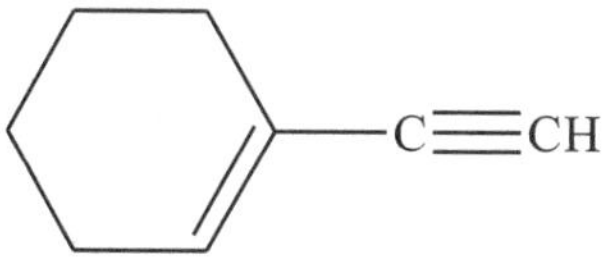

   c. (Z)-4,5-Difluoro-2-octen-6-yne

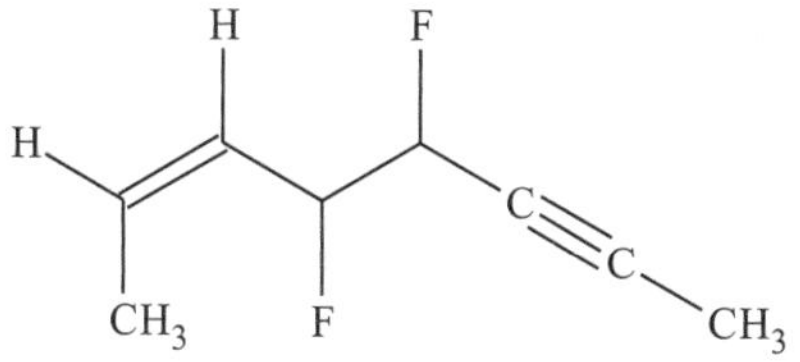

   d. 1,5,9-Cyclododecatriyne

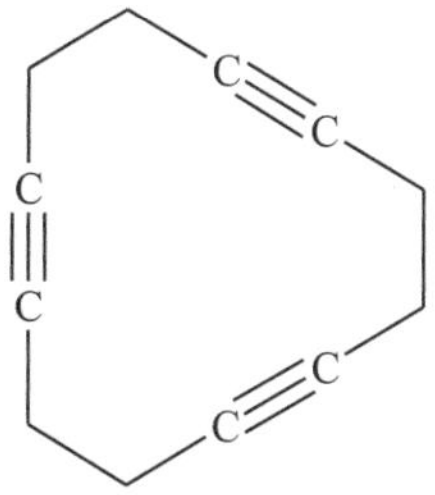

2. a. 3,4-dimethylcycloundeca-1,5-diyne

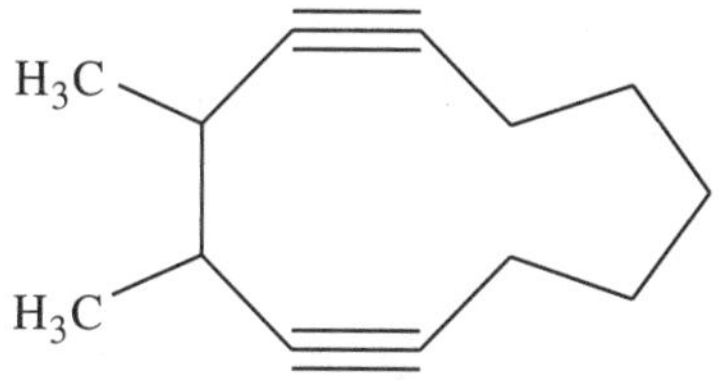

b. (*Z*)-11-methyldodec-4-en-8-yne

3. a.		Two H's add to each carbon of the triple bond.
b.		Two Br's add to each carbon of the triple bond.
c.		Addition of water according to Markovnikov's rule.
d.		NaNH$_2$ removes the *sp* H, and the carbanion then substitutes for the I in ICH$_2$CH$_2$CH$_3$.
e.		Addition of 2 HCl, according to Markovnikov's Rule.

4. a. Hydroboration-oxidation adds water in an anti-Markovnikov fashion to a C≡C.

b. Water adds to a C≡C in a Markovnikov fashion.

c. NaNH₂ pulls of the *sp* H to form a carbanion. The carbanion attacks the CH₂ bonded to the Br, and forms a new C–C bond.

d. Hydrogen with Lindlar Pd adds two hydrogens to the triple bond in a cis fashion.

5.

A "vinylcation"

6.

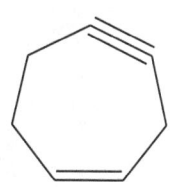

The C≡C carbons are *sp* hybridized and would prefer to have bond angles of 180°, if possible. Putting four *sp* carbons inside of a seven-membered ring would introduce tremendous ring strain, so it is not likely to be prepared easily.

7. a.

b.

Answers to Chapter 6 Review Exercises

1. a.

1°   Strong   S$_N$2

b.

3°   Weak   S$_N$1

c.

1°   Strong S_N2

d.

3°   Strong   E2

2. a. Tertiary alkyl halide + weak nucleophile = S_N1

b. Tertiary alkyl halide + strong nucleophile = E2

c. Primary alkyl halide + strong nucleophile = S_N2

3. a. Cyanide attacks from the front, and pushes the Br out of the molecule.

b. An amine attacks a primary alkyl halide to make the product.

or

4.

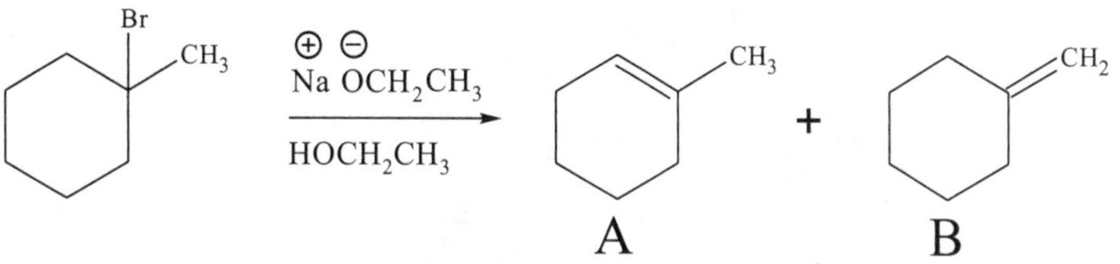

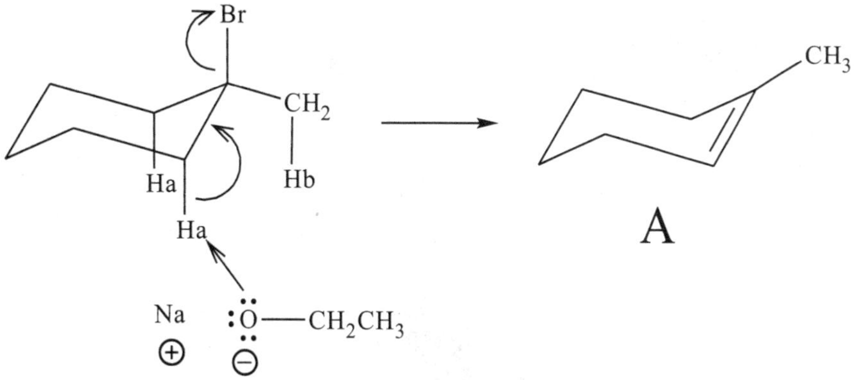

The E2 mechanism is used to form both products. A is formed by the base pulling off either of the hydrogens marked Ha. To form B, the base pulls off the hydrogen marked Hb, as shown below.

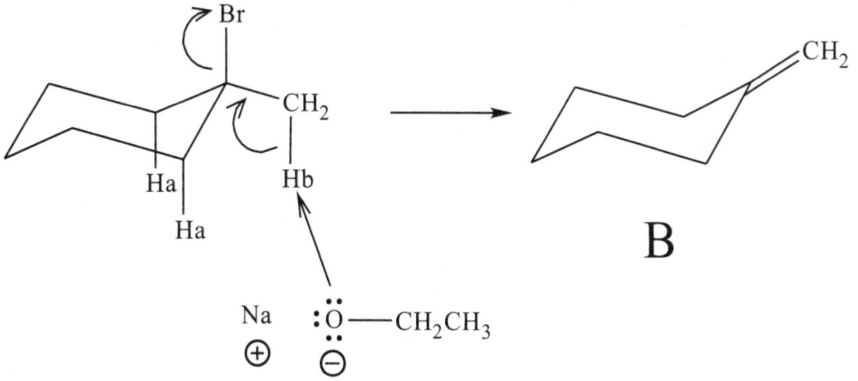

Sodium ethoxide is a relatively small base, so it can easily get to the Ha hydrogens, and pull them off to form A. Potassium *t*-butoxide is a much bulkier base and has difficulty getting to the Ha hydrogens. Therefore, it pulls off the more accessible Hb, to form B as the major product.

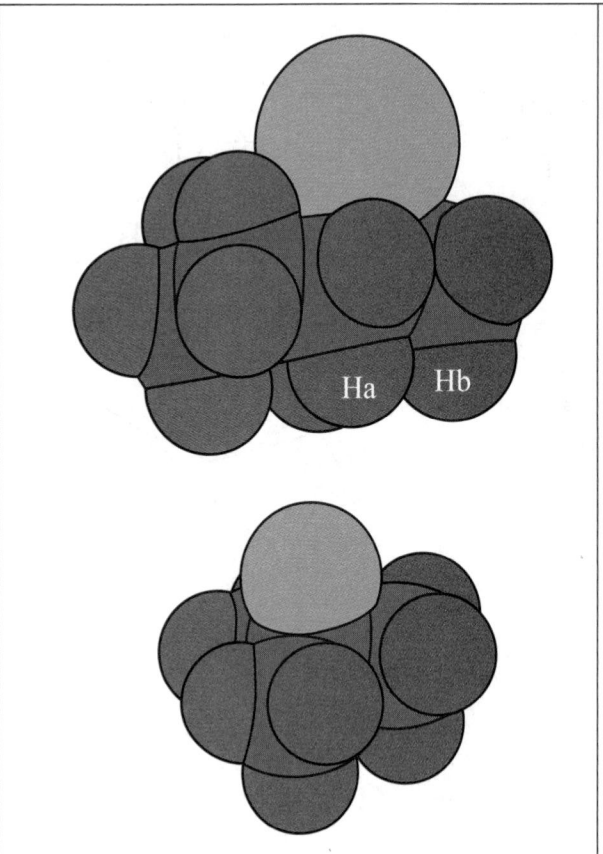

Space-filling models of 1-bromo-1-methylcyclohexane and *t*-butoxide are shown below. Ha is on the inside of the molecule, and is less accessible than Hb, which is on the outer part of the molecule. So Hb can be pulled off more easily by *t*-butoxide than Ha.

Answers to Chapter 7 Review Exercises

1. a.

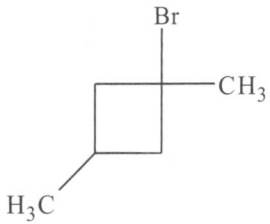

Bromine is much more selective for substitution of a hydrogen at a tertiary carbon.

b.

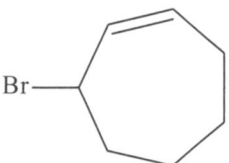

Substitution at the resonance-stabilized secondary radical is much faster than substitution at a nonresonance-stabilized secondary radical.

c.

Substitution at the resonance-stabilized secondary radical is much faster than substitution at a nonresonance-stabilized secondary radical.

d.

All the hydrogens are equivalent, so the chlorine can substitute for any hydrogen.

e.

Bromine is much more selective for substitution of a hydrogen at a tertiary carbon.

2.

Problem 1a.

Generation of bromine radicals.

Bromine radical pulls of a tertiary hydrogen to form a tertiary radical.

The tertiary radical reacts with a bromine molecule to form the product, plus a new bromine radical.

Problem 1b.

A trace amount of bromine forms some radicals.

A bromine radical pulls a hydrogen off on the allylic carbon. This shows resonance stabilization of the radical.

HBr reacts with NBS to form bromine plus succinimide.

Bromine reacts with the radical to form the product, plus more bromine radical.

3.

Cl–Cl ⇌ 2 Cl•	This is initiation, which is fine.
Cl• + CH₃CH₃ ⇌ H• + CH₃CH₂–Cl	$\Delta H°$ for this step is +17 kcal/mol (+71 kJ/mol), which is quite endothermic, and therefore unlikely to occur.
H• + Cl–Cl ⇌ H–Cl + Cl•	$\Delta H°$ for this step is –45 kcal/mol (198 kJ/mol), which is quite exothermic.

4. If you have a high concentration of bromine during the reaction, the bromine can add to the alkene to form a dibromoalkane. With a low concentration of bromine, it is less likely to react with the alkene and find a radical to react with to carry on the free-radical halogenation.

5.

a. Cl₂ + CH₃CH₃ → 2CH₃Cl is exothermic.

Bonds Broken	Bonds Formed	$\Delta H°$
58 + 88 = 146 kcal/mol	2 × –84 = –168 kcal/mol	–22 kcal/mol
243 + 368 = 611 kJ/mol	2 × –351 = –702 kJ/mol	–91 kJ/mol

b. Br₂ + CH₃CH₃ → 2CH₃Br is exothermic.

Bonds Broken	Bonds Formed	$\Delta H°$
46 + 88 = 134 kcal/mol	2 × –70 = –140 kcal/mol	–6 kcal/mol
192 + 368 = 560 kJ/mol	2 × –293 = –586 kJ/mol	–26 kJ/mol

c. (CH₃)₃C–H + I₂ → (CH₃)₃C–I + H–I is endothermic.

Bonds Broken	Bonds Formed	$\Delta H°$
92 + 36 = 128 kcal/mol	–51 + –71 = –122 kcal/mol	6 kcal/mol
385 + 151 = 536 kJ/mol	–213 + –297 = –510 kJ/mol	26 kJ/mol

d. $CH_4 + 2O_2 \rightarrow CO_2 + H_2O$ is exothermic.

Bonds Broken	Bonds Formed	$\Delta H°$
$(4 \times 104) + (2 \times 119) =$ 654 kcal/mol	$(2 \times -127) + (4 \times -119) =$ -730 kcal/mol	-76 kcal/mol
$(4 \times 435) + (2 \times 498) =$ 2736 kJ/mol	$(2 \times -531) + (4 \times -498) =$ -3054 kJ/mol	-318 kJ/mol

Answers to Chapter 8 Review Exercises

1. If the reaction mechanism was an S_N1, we would expect that the secondary cation that forms initially could rearrange to another secondary cation. Rearrangement to a primary cation is energetically unfavorable and would not occur. Therefore, the original cation from 3-pentanol could rearrange from carbon 3 to either carbon 2 or carbon 4. The original cation from 2-pentanol could rearrange from carbon 2 to carbon 3, and then on to carbon 4. Assuming that these rearrangements occur readily, we should end up with two-thirds 2-bromopentane and one-third 3-bromopentane.

2. a. 7,7,7-Trifluoroheptan-2-ol
   b. 1-Cyclohexyl-4-methylpentan-1-ol
   c. 2-*tert*-Butyl-4-phenylcyclopentanol
   d. 5,9-Dimethylundec-10-en-4-ol
   e. Hex-2-yne-1-thiol
   f. 6-Mercapto-2-propylcyclooctanol

3. a.

1.	2.	3.
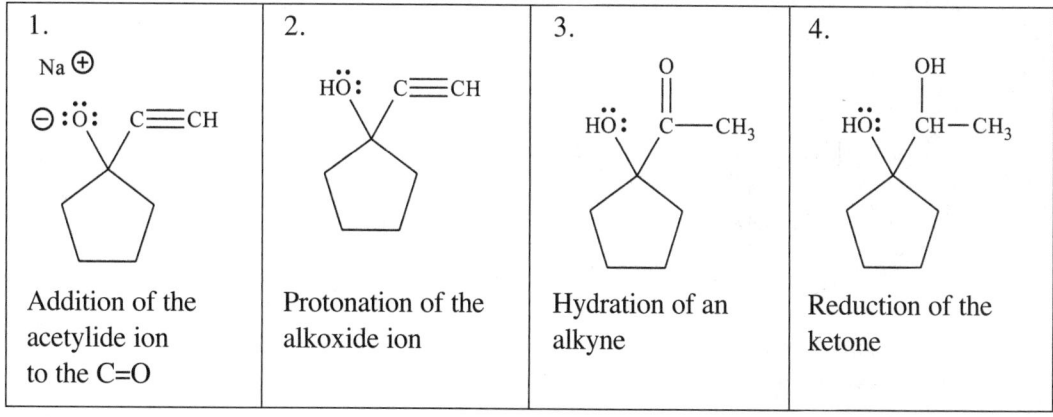		
Preparation of the Grignard reagent	Intermediate after the attack of the Grignard reagent	Alcohol product after protonation of the anion

b.

1.	2.	3.	4.
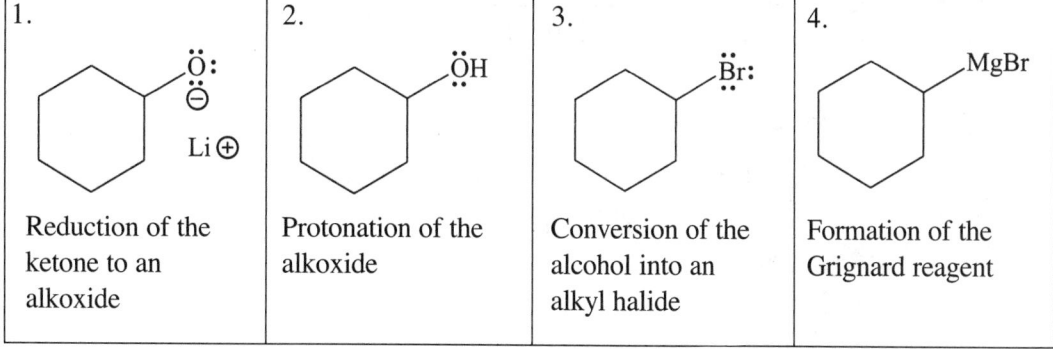			
Addition of the acetylide ion to the C=O	Protonation of the alkoxide ion	Hydration of an alkyne	Reduction of the ketone

c.

1.	2.	3.	4.
Reduction of the ketone to an alkoxide	Protonation of the alkoxide	Conversion of the alcohol into an alkyl halide	Formation of the Grignard reagent

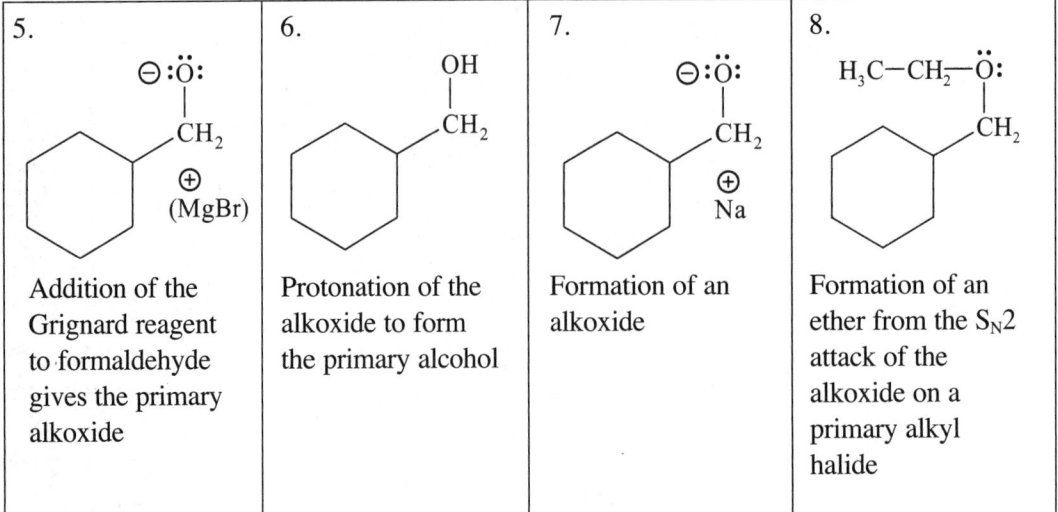

5.	6.	7.	8.
Addition of the Grignard reagent to formaldehyde gives the primary alkoxide	Protonation of the alkoxide to form the primary alcohol	Formation of an alkoxide	Formation of an ether from the S_N2 attack of the alkoxide on a primary alkyl halide

4. a.

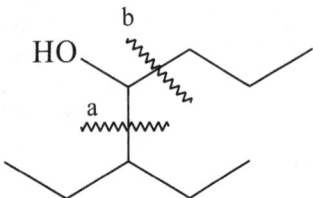

This is a secondary alcohol, so we need to use a Grignard reagent plus an aldehyde. There are two possible combinations. To form bond a,

MgBr

1. O (image)

2. HOH, H$^+$

To form bond b,

1. BrMg

2. HOH, H$^+$

b.

HO a
c
b

Because this is a tertiary alcohol, we use a ketone plus a Grignard reagent. There are three possibilities. Here is possibility a:

Here is possibility b:

Here is possibility c:

5. a. Using a Grignard reaction, you have two choices. This is one.

This is the other.

b. Using a reduction of a C=O compound:

c. From an alkene:

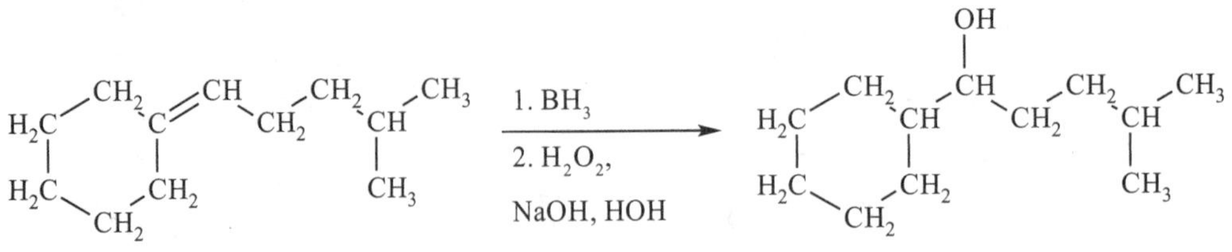

d. Starting with 3-methyl-1-butyne:

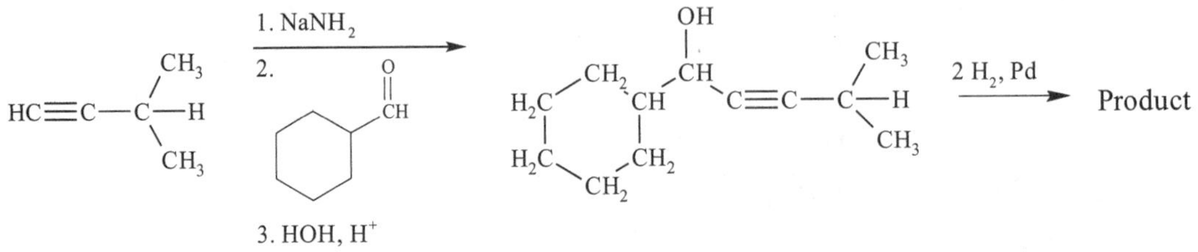

Answers to Chapter 9 Review Exercises

1. a. 4-Cyclopentylhept-5-yn-3-ol. The longest chain is seven, and the alcohol has the priority in numbering.

   b. 8-Ethoxyoct-1-en-4-thiol. The longest chain is eight, and the thiol has the priority in numbering.

   c. 2-Cyclopropyloxirane, or just cyclopropyloxirane.

   d. 2,2-Dichloro-1,1-difluoro-1-methoxyethane.

   e. 1-(Isobutylthio)prop-1-yne. The alkyne is the highest priority functional group.

   f. 6-Cyclobutyl-2-ethoxy-4-(ethylthio)-3-iodocycloheptanol. The highest priority functional group is the alcohol, and we number counterclockwise starting from the carbon the OH group is bonded to.

2. a. Sodium metal makes the alkoxide, which does an S_N2 attack on the alkyl halide.

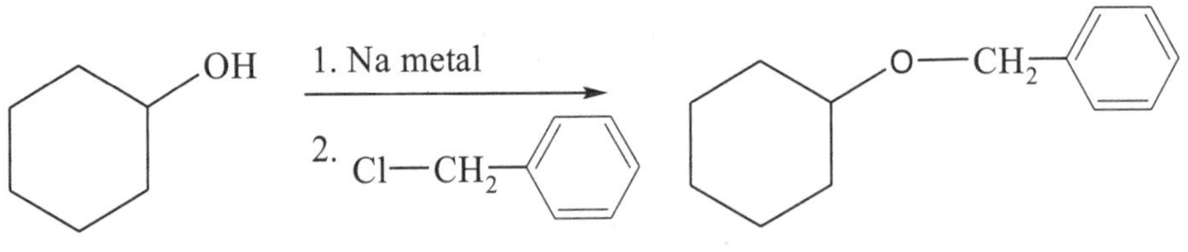

   b. The strong nucleophile attacks the less substituted carbon of the epoxide.

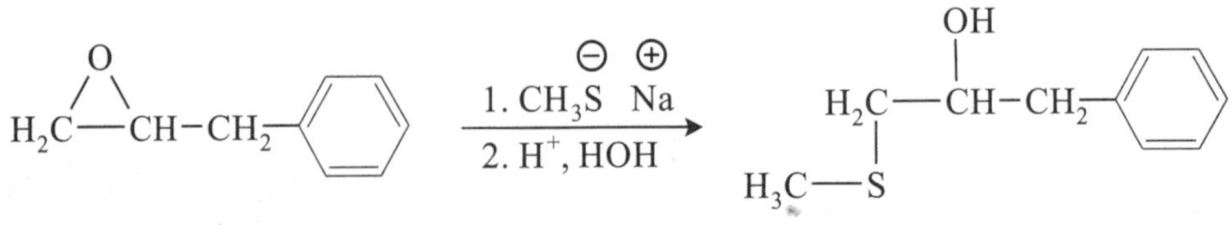

c. The weak nucleophile attacks the more substituted carbon of the protonated epoxide.

d. HBr converts the alcohol into an alkyl halide, which is displaced by the strong nucleophile.

e. MCPBA oxidizes thioethers to sulfoxides.

f. Hydrogen peroxide oxidizes thioethers to sulfones.

g. Concentrated HI cleaves all of the C–O bonds and makes alkyl iodides.

3.
Problem 2c.

Protonation of the epoxide by the acid catalyst.

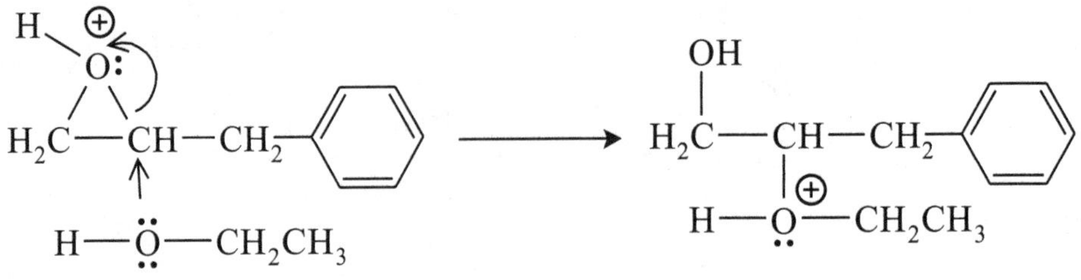

The alcohol attacks the more substituted carbon because it has the most positive character.

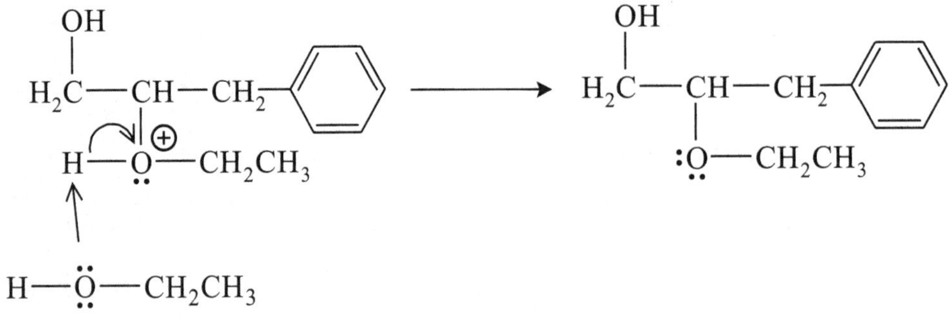

Loss of H⁺ gives us the product.

Problem 2d.

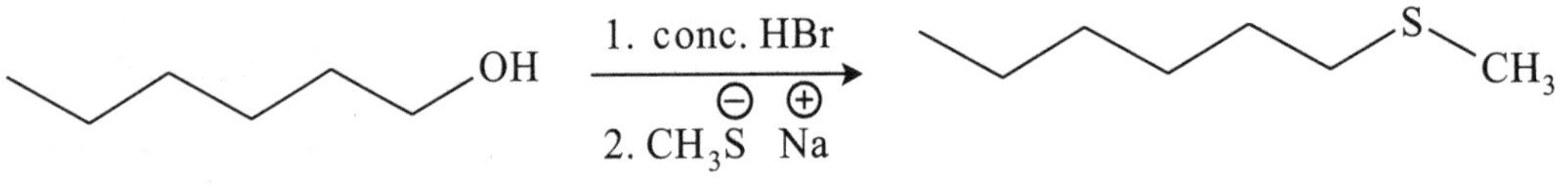

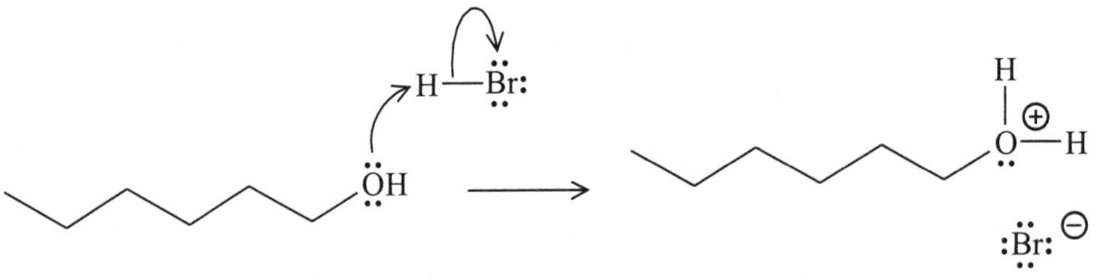

The OH group is protonated by H–Br to form the protonated alcohol, plus a bromide ion.

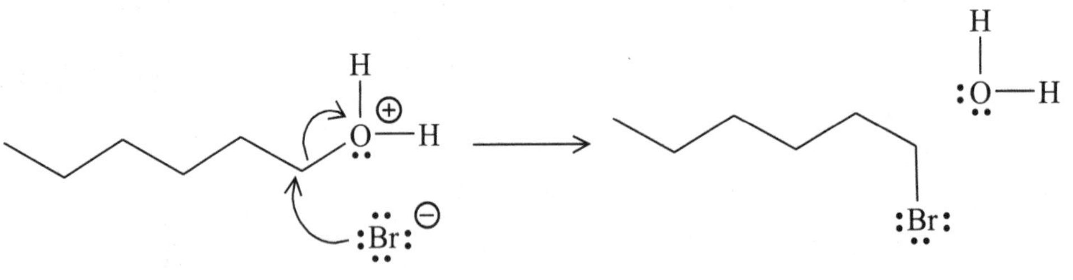

Bromide does an S_N2 attack on the backside of the carbon and pushes out the water.

Methanethiolate ion does an S_N2 attack on the backside of the carbon and pushes out the bromide.

4. a.

b.

c.

5.

The phenol is converted into an ether. The methyl group next to the benzene ring is halogenated, and then the halogen is replaced by the thiolate anion in an S_N2 reaction.

The thioether is oxidized to the sulfide with hydrogen peroxide. Concentrated HBr cleaves the methyl ether and regenerates the phenol.

Answers to Chapter 10 Review Exercises

1.

Compound	Monochloro Derivatives
a.	
b.	
c.	
d.	

Compound	Dichloro Derivatives
a.	
b.	
c.	
d	

2. a. 4-Iodo-2,5-dimethoxybenzoic acid (benzoic acid is a special name).
  b. 1-*t*-Butyl-2-cyclopropyl-4-nitrobenzene (no special names, so alphabetize and give the lowest numbers).
  c. 7-Bromo-2,6-diphenyl-1-hepten-4-yn-3-ol (just to see if you can use all of the priority rules).
  d. 4-dodecylbenzenesulfonic acid (benzenesulfonic acid is a special name).

3.
  a. Alkyl groups are ortho-para directors.

  b. Ketones are meta directors because the C=O carbon is slightly positive.

  c. CH$_3$CH$_2$O is an ortho-para director because the O has unshared pairs of electrons on it.

  d. CF$_3$ is a meta director because the C is slightly positive due to the F's electronegativity.

4. a. Because alkyl groups are ortho-para directors, the alkyl group has to go on first. The alkyl group will direct the ketone group to the para position. If we put the ketone on first, because it is a meta director, it would direct substitution to the meta position. Also, the Friedel–Crafts alkylation wouldn't work because the ketone is a strongly deactivating group.

   b. The sulfonic acid group is a meta director, so we put it on the benzene ring first. Chlorine is an ortho-para director, so putting it on first would direct the sulfonic acid to the para position.

   c. Because both NH_2 and pentyl are ortho-para directing, they can't be used to introduce a group meta. Also, pentyl cannot be added by a Friedel–Crafts alkylation because the intermediate primary cation will rearrange. The only way to add a straight chain group is by Friedel–Crafts acylation and then reduction. Because a ketone is a meta director, a nitro is added after the acylation. The nitro is reduced, and then the ketone is reduced to a CH_2 using the Clemmensen reduction procedure.

d. Direct bromination of isopropylbenzene would give mostly para product. Therefore, the para position is blocked first with a sulfonic acid group, so bromination is forced to go ortho. Removal of the sulfonic acid group gives the product.

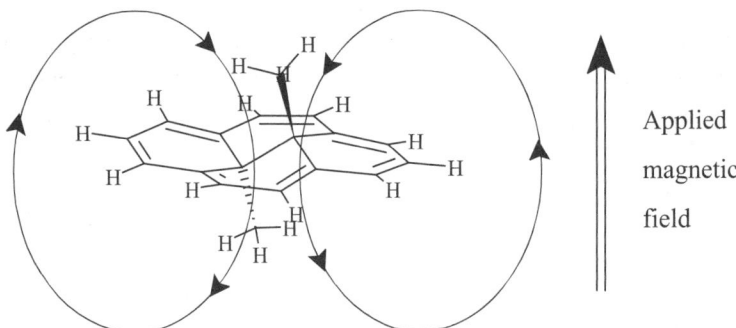

5. a. Antiaromatic: 12 p electrons, planar, cyclic, p orbitals on each ring atom.
   b. Nonaromatic: not planar, due to eight-membered ring.
   c. Aromatic: 14 p electrons, planar, cyclic, p orbitals on each perimeter ring atom.
   d. Aromatic: 10 p electrons, planar, cyclic, p orbitals on each ring atom.

Induced magnetic field

Applied magnetic field

6. The applied magnetic field in the NMR sets up an induced magnetic field in an aromatic ring. The induced field is going the same direction as the applied field outside the ring, but it is going in the opposite direction inside the ring. This means the H's on the outside of the ring experience the induced field in the same direction as the applied field, so less applied field is needed to bring them into resonance. This is what is referred to as deshielding, so the H's on the outside of the ring show up between 8 and 9, which is close to where benzene's H's show up.

The methyls are inside the ring, so the induced field opposes the applied field and cancels part of it—this is referred to as shielding. Therefore, more applied field is needed to bring them into resonance, so these H's show up far to the right of where a "normal" methyl group shows up.

This compound was made specifically to test hypotheses about NMR properties of aromatic compounds. The data are consistent with this compound being aromatic.

Answers to Chapter 11 Review Exercises

1. a. (3R,6S)-6-Methoxy-3-methyl-3-heptanamine
    b. 6-(Dimethylamino)nonan-3-ol
    c. 4-Chloro-2-ethynylpyridine
    d. N,4-Diisopropyl-N-methylaniline

2. a. Form an imine, and then reduce it to the amine.

    b. React the alkyl halide with excess ammonia.

    c. Reduce the amide to an amine with lithium aluminum hydride.

    d. Hofmann rearrangement converts an amide into an amine with one carbon fewer.

3. a.

    b.

    c.

d.

then proceed as in
problem b or c, above.

4. Let R = the phenyl group.

Sodium cyanoborohydride doesn't reduce ketones very well, but it does reduce C=N's,
especially if the N has been protonated, as in the preceding mechanism.

5. a. Treatment of the acid chloride with excess ammonia makes the amide, which can be reduced
by lithium aluminum hydride to make the amine.

b. Because the product has one less carbon than the starting material, we need to do a Curtius
or a Hofmann rearrangement. I am showing a Curtius rearrangement here.

c. To attach benzene ring to the acid chloride, we use the Friedel–Crafts acylation. This gives us a ketone. To convert the ketone into an amine, we use reductive amination. I am showing a one-step process, reducing the imine as it is formed with sodium cyanoborohydride. You could also make the imine first and then reduce it with $NaBH_4$ or H_2/Pd.

d. Because the carbon chain of the primary amine product is one carbon longer than that of the starting material, we need to add a carbon in an appropriate place. Reacting the alkyl halide with cyanide adds a carbon and gives us a nitrile. Reduction of the nitrile gives us the primary amine.

e. We need to add the five-membered nitrogen-containing ring (pyrrolidine). One way to do it is by adding an excess of pyrrolidine.

f. Because we can't fluorinate benzene directly, we have to find another way. We can convert aniline into fluorobenzene using a diazonium reaction. We can make aniline by nitrating benzene and then reducing the nitro group.

g. The key to this is recognizing that an amine can be converted into a nitrile with a diazonium reaction. Amines are ortho-para directors and are strongly activating, so we can put the halogens on one at a time.

Answers to Chapter 12 Review Exercises

1. a. 4-Ethyl-4-phenylhex-5-yn-2-one
    b. 2,3-Diethylhexanal
    c. 2,7-Diisopropyl-4,5-dimethylcycloheptanone

2. a.

(Two different ways, with *different types of reactions*.)

    b.

    c.

    d.

e.

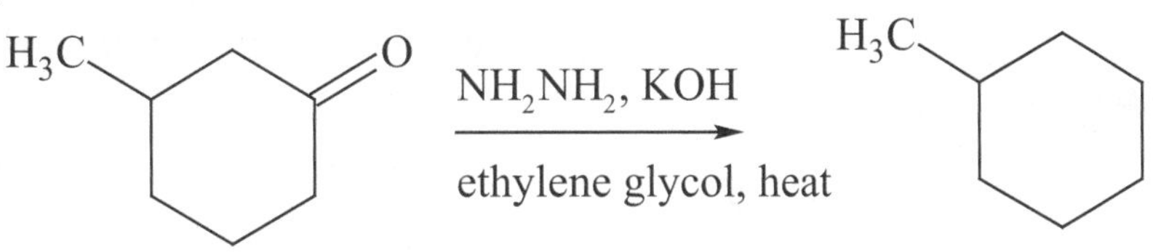

f.

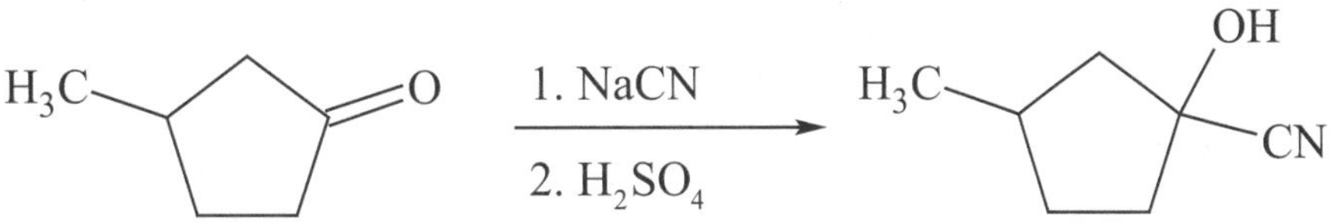

3. a. The key step is making the trans C=C. Dehydration of an appropriate alcohol gives the trans C=C as the major product, so the last step should be the following.

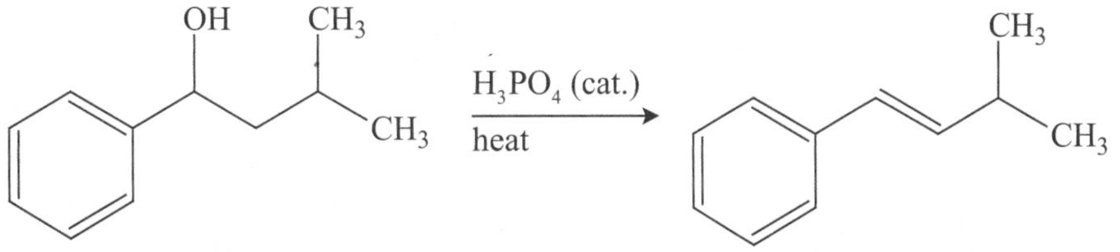

You can make the alcohol using a Grignard reaction on benzaldehyde, as shown below.

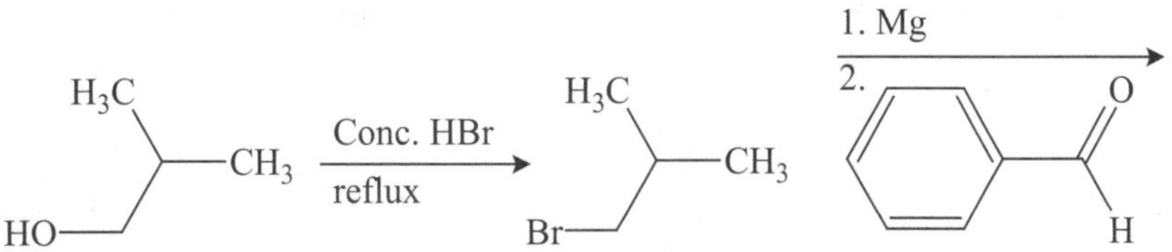

b. To make a cis C=C, you need to use a Wittig reaction that does not involve an ylid that is stabilized by resonance with a benzene ring or a carbonyl. Such a reaction is shown below.

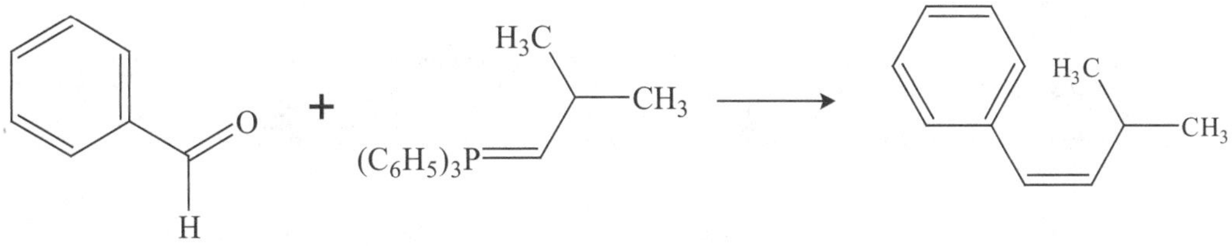

To make the ylid, you convert the alcohol into an alkyl halide, react the alkyl halide with triphenylphosphine, and then react it with a very strong base.

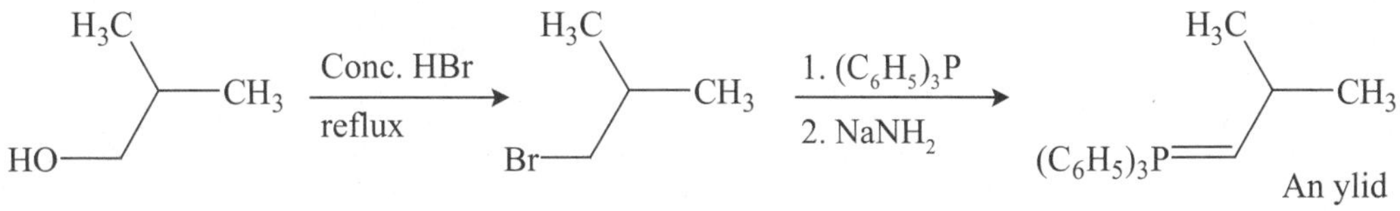

c. The last step is most likely a Grignard reaction, as shown below. Therefore, we need to make both the Grignard reagent and cyclopentanone from cyclopentene by some sequence of reactions.

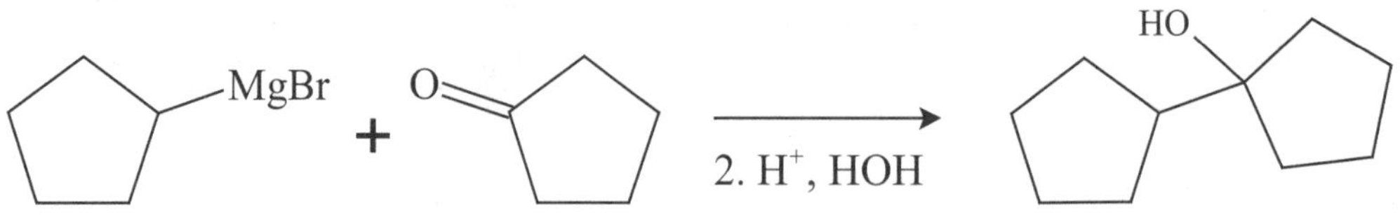

The Grignard reagent and cyclopentanone can each be made in two steps from cyclopentene, as shown below.

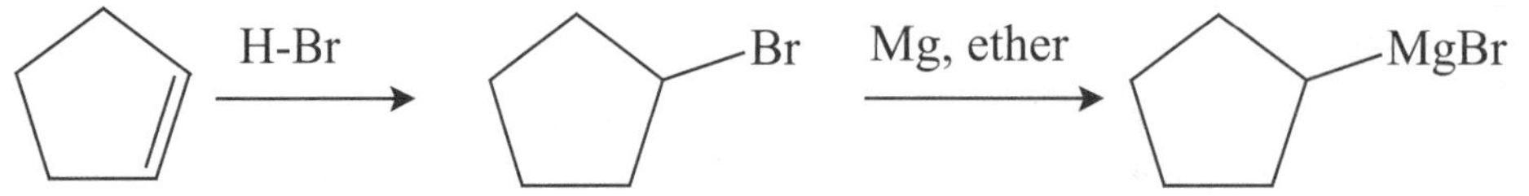

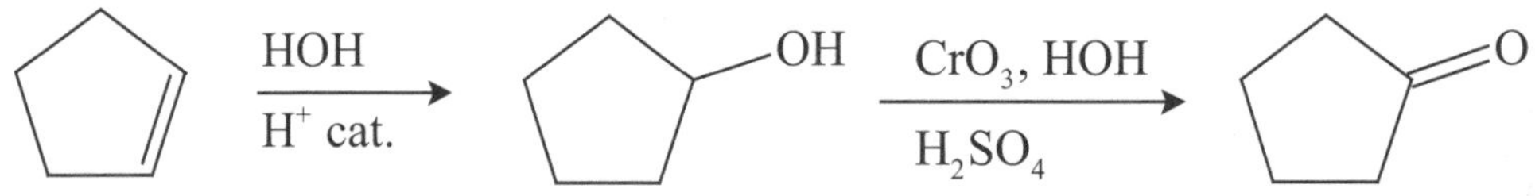

d. This is only one step, if you see it as a Claisen–Schmidt reaction.

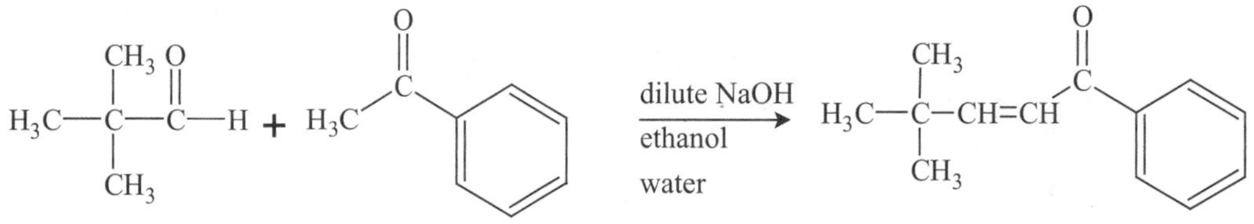

Answers to Chapter 13 Review Exercises

1. a. (*E*)-2-Cyclopentyl-2-nitro-6-phenyl-4-heptenoic acid
   b. Propyl 2-methylhexanoate
   c. 2-Fluoro-3-iodo-4-isopropylbenzoic acid
   d. *N*-Ethyl-4-hydroxy-*N*-propyl-5-octynamide

2.

Product	Reagent
a. 3-Ethylpentanoyl chloride	$SOCl_2$
b. Ethyl 3-ethylpentanoate	Ethanol, H^+ catalyst, heat
c. 3-Ethyl-*N*-phenylpentanamide	(1) $SOCl_2$; (2) aniline, dilute NaOH
d. *N*,3-Diethyl-1-pentanamine	

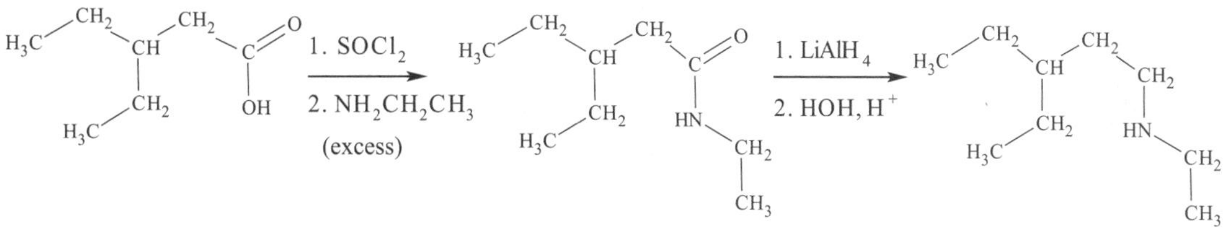

e. 3-Ethyl-1-(4-methoxyphenyl)-1-pentanone

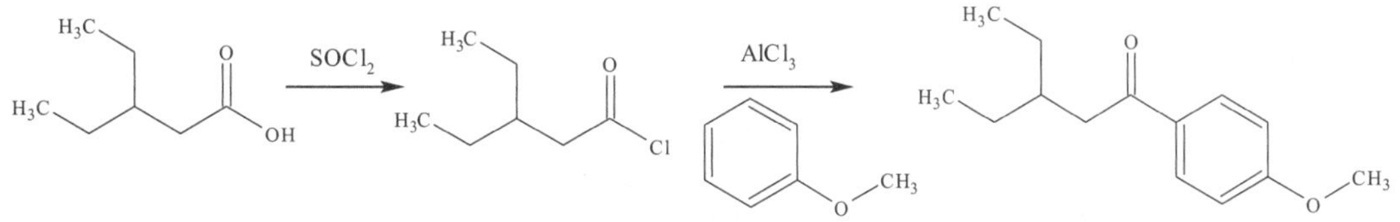

f. 3,5-Diethyl-3-heptanol

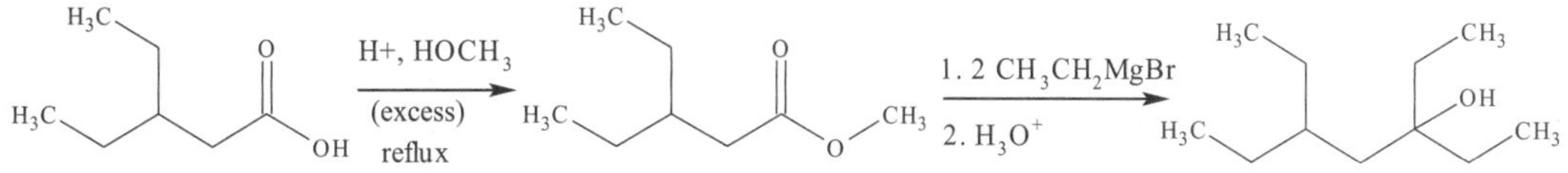

3. a. 3-Ethylpentanoyl chloride: add water carefully!

b. Ethyl 3-ethylpentanoate: reflux in 2 M HCl for a couple hours.

c. 3-Ethyl-*N*-phenylpentanamide: reflux in 6 M HCl for several hours.

d. 3-(Bromomethyl)pentane: NaCN in an ethanol-water mixture at reflux, followed by reflux in 6 M HCl for several hours.

e. 3-Ethylpentanenitrile: reflux in 6 M HCl for several hours.

4.

5. a.

b.

c.

6.

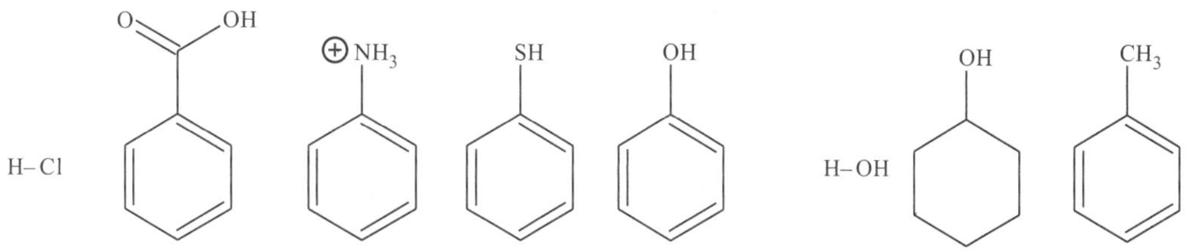

pK$_a$:   –7   4.2   4.6   7   10   15.7   16–17   40

Answers to Chapter 14 Review Exercises

1.

	Starting Material	Product
a. 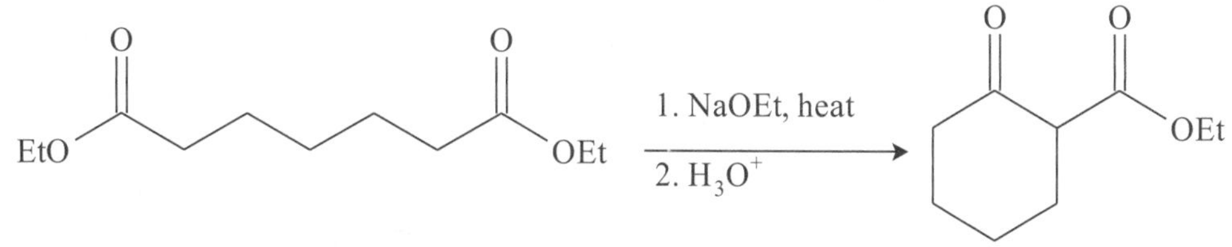	This is a Dieckmann condensation.	
b. 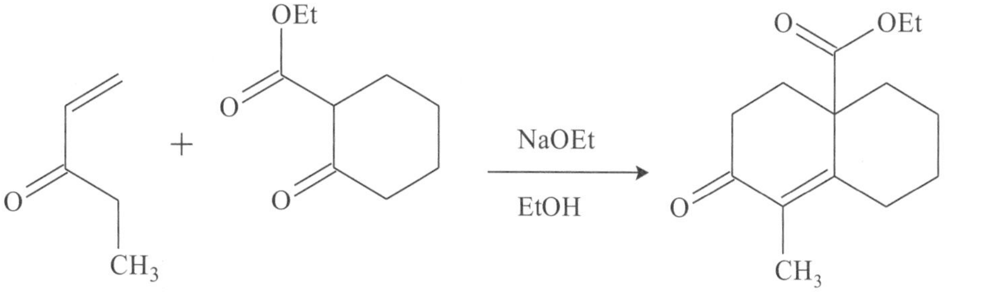	This is a Robinson annelation.	
c. 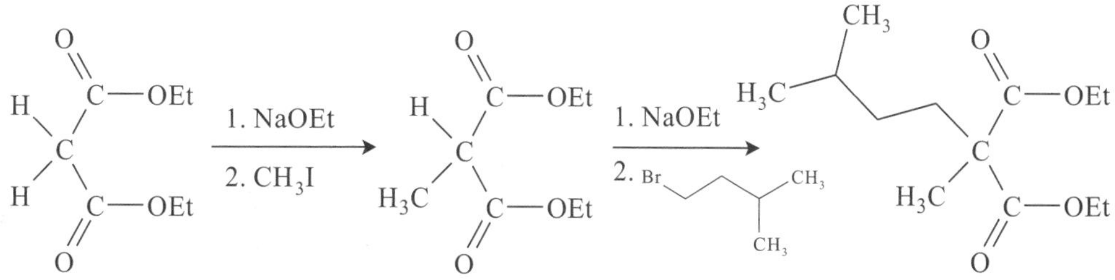	This is the sequential alkylation of diethyl malonate with two different alkyl halides.	

Starting Material	Product

d.

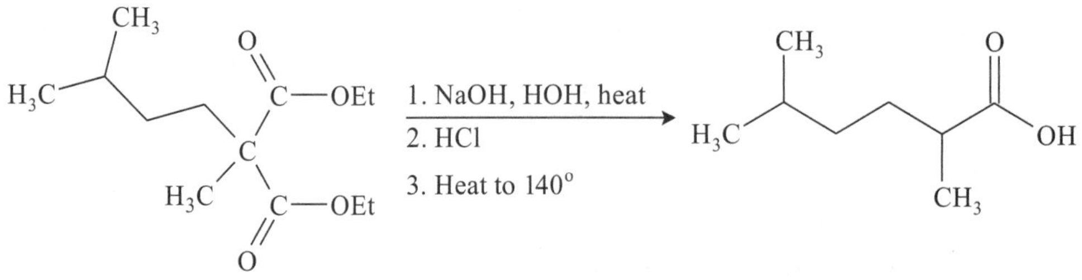

This is the hydrolysis and decarboxylation of a malonic ester to make a substituted acetic acid.

e.

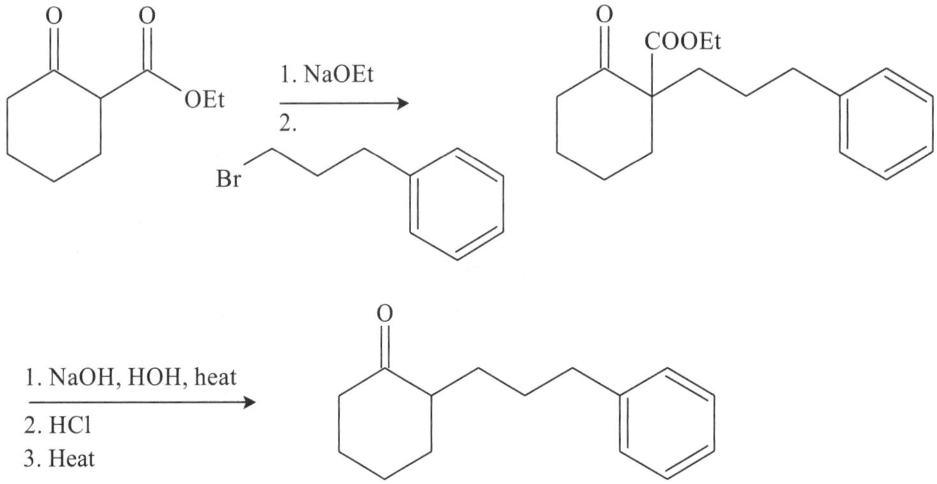

2.

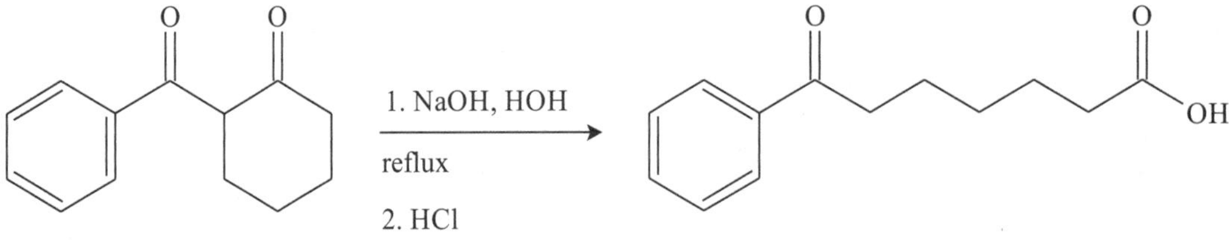

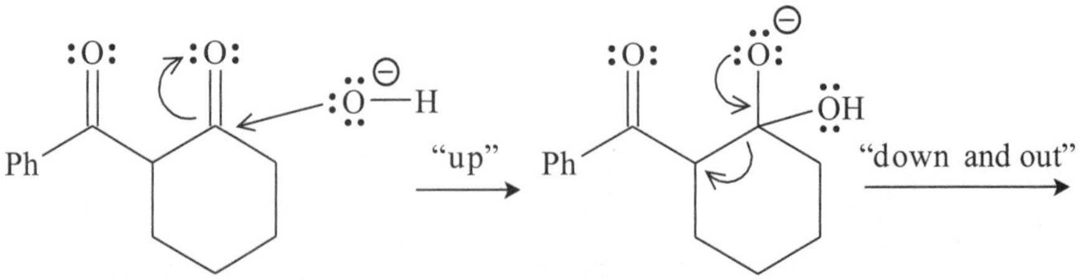

Hydroxide attacks the ring carbonyl, and pushes an electron pair up on to the oxygen. The electron pair comes back down to reform the C=O and pushes out the electron pair onto the alpha-carbon of the other ketone.

The anion formed is resonance-stabilized, so it is a much better leaving group than a normal carbanion. Protonation of the enolate on carbon forms the product. The carboxylic acid is deprotonated under the basic conditions as well, but I didn't show this in the mechanism.

3.

Compound A

4.

Answers to Chapter 15 Review Exercises

1. a. C_8H_{18}, 0.9 ppm	b. C_5H_{10}, 1.5 ppm	c. $C_{12}H_{18}$, 2.2 ppm
d. $C_4H_8O_2$, 3.7 ppm	e. $C_4H_4N_2$, 8.6 ppm	f. $C_4H_4N_2$, 2.9 ppm

2. a. C=O is producing the peak at 1750 cm^{-1}.
   b. C–O stretches the peak at 1200 cm^{-1}.
   c. sp^3 C–H stretches the peak at 3000–2800 cm^{-1}.

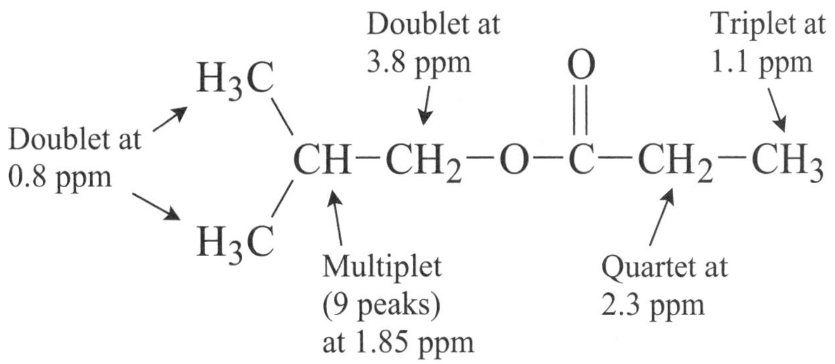

The CH$_2$ next to the oxygen is at 3.8 due to the electronegativity of the oxygen. The CH$_2$ at 2.3 is next to the C=O because it is farther away from an oxygen. The other hydrogens are all on sp^3 carbons bonded to sp^3 carbons, so they appear between 2 and 0.5.

We see only six peaks in the carbon spectrum because the two methyls on the left side of the structure above are equivalent, so there is one fewer type of carbon than total carbons.

The C=O carbon peak is about 175 ppm.

The C–O carbon peak is about 70 ppm.

3. In the mass spectrum, there is an even molecular weight (148, from the molecular ion, M$^+$). There is no indication of a chlorine or bromine. $m/z = 91$ could be a tropylium ion, C$_7$H$_7^+$, which could indicate a monosubstituted benzene with a carbon attached.

The IR spectrum shows sp^2 C–H stretches at about 3100 cm^{-1}, sp^3 C–H stretches at about 2900 cm^{-1}, a C=O as the strong peak about 1700 cm^{-1}, and a potential monosubstituted benzene from the peaks at 750 and 700 cm^{-1}.

The proton NMR spectrum shows two sets of peaks that integrate for a total of five hydrogens at 7.3 and 7.2 ppm. This indicates a monosubstituted benzene ring, that doesn't have a C=O or an electronegative atom attached to it, most likely. The triplets at 2.9 and 2.75 ppm are most likely two adjacent CH$_2$'s that are coupling each other. The singlet at 2.1 ppm is a methyl group that is not adjacent to any other CH's.

The carbon NMR spectrum shows seven peaks. There is a carbonyl at about 209 ppm, and the four benzene-type carbons are between 126 and 142 ppm. The three peaks between 30 and 50 ppm are the sp^3 carbons.

This gives us four pieces: a monosubstituted benzene ring, a C=O, a CH$_2$–CH$_2$, and a methyl that is adjacent to no other hydrogens. The only way to put these together that meets these criteria is as follows:

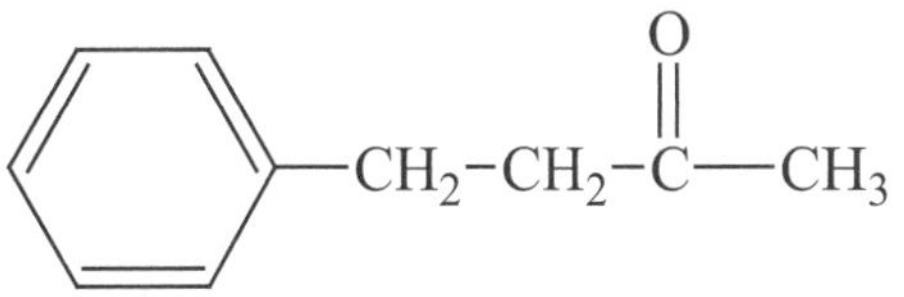

If you put them together in the following way, the methyl is split by the adjacent CH_2, and will not be a singlet. The CH_2 between the methyl and the other CH_2 would have six peaks. The hydrogens ortho to the C=O on the benzene ring would probably be at about 8 ppm, so this structure doesn't fit the data.

$$\text{C}_6\text{H}_5-\overset{\overset{\textstyle O}{\|}}{\text{C}}-CH_2-CH_2-CH_3$$

Final Exam

Write answers on a separate sheet of paper.

1. Give IUPAC names for the following compounds.
   a.

   b.

   c.

   d.

2. Give the major organic product of each of the following reactions.

a.

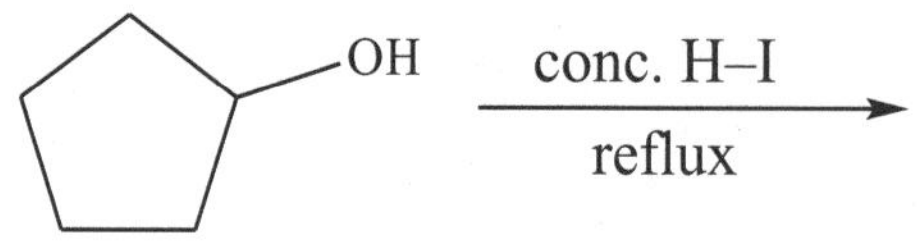

$$\xrightarrow[\text{reflux}]{\text{conc. H–I}}$$

b.

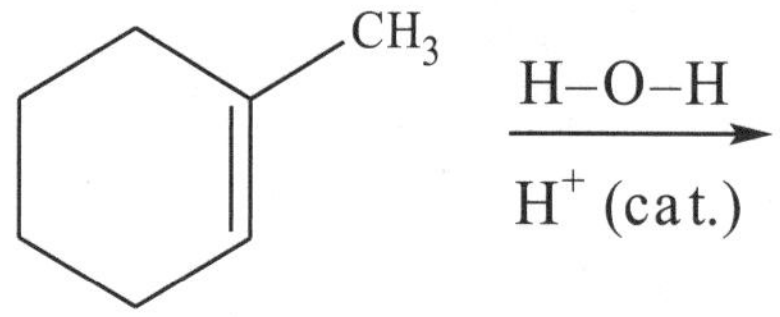

$$\xrightarrow[\text{H}^+ \text{ (cat.)}]{\text{H–O–H}}$$

c.

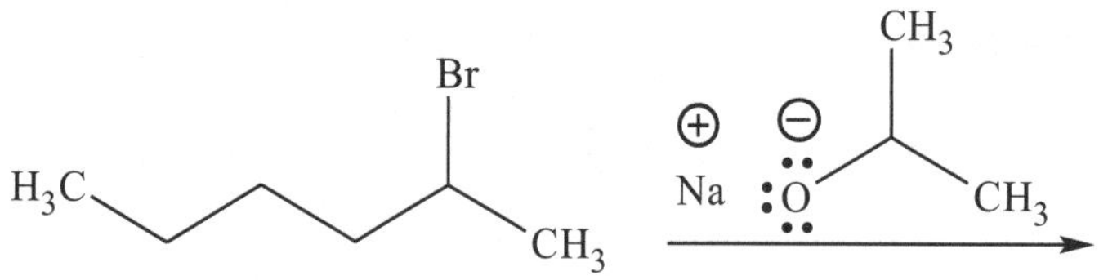

d.

$$\xrightarrow{\text{2H–Br}}$$

e.

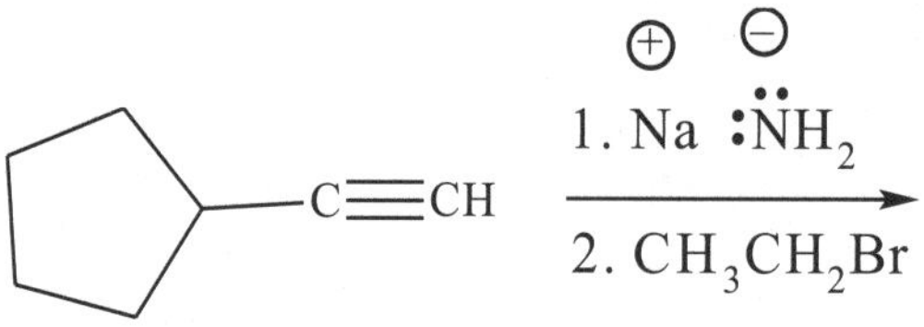

$$\xrightarrow[\text{2. CH}_3\text{CH}_2\text{Br}]{\text{1. Na}\ \overset{\oplus}{}\ \overset{\ominus}{:}\ddot{\text{N}}\text{H}_2}$$

f.

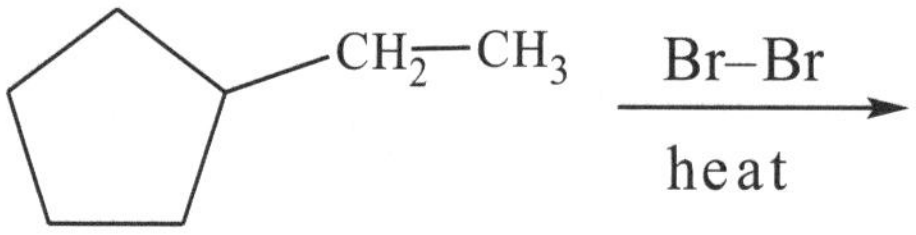

$$\xrightarrow[\text{heat}]{\text{Br–Br}}$$

g.

$$\xrightarrow[\text{HOH}]{\text{CrO}_3, \text{H}^+}$$

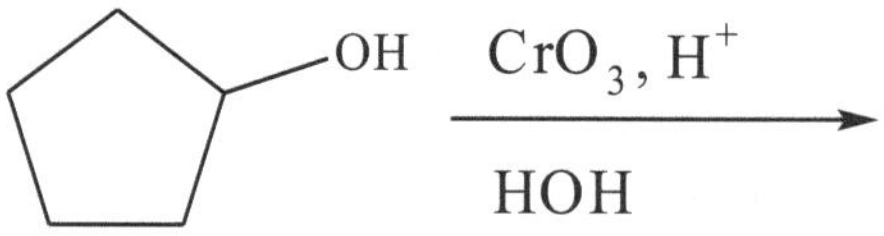

h.

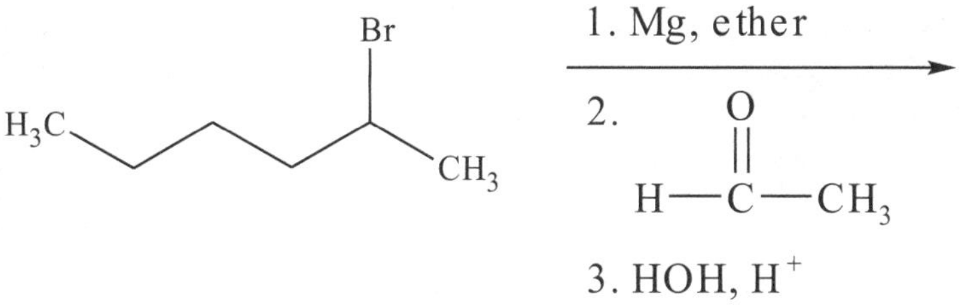
Br
H_3C
CH_3
1. Mg, ether
2.
O
H—C—CH_3
3. HOH, H^+

i.

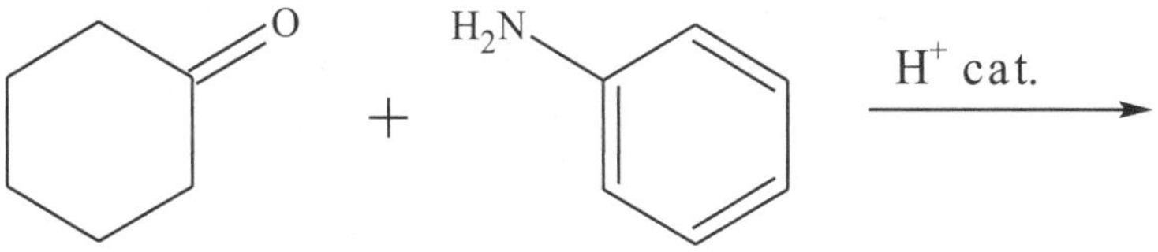
O
+
H_2N
H^+ cat.

j.

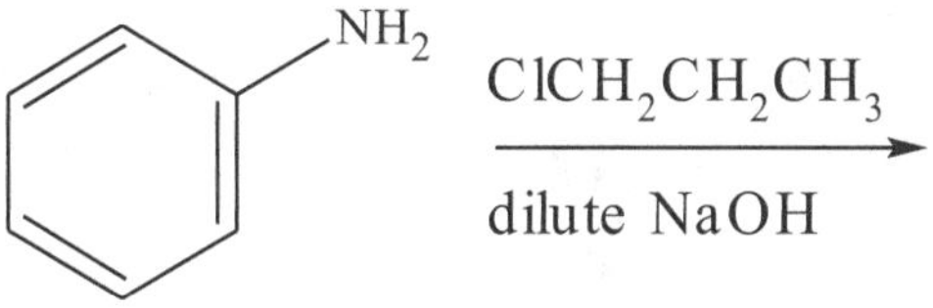
NH_2
$ClCH_2CH_2CH_3$
dilute NaOH

k.

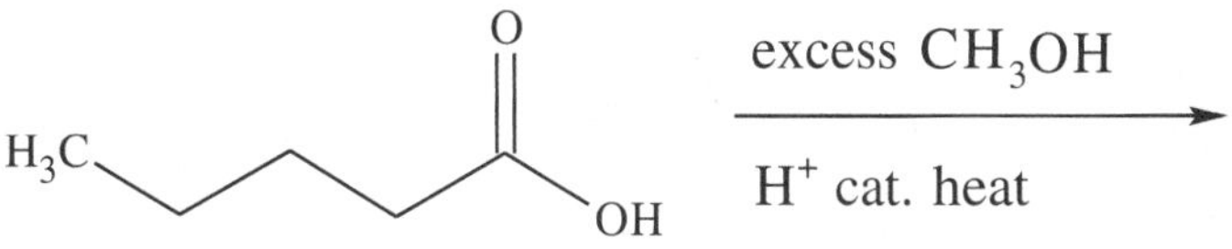
O
H_3C
OH
excess CH_3OH
H^+ cat. heat

l.

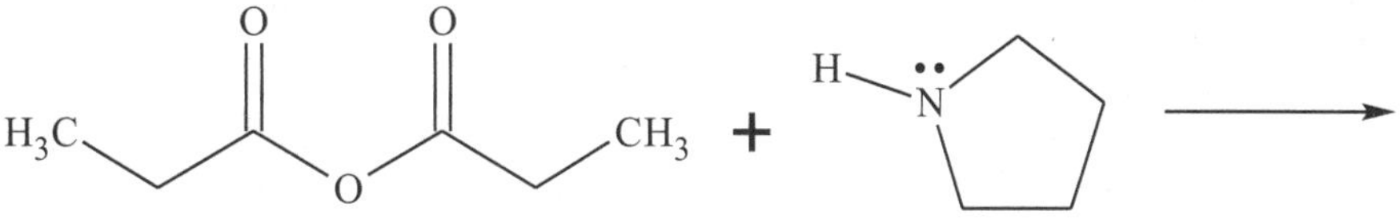
O
O
H_3C
O
CH_3
+
H
N

3. Give reasonable arrow-pushing mechanisms for the following reactions. Show any important resonance forms.

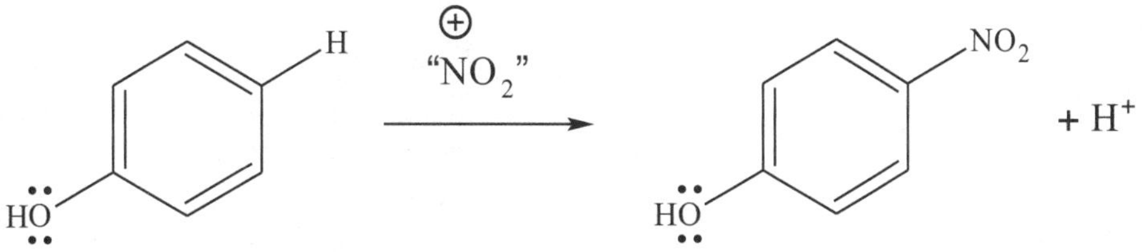
H
⊕
"NO_2"
HO
NO_2
+ H^+
HO

b.

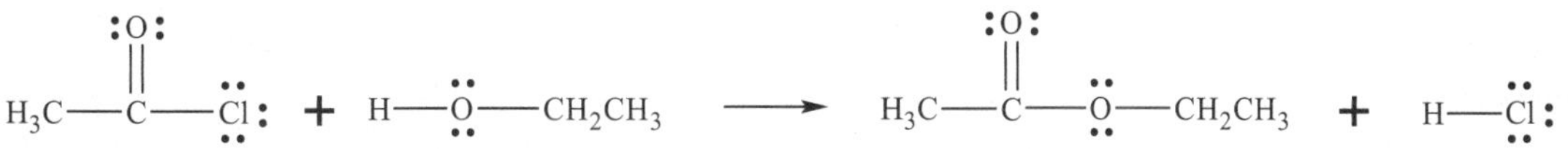
:O:
H_3C—C—Cl:
+
H—O—CH_2CH_3
:O:
H_3C—C—O—CH_2CH_3
+
H—Cl:

4. Provide a reasonable synthesis of each of the following three products from the indicated starting material, plus any other reagents. More than one step is required in each case.

Starting Material   **Product**

a.

b.

$H_3C\!-\!CH_2\!-\!CH_2\!-\!Br$ $H_3C\!-\!CH_2\!-\!CH_2\!-\!\overset{\overset{\displaystyle O}{\|}}{C}\!-\!CH_3$

c.

5. Rewrite the following compounds in order of decreasing boiling point, so that the compound with the highest boiling point is on the left and that with the lowest boiling is on the right. Briefly explain your reasoning.

6. Match the structure with the appropriate pK_a value.

Structure	H₃C—C(=O)OH	H₃C–CH(OH)–CH₃	H–Cl	H–O–H
pKa				

Choose from the following pK_a values:

   –7,   0,   4.7,   7,   10,   15.7,   17,   33

7. Indicate whether each of the following pairs of structures are enantiomers, diastereomers, structural isomers, or identical structures.

   a.

   b.

   c.

8. Rank the groups on the stereogenic center in the molecule below from highest (A) to lowest (D), and classify the stereogenic center as *R* or *S*.

9. a. Draw two structures that correspond to the formula $C_4H_8O_2$. Show all atoms, all bonds, and all unshared pairs of electrons. No atoms may have a formal charge. One of your structures *must* contain a ring, and the other *must not* contain a ring.

   b. Circle and label neatly each occurrence of the following functional groups in the structure below.

Alkene	Ether	Amine	Benzene ring
Alkyne	Alcohol	Ketone	Aldehyde
Ester	Amide	Alkyl halide	Carboxylic acid

10. Propose a reasonable structure that is consistent with the spectra below. Briefly explain your reasoning.

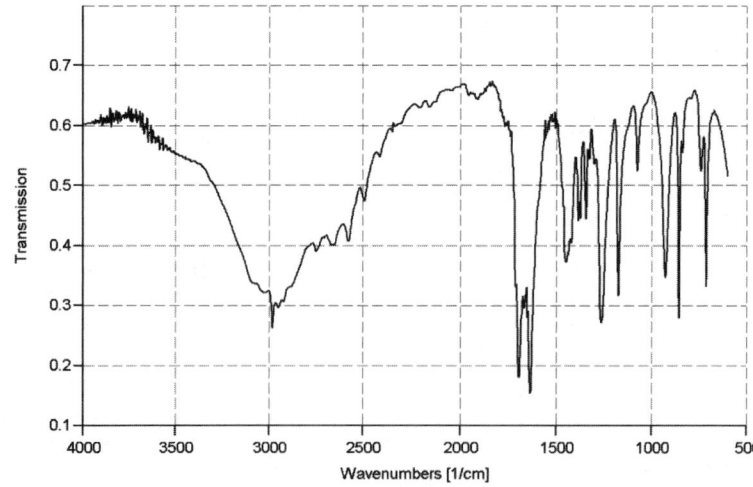

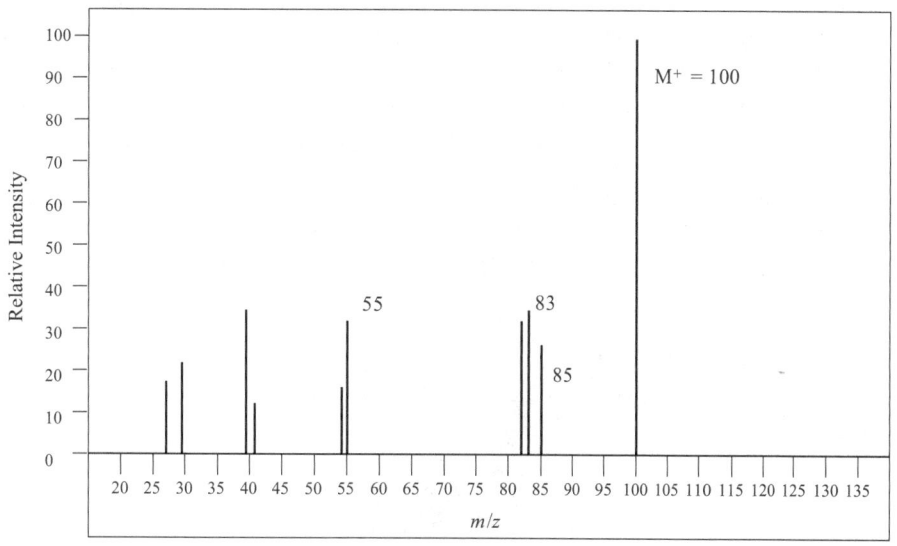

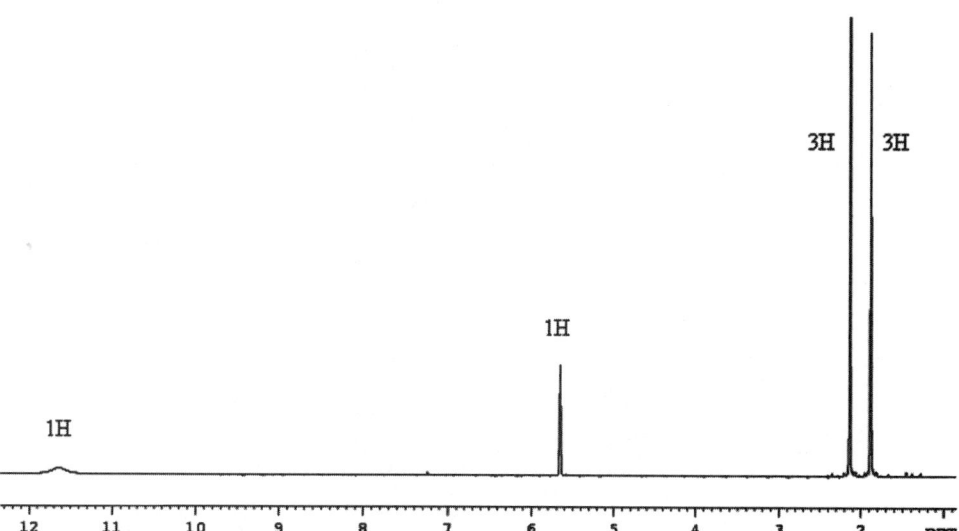

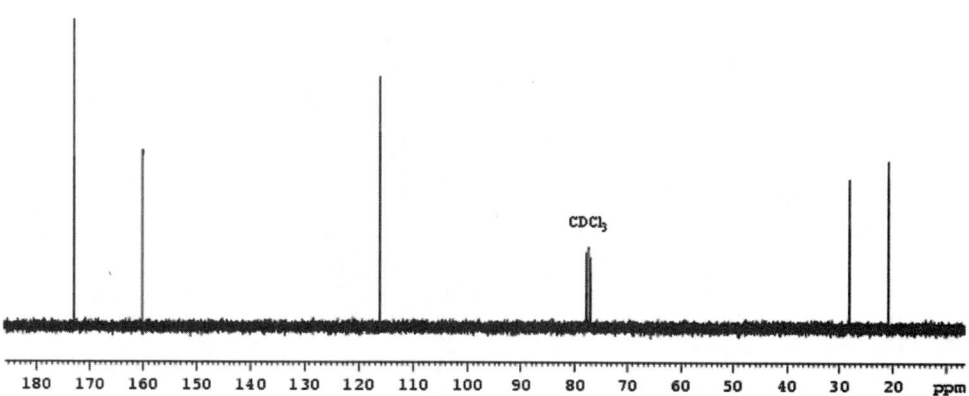

11. a. 2,6-Dimethylcyclohexanone has more than one stereoisomer. When one stereoisomer reacts with NaBH$_4$ in water, the product is a mixture of two stereoisomers. Draw the stereoisomer of 2,6-dimethylcyclohexanone and the two stereoisomer products produced when it reacts with NaBH$_4$ in water.

b. Another stereoisomer of 2,6-dimethylcyclohexanone produces only one product when reacted with NaBH$_4$ in water. Draw this stereoisomer of 2,6-dimethylcyclohexanone and the one product produced when it reacts with NaBH$_4$ in water.

Answers to Final Exam

1. a.   (4Z)-6-Bromo-5-fluoro-10-methyldodec-4-ene
    b.   4-Hydroxycyclooctanone
    c.   Propyl 3-nitro-5-phenylpentanoate
    d.   5-Cyclobutyl-2-methoxyaniline

2. a.

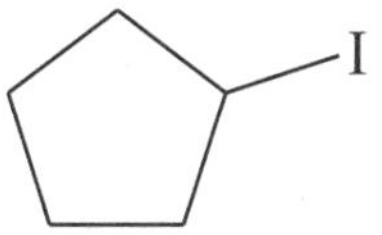

I

b.

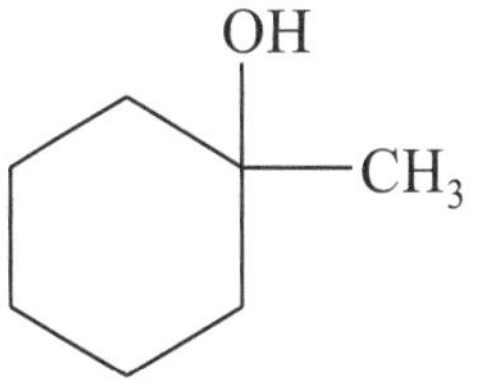

OH
—CH$_3$

c.

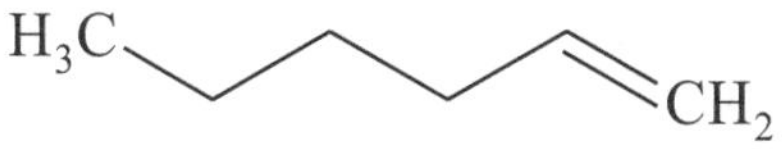

H$_3$C
CH$_2$

d.

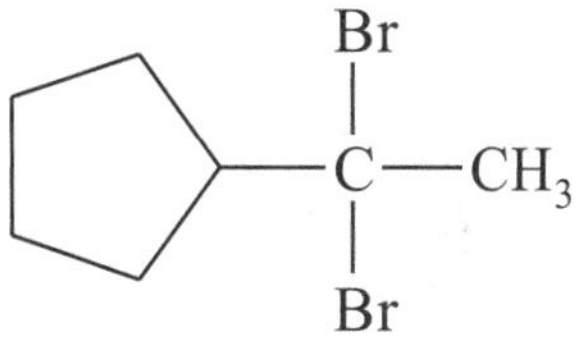

Br
C—CH$_3$
Br

e.

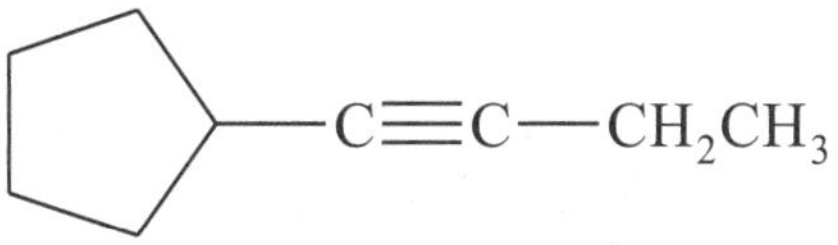

C≡C—CH$_2$CH$_3$

f.

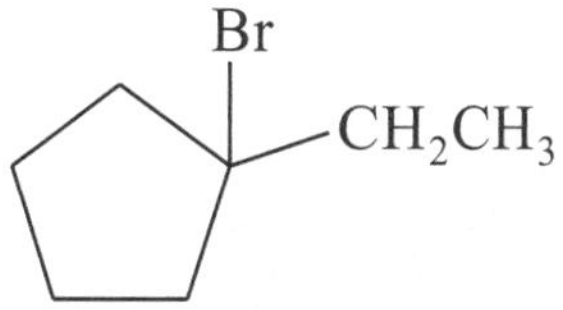

Br
CH$_2$CH$_3$

g.

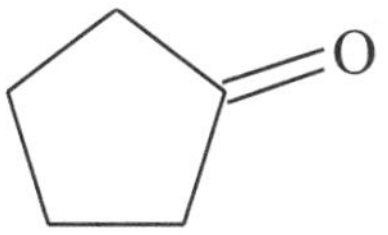

O

h.

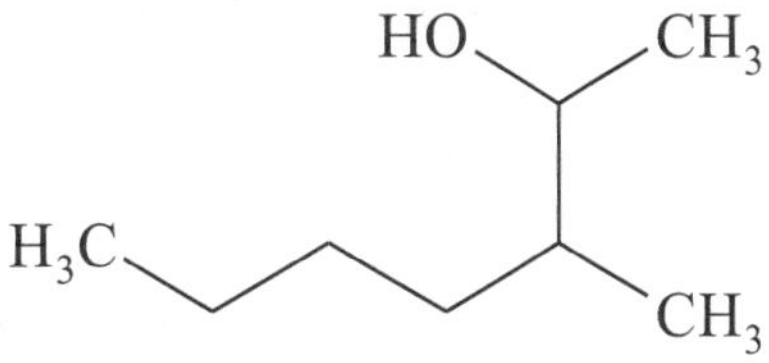

HO   CH$_3$
H$_3$C
CH$_3$

i.

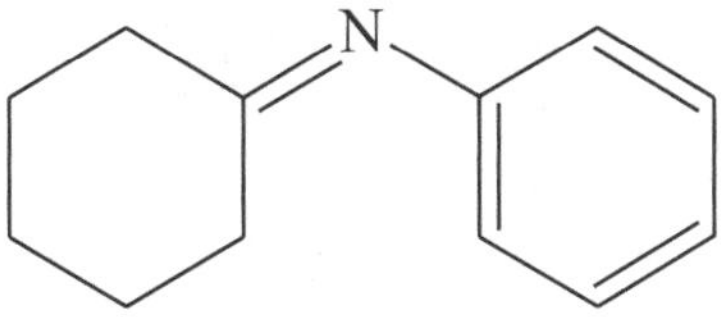

N

j.

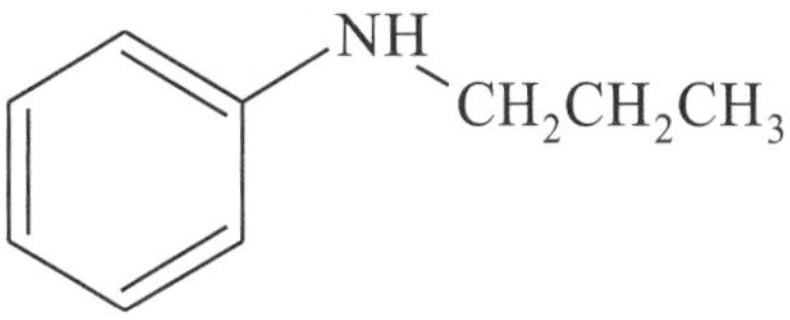

NH
CH$_2$CH$_2$CH$_3$

k.

l.

3. a.

b.

4. a.

b.

c.

5. Carboxylic acids form hydrogen-bonded dimers, so they are very strongly attracted to each other. Alcohols form one hydrogen bond between pairs of molecules. The carbonyl oxygen of a ketone is attracted to the carbonyl oxygen of another ketone, which a weaker attractive force than a hydrogen bond. Ethers are not strongly attracted to other ether molecules.

6.

Structure			H–Cl	H–O–H
pK_a	4.7	17	–7	15.7

7. a.

Identical; there is only one stereocenter.

b.

Diastereomers; the stereocenters are not mirror images.

c.

Structural isomers.

8.

A
D H   O—CH$_3$
HO
CH$_2$ C
C   CH$_2$
B
H$_2$C
CH$_3$

The lowest priority group is going away from us. Looking from A to B to C is counterclockwise, so the stereocenter is *S*.

9. a. Only 2 possibilities are shown: there are many more!

H—Ö: :Ö—H   H
H—C—C—C=C—H
H   H   H

:O:
H   C
H   C   :O:
H   C—C
H H   H

b.

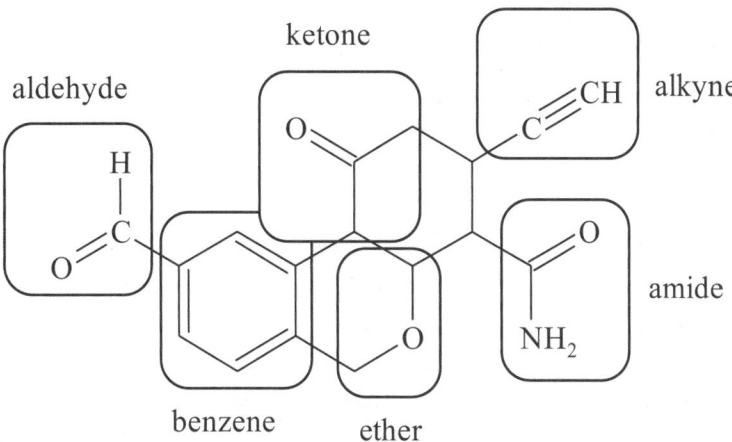

10. The IR spectrum shows a strong carboxylic acid O–H stretch from 3400 to 2400 cm^{-1}, a C=O at 1720 cm^{-1}, and a C=C at about 1620 cm^{-1}.

    The mass spectrum shows the molecular ion at $m/z = 100$. Loss of OH gives the peak at $m/z = 83$.

    The proton NMR spectrum shows the carboxylic acid OH at 11.7 ppm, an alkene hydrogen at 5.7 ppm, and two isolated methyl groups are around 2 ppm. The lack of coupling means there are no adjacent hydrogens.

    The carbon NMR spectrum shows five peaks, for five types of carbons. The peak at 174 ppm is the C=O, the alkene carbons at 160 and 116 ppm, and two alkane carbons at 28 and 21 ppm.

    Five carbons, two oxygens, and eight hydrogens add up to a molecular weight of 100. There is a carboxylic acid, a C=C–H piece, and two methyls. The only way that these can be put together with no adjacent hydrogens is shown below.

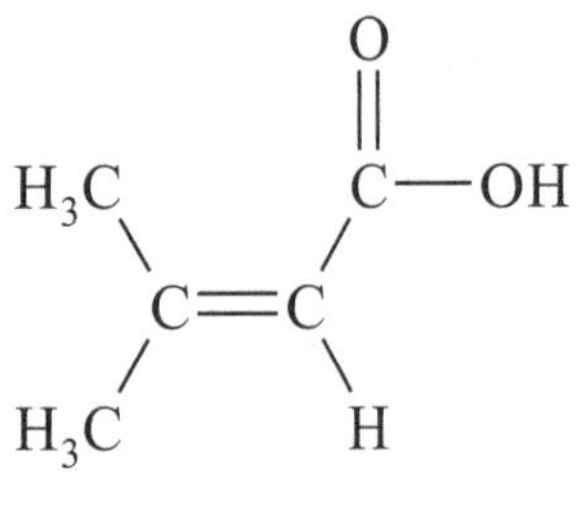

11. a.

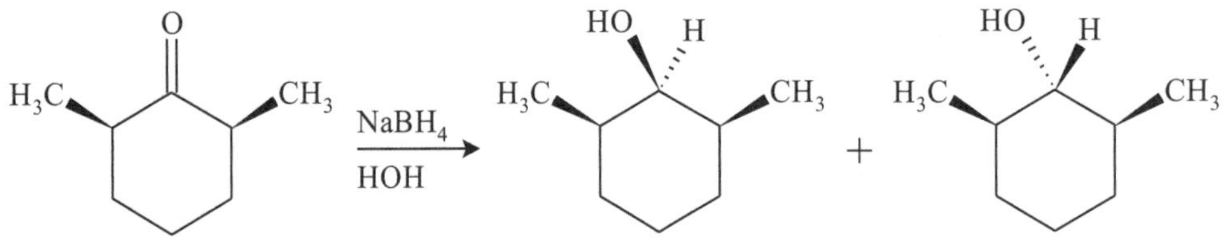

    b.

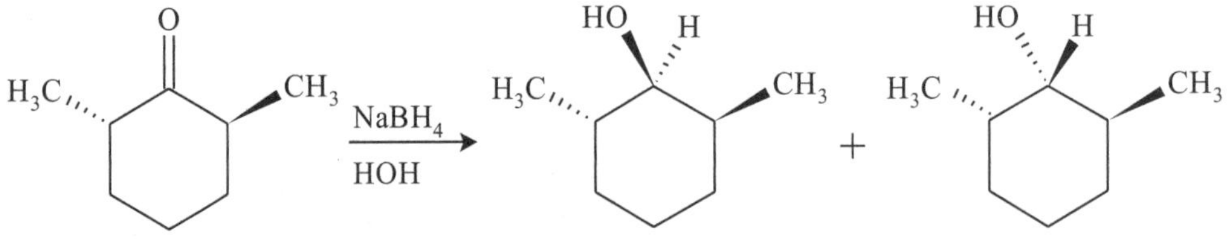

    These molecules are actually identical. If you flip one over, it is the same as the other.

INDEX

MOVE TO THE HEAD OF YOUR CLASS
THE EASY WAY!

Barron's presents **THE E-Z SERIES** (formerly THE EASY WAY SERIES)—specially prepared by top educators, it maximizes effective learning while minimizing the time and effort it takes to raise your grades, brush up on the basics, and build your confidence. Comprehensive and full of clear review examples, **THE E-Z SERIES** is your best bet for better grades, quickly!

ISBN 978-0-7641-4256-7   **E-Z Accounting**—$16.99, *Can$19.99*
ISBN 978-0-7641-4257-4   **E-Z Algebra**—$16.99, *Can$19.99*
ISBN 978-1-4380-0039-8   **E-Z Algebra 2**—$16.99, *Can$19.50*
ISBN 978-0-7641-4258-1   **E-Z American History**—$18.99, *Can$22.99*
ISBN 978-0-7641-4458-5   **E-Z American Sign Language**—$16.99, *Can$19.99*
ISBN 978-0-7641-4468-4   **E-Z Anatomy and Physiology**—$16.99, *Can$19.99*
ISBN 978-0-7641-4466-0   **E-Z Arithmetic**—$16.99, *Can$19.99*
ISBN 978-0-7641-4134-8   **E-Z Biology**—$18.99, *Can$21.99*
ISBN 978-0-7641-4133-1   **E-Z Bookkeeping**—$14.99, *Can$17.99*
ISBN 978-0-7641-4259-8   **E-Z Business Math**—$16.99, *Can$19.50*
ISBN 978-0-7641-4461-5   **E-Z Calculus**—$16.99, *Can$19.99*
ISBN 978-0-7641-4128-7   **E-Z Chemistry**—$16.99, *Can$19.99*
ISBN 978-0-7641-2579-9   **Creative Writing the Easy Way**—$14.99, *Can$17.99*
ISBN 978-0-7641-4464-6   **E-Z Earth Science**—$16.99, *Can$19.99*
ISBN 978-0-7641-3736-5   **English for Foreign Language Speakers the Easy Way**—$18.99, *Can$21.99*
ISBN 978-0-7641-4260-4   **E-Z English**—$14.99, *Can$17.99*
ISBN 978-0-7641-3050-2   **Forensics the Easy Way**—$14.99, *Can$16.99*
ISBN 978-0-7641-4455-4   **E-Z French**—$16.99, *Can$19.99*
ISBN 978-0-7641-2435-8   **French Grammar the Easy Way**—$16.99, *Can$19.99*
ISBN 978-0-7641-3918-5   **E-Z Geometry**—$16.99, *Can$19.99*
ISBN 978-0-7641-4261-1   **E-Z Grammar**—$14.99, *Can$16.99*
ISBN 978-0-7641-4454-7   **E-Z Italian**—$14.99, *Can$17.99*
ISBN 978-0-8120-9627-9   **Japanese the Easy Way**—$18.99, *Can$22.99*
ISBN 978-0-7641-3237-7   **Macroeconomics the Easy Way**—$14.99, *Can$17.99*
ISBN 978-0-7641-9369-9   **Mandarin Chinese the Easy Way**—$23.99, *Can$27.50*
ISBN 978-0-7641-4132-4   **E-Z Math**—$14.99, *Can$17.99*
ISBN 978-0-7641-1871-5   **Math Word Problems the Easy Way**—$16.99, *Can$19.99*
ISBN 978-0-7641-4456-1   **E-Z Microbiology**—$18.99, *Can$22.99*
ISBN 978-0-8120-9601-9   **Microeconomics the Easy Way**—$16.99, *Can$19.99*
ISBN 978-0-7641-4467-7   **E-Z Organic Chemistry**—$16.99, *Can$19.99*
ISBN 978-0-7641-4126-3   **E-Z Physics**—$16.99, *Can$19.99*
ISBN 978-1-4380-0011-4   **E-Z Pre-Algebra**—$14.99, *Can$16.99*
ISBN 978-0-7641-4465-3   **E-Z Precalculus**—$16.99, *Can$19.99*
ISBN 978-0-7641-4462-2   **E-Z Psychology**—$16.99, *Can$19.99*
ISBN 978-0-7641-4129-4   **E-Z Spanish**—$16.99, *Can$19.99*
ISBN 978-0-7641-4249-9   **E-Z Spanish Grammar**—$16.99, *Can$19.99*
ISBN 978-0-7641-4459-2   **E-Z Spelling**—$14.99, *Can$17.99*
ISBN 978-0-7641-3978-9   **E-Z Statistics**—$16.99, *Can$19.99*
ISBN 978-0-7641-4251-2   **E-Z Trigonometry**—$16.99, *Can$19.99*
ISBN 978-0-8120-9765-8   **World History the Easy Way, Vol. One**—$18.99, *Can$22.99*
ISBN 978-0-8120-9766-5   **World History the Easy Way, Vol. Two**—$21.99, *Can$26.50*

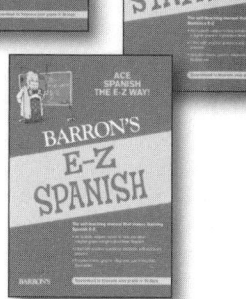

Available at your local book store
or visit **www.barronseduc.com**

Barron's Educational Series, Inc.
250 Wireless Blvd.
Hauppauge, NY 11788
Order toll-free: 1-800-645-3476
Order by fax: 1-631-434-3217

In Canada:
Georgetown Book Warehouse
34 Armstrong Ave.
Georgetown, Ontario L7G 4R9
Canadian orders: 1-800-247-7160
Order by fax: 1-800-887-1594

(#45) R5/13

Prices subject to change without notice.